A SOURCE BOOK IN THE
HISTORY OF PSYCHOLOGY

A SOURCE BOOK

IN THE HISTORY

OF PSYCHOLOGY

EDITED BY

RICHARD J. HERRNSTEIN

EDWIN G. BORING

HARVARD UNIVERSITY

HARVARD UNIVERSITY PRESS

CAMBRIDGE, MASSACHUSETTS · 1965

GENERAL EDITOR'S PREFACE

The Source Books in this series are collections of classical papers that have shaped the structures of the various sciences. Some of these classics are not readily available and many of them have never been translated into English, thus being lost to the general reader and frequently to the scientist himself. The point of this series is to make these texts readily accessible and to provide good translations of the ones that either have not been translated at all or have been translated only poorly.

The series was planned originally to include volumes in all the major sciences from the Renaissance through the nineteenth century. It has been extended to include ancient and medieval Western science and the development of the sciences in the first half of the present century. Many of these books have been published already and several more are in various stages of preparation.

The Carnegie Corporation originally financed the series by a grant to the American Philosophical Association. The History of Science Society and the American Association for the Advancement of Science have approved the project and are represented on the Editorial Advisory Board. This Board at present consists of the following members:

Marshall Clagett, History of Science, University of Wisconsin

I. Bernard Cohen, History of Science, Harvard University

C. J. Ducasse, Philosophy, Brown University

Ernst Mayr, Zoology, Harvard University

Ernest A. Moody, Philosophy, University of California at Los Angeles

Ernest Nagel, Philosophy, Columbia University

Harlow Shapley, Astronomy, Harvard University

Harry Woolf, History of Science, Johns Hopkins University

The series was begun and sustained by the devoted labors of Gregory D. Walcott and Everett W. Hall, the first two General Editors. I am indebted to them, to the members of the Advisory Board, and to Joseph D. Elder, Science Editor of Harvard University Press, for their indispensable aid in guiding the course of the Source Books.

EDWARD H. MADDEN
General Editor

Department of Philosophy
State University of New York at Buffalo
Buffalo, New York

PREFACE

A historical source book represents an intermediate stage between the past as it actually was and as it is recounted in a historical narrative. It is a selection from the original documents, a selection that any historian unavoidably makes as he writes a scholarly history. It would be hard to say how close to completion the historian is when he reaches the source-book stage, but let us say he is halfway there. By this stage, he has selected from the innumerable events of the past the decisive ones for an understanding of subsequent events, but he has not yet set down his interpretations and justifications. The cautious reader will bear in mind that a source book is thus only little less arbitrary than a history. A source book may be, however, more than merely a half-completed history, for it contains the very words in which the important ideas were stated, the raw data, as it may be said, of historical study from which any reader may draw conclusions.

Instead of organizing our source book chronologically, we have sorted the great writings of the past into fifteen separate chronologies, corresponding to the fifteen chapters. Modern psychology is multidimensional, and so is its past. Fourteen of our fifteen chapters deal each with a specific topic in psychology; the fifteenth traces opinion about the nature of psychology itself. We have limited ourselves to topics in the history of experimental or quantitative psychology, omitting the history of social and clinical psychology and recognizing that even within our chosen field we have been neither exhaustive nor unbiased. Because our source book is topically organized, certain outstanding writers inevitably turn up in more than one chapter. We have tried to come no closer to the present than 1900, but occasionally more recent work has made it seem essential to push a little way into the present century.

In general, we have arranged each section chronologically, but occasionally when a topic is bifurcated we have followed along one branch and then gone back to complete the other. We have tried to use the more important edition of a work, the first or a later one, according to which is the more appropriate. For each chapter we have provided an introduction which is to some extent a preview of the topic considered.

We have sought then to make each trend continuous, that is to say, we have given every excerpt an introduction which indicates the nature of its significance and also its connection with the preceding excerpt. Sometimes there is a gap between two excerpts, occasioned by the fact that prolific intervening writers have provided nothing compact enough for excerpting. In that case we name the intervening writers and say how they form the linkage between the two excerpts.

The selections that were originally in English we have reproduced with the original orthography and punctuation, hoping thus to induce some of the flavor of the past, but alas, modernity does not permit the old-fashioned tailed *s*. Of the 116 excerpts there are 46 which were originally in French or German, three in Greek, three in Latin, and two in Russian. Half of these are borrowed from published English translations, most of which are well known, but 27—a quarter of them all—are new in this book. Each excerpt indicates who translated it.

We have edited the excerpts only in the sense of selecting material and of deleting what we thought was unessential. Small omissions are denoted by the usual three dots, longer ones by a line of dots. While we have tried to be faithful to the author excerpted, we have felt less responsibility to the English translator, where our duty is primarily to the reader. In the translations we have sometimes made minor adjustments to increase clarity, often by small modernizations. And in a number of selections we have omitted the author's notes and references where they did not seem to contribute substantially to the text. It was only by extensive deletions that we were able to condense so much of the past into a volume of this size, for great men do not always write succinctly. Permissions to reprint excerpts are indicated in the List of Excerpted Works following the text. We are grateful to authors and publishers who so readily lent us their work.

Our task was after all little more than sampling the wealth of material in the Harvard College Library, the Harvard Medical Library, and the Boston Medical Library. What civilization and scholarship owe to libraries! Professor Curt P. Richter of the Johns Hopkins University lent his volume of Sechenov and the Clark University Library their first edition of Mach's *Analyse*. We are grateful. Academia is like the sea; a cry of distress seldom goes unanswered. The Carnegie Fund of the American Philosophical Association helped us to get our translations in hand and also our minds in order. A gracious institution that. In connection with the excerpts from Sechenov and Pavlov our incompetence with Russian was mitigated by Mr. William Baum of Harvard and Mr. John Molino of Columbia, who leapt this barrier for us. We thank them. And then there is Miss Edith L. Annin, who has brought order into the chaos of so many publications from the Harvard Psychological Laboratories in recent years that it seems redundant to thank her, and Miss Judith Schiek, who with skill, patience, and insight has borne half the load of making thought orderly. We thank them both. And then at the Harvard University Press there is that ever patient, immutable arbitress, not only of order and consistency, but also of unpredictable wisdom, Miss Blair McElroy; how fortunate we were to be under her command.

Boldly we date our preface on Descartes's 369th birthday. How much of him is still with us!

Cambridge, Massachusetts   R. J. H.
31 March 1965   E. G. B.

CONTENTS

A SOURCE BOOK IN THE
HISTORY OF PSYCHOLOGY

I SENSORY SPECIFICATION

Sensation—learning—motivation: it is not too great an oversimplification to say that the genetic lines in the history of thinking that led eventually to modern scientific psychology had to do with those three great topics and that they, in spite of overlapping and developmental irregularities, emerged and grew in that order, with sensation first. That is the fact. It is hard to say whether there may have been genetic necessity for it. At any rate it was natural for experimental psychology to emerge from British empiricism, and equally natural for empiricism to concern itself with sensation, the avenue by which experience gets into the mind. For philosophers to understand experience it was necessary for them to describe it, and description is, of course, analytical; one needs eventually to know the inventory of sensory elements. In some such way British empiricism became sensationistic and elementistic, and that process was abetted by the coincidental emergence of sensory physiology, which, having established the distinction between sensory and motor nerves, became involved in the specification of the qualities that categorize the departments of sense. Elementism in psychology was, moreover, reinforced by the discovery of chemical elements and the great success of atomistic chemistry. Might not psychology be equally successful by the specification of the sensory elements and the description of the many perceptual and cognitive compounds that are formed from them by association? That question represents the outstanding aspiration of the new experimental psychology at the end of the nineteenth century.

The present chapter, with fourteen excerpts from the contributions of philosophers and physiologists, shows how the list of sensory qualities became more and more specific up to the end of the nineteenth century, sometimes by introspective analysis, sometimes by physiological specification, but more often by a combination of these two modes of analysis. The chapter begins with Aristotle's classical specification of the five senses (ca. 350 B.C.), and later presents John Locke's further differentiation between the primary and secondary qualities of sensation (1690). It shows how Newton divided the spectrum into seven colors (1675), and how Thomas Young, a century later, used Newton's law of color mixture to argue for three physiologically elemental colors (1802). Bell and Magendie are cited for the law of the spinal nerve roots, the discovery that separated the sensory from the motor problems of the nervous system (1811, 1822), and Bell and Johannes Müller are cited on the specific energies of nerves, the principle that in its first formulation differentiated the five senses (1811, 1838). Next E. H. Weber is quoted on the tactual sense and common sensibility (1846). After that comes Helmholtz's extension of the theory of specific energies to the separate fibers within the optic and auditory nerves and the physiological analysis of these two senses into three colors and a large number of pitches (1860, 1863). The work on the differentiation of the tactual sense is summarized by von Frey, whose researches had contributed to the generally accepted analysis of touch into four fundamental qualities (1904). The chapter ends with a brief excerpt from Titchener, summarizing in a way the current state of elementistic psychology at the end of the century by attempting to tabulate the number of sensory elements then known to exist (1896). The analogy with chemistry was still uppermost in the thinking of these systematic psychol-ogists, and Titchener's table of known sensory elements had in it blanks where newly discovered items could be fitted in later.

1 ARISTOTLE (384–322 B.C.) ON THE FIVE SENSES, ca. 350 B.C.

Aristotle, *De anima,* bk. II. Translated by J. A. Smith in W. D. Ross, ed., *The Works of Aristotle* (Oxford: Clarendon Press, 1931), III, 417b–424a.

These passages constitute the classical differentiation of the five senses, a differentiation that has never been called into question. Aristotle was al-ready aware that the fifth sense, touch, is more complex than the other four senses, includes perhaps more than one sense within itself.

In dealing with each of the senses we shall have first to speak of the objects which are perceptible by each. The term 'object of sense' covers three kinds of objects, two kinds of which are, in our language, directly perceptible, while the remaining one is only incidentally perceptible. Of the first two kinds one (*a*) consists of what is perceptible by a single sense, the other (*b*) of what is perceptible by any and all of the senses. I call by the name of special object of this or that sense that which cannot be perceived by any other sense than that one and in respect of which no error is possible; in this sense colour is the special object of sight, sound of hearing, flavour of taste. Touch, indeed, discriminates more than one set of different qualities. Each sense has one kind of object which it discerns, and never errs in reporting that what is before it is colour or sound (though it may err as to what it is that is coloured or where that is, or what it is that is sounding or where that is). Such objects are what we propose to call the special objects of this or that sense.

'Common sensibles' are movement, rest, number, figure, magnitude; these are not peculiar to any one sense, but are common to all. There are at any rate certain kinds of movement which are perceptible both by touch and by sight.

We speak of an incidental object of sense where e.g. the white object which we see is the son of Diares; here because 'being the son of Diares' is incidental to the directly visible white patch we speak of the son of Diares as being (incidentally) perceived or seen by us. Because this is only incidentally an object of sense, it in no way as such affects the senses. Of the two former kinds, both of which are in their own nature perceptible by sense, the first kind—that of special objects of the several senses—constitute *the* objects of sense in the strictest sense of the term and it is to them that in the nature of things the structure of each several sense is adapted.

The object of sight is the visible, and what is visible is (a) colour and (b) a certain kind of object which can be described in words but which has no single name; what we mean by (b) will be abundantly clear as we proceed. Whatever is visible is colour and colour is what lies upon what is in its own nature visible; 'in its own nature' here means not that visibility is involved in the definition of what thus underlies colour, but that that substratum contains in itself the cause of visibility. Every colour has in it the power to set in movement what is actually transparent; that power constitutes its very nature. That is why it is not visible except with the help of light; it is only in light that the colour of a thing is seen.

·   ·   ·   ·   ·

The following experiment makes the necessity of a medium clear. If what has colour is placed in immediate contact with the eye, it cannot be seen. Colour sets in movement not the sense organ but what is transparent, e.g. the air, and that, extending continuously from the object of the organ, sets the latter in movement. Democritus misrepresents the facts when he expresses the opinion that if the interspace were empty one could distinctly see an ant on the vault of the sky; that is an impossibility. Seeing is due to an affection or change of what has the perceptive faculty, and it cannot be affected by the seen colour itself; it remains that it must be affected by what comes between. Hence it is indispensable that there be *something* in between—if there were nothing, so far from seeing with greater distinctness, we should see nothing at all.

·   ·   ·   ·   ·

The same account holds also of sound and smell; if the object of either of these senses is in immediate contact with the organ no sensation is produced. In both cases the object sets in movement only what lies between, and this in turn sets the organ in movement: if what sounds or smells is brought into immediate contact with the organ, no sensation will be produced. The same, in spite of all appearances, applies also to touch and taste; why there is this apparent difference will be clear later. What comes between in the case of sounds is air; the corresponding medium in the case of smell has no name. But, corresponding to what is transparent in the case of colour, there is a quality found both in air and water, which serves as a medium for what has smell—I say 'in water' because animals that live in water as well as those that live on land seem to possess the sense of smell, and 'in air' because man and all other land animals that breathe, perceive smells only when they breathe air in. The explanation of this too will be given later.

Now let us, to begin with, make certain distinctions about sound and hearing.

Sound may mean either of two things—(a) actual, and (b) potential, sound. There are certain things which, as we say, 'have no sound', e.g. sponges or wool, others which have, e.g. bronze and in general all things which are smooth and solid—the latter are said to have a sound because they can make a sound, i.e. can generate actual sound between themselves and the organ of hearing.

Actual sound requires for its occurrence (i,ii) two such bodies and (iii) a space between them; for it is generated by an impact. Hence it is impossible for one body only to generate a sound—there must be a body impinging and a body impinged upon; what sounds does so by striking against something else, and this is impossible without a movement from place to place.

· · · · ·

Further, we must remark that sound is heard both in air and in water, though less distinctly in the latter. Yet neither air nor water is the principal cause of sound. What is required for the production of sound is an impact of two solids against one another and against the air. The latter condition is satisfied when the air impinged upon does not retreat before the blow, i.e. is not dissipated by it.

· · · · ·

Smell and its object are much less easy to determine than what we have hitherto discussed; the distinguishing characteristic of the object of smell is less obvious than those of sound or colour. The ground of this is that our power of smell is less discriminating and in general inferior to that of many species of animals; men have a poor sense of smell and our apprehension of its proper objects is inseparably bound up with and so confused by pleasure and pain, which shows that in us the organ is inaccurate . . . It seems that there is an analogy between smell and taste, and that the species of tastes run parallel to those of smells—the only difference being that our sense of taste is more discriminating than our sense of smell, because the former is a modification of touch, which reaches in man the maximum of discriminative accuracy. While in respect of all the other senses we fall below many species of animals, in respect of touch we far excel all other species in exactness of discrimination. That is why man is the most intelligent of all animals. This is confirmed by the fact that it is to differences in the organ of touch and to nothing else that the differences between man and man in respect of natural endowment are due; men whose flesh is hard are ill-endowed by nature, men whose flesh is soft, well-endowed.

As flavours may be divided into (*a*) sweet, (*b*) bitter, so with smells. In some things the flavour and the smell have the same quality, i.e. both are sweet or both bitter, in others they diverge. Similarly a smell, like a flavour, may be pungent, astringent, acid, or succulent. But, as we said, because smells are much less easy to discriminate than flavours, the names of these varieties are applied to smells only metaphorically; for example 'sweet' is extended from the taste to the smell of saffron or honey, 'pungent' to that of thyme, and so on.

. . . . .

What can be tasted is always something that can be touched, and just for that reason it cannot be perceived *through* an interposed foreign body, for touch means the absence of any intervening body. Further, the flavoured and tasteable body is suspended in a liquid matter, and this is tangible. Hence, if we lived in water, we should perceive a sweet object introduced into the water, but the water would not be the medium *through* which we perceived; our perception would be due to the solution of the sweet substance in what we imbibed, just as if it were mixed with some drink.

. . . . .

The species of flavour are, as in the case of colour, (*a*) simple, i.e. the two contraries, the sweet and the bitter, (*b*) secondary, viz. (i) on the side of the sweet, the succulent, (ii) on the side of the bitter, the saline, (iii) between these come the pungent, the harsh, the astringent, and the acid; these pretty well exhaust the varieties of flavour. It follows that what has the power of tasting is what is potentially of that kind, and that what is tasteable is what has the power of making it actually what it itself already is.

Whatever can be said of what is tangible, can be said of touch, and vice versa; if touch is not a single sense but a group of senses, there must be several kinds of what is tangible. It is a problem whether touch is a single sense or a group of senses. It is also a problem, what is the organ of touch; is it or is it not the flesh (including what in certain animals is homologous with flesh)? On the second view, flesh is 'the medium' of touch, the real organ being situated farther inward. The problem arises because the field of each sense is according to the accepted view determined as the range between a single pair of contraries,

5

white and black for sight, acute and grave for hearing, bitter and sweet for taste; but in the field of what is tangible we find several such pairs, hot cold, dry moist, hard soft, &c. This problem finds a partial solution, when it is recalled that in the case of the other senses more than one pair of contraries are to be met with, e.g. in sound not only acute and grave but loud and soft, smooth and rough, &c.; there are similar contrasts in the field of colour. Nevertheless we are unable clearly to detect in the case of touch what the single subject is which underlies the contrasted qualities and corresponds to sound in the case of hearing.

To the question whether the organ of touch lies inward or not (i.e. whether we need look any farther than the flesh), no indication in favour of the second answer can be drawn from the fact that if the object comes into contact with the flesh it is at once perceived. For even under present conditions if the experiment is made of making a web and stretching it tight over the flesh, as soon as this web is touched the sensation is reported in the same manner as before, yet it is clear that the organ is not in this membrane. If the membrane could be *grown* on to the flesh, the report would travel still quicker. The flesh plays in touch very much the same part as would be played in the other senses by an air-envelope growing round our body; had we such an envelope attached to us we should have supposed that it was by a single organ that we perceived sounds, colours, and smells, and we should have taken sight, hearing, and smell to be a single sense. But as it is, because that through which the different movements are transmitted is not naturally attached to our bodies, the difference of the various sense-organs is too plain to miss. But in the case of touch the obscurity remains.

· · · · ·

In general, flesh and the tongue are related to the real organs of touch and taste, as air and water are to those of sight, hearing, and smell. Hence in neither the one case nor the other can there be any perception of an object if it is placed immediately upon the organ, e.g. if a white object is placed on the surface of the eye. This again shows that what has the power of perceiving the tangible is seated inside. Only so would there be a complete analogy with all the other senses. In their case if you place the object on the *organ* it is not perceived, here if you place it on the flesh it *is* perceived; therefore flesh is not the organ but the *medium* of touch.

What can be touched are distinctive qualities of body *as* body; by such differences I mean those which characterize the elements, viz. hot cold, dry moist, of which we have spoken earlier in our treatise on the elements. The organ for the perception of these is that of touch—that part of the body in which primarily the sense of touch resides.

.   .   .   .   .

The following results applying to any and every sense may now be formulated.

(A) By a 'sense' is meant what has the power of receiving into itself the sensible forms of things without the matter. This must be conceived of as taking place in the way in which a piece of wax takes on the impress of a signet-ring without the iron or gold; we say that what produces the impression is a signet of bronze or gold, but its particular metallic constitution makes no difference: in a similar way the sense is affected by what is coloured or flavoured or sounding, but it is indifferent what in each case the *substance* is; what alone matters is what *quality* it has, i.e. in what *ratio* its constituents are combined.

(B) By 'an organ of sense' is meant that in which ultimately such a power is seated.

2   ISAAC NEWTON (1642–1727) ON THE SEVEN COLORS OF THE SPECTRUM, 1675

Thomas Birch, *History of the Royal Society of London* (London, 1757), III, 262–263. On 9 December 1675 the Royal Society received from Newton a paper which he called "An hypothesis explaining the properties of light, discoursed in my several papers," and they read the second part, "explaining colours," at the next meeting, on 16 December. The excerpt is from this manuscript.

Violet, indigo, blue, green, yellow, orange, and red: into these seven colors Newton analyzed the spectrum. Musical theory—the existence of seven notes in the musical scale—influenced his judgment. For colors that is a very hard judgment to make. The modern view is that there are four principal hues—blue, green, yellow, and red—but it takes better experimental control than merely the direct observation of the spectrum to establish this tetrachromatic analysis.

And now to explain colours; I suppose, that as bodies of various sizes, densities, or sensations, do by percussion or other action excite sounds of various tones, and consequently vibrations in the air of various bigness; so when the rays of light, by impinging on the stiff refracting superficies, excite vibrations in the æther, those rays, whatever they be, as they happen to differ in magnitude, strength or vigour, excite vibrations of various bigness; the biggest, strongest, or most potent rays, the largest vibrations; and others shorter, according to their bigness, strength, or power: and therefore the ends of the capillamenta of the optic nerve, which pave or face the retina, being such refracting superficies, when the rays impinge upon them, they must there excite these vibrations, which vibrations (like those of sound in a trunk or trumpet) will run along the aqueous pores or crystalline pith of the capillamenta through the optic nerves into the sensorum (which light itself cannot do) and there, I suppose, affect the sense with various colours, according to their bigness and mixture; the biggest with the strongest colours, reds and yellows; the least with the weakest, blues and violets; the middle with green, and a confusion of all with white, much after the manner, that in the sense of hearing, nature makes use of aereal vibrations of several bignesses to generate sounds of divers tones; for the analogy of nature is to be observed. And further, as the harmony and discord of sounds proceed from the proportions of the aereal vibrations, so may the harmony of some colours, as of golden and blue, and the discord of others, as of red and blue, proceed from the proportions of the æthereal. And possibly colour may be distinguished into its principal degrees, red, orange, yellow, green, blue, indigo, and deep violet, on the same ground, that sound within an eighth is graduated into tones. For, some years past, the prismatic colours being in a well darkened room cast perpendicularly upon a paper about two and twenty foot distant from the prism, I desired a friend to draw with a pencil lines cross the image, or pillar of colours, where every one of the seven aforenamed colours was most full and brisk, and also where he judged the truest confines of them to be, whilst I held the paper so, that the said image might fall within a certain compass marked on it. And this I did, partly because my own eyes are not very critical in distinguishing colours, partly because another, to whom I had not communicated my thoughts about this matter, could have nothing but his eyes to determine his fancy in making those marks. This observation we repeated divers times, both in the same and divers days, to see how the marks on several papers would agree; and comparing the observations, though the just confines of the colours are hard to be assigned, because they pass into one another by insensible gradation; yet the *differences* of the observations were but little, especially towards the red end, and taking means between those differences, that were, the length of the image (reckoned not by the distance of the verges of the semicircular ends, but by the distance of the centres of those semicircles, or length of the strait sides as it ought to be) was divided in about the same proportion that a string is, between the end and the middle, to sound the tones in the eighth. You will understand me best by viewing the annexed figure, in which AB and CD represent the strait sides, about ten inches long, APC and BTD the semicircular ends, X and Y the centres of those semicircles, XZ the length of a musical string double to XY, and

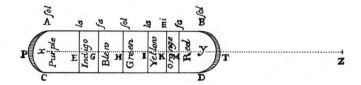

divided between *X* and *Y*, so as to sound the tones expressed at the side (that is *XH* the half, *XG* and *GI* the third part, *YK* the fifth part, *YM* the eighth part, and *GE* the ninth part of *XY*) and the intervals between these divisions express the spaces which the colours written there took up, every colour being most briskly specific in the middle of those spaces.

3   NEWTON ON THE COLOR CIRCLE, 1704

Isaac Newton, *Opticks* (London, 1704), bk. I, pt. 2, propositions 5, 6. The more available fourth edition (London, 1730) is like the first edition in this section. The substance of this report, but without the description of the color circle, appears in Newton's letters of 1672 to the Royal Society.

Newton's color circle is the basis for all the color diagrams (circular, square, triangular) that have since been proposed, but Newton's figure omits variation in brightness and provides only for difference in hue and saturation. Newton used the circle to explain color mixture when, in the mixture of two colors, the resultant hue is intermediate and the saturation less the more remote the two components. Newton is quite clear that the circle is completed by the purples, which lie between violet and red and which arise only from mixtures, having no corresponding 'homogeneal' light, although the circle as shown does not seem to take account of this peculiarity.

PROP. V. THEOR. IV.

Whiteness and all grey Colours between white and black, may be compounded of Colours, and the whiteness of the Sun's Light is compounded of all the primary Colours mix'd in a due Proportion.

. . . . .

Now, considering that these grey and dun Colours may be also produced by mixing Whites and Blacks, and by consequence differ from perfect Whites, not in Species of Colours, but only in degree of Luminousness, it is manifest that there is nothing more requisite to make them perfectly white than to increase their Light sufficiently; and, on the contrary, if by increasing their Light they can be brought to perfect Whiteness, it will thence also follow, that they are of the same Species of Colour with the best Whites, and differ from them only in the Quantity of Light. And this I tried as follows. I took the third of the above-mention'd grey Mixtures, (that which was compounded of Orpiment, Purple, Bise, and *Viride Æris*) and rubbed it thickly upon the Floor of my Chamber, where the Sun shone upon it through the opened Casement; and by it, in the shadow, I laid a Piece of white Paper of the same Bigness. Then going from them to the distance of 12 or 18 Feet, so that I could not discern the Unevenness of the Surface of the Powder, nor the little Shadows let fall from the gritty Particles thereof; the Powder appeared intensely white, so as to transcend even the Paper it self in Whiteness, especially if the Paper were a little shaded from the Light of the Clouds, and then the Paper compared with the Powder appeared of such a grey Colour as the Powder had done before. But by laying the Paper where the Sun shines through the Glass of the Window, or by shutting the Window that the Sun might shine through the Glass upon the Powder, and by such other fit Means of increasing or decreasing the Lights wherewith the Powder and Paper were illuminated, the Light wherewith the Powder is illuminated may be made stronger in such a due Proportion than the Light wherewith the Paper is illuminated, that they shall both appear exactly alike in Whiteness. For when I was trying this, a Friend coming to visit me, I stopp'd him at the Door, and before I told him what the Colours were, or what I was doing; I asked him, Which of the two Whites were the best, and wherein they differed? And after he had at that distance viewed them well, he answer'd, that they were both good Whites, and that he could not say which was best, nor wherein their Colours differed. Now, if you consider, that this White of the Powder in the Sun-shine was compounded of the Colours which the component Powders (Orpiment, Purple, Bise, and *Viride Æris*) have in the same Sun-shine, you must acknowledge by this Experiment, as well as by the former, that perfect Whiteness may be compounded of Colours.

· · · · · ·

PROP. VI. PROB. II.

In a mixture of Primary Colours, the Quantity and Quality of each being given, to know the Colour of the Compound.

With the Center *O* [in Fig. 11] and Radius *OD* describe a Circle *ADF*, and distinguish its Circumference into seven Parts *DE, EF, FG, GA, AB, BC, CD,* proportional to the seven Musical Tones or Intervals of the eight Sounds, *Sol, la, fa, sol, la, mi, fa, sol,* contained in an eight, that is, proportional to the Number $\frac{1}{9}, \frac{1}{16}, \frac{1}{10}, \frac{1}{9}, \frac{1}{16}, \frac{1}{16}, \frac{1}{9}$. Let the first Part *DE* represent a red Colour, the second *EF* orange, the third *FG* yellow, the fourth *GA* green, the fifth *AB* blue, the sixth *BC* indigo, and the seventh *CD* violet. And conceive that these are all the Colours of uncompounded Light gradually passing into one another, as they do when made by Prisms; the Circumference *DEFGABCD*, representing the whole Series of Colours from one end of the Sun's colour'd Image to the other, so that from *D* to *E* be all degrees of red, at *E* the mean Colour between red and orange, from *E* to *F* all degrees of orange, at *F* the mean between orange and yellow, from *F* to *G* all degrees of yellow, and so on. Let *p* be the Center of Gravity of the Arch *DE*, and *q, r, s, t, u, x*, the Centers

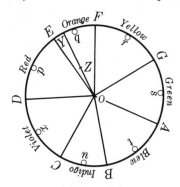

Fig. 11.

of Gravity of the Arches *EF, FG, GA, AB, BC*, and *CD* respectively, and about those Centers of Gravity let Circles proportional to the Number of Rays of each Colour in the given Mixture be describ'd: that is, the Circle *p* proportional to the Number of the red-making Rays in the Mixture, the Circle *q* proportional to the Number of the orange-making Rays in the Mixture, and so of the rest. Find the common Center of Gravity of all those Circles, *p, q, r, s, t, u, x*. Let that Center be *Z*; and from the Center of the Circle *ADF*, through *Z* to the Circumference, drawing the Right Line *OY*, the Place of the Point *Y* in the Circumference shall shew the Colour arising from the Composition of all the Colours in the given Mixture, and the Line *OZ* shall be proportional to the Fulness or Intenseness of the Colour, that is, to its distance from Whiteness. As if *Y* fall in the middle between *F* and *G*, the compounded Colour shall be the best yellow; if *Y* verge from the middle towards *F* or *G*, the compound Colour shall accordingly be a yellow, verging towards orange or green. If *Z* fall upon the Circumference, the Colour shall be intense and florid in the highest Degree; if it fall in the mid-way between the Circumference and Center, it shall be but half so intense, that is, it shall be such a Colour as would be made by diluting the intensest yellow with an equal quantity of whiteness; and if it fall upon the center *O*, the Colour shall have lost all its intenseness, and become a white. But it is to be noted, That if the point *Z* fall in or near the line *OD*, the main ingredients being the red and violet, the Colour compounded shall not be any of the prismatick Colours, but a purple, inclining to red or violet, accordingly as the point Z lieth on the side of the line DO towards E or towards C, and in general the compounded violet is more bright and more fiery than the uncompounded.

4   THOMAS YOUNG (1773–1829) ON NEWTON AND THE EXCITATION OF THE RETINA BY COLORS, 1802

Thomas Young, "On the theory of light and colours," *Philosophical Transactions of the Royal Society of London 92,* 18–21 (1802); also in George Peacock, ed., *Miscellaneous Works of the Late Thomas Young,* vol. I (London, 1855). This lecture was delivered on 12 November 1801. In it Young quotes a letter of Newton's, written on 18 November 1672 and published in *Philosophical Transactions 7,* 5088–5089 (1672).

Theory about the differentiation of colors matured slowly. It was 130 years from Newton's letter, which Young quotes, to his own similar statement of the problem, and fifty years more until Helmholtz (1852) made this view clear and secure. Newton had said that the biggest (longest) vibrations of light excite red at the retina and the shortest violet. In his scholium Young takes over Newton's laws of color mixture to conclude that all colors can be produced by a mixture of the proper three and thus to suggest that there may be in the retina three kinds of nerve fibers that respond differentially to light vibrations of different frequencies. It could be assumed that the many other colors were occasioned by different proportions of the excitations of the three fundamental kinds of fibers. This, then, is the origin in Newton and Young of the "Young-Helmholtz" theory of color vision. Actually this idea of Newton's and Young's constitutes an anticipation of the theory of specific energies of separate nerve fibers long before the general theory (Charles Bell in 1811, Johannes Müller in 1826) and the extension of the theory to specific fibers (Helmholtz in 1852) were thought of.

HYPOTHESIS III

The Sensation of different Colours depends on the different frequency of Vibrations, excited by Light in the Retina.

Passages from Newton

"The objector's [Robert Hooke's] hypothesis, as to the fundamental part of it, is not against me. That fundamental supposition is, that the parts of bodies, when briskly agitated, do excite vibrations in the ether, which are propagated every way from those bodies in straight lines, and cause a sensation of light by beating and dashing against the bottom of the eye, something after the manner that vibrations in the air cause a sensation of sound by beating against the organs of hearing. Now, the most free and natural application of this hypothesis to the solution of phenomena, I take to be this: that the agitated parts of bodies, according to their several sizes, figures, and motions, do excite vibrations in the ether of various depths or bignesses, which, being promiscuously propagated through that medium to our eyes, effect in us a sensation of light of a white colour; but if by any means those of unequal bignesses be separated from one another, the largest beget a sensation of a red colour, the least or shortest of a deep violet, and the intermediate ones of intermediate colours; much after the manner that bodies, according to their several sizes, shapes, and motions, excite vibrations in the air of various bignesses, which, according to those bignesses, make several tones in sound: that the largest vibrations are best able to overcome the resistance of a refracting superficies, and so break through it with least refraction; whence the vibrations of several bignesses, that is, the rays of several colours, which are blended together in light, must be parted from one another by refraction, and so cause the phenomena of prisms, and other refracting substances; and that it depends on the thickness of a thin transparent plate or bubble, whether a vibration shall be reflected at its further superficies, or transmitted; so that, according to the number of vibrations, interceding the two superficies, they may be reflected or transmitted for many successive thicknesses. And, since the vibrations which make blue and violet, are supposed shorter than those which make red and yellow, they must be reflected at a less thickness of the plate: which is sufficient to explicate all the ordinary phenomena of those plates or bubbles, and also of all natural bodies, whose parts are like so many fragments of such plates."

·   ·   ·   ·   ·

Scholium. Since, for the reason here assigned by Newton, it is probable that the motion of the retina is rather of a vibratory than of an undulatory nature, the frequency of the vibrations must be dependent on the constitution of this substance. Now, as it is almost impossible to conceive each sensitive point of the retina to contain an infinite number of particles, each capable of vibrating in perfect unison with every possible undulation, it becomes necessary to suppose the number limited, for instance, to the three principal colours, red, yellow, and blue, of which the undulations are related in magnitude nearly as the numbers 8, 7, and 6; and that each of the particles is capable of being put in motion less or more forcibly, by undulations differing less or more from a perfect unison; for instance, the undulations of green light being nearly in the ratio of $6\frac{1}{2}$, will affect equally the particles in unison with yellow and blue, and produce the same effect as a light composed of those two species: and each sensitive filament of the nerve may consist of three portions, one for each principal colour. Allowing this statement, it appears that any attempt to produce a musical effect from colours, must be unsuccessful, or at least that nothing more than a very simple melody could be imitated by them; for the period, which in fact constitutes the harmony of any concord, being a multiple of the periods of the single undulations, would in this case be wholly without the limits of sympathy of the retina, and would lose its effect; in the same manner as the harmony of a third or a fourth is destroyed, by depressing it to the lowest notes of the audible scale. In hearing, there seems to be no permanent vibration of any part of the organ.

5   JOHN LOCKE (1632–1704) ON PRIMARY AND SECONDARY QUALITIES, 1690

John Locke, *An Essay concerning Humane Understanding: In Four Books* (London, 1690), bk. II, chap. 8.

The distinctions between primary and secondary qualities in these sections anticipate the later doctrine of sensory attributes in which quality is equivalent to Locke's secondary qualities. In a sense the doctrine of secondary qualities anticipates Johannes Müller's doctrine of specific nerve energies in that it asserts that what the mind perceives is different from the property of the object that arouses the quality.

7. To discover the nature of our *Ideas* the better, and to discourse of them intelligibly, it will be convenient to distinguish them, as they are *Ideas,* or Perceptions in our Minds; and as they are in the Bodies, that cause such Perceptions in us: that so we *may not* think (as perhaps usually is done) that they are exactly the Images and *resemblances* of something inherent in the subject; most of those of Sensation being in the Mind no more the likeness of something existing without us, than the Names that stand for them, are the likeness of our *Ideas,* which yet upon hearing, they are apt to excite in us.

8. Whatsoever the Mind perceives in it self, or is the immediate object of Perception, Thought, or Understanding, that I call *Idea;* and the power to produce any *Idea* in our mind, I call *Quality* of the Subject wherein that power is. Thus a Snow-ball having the power to produce in us the *Ideas* of *White, Cold,* and *Round,* the powers to produce those *Ideas* in us, as they are in the Snow-ball, I call *Qualities;* and as they are Sensations, or Perceptions, in our Understandings, I call them *Ideas:* which *Ideas,* if I speak of sometimes, as in the things themselves, I would be understood to mean those Qualities in the Objects which produce them in us.

9. Concerning these *Qualities,* we may, I think, observe these *primary* ones in Bodies, that produce simple *Ideas* in us, viz. Solidity, Extension, Motion or Rest, Number and Figure.

10. These which I call *original* or *primary Qualities* of Body, are wholly inseparable from it; and such as in all the alterations and changes it suffers, all the force can be used upon it, it constantly keeps; and such as Sense constantly finds in every particle of Matter, which has bulk enough to be perceived, and the Mind finds inseparable from every particle of Matter, though less than to make it self singly be perceived by our Senses. *e.g.,* Take a grain of Wheat, divide it into two parts, each part has still *Solidity, Extension, Figure,* and *Mobility;* divide it again, and it retains still the same qualities; and so divide it on, till the parts become insensible, they must retain still each of them all those qualities. For division (which is all that a Mill, or Pestle, or any other Body, does upon another, in reducing it to insensible parts) can never take away either Solidity, Extension, Figure, or Mobility from any Body, but only makes two distinct Bodies, or more, of one, which altogether after division have their certain number.

11. The next thing to be considered, is, how *Bodies operate* one upon another, and that is manifestly *by impulse,* and nothing else. It being impossible to conceive, that Body should operate on what it does not touch, (which is all one as to imagine it can operate where it is not) or when it does touch, operate any other way than by Motion.

12. If then Bodies cannot operate at a distance; if external Objects be not united to our Minds, when they produce *Ideas* in it; and yet we perceive *these original Qualities* in such of them, as singly fall under our Senses, 'tis evident that some motion must be thence continued by our Nerves, or animal Spirits, by some parts of our Bodies, to the Brains, the seat of Sensation, there to *produce in our Minds the particular Ideas we have of them.* And since the Extension, Figure, Number, and Motion of Bodies of an observable bigness, may be perceived at a distance *by* the sight, 'tis evident some singly imperceptible Bodies must come from them to the Eyes, and thereby convey to the Brain some *Motion,* which produces these *Ideas* we have of them in us.

13. After the same manner, that the *Ideas* of these original Qualities are produced in us, we may conceive, that the *Ideas of secundary Qualities* are also *produced,* viz. *by the operation of insensible particles on our Senses.* For it being manifest, that there are Bodies, and good store of Bodies, each whereof is so small, that we cannot, by any of our Senses, discover either their bulk, figure, or motion, as is evident in the Particles of the Air and Water, and other extreamly smaller than those, perhaps, as much less than the Particles of Air, or Water, as the Par-

ticles of Air, or Water, are smaller than Pease or Hail-stones. Let us suppose at present, that the different Motions and Figures, Bulk, and Number of such Particles, affecting the several Organs of our Senses, produce in us those different Sensations, which we have from the Colours and Smells of Bodies, *e.g.* a Violet, by which impulse of those insensible Particles of Matter of different figures and bulks, and in a different Degree and Modification, we may have the *Ideas* of the blue Colour, and sweet Scent of a Violet produced in our Minds. It being no more conceived impossible, to conceive, that God should annex such *Ideas* to such Motions, with which they have no similitude; than that he should annex the *Idea* of Pain to the motion of a peice of Steel, dividing our Flesh, with which that *Idea* hath no resemblance.

14. What I have said concerning *Colours* and *Smells,* may be understood also of *Tastes,* and *Sounds, and other the like sensible Qualities;* which, whatever reality we by mistake attribute to them, are in truth nothing in the Objects themselves, but Powers to produce various Sensations in us, and *depend on those primary Qualities, viz.* Bulk, Figure, Texture, and Motion of Parts; and therefore *I call* them *Secondary Qualities.*

15. From whence, I think, it is easie to draw this Observation, That the *Ideas of primary Qualities* of Bodies, *are Resemblances* of them, and their Patterns do really exist in the Bodies themselves; but the *Ideas, produced* in us *by* these *Secondary Qualities, have no resemblance* of them at all. There is nothing like our *Ideas,* existing in the Bodies themselves. They are in the Bodies, we denominate from them, only a Power to produce those Sensations in us: And what is Sweet, Blue, or Warm in *Idea,* is but the certain Bulk, Figure, and Motion of the insensible Parts, in the Bodies themselves we call so.

.   .   .   .   .

23. The *Qualities* then that are *in Bodies* rightly considered, are of *Three sorts:*

First, The *Bulk, Figure, Number, Situation,* and *Motion, or Rest* of their solid Parts; these are in them, whether we perceive them or no; and when they are of that size, that we can discover them, we have by these an *Idea* of the thing, as it is in it self, as is plain in artificial things. These I call *primary Qualities.*

Secondly, The *Power* that is in any Body, *by* Reason of *its* insensible *primary Qualities,* to operate after a peculiar manner on any of our Senses, and thereby *produce in us* the *different Ideas* of several Colours, Sounds, Smells, Tastes, &c. These are usually called sensible Qualities.

Thirdly, The *Power* that is in any Body, *by* Reason of the particular Constitution of *its primary Qualities, to* make such a *change* in the *Bulk, Figure, Texture, and Motion of another Body,* as to make it operate on our Senses, differently from what it did before. Thus the Sun has a Power to make Wax white, and Fire to make Lead fluid.

The First of these, as has been said, I think, may be properly called *real, Original,* or *primary Qualities,* because they are in the things themselves, whether they are perceived or no: and upon their different Modifications it is, that the secundary Qualities depend.

The other two, are only Powers to act differently upon other things, which Powers result from the different Modifications of those primary Qualities.

6   CHARLES BELL (1774–1842) ON SPINAL NERVE ROOTS, 1811

Charles Bell, *Idea of a New Anatomy of the Brain: Submitted for the Observation of His Friends* (privately printed, London, 1811), pp. 21–24, 28–29, 34–37. Variously reprinted, most recently in Wayne Dennis, ed., *Readings in the History of Psychology* (New York: Appleton-Century-Crofts, 1948), pp. 113–124.

That the motor nerves leave the spinal cord by the anterior roots and the sensory nerves enter it by the posterior roots is the Bell-Magendie law of the spinal nerve roots. Bell made the discovery first as reported in this pamphlet, that is to say, he found that direct excitation of the anterior roots of the spinal cord of a rabbit produces muscular convulsions of the back, whereas direct stimulation of the posterior roots, having no observable effect, may be assumed to be sensory in function. Magendie came to this same conclusion eleven years later as a result of more carefully controlled and thus more convincing experiments. There was controversy about which investigator should receive credit for the discovery, and so the law was finally named for both men. This excerpt shows Bell distinguishing between sensory and motor nerves and tracts with a clarity that had been impossible before this separation of the two neural functions was established.

The *medulla spinalis* has a central division, and also a distinction into anterior and posterior fasciculi, corresponding with the anterior and posterior portions of the brain. Further we can trace down the crura of the *cerebrum* into the anterior fasciculus of the spinal marrow, and the crura of the *cerebellum* into the posterior fasciculus. I thought that here I might have an opportunity of touching the *cerebellum,* as it were, through the posterior portion of the spinal marrow, and the cerebrum by the anterior portion. To this end I made experiments which, though they were not conclusive, encouraged me in the view I had taken.

I found that injury done to the anterior portion of the spinal marrow, convulsed the animal more certainly than injury done to the posterior portion; but I found it difficult to make the experiment without injuring both portions.

Next considering that the spinal nerves have a double root, and being of opinion that the properties of the nerves are derived from their connections with the parts of the brain, I thought that I had an opportunity of putting my opinion to the test of experiment, and of proving at the same time that nerves of different endowments were in the same cord, and held together by the same sheath.

On laying bare the roots of the spinal nerves, I found that I could cut across the posterior fasciculus of nerves, which took its origin from the posterior portion of the spinal marrow without convulsing the muscles of the back; but that on touching the anterior fasciculus with the point of the knife, the muscles of the back were immediately convulsed.

Such were my reasons for concluding that the cerebrum and the cerebellum were parts distinct in function, and that every nerve possessing a double function obtained that by having a double root. I now saw the meaning of the double connection of the nerves with the spinal marrow; and also the cause of that seeming intricacy in the connections of nerves throughout their course, which were not double at their origins.

The spinal nerves being double, and having their roots in the spinal marrow, of which a portion comes from the cerebrum and a portion from the cerebellum, they convey the attributes of both grand divisions of the brain to every part; and therefore the distribution of such nerves is simple, one nerve supplying its destined part. But the nerves which come directly from the brain, come from parts of the brain which vary in operation; and in order to bestow different qualities on the parts to which the nerves are distributed, two or more nerves must be united in their course or at their final destination.

· · · · ·

The nerves of sense, the olfactory, the optic, the auditory, and the gustatory nerve, are traced backwards into certain tubercles or convex bodies in the base of the brain. And I may say, that the nerves of sense either form tubercles before entering the brain, or they enter into those convexities in the base of the *cerebrum*. These convexities are the constituent parts of the cerebrum, and are in all animals necessary parts of the organs of sense: for as certainly as we discover an animal to have an external organ of sense, we find also a medullary tubercle; whilst the superiority of animals in intelligence is shewn by the greater magnitude of the hemispheres or upper part of the cerebrum.

· · · · ·

All ideas originate in the brain: the operation producing them is the remote effect of an agitation or impression on the extremities of the nerves of sense; directly they are consequences of a change or operation in the proper organ of the sense which constitutes a part of the brain, and over these organs, once brought into action by external impulse, the mind has influence. It is provided, that the extremities of the nerves of the senses shall be susceptible each of certain qualities in matter; and betwixt the impression of the outward sense, as it may be called, and the exercise of the internal organ, there is established a connection by which the ideas excited have a permanent correspondence with the qualities of bodies which surround us.

· · · · ·

From the cineritious matter, which is chiefly external, and forming the surface of the cerebrum; and from the grand center of medullary matter of the cerebrum, what are called the *crura* descend. These are fasciculated processes of the cerebrum, from which go off the nerves of motion, the nerves governing the muscular frame. Through the nerves of sense, the *sensorium* receives impressions, but the will is expressed through the medium of the nerves of motion. The secret operations of the bodily frame, and the connections which unite the parts of the body into a system, are through the cerebellum and nerves proceeding from it.

7   FRANÇOIS MAGENDIE (1783–1855) ON SPINAL NERVE ROOTS, 1822

François Magendie, "Expériences sur les fonctions des racines des nerfs rachidiens," *Journal de physiologie expérimentale et pathologique 2,* 276–279 (1822); "Expériences sur les fonctions des nerfs qui naissent de la moëlle épinière," *ibid.,* pp. 366–371. Translated for this book by Mollie D. Boring.

Here is the account of Magendie's   of the spinal nerve roots.
experiments which established the law

EXPERIMENTS ON THE FUNCTIONS OF THE SPINAL NERVE ROOTS

For a long time I have wanted to do the experiment of cutting the posterior roots of the nerves emanating from the spinal cord in an animal. I have made many attempts to do this, but without success, since it is difficult to open the vertebral canal without injuring the cord and causing the death of the animal, or at any rate gravely wounding it. But last month a litter of eight puppies, six weeks old, was brought into my laboratory; these animals seemed to me eminently suited for a new attempt at opening the vertebral canal. And, in fact, by using a very sharp scalpel I was able with a single stroke, so to speak, to lay bare the posterior half of the spinal cord within its envelopes. With this organ now all but bared, I had only to cut the surrounding *dura mater,* and this I did easily; I now had the posterior roots of the lumbar and sacral pairs before my eyes, and, lifting them successively with the blades of a small pair of scissors, I was able to cut them on one side and leave the cord intact. I did not know what would result from this operation. I stitched up the wound by suturing the skin and observed the animal. At first I believed the limb corresponding to the cut nerves to be completely paralyzed; it was insensitive to pricking and the hardest pressures and, further, it seemed immobile; but soon, to my very great surprise, I clearly saw it move, although sensibility remained completely absent. A second and a third experiment gave me exactly the same result; it began to seem to me probable that the posterior roots of the spinal nerves could have different functions from the anterior roots and that these pertained most particularly to sensibility.

It naturally occurred to me now to cut the anterior roots while leaving the posterior ones intact, but such a project was easier to entertain than to carry out. How could one get around the posterior roots to get at the anterior part of the cord? I confess the problem at first seemed to me insurmountable; however, I kept thinking about it for two days, and I finally decided to use a sort of cataract knife, with a blade narrow enough to get in under the posterior roots and to cut the anterior roots by pressing them against the posterior surface of the vertebrae; but I was obliged to give up this tactic because of the large veins in the canal on the side that I opened with each movement forward. In making these attempts, I noticed that, by pulling up the vertebral *dura mater,* I could see the anterior roots joined into bundles at their points of entry into the membrane. I could not have asked for anything better, and in a few minutes I had cut all the pairs I wished to divide. As in the previous experiments, I made the section on one side only, so as to have a measure of comparison. You can imagine with what curiosity I followed the results of this section: there could be no doubt whatsoever; the limb was completely immobile and flaccid, although it retained an unequivocal sensibility. Finally, for completeness' sake, I cut both the anterior and the posterior roots; there was an absolute loss of feeling and movement.

I repeated and varied these experiments on several species of animals; the results I have just described were verified in every way, both for the anterior and the posterior limbs. I am continuing this research . . . It is at present sufficient for me to be able to state here positively that the anterior and posterior roots of the nerves emanating from the spinal cord have different functions, that the posterior seem to pertain more particularly to sensibility, whereas the anterior seem especially linked with movement.

.   .   .   .   .

Experiments on the Functions of Nerve Roots Emanating from the Spinal Cord

The discoveries I announced . . . are too important for me not to have tried clarifying them in further research.

First, I wanted to make sure it was not possible to cut the anterior or posterior roots of the spinal nerves without opening the large canal of the vertebral *dura mater;* for, in exposing the spinal cord to air and cold temperature, one appreciably weakens nervous action and, perforce, reaches one's conclusions in a roundabout manner.

The anatomical arrangement of these parts did not make the task impossible, for each bundle of spinal roots follows on for a little within a particular canal before rejoining and fusing with the other bundle. Hence, using scissors with damp points, I found it possible to remove enough of the vertebral lamina and lateral parts to expose the ganglion of each lumbar pair; it is then not too difficult—using a small probe—to separate the canal containing the posterior roots, and there is no further difficulty in making the section. This method of doing the experiment yielded the same results I had observed previously; but, as the experiment is much longer and more laborious than in the procedure of opening the large canal of the spinal *dura mater,* I see no reason to prefer it to the first method.

I next wanted to submit my earlier results to a further proof. Everyone knows that nux vomica will cause very violent general tetanic convulsions in man and animals. I was curious to know whether these convulsions would take place in a limb in which the nerves of movement had been severed, and whether they would be just as violent with the nerves of feeling severed. The results agreed completely with my earlier ones; that is to say, in an animal whose posterior roots had been cut the tetanus was complete and as intense as if the spinal roots were intact; on the other hand, with an animal in which I had cut the nerves of movement of one posterior limb, this limb remained supple and immobile at the moment when, under the influence of the poison, all the other body muscles showed the most pronounced tetanic contractions.

By irritating the nerves of feeling directly, or the posterior spinal roots, could one produce contractions? Would a direct irritation to the nerves of movement evoke pain? Such were the questions I asked myself, and experimentation alone could answer them.

With these in mind, I began by examining the posterior roots, the nerves of feeling. Here is what I observed: in pinching, plucking, or pricking these roots, the animal manifests pain; but this pain is nothing compared to the intensity that develops if one touches, even lightly, the spinal cord at the origin of these roots. Almost every time that the posterior roots are excited contractions are produced in the corresponding muscles; however, these contractions are not very marked and are infinitely weaker than if one touches the spinal cord itself. If one severs an entire bundle of posterior roots, a gross movement is produced in the limb corresponding to the bundle.

I repeated these operations on the anterior bundles and obtained analagous, though opposite, results: for here the contractions induced by pinching, jabbing, and the like are very strong and even convulsive, whereas the signs of sensibility are scarcely visible. These facts, then, confirm my earlier statements, though they seem to establish the fact that feeling is not exclusively in the posterior roots, any more than movement is exclusively in the anterior ones. However, one problem may arise. When, in the preceding experiment, the roots had been cut, they were still continuous with the spinal cord. Could not the disturbance communicated to the cord itself be the real source of the contractions, or even of the pain, experienced by the animals? To remove this doubt, I redid the experiments after separating the roots from the cord; and I must say that in all cases—with the exception of two animals in which I saw contractions when I pinched or pulled the anterior and posterior bundles—I observed no appreciable effect of the irritation of the anterior or posterior roots thus separated from the cord.

I had still another type of proof to which to submit the spinal roots: namely, galvanism. Therefore I excited the spinal roots by this method, at first leaving them in their normal state and then cutting them at the spine in order to isolate them. In these different cases, I obtained contractions with both sorts of roots; but the contractions that followed excitation of the anterior roots were in general stronger, more nearly total, than those which came when electric current was introduced into the posterior roots. The phenomena were the same whether one placed the zinc pole or the copper pole on the nerve.

8   BELL ON THE SPECIFICITY OF SENSORY NERVES, 1811

Charles Bell, *Idea of a New Anatomy of the Brain: Submitted for the Observation of His Friends* (privately printed, London, 1811). Variously reprinted, most recently in Wayne Dennis, ed., *Readings in the History of Psychology* (New York: Appleton-Century-Crofts, 1948), pp. 113–124.

These introductory paragraphs to Sir Charles Bell's pamphlet of 1811 constitute the claim that he anticipated Johannes Müller's doctrine of the specific energies of nerves, a conception which, as this chapter has already shown in connection with John Locke's secondary qualities and Thomas Young's guess about the three kinds of fibers in the optic nerve, had already found its way into scientific thinking.

The want of any consistent history of the Brain and Nerves, and the dull unmeaning manner which is in use of demonstrating the brain, may authorize any novelty in the manner of treating the subject.

I have found some of my friends so mistaken in their conception of the object of the demonstrations which I have delivered in my lectures, that I wish to vindicate myself at all hazards. They would have it that I am in search of the seat of the soul; but I wish only to investigate the structure of the brain, as we examine the structure of the eye and ear.

It is not more presumptuous to follow the tracts of nervous matter in the brain, and to attempt to discover the course of sensation, than it is to trace the rays of light through the humours of the eye, and to say, that the retina is the seat of vision. Why are we to close the investigation with the discovery of the external organ?

．　　　．　　　．　　　．　　　．

The prevailing doctrine of the anatomical schools is, that the whole brain is a common sensorium; that the extremities of the nerves are organized, so that each is fitted to receive a peculiar impression; or that they are distinguished from each other only by delicacy of structure, and by a corresponding delicacy of sensation: that the nerve of the eye, for example, differs from the nerves of touch only in the degree of its sensibility.

It is imagined that impressions, thus differing in kind, are carried along the nerves to the sensorium, and presented to the mind; and that the mind, by the same nerves which receive sensation, sends out the mandate of the will to the moving parts of the body.

．　　　．　　　．　　　．　　　．

I have to offer reasons for believing: . . . That the external organs of the senses have the matter of the nerves adapted to receive certain im-

23

pressions, while the corresponding organs of the brain are put in activity by the external excitement: That the idea or perception is according to the part of the brain to which the nerve is attached, and that each organ has a certain limited number of changes to be wrought upon it by the external impression.

. . . . .

When this whole was created, (of which the remote planetary system, as well as our bodies, and the objects more familiar to our observation, are but parts,) the mind was placed in a body not merely suited to its residence, but in circumstances to be moved by the materials around it; and the capacities of the mind, and the powers of the organs, which are as a medium betwixt the mind and the external world, have an original constitution framed in relation to the qualities of things.

It is admitted that neither bodies nor the images of bodies enter the brain. It is indeed impossible to believe that colour can be conveyed along a nerve; or the vibration in which we suppose sound to consist can be retained in the brain: but we can conceive, and have reason to believe, that an impression is made upon the organs of the outward senses when we see, or hear, or taste.

In this inquiry it is most essential to observe, that while each organ of sense is provided with a capacity of receiving certain changes to be played upon it, as it were, yet each is utterly incapable of receiving the impressions destined for another organ of sensation.

It is also very remarkable that an impression made on two different nerves of sense, though with the same instrument, will produce two distinct sensations; and the ideas resulting will only have relation to the organ affected.

As the announcing of these facts forms a natural introduction to the Anatomy of the Brain, which I am about to deliver, I shall state them more fully.

There are four kinds of Papillae on the tongue, but with two of those only we have to do at present. Of these, the Papillae of one kind form the seat of the sense of taste; the other Papillae (more numerous and smaller) resemble the extremities of the nerves in the common skin, and are the organs of touch in the tongue. When I take a sharp steel point, and touch one of *these* Papillae, I feel the sharpness. The sense of *touch* informs me of the shape of the instrument. When I touch a Papilla of taste, I have no sensation similar to the former. I do not know that a point touches the tongue, but I am sensible of a metallic taste, and the sensation passes backward on the tongue.

In the operation of couching the cataract, the pain of piercing the retina with a needle is not so great as that which proceeds from a grain of sand under the eyelid. And although the derangement of the stomach sometimes marks the injury of an organ so delicate, yet the pain is occasioned by piercing the outward coat, not by the affection of the expanded nerve of vision.

If the sensation of light were conveyed to us by the retina, the organ of vision, in consequence of that organ being as much more sensible than the surface of the body as the impression of light is more delicate than that pressure which gives us the sense of touch; what would be the feelings of a man subjected to an operation in which a needle were pushed through the nerve. Life could not bear so great a pain.

But there is an occurrence during this operation on the eye, which will direct us to the truth: when the needle pierces the eye, the patient has the sensation of a spark of fire before the eye.

This fact is corroborated by experiments made on the eye. When the eye-ball is pressed on the side, we perceive various coloured light. Indeed the mere effect of a blow on the head might inform us, that sensation depends on the exercise of the organ affected, not on the impression conveyed to the external organ; for by the vibration caused by the blow, the ears ring, and the eye flashes light, while there is neither light nor sound present.

It may be said, that there is here no proof of the sensation being in the brain more than in the external organ of sense. But when the nerve of a stump is touched, the pain is as if in the amputated extremity. If it be still said that this is no proper example of a peculiar sense existing without its external organ, I offer the following example: Quando penis glandem exedat ulcus, et nihil nisi granulatio maneat, ad extremam tamen nervi pudicae partem ubi terminatur sensus supersunt et exquisitissima sensus gratificatio.

If light, pressure, galvanism, or electricity produce vision, we must conclude that the idea in the mind is the result of an action excited in the eye or in the brain, not of any thing received, though caused by an impression from without. The operations of the mind are confined not by the limited nature of things created, but by the limited number of our organs of sense. By induction we know that things exist which yet are not brought under the operation of the senses. When we have never known the operation of one of the organs of the five senses, we can never know the ideas pertaining to that sense; and what would be the effect on our minds, even constituted as they now are, with a super-added organ of sense, no man can distinctly imagine.

As we are parts of the creation, so God has bound us to the material world by this law of our nature, that it shall require excitement from without, and an operation produced by the action of things external to rouse our faculties: But that once brought into activity, the organs can be put in exercise by the mind, and be made to minister to the memory and imagination, and all the faculties of the soul.

I shall hereafter shew, that the operations of the mind are seated in the great mass of the cerebrum, while the parts of the brain to which the nerves of sense tend, strictly form the seat of the sensation, being the internal organs of sense. These organs are operated upon in two directions. They receive the impression from without, as from the eye and ear: and as their action influences the operations of the brain producing perception, so are they brought into action and suffer changes similar to that which they experience from external pressure by the operation of the will; or, as I am now treating of the subject anatomically by the operation of the great mass of the brain upon them.

In all regulated actions of the muscles we must acknowledge that they are influenced through the same nerves, by the same operation of the sensorium. Now the operations of the body are as nice and curious, and as perfectly regulated before Reason has sway, as they are at any time after, when the muscular frame might be supposed to be under the guidance of sense and reason. Instinctive motions are the operations of the same organs, the brain and nerves and muscles, which minister to reason and volition in our mature years. When the young of any animal turns to the nipple, directed by the sense of smelling, the same operations are performed, and through the same means, as afterwards when we make an effort to avoid what is noxious, or desire and move towards what is agreeable.

9 JOHANNES MÜLLER (1801–1858) ON THE SPECIFIC ENERGIES OF NERVES, 1838

Johannes Müller, *Handbuch der Physiologie des Menschen,* bk. V (Coblenz, 1838), Introduction. Translated by William Baly as *Elements of Physiology,* vol. II (London, 1842).

Johannes Müller, sometimes called the father of experimental physiology, first formulated his principle of the specific energies of nerves in 1826, but the elaborated argument for the doctrine, as reprinted here in part, appeared in his *Handbuch* twelve years later. The importance of this doctrine for psychology was later said by Helmholtz to be comparable to the importance of the theory of the conservation of energy for physics. One should note that Müller, although he preferred to think of the specificity as residing in the nerves, entertained the possibility that it might reside in their terminations in the brain, a view which anticipates the theory of cerebral sensory centers.

The senses, by virtue of the peculiar properties of their several nerves, make us acquainted with the states of our own body, and they also inform us of the qualities and changes of external nature as far as these give rise to changes in the condition of the nerves. Sensation is a property common to all the senses; but the kind of sensation is different in each: thus we have the sensations of light, of sound, of taste, of smell, and of feeling, or touch. By feeling, or touch, we understand the peculiar kind of sensation of which the ordinary sensitive nerves generally—as, the nervus trigeminus, vagus, glossopharyngeus, and the spinal nerves,—are susceptible; the sensations of itching, of pleasure and pain, of heat and cold, and those excited by the act of touch in its more limited sense, are varieties of this mode of sensation. That which through the medium of our senses is actually perceived by the sensorium, is indeed merely a property or change of condition of our nerves; but the imagination and reason are ready to interpret the modifications in the state of the nerves produced by external influences as properties of the external bodies themselves. This mode of regarding sensations has become so habitual in the case of the senses which are more rarely affected by internal causes, that it is only on reflection that we perceive it to be erroneous. In the case of the sense of feeling or touch, on the contrary, where the peculiar sensations of the nerves perceived by the sensorium are excited as frequently by internal as by external causes, it is easily conceived that the feeling of pain or pleasure, for example, is a condition of the nerves, and not a property of the things which excite it. This leads us to the consideration of some general laws, a knowledge of which is necessary before entering on the physiology of the separate senses.

I. In the first place, it must be kept in mind that *external agencies can give rise to no kind of sensation which cannot also be produced by internal causes, exciting changes in the condition of our nerves.*

In the case of the sense of touch, this is at once evident. The sensations of the nerves of touch (or common sensibility) are those of cold and heat, pain and pleasure, and innumerable modifications of these, which are neither painful nor pleasurable, but yet have the same kind of sensation as their element, though not in an extreme degree. All these sensations are constantly being produced by internal causes in all parts of our body endowed with sensitive nerves; they may also be excited by causes acting from without, but external agencies are not capable of adding any new element to their nature. The sensations of the nerves of touch are therefore states or qualities proper to themselves, and merely rendered manifest by exciting causes external or internal. The sensation of smell also may be perceived independently of the application of any odorous substance from without, the nerve of smell being thrown by an internal cause into the condition requisite for the production of the sensation. This perception of the sensation of odours without an external exciting cause, though not of frequent occurrence, has been many times observed in persons of an irritable nervous system; and the sense of taste is probably subject to the same affection, although it would always be difficult to determine whether the taste might not be owing to a change in the qualities of the saliva or mucus of the mouth; the sensation of nausea, however, which belongs to the sensations of taste, is certainly very often perceived as the result of a merely internal affection of the nerves. The sensations of the sense of vision, namely, colour, light, and darkness, are also perceived independently of all external exciting cause. In the state of the most perfect freedom from excitement, the optic nerve has no other sensation than that of darkness. The excited condition of the nerve is manifested, even while the eyes are closed, by the appearance of light, or luminous flashes, which are mere sensations of the nerve, and not owing to the presence of any matter of light, and consequently are not capable of illuminating any surrounding objects. Every one is aware how common it is to see bright colours while the eyes are closed, particularly in the morning when the irritability of the nerves is still considerable. These phenomena are very frequent in children after waking from sleep. Through the sense of vision, therefore, we receive from external nature no impressions which we may not also experience from internal excitement of our nerves; and it is evident that a person blind from infancy in consequence of opacity of the transparent media of the eye, must have a perfect internal conception of light and colours, provided the retina and optic nerve be free from lesion. The prevalent notions with regard to the wonderful sensations supposed to be experienced by persons blind from birth when their sight is restored by operation, are exaggerated and incorrect. The elements of the sensation of vision, namely, the sensations of light, colour, and darkness, must have been previously as well known to such persons as to those of whom the sight has always been perfect. If, moreover, we imagine a man to be from his birth surrounded merely by external objects destitute of all variety of colours, so that he could never receive the impressions of colours from without, it is evident that the sense of vision might nevertheless have been no less perfect in him than in other men; for light and colours are innate endowments of his nature, and require merely a stimulus to render them manifest.

The sensations of hearing also are excited as well by internal as by external causes; for, whenever the auditory nerve is in a state of excitement, the sensations peculiar to it, as the sounds of ringing, humming, &c. are perceived. It is by such sensations that the diseases of the auditory nerve manifest themselves; and, even in less grave, transient affections of the nervous system, the sensations of humming and ringing in the ears afford evidence that the sense of hearing participates in the disturbance.

No further proof is wanting to show, that external influences give rise in our senses to no other sensations, than those which may be excited in the corresponding nerves by internal causes.

II. *The same internal cause excites in the different senses different sensations, in each sense the sensations peculiar to it.*

One uniform internal cause acting on all the nerves of the senses in the same manner, is the accumulation of blood in the capillary vessels of the nerve, as in congestion and inflammation. This uniform cause excites in the retina, while the eyes are closed, the sensation of light and luminous flashes; in the auditory nerve, humming and ringing sounds; and in the nerves of feeling, the sensation of pain. In the same way, also, a narcotic substance introduced into the blood excites in the nerves of each sense peculiar symptoms; in the optic nerves the appearance of luminous sparks before the eyes; in the auditory nerves, "tinnitus aurium;" and in the common sensitive nerves the sensation of ants creeping over the surface.

III. *The same external cause also gives rise to different sensations in each sense, according to the special endowments of its nerve.*

The mechanical influence of a blow, concussion, or pressure excites, for example, in the eye the sensation of light and colours. It is well known that by exerting pressure upon the eye, when the eyelids are closed, we can arouse the appearance of a luminous circle; by more gentle pressure the appearance of colours may be produced, and one colour may be made to change to another. Children, waking from sleep before daylight, frequently amuse themselves with these phenomena. The light thus produced has no existence external to the optic nerve; it is merely a sensation excited in it. However strongly we press upon the eye in the dark, so as to give rise to the appearance of luminous flashes, these flashes, being merely sensations, are incapable of illuminating external objects. Of this any one may easily convince himself by experiment. I have in repeated trials never been able, by means of these luminous flashes in the eye, to recognise in the dark the nearest objects, or to see them better than before; nor could another person, while I produced by pressure on my eye the appearance of brilliant flashes, perceive in it the slightest trace of real light.

.   .   .   .   .

A mechanical influence excites also peculiar sensations of the auditory nerve: at all events, it has become a common saying, "to give a person what will make his ears ring," or "what will make his eyes flash fire," or "what will make him feel;" so that the same cause, a blow, produces in the nerves of hearing, sight, and feeling, the different sensations proper to these senses. It has not become a part of common language that a blow shall be given which will excite the sense of smell, or of taste; nor would such sayings be correct; yet mechanical irritation of the soft palate, of the epiglottis and root of the tongue, excites the sensation of nausea. The action of sonorous bodies on the organ of hearing is entirely mechanical. A sudden mechanical impulse of the air upon the organ of hearing produces the sensation of a report of different degrees of intensity according to the violence of the impulse, just as an impulse upon the organ of vision gives rise to the sensation of light. If the action of the mechanical cause on the organ of hearing be of continued duration, the sound is also continued; and when caused by a rapid succession of uniform impulses, or vibrations, it has a musical character. If we admit that the matter of light acts on bodies by mechanical oscillation (the undulation theory), we shall have another example of a mechanical influence's producing different effects on different senses. These undulations, which produce in the eye the sensations of light and sound, which are well known to result from the action of narcotics.

IV. *The peculiar sensations of each nerve of sense can be excited by several distinct causes internal and external.*

The facts on which this statement is founded have been already mentioned; for we have seen that the sensation of light in the eye is excited:

1. By the undulations or emanations which from their action on the eye are called light, although they have many other actions than this; for instance, they effect chemical changes, and are the means of maintaining the organic processes in plants.

2. By mechanical influences; as concussion, or a blow.

3. By electricity.

4. By chemical agents, such as narcotics, digitalis, &c. which, being absorbed into the blood, give rise to the appearance of luminous sparks, &c. before the eyes independently of any external cause.

5. By the stimulus of the blood in the state of congestion.

The sensation of sound may be excited in the auditory nerve:

1. By mechanical influences, namely, by the vibrations of sonorous bodies imparted to the organ of hearing through the intervention of media capable of propagating them.

2. By electricity.

3. By chemical influences taken into the circulation, such as the narcotics, or alterantia nervina.

4. By the stimulus of the blood.

The sensation of odours may be excited in the olfactory nerves:

1. By chemical influences of a volatile nature,—odorous substances.

2. By electricity.

The sensation of taste may be produced:

1. By chemical influences acting on the gustatory nerves either from without or through the medium of the blood, for, according to Magendie, dogs taste milk injected into their blood-vessels, and begin to lap with their tongues.

2. By electricity.

3. By mechanical influences; for we must refer to taste the sensation of nausea produced by mechanically irritating the velum palati, epiglottis, and root of the tongue.

The sensations of the nerves of touch or feeling are excited:

1. By mechanical influences; as sonorous vibrations, and contact of any kind.

2. By chemical influences.

3. By heat.

4. By electricity.

5. By the stimulus of the blood.

V. *Sensation consists in the sensorium's receiving through the medium of the nerves, and as the result of the action of an external cause, a knowledge of certain qualities or conditions, not of external bodies, but of the nerves of sense themselves; and these qualities of the nerves of sense are in all different, the nerve of each sense having its own peculiar quality or energy.*

The special susceptibility of the different nerves of sense for certain influences—as of the optic nerve for light, of the auditory nerve for vibrations, and so on—was formerly attributed to these nerves having each a specific irritability. But this hypothesis is evidently insufficient to explain all the facts. The nerves of the senses have assuredly a specific irritability for certain influences; for many stimuli, which exert a violent action upon one organ of sense, have little or no effect upon another: for example, light, or vibrations so infinitely rapid as those of light, act only on the nerves of vision and common sensation; slower vibrations, on the nerves of hearing and common sensation, but not upon those of vision; odorous substances only upon the olfactory nerves. The external stimuli must therefore be adapted to the organ of sense—must be "homogeneous:" thus light is the stimulus adapted to the nerve of vision;

while vibrations of less rapidity, which act upon the auditory nerve, are not adapted to the optic nerve, or are indifferent to it; for if the eye be touched with a tuning-fork while vibrating, a sensation of tremours is excited in the conjunctiva, but no sensation of light. We have seen, however, that one and the same stimulus, as electricity, will produce different sensations in the different nerves of the senses; all the nerves are susceptible of its action, but the sensations in all are different. The same is the case with other stimuli, as chemical and mechanical influences. The hypothesis of a specific irritability of the nerves of the senses for certain stimuli, is therefore insufficient; and we are compelled to ascribe, with Aristotle, peculiar energies to each nerve—energies which are vital qualities of the nerve, just as contractility is the vital property of muscle . . .

The sensation of sound, therefore, is the peculiar "energy" or "quality" of the auditory nerve; the sensation of light and colours that of the optic nerve; and so of the other nerves of sense. An exact nature it is always our own sensations that we become acquainted with, and from them we form conceptions of the properties of external objects, which may be relatively correct; but we can never submit the nature of the objects themselves to that immediate perception to which the states of the different parts of our own body are subjected in the sensorium.

VI. *The nerve of each sense seems to be capable of one determinate kind of sensation only, and not of those proper to the other organs of sense; hence one nerve of sense cannot take the place and perform the function of the nerve of another sense.*

The sensation of each organ of sense may be increased in intensity till it becomes pleasurable, or till it becomes disagreeable, without the specific nature of the sensation being altered, or converted into that of another organ of sense. The sensation of dazzling light is an unpleasant sensation of the organ of vision; harmony of colours, an agreeable one. Harmonious and discordant sounds are agreeable and disagreeable sensations of the organ of hearing. The organs of taste and smell have their pleasant and unpleasant tastes and odours; the organ of touch its pleasurable and painful feelings. It appears, therefore, that, even in the most excited condition of an organ of sense, the sensation preserves its specific character.

·   ·   ·   ·   ·

VII. *It is not known whether the essential cause of the peculiar "energy" of each nerve of sense is seated in the nerve itself, or in the parts of the brain and spinal cord with which it is connected; but it is certain that the central portions of the nerves included in the encephalon are susceptible of their peculiar sensations, independently of the more peripheral portion of the nervous cords which form the means of communication with the external organs of sense.*

The specific sensibility of the individual senses to particular stimuli—owing to which vibrations of such rapidity or length as to produce sound are perceived, only by the senses of hearing and touch, and mere mechanical influences, scarcely at all by the sense of taste—must be a property of the nerves themselves; but the peculiar mode of reaction of each sense, after the excitement of its nerve, may be due to either of two conditions. Either the nerves themselves may communicate impressions different in quality to the sensorium, which in every instance remains the same; or the vibrations of the nervous principle may in every nerve be the same and yet give rise to the perception of different sensations in the sensorium, owing to the parts of the latter with which the nerves are connected having different properties. The proof of either of these propositions I regard as at present impossible.

.   .   .   .   .

VIII. *The immediate objects of the perception of our senses are merely particular states induced in the nerves, and felt as sensations either by the nerves themselves or by the sensorium; but inasmuch as the nerves of the senses are material bodies, and therefore participate in the properties of matter generally occupying space, being susceptible of vibratory motion, and capable of being changed chemically as well as by the action of heat and electricity, they make known to the sensorium, by virtue of the changes thus produced in them by external causes, not merely their own condition, but also properties and changes of condition of external bodies. The information thus obtained by the senses concerning external nature, varies in each sense, having a relation to the qualities or energies of the nerve.*

Qualities which are to be regarded rather as sensations or modes of reaction of the nerves of sense, are light, colour, the bitter and sweet tastes, pleasant and unpleasant odours, painful and pleasant impressions on the nerves of touch, cold and warmth: properties which may belong wholly to external nature are "extension," progressive and tremulous motion, and chemical change.

All the senses are not equally adapted to impart the idea of "extension" to the sensorium. The nerve of vision and the nerve of touch, being capable of an exact perception of this property in themselves, make us acquainted with it in external bodies. In the nerves of taste, the sensation of extension is less distinct, but is not altogether deficient; thus we are capable of distinguishing whether the seat of a bitter or sweet taste be the tongue, the palate, or the fauces. In the sense of touch and sight, however, the perception of space is most acute. The retina of the optic nerve has a structure especially adapted for this perception.

33

10   ERNST HEINRICH WEBER (1795–1878) ON THE SENSE OF TOUCH AND COMMON SENSIBILITY, 1846

E. H. Weber, "Der Tastsinn und das Gemeingefühl," in Rudolph Wagner, ed., *Handwörterbuch der Physiologie* (Brunswick, 1846), vol. III, pt. 2, pp. 481–588. Translated for this book by Barbara Haupt.

Weber's first contribution to the sense physiology of touch (*Tastsinn* and *Ortsinn*) was in his booklet of 1834 in Latin, where he also first propounded the relation that we nowadays call Weber's law. This later exposition of 1846 is the standard reference.

On the conditions under which one is induced to relate many sensations to external objects:

Careful examination of the tactual sense and "common sensibility" of the skin and the muscles, when aided by measurement, is of especial interest, because for no other sensory organ can we, without harming ourselves, conduct such varied experiments and carry out so many kinds of measurement. Much of what we observe about the skin in this way can subsequently be applied also to the visual and other senses, as well as to "common sensibility."

Since all the forces that act on the body to arouse sensations are penetrating movements that bring about an alteration in the nerves, one would assume that the object of the sensations would always seem to lie in the sense organs, and indeed such seems to be the case for many sensations. For example, when we suffer from headache, pains in the eyes, pressure on the ears, toothache, or other such pains, we perceive that a particular part of the body is hurt, and we think we are having the sensations where our nerves are affected. We do not distinguish what affects us from the organs affected; on the contrary, we feel only the change that is brought about in the state of the particular part affected. When the surgeon's knife penetrates the skin, it is not sensed as an object that comes into contact with the body; instead, we feel pain in the area injured.

In the areas that are not particular organs of sense one has only sensations of this sort. By means of developed sense organs, on the other hand, one receives, in addition to these sensations, others that lead him to think he perceives an object existing outside the organ.

Thus, for example, we assume that we see things at a certain distance from us, spatially separated from us, and yet it is certain that the energies of our nerves do not extend beyond the surfaces of our bodies and that we can see these things only because the light they emit penetrates the neural skin [retina] of our eyes, there arousing a small image of the visible objects. We do not, however, have the slightest awareness of this contact with the neural skin of the eyes, not even when we turn our attention to the matter. Nor are we conscious even of the fact that we are directing our attention to a particular part of the neural skin in the eye. Instead, we believe that we are capable of paying attention to a visible object out in space and that, in fact, we are compelled to do so.

For every sensation we must distinguish the sensation itself from our interpretation of it. The sensations of light and dark, and of colors, are pure sensations; that something light or dark and colored is either within us or out in space, has a shape, and is resting or moving is but an interpretation of the sensation. That interpretation is, however, associated so closely with the sensation that it is inseparable from it and we take it as a part of the sensation, even though it is only an idea that is derived from the sensation. In many cases, however, false as well as correct interpretations of the sensations merge so completely with the sensations that the two cannot be distinguished at all, even when one has recognized the error and its cause.

To everyone, even the astronomer, the sun and moon seem to have a greater diameter when rising and setting than when they are high in the sky. It is known, however, that this deception is not based on a refraction of the light by the atmosphere that gives rise to a larger image on the neural skin in the eyes; in fact, the visual angle from which we see these heavenly bodies is, as measurement proves, in both cases exactly the same. The delusion is, rather, based on a false interpretation that circumstances oblige everyone to make, so that probably no one has ever freed himself from it; and it is so inseparably bound to the sight of the rising moon and sun that we are incapable of distinguishing it from what we actually sense. We assume that we directly perceive the rising sun and moon to have a greater diameter than when they are high in the sky; yet we are not at all conscious of the cause that leads us to this false interpretation of our sensations, which lies in the fact that the rising sun and moon seem to be farther away from us than when they are high in the sky—for bodies that are seen within the same visual angle appear larger when we consider them farther away, and vice versa.

· · · · ·

THE TACTUAL SENSE

Sensations of location, of pressure, and of temperature:

The tactual sense provides us with two kinds of sensations that are peculiar to it, *sensations of pressure* and *sensations of temperature,* and at the same time the tactual organ and its nerves are so arranged that the same kind of sensations can be distinguished from one another when they occur at different parts of the skin. Therefore we can distinguish the *sense of location,* the *sense of pressure,* and the *sense of temperature* as three capacities of the tactual sense. Compression and expansion of the sensitive organs, or tension—for example, the pressure of a weight on the skin or the extension of the hair follicle on pulling a hair—both excite sensations that we can concisely combine under the name of sensations of pressure. Sensations of temperature are either *positive*—that is to say, *sensations of warmth,* when the temperature in sensitive areas rises as a result of the conduction of heat to them—or *negative* sensations of heat, that is to say, *sensations of cold,* when their temperature sinks because heat is withdrawn from them. Only the areas that have tactual organs provide us with sensations of pressure and temperature. The internal regions that possess no tactual organs can be pressed, warmed, or cooled, but sensations of pressure, warmth, and cold never arise. All other sensations, in addition to those with which the tactual organs provide us, belong to "common sensibility." Therefore one must not confuse the pains that arise from pressure, warmth, and cold with sensations of pressure, warmth, and cold themselves. Moreover, one must distinguish these pure sensations from the ideas to which they give rise . . .

The sensations of pressure and of warmth and cold are so different that it may appear doubtful whether they can be regarded as various modifications of the same sensation. Nevertheless, just as the tongue is at once the seat of two senses, taste and touch, so the question may be raised whether the skin is perhaps also the seat of two senses, pressure and temperature. Only a single tactual sense would need to be assumed to belong to the skin if the same microscopically small sensory organs placed at the ends of the tactual nerves were to serve both purposes, the perception of pressure and its graded variations as well as the perception of hot and cold and their graded variations, and, further, if the sensation of warmth-and-cold had its source in the perception of the pressure that would arise because heat expands bodies but cold contracts them. Then one might possibly suspect that the pressure and pulling affecting an area of the skin in a certain direction would cause the sensations of pressure and pulling, but that compression and expansion taking place in many directions would arouse the sensations of cold and warmth. On the other hand, if two different kinds of organs existed in the skin, of which the one was set in motion by pressure thereby arousing a change in the nerves connected with them, and the other was set in motion by changes in temperature inducing impressions in the nerves connected with them, then one would have to assume the existence of a sense of pressure separate from a sense of temperature in the skin. The first assumption seems better grounded than the second.

In saying this I base my opinion on observations I have made that seem to deserve the attention of physiologists: *cold bodies resting on the skin seem heavier to us, and warm ones lighter, than they should.* The sensation of cold seems accordingly to add to the sensation of pressure, whereas that of warmth does not seem to do so. It may perhaps even act as a negative pressure, thus lessening the simultaneous sensation of pressure. Take equal weights of exactly the same shape that can easily be superimposed on each other. Very good new gold coins [Thaler] are suitable. Cool one of them to below the freezing point, for example down to $-7°$ C or $-4°$ C and heat the others up to $+37°$ or $38°$ C and lay a cold coin on the forehead of an observer who is lying with his head fully supported, his forehead horizontal, and his eyes closed. Then immediately remove the coin and lay two warm superimposed ones on exactly the same spot; then remove them and very quickly put a cold one there; and after removing it again, once more lay two warm coins there, until the observer is able to make a judgment of whether the first or the second weight laid on his forehead is heavier. He will be sure that both weights were equally heavy, or even that the one consisting of two warmed coins was the lighter. This experiment, which I have conducted on several persons with the same success, proves that the sensation of cold very considerably strengthens the sensation of pressure, since the cold weight, not only when it is as great as the warm weight but even when it is only half as great, is considered heavier.

·   ·   ·   ·   ·

THE SENSE OF LOCATION ON THE SKIN

The existence of the local sense depends on the fact that two sensations, even though they be identical, can be distinguished simply by the fact that they are aroused on different parts of the body or of the same sense organ. If pressure or application of heat or cold induces a sensation, one can indicate approximately the spot where the influence inducing the sensation on the skin was exerted; and, when we receive an impression by means of heat, cold, or pressure at two places on our skin that are not too close together, either simultaneously or successively, we can distinguish the two places at which the influences operated, as well as whether these places are at a greater or lesser distance from each other, and one can also cite approximately the direction of the line by which he considers the two places to be connected.

Twenty years ago in a series of experiments I showed to what extent one possesses this ability, finding it present in various parts of the skin in such greatly varying degrees that at tip of the tongue it is more than fifty times as precise as on the skin that covers the center of the upper arm or the thigh. The method of investigation was the following: I touched various persons who had their eyes closed or averted on two parts of the skin simultaneously with two small identically shaped objects and asked them whether they felt that one or more objects were touching them and in which direction the line ran by which they imagined the touched parts of their bodies were connected—lengthwise or crosswise. For this purpose I ground down the points of a compass with cylindrical legs so that the ends, having a diameter of one third of a Paris line [0.75 mm], did not prick when the skin was touched with them but evoked a distinct tactual impression. As soon as touch evokes pain, observation becomes much less accurate because pain is never so well localized as is an adequately strong touch with a surface not too sharp. As I opened the compass, at first more, but then less and less, I [presently] reached that distance between the ends of the legs where the two impressions began to be felt as a single impression. Even then the observer could often still determine whether the line that joined the ends of the compass lay lengthwise or crosswise on his body or his limbs, although he felt indeed only a single impression, but the part of the skin that was touched seemed then to have an elongated shape, so that he could say in which direction the greater and the smaller diameters of this elongated area of the skin extended. I recorded at what distance between the ends of the compass two touches were still distinguished.

* * * * *

THE SENSE OF TEMPERATURE

The sensations of warmth and cold do not function in the same way as the sensations of light and dark, for they are positive and negative quantities and between them lies a zero point, which is determined by the sources of heat that we have within us. If objects surrounding and touching the skin are of such a temperature that the temperature of the skin (regardless of the fact that we have a source of heat within us) neither rises nor sinks, then these objects seem to us to be neither warm nor cold. If they raise the temperature of the skin, then they seem warm, and if they cause the temperature of the skin to sink, we regard them as cold. [Matters are quite different with vision, where] absolute darkness is the zero point of illumination, and the varying degrees of illumination, from darkness to the greatest brightness, are all positive quantities.

A thermometer indicates the temperature of mercury at every moment, whether it is rising, falling, or [standing still]. The situation for the tactual sense is different, for it seems here as if we could perceive the rising or falling of the temperature of the skin as [if it were] the degree to which the [external] temperature has risen or fallen. For example, one does not feel whether his forehead or his hand is the warmer until he lays his hand on his forehead; then he often perceives a great difference between the two, sometimes finding the hand to be warmer, sometimes the forehead. If he lays his hand on his forehead, the colder of the two areas of skin lowers the temperature of the warmer, and vice versa, and this lowering of the temperature in the one or in the other part is what we feel. It is the reciprocal influence that enables us to compare the temperatures in the various parts of the skin directly . . . If we dip one hand into moderately cold water [and keep it there] and dip the other hand into the water repeatedly but only for a moment, we have the impression that the water is colder to the second hand than to the first, although the temperature of the skin of the first hand has fallen more than the temperature of the second. [That is because] no heat is taken from the second hand during the time that it is not in the water and because a part of the heat lost has been replaced by the internal source of heat.

· · · · ·

COMMON SENSIBILITY

Coenaesthesis

The majority of physiologists use the term "common sensibility" to characterize one's ability to perceive his own general state of sensation—for example, pain—and thus distinguish it from the ability to have a sensation that he understands as an object different from his own sensory state, as is the case with a sensation of color or sound. This latter ability thus cannot be considered an actual sense. In addition, many persons have assumed that *all nerves of sense* can provide us with the same sensations under certain conditions, but that there are also sensory nerves which, because they are joined to no particular sense organ, can provide us with no actual sensual impression, but only with "common sensibility." Many are even of the opinion that, upon our first use of our senses, we feel all impressions only as a change of our own total sensory state and that only gradually, by means of a comparison and an interpretation of these sense impressions, do we learn to comprehend certain sensations as objects.

Common sensibility and sensory perception often arise simultaneously and are then only various effects of one and the same impression, as is the case with nausea, which is aroused by a smell, or in general the pleasantness or unpleasantness of sensations that is perceived directly and simultaneously with the sensations and does not first result from a comparison of the sensations.

Therefore the name *common sensibility* is used to characterize the consciousness of our general sensory state, which all parts provided with sensory nerves mediate.

39

11   HERMANN LUDWIG FERDINAND VON HELMHOLTZ (1821–1894) ON THE THREE-COLOR THEORY OF VISION AND VISUAL SPECIFIC NERVE ENERGIES, 1860

H. L. F. von Helmholtz, *Handbuch der physiologischen Optik,* vol. II (Leipzig, 1860), sect. 20. Translated by J. P. C. Southall as *Helmholtz' Treatise on Physiological Optics,* vol. II ([Rochester, N. Y.]: Optical Society of America, 1924).

Helmholtz's espousal of Thomas Young's three-fiber theory of color vision dates from his papers of 1852 and 1854, but his definitive argument is in the *Optik.* Here Johannes Müller's doctrine of the specific energies of nerves stands him in good stead, for the optic nerve has insufficient fibers to transmit both the spatial pattern and the enormous variety of hues, brightnesses, and saturations of the visual stimulus object. Helmholtz followed Thomas Young in his solution of this problem.

Three fundamentally elementary kinds of nerve fibers can be excited in an enormous number of different proportions, and the Young-Helmholtz theory supposes that each of the many discriminable colors is aroused by its proper ratio of the excitations of the three basic fibers, each of which has its own specific energy. The theory thus constitutes an extension of the general theory of the specific energy of nerves, positing three kinds of specific visual energies.

Every difference of impression made by light . . . may be regarded as a function of three independent variables; and the three variables which have been chosen thus far were (1) the luminosity, (2) the hue, and (3) the saturation, or (1) the quantity of white, (2) the quantity of some colour of the spectrum, and (3) the wave-length of this colour. However, instead of these variables, three others may also be employed; and in fact this is what it amounts to, when all colours are regarded as being mixtures of variable amounts of *three so-called fundamental colours,* which are generally taken to be *red, yellow* and *blue.* To conceive this theory objectively, and to assert that there are simple colours in the spectrum which can be combined to produce a visual impression that will be the same as that produced by any other simple or compound light, would not be correct. There are no such three simple colours that can be combined to match the other colours of the spectrum even fairly well, because the colours of the spectrum invariably appear to be more saturated than the composite colours. Least suited for this purpose are red, yellow and blue; for if we take for blue a colour like the hue of the sky, and not a more greenish blue, it will be impossible to get green at all by mixing these colours. By taking a greenish yellow and a greenish blue, the best we can get is a very pale green. These three colours would not have been selected, had it not been that most persons, relying on the mixture of pigments, made the mistake of thinking that a mixture of yellow and blue light gives green. It would be rather better to take *violet, green* and *red* for fundamental colours. Blue can be obtained by mixing violet and green, but it is not the saturated blue of the spectrum; and a dead yellow can be made with green and red, which is not at all like the brilliant yellow in the spectrum.

If we think of the colours as plotted on a colour-chart by the method sketched above, it is evident from the rules given for the construction that all colours that are to be made by mixing three colours must be contained within the triangle whose vertices are the places in the chart where the three fundamental colours are. Thus, in the adjoining colour circle (Fig. 20), where the positions of the colours are indicated by the initial

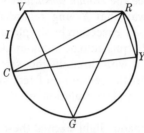

Fig. 20.

letters of their names (I = indigo-blue, C = cyan-blue, Y = yellow, G = green, etc.), all the colours that can be made by mixing red, cyan-blue and yellow are comprised within the triangle RCY. Thus, as we see, two large pieces of the circle are missing, and all that could be obtained would be a very pale violet and a very pale green. But if, instead of cyan-blue, the colour of the blue sky, indigo-blue, were taken, green would be missing entirely. The triangle VRG comprises the colours obtained by mixing violet, red and green, and a larger number of the existing colours would indeed be represented. But, as the diagram shows, large portions of the circle are still missing, as must always be the case according to the results of experiments on the mixture of the colours of the spectrum. The conclusion is that the boundary of the colour chart must be a curved line which differs considerably from the perimeter of the triangle.

Brewster, endeavouring to defend the objective nature of three fundamental colours, maintained that for every wave-length there were three different kinds of light, red, yellow and blue, mixed merely in different proportions so as to give the different colours of the spectrum. Thus, the colours of the spectrum were considered as being compound colours consisting of three kinds of light of different quality; although the degree of refrangibility of the rays was the same for each individual simple colour. Brewster's idea was that light of all three fundamental colours could be proved to exist in the different simple colours by the absorption of light by coloured media. His entire theory is based on this conception, which [has been] shown . . . to be erroneous.

Apart from Brewster's hypothesis, the notion of three fundamental colours as having any objective significance has no meaning anyhow. For as long as it is simply a question of physical relations, and the human eye is left out of the game, the properties of the compound light are dependent only on the relative amounts of light of all the separate wavelengths it contains. When we speak of reducing the colours to three fundamental colours, this must be understood in a subjective sense and as being an attempt to trace the *colour sensations* to three *fundamental sensations.* This was the way that Young regarded the problem; and, in fact, his theory affords an exceedingly simple and clear explanation of all the phenomena of the physiological colour theory. He supposes that:

1. The eye is provided with three distinct sets of nervous fibres. Stimulation of the first excites the sensation of red, stimulation of the second the sensation of green, and stimulation of the third the sensation of violet.

2. Objective homogeneous light excites these three kinds of fibres in various degrees, depending on its wave-length. The red-sensitive fibres are stimulated most by light of longest wave-length, and the violet-sensitive fibres by light of shortest wave-length. But this does not mean that each colour of the spectrum does not stimulate all three kinds of fibres, some feebly and others strongly; on the contrary, in order to explain a series of phenomena, it is necessary to assume that that is exactly what does happen. Suppose that the colours of the spectrum are plotted horizontally in Fig. 21 in their natural sequence, from red to violet, the three curves

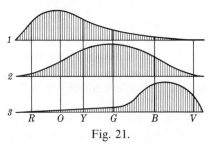

Fig. 21.

may be taken to indicate something like the degree of excitation of the three kinds of fibres, No. 1 for the red-sensitive fibres, No. 2 for the green-sensitive fibres, and No. 3 for the violet-sensitive fibres.

Pure *red* light stimulates the red-sensitive fibres strongly and the two other kinds of fibres feebly; giving the sensation red.

Pure *yellow* light stimulates the red-sensitive and green-sensitive fibres moderately and the violet-sensitive fibres feebly; giving the sensation yellow.

Pure *green* light stimulates the green-sensitive fibres strongly, and the two other kinds much more feebly; giving the sensation green.

Pure *blue* light stimulates the green-sensitive and violet-sensitive fibres moderately, and the red-sensitive fibres feebly; giving the sensation blue.

Pure *violet* light stimulates the violet-sensitive fibres strongly, and the other fibres feebly; giving the sensation violet.

When all the fibres are stimulated about equally, the sensation is that of *white* or pale hues.

It might be natural to suppose that on this hypothesis the number of nervous fibres and nerve-endings would have to be trebled, as compared with the number ordinarily assumed when each single fibre is made to conduct all possible colour stimulations. However, in the writer's opinion there is nothing in Young's hypothesis that is opposed to the anatomical facts in this respect; because we are entirely ignorant as to the number of conducting fibres, and there are also quantities of other microscopical elements (cells, nuclei, rods) to which hitherto no specific functions could be ascribed. But this is not the essential thing in Young's hypothesis. That appears to the writer to consist rather in the idea of the colour sensations being composed of three processes in the nervous substance that are perfectly independent of one another. This independence is manifested not merely in the phenomena which are being considered at present but also in those of fatigue of the nervous mechanism of vision. It would not be absolutely necessary to assume different nervous fibres for these different sensations. So far as mere explanation is concerned, the same advantages that are afforded by Young's hypothesis could be gained by supposing that within each individual fibre there might occur three activities all different from and independent of one another. But the form of this hypothesis as originally proposed by Young is clearer in both conception and expression than it would be if it were modified as suggested, and hence it will be retained in its original concrete form, for the sake of exposition if for nothing else. Nowhere in the physical (electrical) phenomena of nervous stimulation either in the sensory or motor nerves can there be detected any such differentiation of activity as must exist if each fibre of the optic nerve has to transmit all the colour sensations. By Young's hypothesis it is possible even in this connection to transfer directly to the optic nerve the simple conceptions as to the mechanism of the stimulation and its conduction which we were led to form at first by studying the phenomena in the motor nerves. This would not be the case on the assumption that each fibre of the optic nerve has to sustain three different kinds of states of stimulation which do not mutually interfere with one another. Young's hypothesis is only a more special application of the law of specific sense energies. Just as tactile sensation and visual sensation in the eye are demonstrably affairs of different nervous fibres, the same thing is assumed here too with respect to the various sensations of the fundamental colours.

12 HELMHOLTZ ON THE RESONANCE THEORY OF HEARING AND AUDITORY SPECIFIC NERVE ENERGIES, 1863

H. L. F. von Helmholtz, *Die Lehre von den Tonempfindungen* (Brunswick, 1863; 3rd ed., 1870), chap. 6. Third edition translated by A. J. Ellis as *Sensations of Tone* (London, 1875).

Helmholtz's resonance theory of hearing constitutes a bold extension of the theory of specific energies of nerves in that it assigns a different nerve fiber (specific energy) in the auditory nerve to each of the thousand-odd discriminably different pitches. This is the theory that persisted in spite of many opposing proposals for over fifty years. The first theory, in 1863, fixed on the arches of Corti as the resonators in the inner ear. In 1869 new anatomical data suggested that the fibers of the basilar membrane might act as resonators, and Helmholtz amended the account in his 1870 edition, from which these excerpts are taken. The discussion is changed only a little from the one of 1863, by addition of the reference to the basilar membrane.

Now we cannot precisely ascertain what parts of the ear actually vibrate sympathetically with individual tones. We can only conjecture what they are at present in the case of human beings and mammals. The whole construction of the partition of the cochlea, and of Corti's arches which rest upon it, appears most suited for executing independent vibrations. We do not need to require of them the power of continuing their vibrations for a long time without assistance.

But if these formations are to serve for distinguishing tones of different pitch, and if tones of different pitch are to be equally well perceived in all parts of the scale, the elastic formations in the cochlea, which are connected with different nerve fibres, must be differently tuned, and their proper tones must form a regularly progressive series of degrees through the whole extent of the musical scale.

According to the recent anatomical researches of V. Hensen and C. Hasse, it is probably the breadth of the membrana basilaris in the cochlea, which determines the tuning. At its commencement opposite the oval window, it is comparatively narrow, and it continually increases in width as it approaches the apex of the cochlea. The following measurements of the membrane in a newly born child, from the line where the nerves pass through on the inner edge, to the attachment to the ligamentum spirale on the outer edge, are given by V. Hensen [in Table 1].

[TABLE 1.]

Place of section	Breadth of membrane or length of transverse fibres	
	millimetre	inch
0.2625 mm. [= 0.010335 in.] from root	0.04125	0.00162
0.8626 mm. [= 0.033961 in.] from root	0.0825	0.00325
Middle of the first spire	0.169	0.00665
End of first spire	0.3	0.01181
Middle of second spire	0.4125	0.01624
End of second spire	0.45	0.01772
At the hamulus	0.495	0.01949

The breadth therefore increases more than twelvefold from the beginning to the end.

Corti's rods also exhibit an increase of size as they approach the vertex of the cochlea, but in a much less degree than the membrana basilaris . . . Hensen's measurements [are given in Table 2].

[TABLE 2.]

	At the round window		At the hamulus	
	mm.	inch	mm.	inch
Length of inner rod	0.048	0.00189	0.0855	0.00337
Length of outer rod	0.048	0.00189	0.098	0.00386
Span of the arch	0.019	0.00075	0.085	0.00335

Hence it follows, as Henle has also proved, that the greatest increase of breadth falls on the outer zone of the basilar membrane, beyond the line of the attachment of the outer rods. This increases from 0.023 mm. to 0.41 mm. or nearly twentyfold.

In accordance with these measures, the two rows of Corti's rods are almost parallel and upright near to the round window, but they are bent much more strongly towards one another near the vertex of the cochlea.

It has been already mentioned that the membrana basilaris of the cochlea breaks easily in the radial direction, but that its radial fibres have considerable tenacity. This seems to me to furnish a very important mechanical relation, namely that this membrane in its natural connection admits of being tightly stretched in the transverse direction from the modiolus to the outer wall of the cochlea, but can have only little tension in the direction of its length, because it could not resist a strong pull in this direction.

Now the mathematical theory of the vibration of a membrane with different tensions in different directions shews that it behaves very differently from a membrane which has the same tension in all directions. On the latter, vibrations produced in one part, spread uniformly in all directions, and hence if the tension were uniform it would be impossible to set one part of the basilar membrane in vibration, without producing nearly as strong vibrations (disregarding individual nodal lines) in all other parts of the membrane.

But if the tension in direction of its length is infinitesimally small in comparison with the tension in direction of the breadth, then the radial fibres of the basilar membrane may be approximatively regarded as forming a system of stretched strings, and the membranous connection as only serving to give a fulcrum to the pressure of the fluid against these strings. In that case the laws of their motion would be the same as if every individual string moved independently of all the others, and obeyed, by itself, the influence of the periodically alternating pressure of the fluid of the labyrinth contained in the vestibule gallery. Consequently any exciting tone would set that part of the membrane into sympathetic vibration, for which the proper tone of one of its radial fibres that are stretched and loaded with the various appendages already described, corresponds most nearly with the exciting tone; and thence the vibrations will extend with rapidly diminishing strength on to the adjacent parts of the membrane . . .

The strongly vibrating parts of the membrane would, as has been explained in respect to all bodies which vibrate sympathetically, be more or less limited, according to the degree of damping power in the adjacent parts, by friction against the fluid in the labyrinth and in the soft gelatinous parts of the nerve fillet.

Under these circumstances the parts of the membrane in unison with higher tones must be looked for near the round window, and those with the deeper, near the vertex of the cochlea, as Hensen also concluded from his measurements. That such short strings should be capable of corresponding with such deep tones, must be explained by their being loaded in the basilar membrane with all kinds of solid formations; the fluid of both galleries in the cochlea must also be considered as weighting the membrane, because it cannot move without a kind of wave motion in that fluid.

．　　　．　　　．　　　．　　　．

According to Waldeyer there are about 4500 outer arch fibres in the human cochlea. If we deduct 300 for the simple tones which lie beyond musical limits, and cannot have their pitch perfectly apprehended, there remain 4200 for the seven octaves of musical instruments, that is, 600 for every Octave, 50 for every Semitone; certainly quite enough to explain the power of distinguishing small parts of a Semitone. According to Prof. W. Preyer's investigations, practised musicians can distinguish with certainty a difference of pitch arising from half a vibration in a second, in the doubly accented Octave. This would give 1000 distinguishable degrees of pitch in the Octave, from 500 to 1000 vibrations in the second. Towards the limits of the scale the power to distinguish differences diminishes. The 4200 Corti's arches appear then, in this respect, to be enough to apprehend distinctions of this amount of delicacy. But even if it should be found that many more than 4200 degrees of pitch could be distinguished in the Octave, it would not prejudice our assumption. For if a simple tone is struck having a pitch between those of two adjacent Corti's arches, it would set them both in sympathetic vibration, and that arch would vibrate the more strongly which was nearest in pitch to the proper tone. The smallness of the interval between the pitches of two fibres still distinguishable, will therefore finally depend upon the delicacy with which the different forces of the vibrations excited can be compared. And we have thus also an explanation of the fact that as the pitch of an external tone rises continuously, our sensations also alter continuously and not by jumps, as must be the case if only one of Corti's arches were set in sympathetic motion at once.

To draw further conclusions from our hypothesis, when a simple tone is presented to the ear, those Corti's arches which are nearly or exactly in unison with it will be strongly excited, and the rest only slightly or not at all. Hence every simple tone of determinate pitch will be felt only by certain nerve fibres, and simple tones of different pitch will excite different fibres. When a compound musical tone or chord is presented to the ear, all those elastic bodies will be excited, which have a proper pitch corresponding to the various individual simple tones contained in the whole mass of tones, and hence by properly directing attention, all the individual sensations of the individual simple tones can be perceived. The chord must be resolved into its individual compound tones, and the compound tone into its individual harmonic partial tones.

This also explains how it is that the ear resolves a motion of the air into pendular vibrations and no other. Any particle of air can of course execute only one motion at one time. That we considered such a motion mathematically as a sum of pendular vibrations, was in the first instance merely an arbitrary assumption to facilitate theory, and had no mean-

47

ing in nature. The first meaning in nature that we found for this resolution came from considering sympathetic vibration, when we discovered that a motion which was not pendular, could produce sympathetic vibrations in bodies of those different pitches, which corresponded to the harmonic upper partial tones. And now our hypothesis has also reduced the phenomenon of hearing to that of sympathetic vibration, and thus furnished a reason why an originally simple periodic vibration of the air produces a sum of different sensations, and hence also appears as compound to our perceptions.

The sensation of different pitch would consequently be a sensation in different nerve fibres. The sensation of a quality of tone would depend upon the power of a given compound tone to set in vibration not only those of Corti's arches which correspond to its prime tone, but also a series of other arches, and hence to excite sensation in several different groups of nerve fibres.

Physiologically it should be observed that the present assumption reduces sensations which differ qualitatively according to pitch and quality of tone, to a difference in the nerve fibres which are excited. This is a step similar to that taken in a wider field by Johannes Müller in his theory of the specific energies of sense. He has shewn that the difference in the sensations due to various senses, does not depend upon the actions which excite them, but upon the various nervous arrangements which receive them. We can convince ourselves experimentally that in whatever manner the optic nerve and its expansion, the retina of the eye, may be excited, by light, by twitching, by pressure, or by electricity, the result is never anything but a sensation of light, and that the tactual nerves, on the contrary, never give us sensations of light or of hearing or of taste. The same solar rays which are felt as light by the eye, are felt by the nerves of the hand as heat; the same agitations which are felt by the hand as twitterings, are tone to the ear.

Just as the ear apprehends vibrations of different periodic time as tones of different pitch, so does the eye perceive luminiferous vibrations of different periodic time as different colours, the quickest giving violet and blue, the mean green and yellow, the slowest red. The laws of the mixture of colours led Thomas Young to the hypothesis that there were three kinds of nerve fibres in the eye, with different powers of sensation, for feeling red, for feeling green, and for feeling violet. In reality this assumption gives a very simple and perfectly consistent explanation of all the optical phenomena depending on colour. And by this means the qualitative differences of the sensations of sight are reduced to differences in the nerves which receive the sensations. For the sensations of each individual fibre of the optic nerve there remains only the quantitative differences of greater or less irritation.

The same result is obtained for hearing by the hypothesis to which our investigation of quality of tone has led us. The qualitative difference of pitch and quality of tone is reduced to a difference in the fibres of the nerve receiving the sensation, and for each individual fibre of the nerve there remains only the quantitative differences in the amount of excitement.

13   MAX VON FREY (1852–1932) ON THE FOUR CUTANEOUS SENSES, 1904

Max von Frey, *Vorlesungen über Physiologie* (Berlin: Springer, 1904), pp. 308–326. Translated for this book by Don Cantor.

The doctrine of specific energies of nerves led Magnus Blix in 1882–83 and Alfred Goldscheider in 1884–85 to use very small stimuli on the skin and thus to discover separate sensitive spots for warmth and cold (Blix) and for pressure (Goldscheider). Von Frey in 1894–1896 added the pain spots and developed the classical argument for four modalities in the skin: warmth, cold, pressure, and pain. He also discussed the question of the identity of the end organs that might mediate these four senses. His own best summary of the classical theory is in his *Vorlesungen*.

THE CUTANEOUS DEPARTMENT OF SENSE

1. *Sensations of temperature*

The best way to obtain a summary view of these sensations is to consider what reports about the character of external objects are obtained by stimulating the skin alone, without the support of other sensory organs. In order to exclude the latter entirely, information must not be communicated by so-called active touch (the movement of the sensory area itself) but only by passive touch (the movement of objects upon the resting skin).

Even under such limited experimental conditions, the incoming information is very diverse. It includes size, form and volume, absolute and specific weight, texture, general state, elasticity, temperature, and other factors, and, besides these sensory complexes, there are still others that are in general related not to external objects but to the subject's own body—tickling, well-being, itching, shuddering, and so forth.

It is not difficult to observe that in objective reports certain elements always recur, even though they appear in such a variety of contexts; thus it is possible to infer which of the items are fundamental elements.

The analysis is especially trustworthy when one succeeds in bringing out individual elements experimentally and placing them in the order of their intensity. It then becomes clear that the objective reports refer consistently only to sensations of cold, warmth, pressure, and pain and, in addition, to their variation in space and time.

The subjective sensory complexes, which cannot with certainty be called forth and graded experimentally and therefore cannot accurately be observed, are much more difficult to analyze. They include, besides the sensations already mentioned, many forms of pain. Still, one can show that many varieties of pain are probably not pure cutaneous sensations, inasmuch as stimuli from deeper parts, especially from the motor apparatus, contribute to them. This is certainly true of the sensation of well-being, and, as can rather easily be demonstrated, of shuddering. The appearance of gooseflesh, moreover, is connected with characteristic sensations that come in part from the fixed hair follicles and perhaps in part directly from the contracted smooth-muscle cells. Vascular contraction is also a basic factor in these mixed sensations, and is probably what causes the pain associated with the effect of cold on large surfaces of the body.

If the skin were a uniform sensory surface through which various kinds of sensations could be communicated within one sensory area by various anatomical structures, in contradiction to the preceding supposition, one would expect that those skin surfaces which, like the fingertips, possess the greatest sensitivity to pressure would also be especially well qualified to perceive temperature. This inference contradicts everyday experience. Everyone is familiar with the fact that the skin of the face and torso possesses a much livelier sensitivity to temperature than the skin of the extremities, especially of their distal parts. If one fills two small containers with water having temperatures of 25° and 35° C respectively, they will scarcely be perceived as differing in temperature when touched by the fingertips, although this difference is quite striking when they are touched by the eyelid or the corner of the mouth. That this difference in perception is not solely a function of the unequal thickness of the [various parts of the] epidermis is clear from the fact that it is poorly perceived through the thin skin of the wrist. Small metal objects at room temperature can take only a very little heat from the body and are therefore not felt as differing [in temperature] when placed upon many parts of the skin. The eyelid or the lip, however, clearly feels them as cold.

When one uses cold or warm objects having small surfaces for thermal stimulation of the skin, it becomes clear from even the most cursory examination that corresponding sensations cannot be produced on each surface element of the skin. Sensitivity to temperature appears only in isolated places. Blix, who first discovered this fact, remarked also that stimuli for cold and warmth are not effective in the same places. In relating receptivity to thermal stimulation one must, therefore, distinguish between three surface conditions of the skin: (1) insensitivity, (2) sensitivity to cold, and (3) sensitivity to warmth. The areas for warmth are always in the minority. Blix called the surface elements of the second type *cold spots* and those of the third type *warm spots* . . . Stimulation with objects of small surface fails on portions of the skin where the epidermis is thick, but even there unequal sensitivity from place to place remains demonstrable.

When one has determined the place and number of the sensitive spots for a large skin surface, it becomes clear that these can be found again and again after a suitable interval of time. There is, therefore, no doubt that we are dealing with fixed anatomical structures—the ends of centripetal nerves, which give the skin its sensitivity to temperature and whose terminations on its outer surface are represented by the experimentally fixed warm and cold spots. Through interesting experiments, Thunberg has shown it to be very likely that cold spots are nearer the outer surface than warm spots . . .

Analyses of the density of these structures have been made by Sommer for the majority of skin surfaces. He found an average density of 13 cold spots per square centimeter, a figure that yields a total of 250,000 for the entire surface of the body. For the warm spots, on the other hand, he found an average of only 1.5 per square centimeter, or about 30,000 for the entire body.

The distribution over the skin of the spots sensitive to temperature is very irregular. Only rarely are they entirely isolated, and for the most part they are so clustered together that between the groups gaps as large as 1 square centimeter may occur, gaps in which sensitivity to warmth, to cold, or to both is lacking.

On many parts of the skin that have keen or even extraordinary sensitivity to cold, sensitivity to warmth is quite lacking. Among these parts are the conjunctiva and cornea of the eye, the glans penis, and the nipples. The mucous membranes of the nose and throat lack, for the most part, sensitivity to temperature, as has long been known of the stomach and intestine.

The supposition, previously discussed, of the dissimilarity of apparently homogeneous sensory areas is, then, fully confirmed in respect of the excitability of the skin by temperature . . .

Having established the distribution over the skin of the organs sensitive to temperature, one is in a better position to examine the second hypothesis put forth above, according to which the sensation caused by the stimulation of certain structural elements within a sensory area should vary only in quantity of intensity, and not in quality. This supposition, too, can be fully confirmed with reference to those elements of the skin that are sensitive to temperature. As Blix has shown, they may be stimulated electrically or mechanically as well as thermally and yield only sensations of the quality proper to them [—warmth or cold]. Finally, the cold spots also react with the "cold" sensation for temperatures above 45°, a phenomenon called the paradoxical sensation of cold.

Physical and physiological scales of temperature. In physics the series of various temperatures makes a single extended scale. One can go from one temperature, *a,* to another, *b,* in one way only, that is to say, through all the temperatures that lie in between. If the zero point of the scale and the size of the unit of measurement are understood, then specifying the degrees clearly establishes the temperatures. For the physical measurement of temperature, differentiating between cold (or negative) and warm (or positive) temperatures is, therefore, meaningless, and is only a historical remainder of the original physiological measurement [of warmth and cold]. This fact is expressed in the arbitrary standardization of the physical zero point, which is set either at exactly the melting temperature of ice or at $-17.8°$ C.

Sensations of temperature, on the other hand, show a much greater variety [than does the thermometer]. Sensations of cold and warmth, taken separately or together, can be aroused in many intensities, and under certain circumstances temperature causes pain. One may best show that the sensation of pain does not occur in temperature spots of the skin by using an appropriately shaded convex lens to project sunlight upon the skin. If the rays fall on a warm spot, the sensation of warmth occurs. If weak rays fall on one of the surface elements that is incapable of sensitivity to temperature, no sensation is felt, but if the rays are strong, pain appears. Insensitivity to pain, analgesia, of temperature spots was first observed by Goldscheider. It is not found, incidentally, for all temperature spots . . . Finally, if one projects the sunlight onto a cold spot, the paradoxical sensation of cold appears. Temperatures that excite both the warm and the cold spots [simultaneously], are [called] hot.

There are, then, three different nerve structures in the skin which can be stimulated by high temperatures. Large-surface stimulation brings all three into action at the same time. Low temperatures, such as the freezing point, can also produce pain. One calls such temperatures cutting, biting, or burning cold. In this kind of pain, two nerve structures are simultaneously brought into action.

The verbal designations of sensations of temperature show, then, the following relations to the kind and number of the structures brought into action.

	Warmth nerves	Cold nerves	Pain nerves
Cutting or biting cold	0	+	+
Cold	0	+	0
Cool	0	+	0
Indifferent	0	0	0
Lukewarm	+	0	0
Warm	+	0	0
Hot	+	+	0
Burning hot	+	+	+

A plus sign indicates stimulation and zero the absence of stimulation.

. . . . .

The point of indifference, or the temperature at which all three apparatuses are at rest, is arbitrarily set at 33° C. Rising temperatures stimulate first the warmth nerves, then the cold nerves, and finally the pain nerves . . .

The alteration of the cutaneous temperature nerves by the temperature of their surroundings cannot be regarded as a function of exhaustion, for their excitability does not suffer at all from fatigue. An understanding of this process will come first from a theory of stimulation through temperature, which is at present lacking. One can only say that all changes that lower the temperature of a skin surface that is [at first] thermally inactive, whether the change be a lowering of the outside temperature, evaporation from the skin, or a diversion of the blood stream from it, call forth the sensation of cold, while the opposite changes elicit the sensation of warmth.

The more abrupt the changes in temperature the more quickly their effects are felt (for example, with metal [stimulus objects] because of their great ability to conduct heat). The greater the skin surface they affect, the clearer the sensations.

2. Sensations of pressure

The analysis of perceptions transmitted by the skin shows that the sensation of pressure, as well as of warmth and cold, is to be considered among those which are simple and unanalyzable. Statements about the size, form, weight, and texture of objects can be made quite independently of their temperature. The assertion by E. H. Weber that when two equal weights are laid on the skin the colder appears heavier can probably be explained by an error in judgment, for a weight cooled to well under 0° does lead to the conclusion that a larger volume of metal is present.

It has already been said that the greatest sensitivity to stimulation by temperature does not occur at the places of greatest sensitivity to touch. A quantitative comparison of the sensitivities to touch of various skin surfaces is, however, not at all easy. The problem of distributing weight uniformly on the human skin, which is everywhere curved, would best be met by the use of hydrostatic pressure. Experience teaches, however, that such pressure is not felt. If one immerses a part of the body in 1 meter of water or 7.6 cm of mercury, the pressure then exerted on it, $\frac{1}{10}$ that of atmospheric pressure, will not be perceived. If, on the other hand, one places a weight of 100 grams on a skin surface which is as level as possible, placing a cork tablet under the weight to make sure that the area pressed is exactly 1 cm square, the pressure, which amounts once again to $\frac{1}{10}$ atmospheric pressure, will very probably be perceived.

[It appears that] the physiological effect depends not only on the amount of pressure but also on the number of surface elements on which it is exerted, and the smaller the area pressed, the greater the effect.

Stimulation of a very small surface can be produced by pushing a hair or bristle, cut directly across, vertically against the skin . . .

By using a stimulus of this kind it is not difficult to show that the tip of the tongue, the lips, and the fingertips are exceptional among the parts of the body in their great sensitivity to mechanical stimulation, although for these, too, there is a threshold that must be passed if sensation is to occur . . . In reality, the places of greatest thermal sensitivity are not those of greatest sensitivity to touch. This fact suggests that the two sensations are communicated by different organs.

Stimulating the skin with bristles or hairs in the manner described shows that there is a great difference between single surface elements. If one holds the stimulation near the threshold for the skin surfaces under examination, one always finds the major portion of the surfaces insensitive and only single surface elements sensitive. These are called pressure or touch spots. Their position is constant and is different from the positions of the cold and warm spots; they are also more numerous. Their dispersion over the surface is, unlike that of the temperature spots, very even. A touch spot lies near each hair of the hairy parts of the skin. The number of hairs and touch spots is, then, generally the same. If one projects the hair follicle, which always slants in relation to the skin, onto the skin surface, the touch spot is found to be over the projection. We are dealing, then, with a nervous apparatus which has a relation to the hair follicle. On the hairless surfaces of the skin (about 5 per cent of the surface area of the body) the dispersion is similar, not quite so regular, and often more dense . . .

3. *Sensations of pain*

Pain belongs to the group of sensations that can be produced from the skin; it must, like the sensations of cold, warmth, and pressure, be regarded as an elementary, fundamental part of consciousness, for, although it often occurs along with other sensations, it is nonetheless not dependent on them for its intensity and under certain circumstances appears without them.

Pain is distinguished from the other sensations produced in the skin not only by its special character but also by the fact that it is to a much greater degree accompanied by feeling, especially feeling of aversion. It is physiological, nevertheless, for it can be produced not only in the skin but also in a great number of internal organs. A further characteristic of the sensation of pain is its relation to reflex action.

Not only the skin is sensitive to pain, and that apparently in all its layers; the muscles as well, both the smooth and the striated [mediate pain]. Bruising or tearing of muscles and pathological changes in them, like the rheumatic changes, are very painful. Strong contractions of the smooth muscles arouse colic pains, or rather colic aches. The other parts of the apparatus of movement, such as tendons (or their sheaths), joints, and surface membranes of bones (periosteum) are also sensitive to pain. Among the glands, those of sex are surely subject to pain, but this statement cannot so certainly be made of the rest. It is remarkable that widespread changes can take place in the lungs without producing pain. The brain is also among the organs that are not sensitive to pain, especially its cortex, although the meninges can be sensitive. The mucous membranes of the intestine and stomach are not sensitive to pain, and, according to Lenander, neither is the visceral peritoneum, whereas the parietal membrane is said to be very sensitive. The sensitivity of the mouth to pain (except the teeth, the tip of the tongue, and the mucous membrane of the lips) is very slight. Kiesow has shown it to be likely that a large area of the surface of the mucous membrane of the cheeks is entirely insensitive to pain.

The sensations of pain in internal organs probably differ in character from those in the skin. Weak sensations of pain in the skin are called itching, stronger ones pricking, burning, biting, and cutting. The description of the sensation depends partly on the temporal course of the pain and its extent and partly on accompanying sensations of other kinds. Internal pains are not accompanied by other sensations, are not so sharply circumscribed, and generally also last longer. One calls such pains muffled.

Sensations of pain can be aroused in ways that differ greatly: mechanically, thermally, electrically, and chemically. To cause pain by mechanical and thermal stimulation requires, in general, higher intensities of stimulation than are necessary to produce sensations of touch and temperature, a circumstance which has given rise to the opinion that nerves that are sufficiently strongly stimulated themselves communicate pain. With electrical stimulation, however, this difference of threshold does not hold. The use of different small electrodes, especially on the hairy parts of the body, leads to sensations of pain as well as of touch, even though the intensity be the same or even less. Chemical stimulation, by placing small drops of acid on the skin, for example, has an exclusively painful effect. Here one could raise the objection . . . that the nerves sensitive to pain are the ones nearest the surface and thus the first to be stimulated . . . Sensations solely of pain may be caused, incidentally, by the application of heat over a very small surface if one skillfully stimulates such areas as the skin of the leg, where the temperature spots are especially far from one another. If one concentrates the sunlight refracted by a convex lens onto parts of the surface that are not sensitive to temperature, the effect is painful, without any other sensation.

The experiment just described proves further that organs for sensing pain are present where those for temperature are lacking, and the same is true of organs for sensing touch. Electrical stimulation of a small surface of the skin by faradic current or by the sparks from an electric machine shows the places sensitive to pain to be quite independent of the location of the touch spots. An attempt has been made to prove that certain points on the skin are especially sensitive to pain, and these have been called pain spots, in accordance with the terminology used for the other sensory apparatuses of the skin. This problem presents especially great difficulties, however, because in most places the density of these spots is so much greater than that of the others that the usual methods for isolated stimulation are of little use. Still, experiments have demonstrated decisively that the skin's sensitivity to pain is not continuous. This fact is shown especially clearly by the existence of real gaps, such as, for example, the Kiesow area in the mucous membrane of the cheeks.

The apparatus for sensing pain can also be distinguished from the others discussed earlier by its characteristic ways of reacting. With weak painful stimulation, the sensation occurs quite late, so that a possibly simultaneous thermal or tactile stimulation can be distinguished from the pain by an interval free of sensation.

. . . . .

The sensory function of the skin and the form of the nerve endings. The supposition stated at the beginning was that in all cases where different kinds of sensations can be produced in an apparently homogeneous sensory area various kinds of nerves and nerve endings are present. This view has been shown to be thoroughly well founded. [Let us now consider] a selection of the various anatomically traceable forms of nerve endings in the skin, with the intention of establishing what form belongs to each sensory quality.

The matter is simplest for the sense of touch. Its relation to the hairs of the body has been experimentally established. One finds, in fact, as Bonnet first showed, a characteristic form of nerve ending on each hair, which has a special development on the so-called feeling hairs [vibrissae] of carniverous animals and the anatomy of which can therefore be minutely examined. These forms of nerve endings, which one may call the nerve rings of the hairs, are therefore to be considered the ends of touch nerves where the skin is hairy. A great number of differing forms may be considered for the hairless touch surfaces of the extremities [palms and soles]: Vater-Pacini and Golgi-Mazzoni corpuscles, Krause's end bulbs, Meissner's corpuscles, Ruffini's nerve brushes, Merkel's touch cells, and free nerve endings.

In choosing, the criterion of sufficient profusion must above all be satisfied, for the density of the organs of touch must, of course, reflect at least the density of the experimentally demonstrable touch spots. There is but one known form that satisfies this demand, namely, Meissner's corpuscles. Their superficial position in the skin reflects the ease with which the touch spots may be experimentally outlined . . .

For the organs that sense temperature, it is best to look for specific forms of nerve endings in those places where sensitivity is well developed and where the other qualities of sensitivity are as much as possible excluded. In this respect, the eye has for us a special interest. The sense of touch and sensitivity for warmth are lacking in the conjunctiva and the cornea, whereas sensitivity to cold and pain are very well represented there, although sensitivity to cold is limited to the conjunctiva and the border of the cornea and is especially well developed in the latter. Only in these places does one find, as Dogiel has shown, great numbers of the so-called end bulbs, so that they may be designated as organs for sensing cold . . .

The middle of the cornea possesses sensitivity to pain only, and here is found, according to present knowledge, only one kind of nerve ending, the so-called free intraepithelial endings between undifferentiated cells, first described by J. Cohnheim. This type of ending has recently been shown to exist [not only] in the most varying epithelia but also inside of structures that are not ectodermal in origin. The supposition that they are related to the several kinds of sensitivity to pain gains in probability precisely because of their extraordinarily wide distribution. Furthermore, von Frey's experiments show that the pain nerves in the skin must have their endings very near the surface.

A deeper position must, however, be assumed for the nerve endings that communicate the sensation of warmth. The difficulty of mapping exactly the positions of the warm spots, as well as the results of Thunberg's experiments, argue for this location. Accordingly, the Pacini and Golgi-Mazzoni corpuscles, as well as the endings newly described by Ruffini, come under our consideration. The Pacini and Golgi-Mazzoni corpuscles are extraordinarily similar in structure and are therefore perhaps also identical in function. The dispersion of the Pacini corpuscles does not meet the conditions that one must set for an organ for the perception of warmth. How the two other forms are distributed is as yet too little known. Yet the fact that Ruffini's endings are found in the eyelid, where sensitivity to warmth is very great, argues for their mediation of this quality.

14 EDWARD BRADFORD TITCHENER (1867–1927) ON THE NUMBER OF SENSORY ELEMENTS, 1896

E. B. Titchener, *An Outline of Psychology* (New York: Macmillan, 1896), pp. 74–75.

In 1896 Titchener was fresh from Wundt's laboratory in Leipzig, where introspective atomism (mental chemistry, really) was thought to provide the key to psychology's scientific future. The question of how many mental elements there are was an important one. Oswald Külpe, Titchener's close associate at Leipzig, was about to found the Würzburg school and to discover the new element of imageless thought. It was natural for Titchener to try to get up the list of discovered elements, a sort of Mendeléyev table for psychology. He was then a youth of twenty-nine. In his maturity (in 1910, say) he would have been likely to characterize so rigid a commitment to sensory elements as naive.

THE TOTAL NUMBER OF ELEMENTARY SENSATIONS

Putting together the results of the foregoing Sections, we obtain the following list of sensation qualities:

Eye	32,820	Alimentary canal	3?
Ear (audition)	11,600	Blood-vessels	?
Nose	?	Lungs	1?
Tongue	4	Sex organs	1
Skin	4	Ear (static sense)	1
Muscle	2		
Tendon	1	More than	44,435
Joint	1		

Each one of these forty thousand qualities is a *conscious element,* distinct from all the rest, and altogether simple and unanalysable. Each one may be blended or connected with others in various ways, to form *perceptions* and *ideas.* A large part of psychology is taken up with the determination of the laws and conditions which govern the formation of these sensation complexes.

The above list represents the full resources of the normal mind. It must not be supposed, however, that every normal individual has had experience of all the qualities enumerated. It is safe to say that no one, not even the most experienced psychologist, has seen all the possible visual qualities, heard all the possible tones, smelled all the possible scents, etc. The list is a summary of the results obtained by many observers in the course of minute investigations of our capacity of discrimination in the various fields of sense.

Apart from this, a slight abnormality is much more common than is ordinarily supposed. Very many people are more or less colour-blind; they confuse red with green, or have a shortened spectrum, *i.e.,* do not see the full number of red and violet qualities. Very many are partially tone-deaf, have a defective sense of smell, etc. But when all allowances are made, the average number of conscious elements must run into the tens of thousands.

Psychophysics, both the word and the field of scientific investigation, were originated by G. T. Fechner (1860), who worked out and formalized the methods of determining thresholds. By assuming that the differential threshold, the "just noticeable difference," can be taken as a unit of sensory magnitude, he was able to convert Weber's law into a means for measuring the magnitude of sensation. In this way he founded psychophysics—inadvertently, since he was really waging a philosophical battle against materialism by showing that the material world could be reduced by his transformation equation to the psychic —and, in founding psychophysics, Fechner also almost founded scientific psychology itself, which was then waiting only for the demonstration that mind can actually be measured. Psychophysics thus began by measuring thresholds, but the measurement of thresholds had not waited for Fechner. The problem was too obvious.

This chapter begins with Bouguer's measurement (1760) of the differential threshold for visual brightness and Delezenne's measurement (1827) of the differential threshold for tonal pitch. Then follows an excerpt from Weber's monograph in Latin (1834), giving his first account of the law which later was to bear his name. Next comes Fechner (1860) with his discussion of the fundamental principles that led to the law that is named for him. The text skips over the wearisome argument of the forty years after Fechner as to whether Fechner's law (which Fechner called Weber's) was true or only an approximation, and continues with the problem of sensory measurement. Plateau (1872) described how he had arranged to have painters bisect the sensory distance between two grays by painting an intermediate gray that, contrasting equally with the two others, might be regarded as lying midway between them. Delboeuf (1883) expanded this view into a theory of sensory measurement in terms of the sensed contrasts between sensations, and Titchener (1905), substituting the concept of sense distance for Delboeuf's sensed contrast, developed the theory of supraliminal sensory interval measurement, mostly by the bisection of sense distances. With the substitution of the ratio scaling of sensory magnitudes for interval scaling in the mid-twentieth century, Plateau's disparagement of Fechner's law may seem to have gained strength, but this book does not undertake to penetrate so far into the present century.

15 PIERRE BOUGUER (1698–1758) ON THE DIFFERENTIAL THRESHOLD FOR ILLUMINATION, 1760

Pierre Bouguer, *Traité d'optique sur la gradation de la lumière* (Paris, 1760), bk. I, sect. 2, art. l, pp. 51–56. Translated for this book by Mollie D. Boring.

Bouguer, a French physicist and mathematician, was a pioneer in the development of photometric methods of measuring illumination (he found that the sun is about 300,000 times as bright as the moon) and the inventor of the heliometer, for measuring the sun's diameter and the distance between stars. Many of his important contributions to optics are contained in his *Essai d'optique* (1729), of which the work cited above is a greatly enlarged, slightly posthumous, revised edition, but the determination of the differential threshold for illumination at about $\frac{1}{64}$ is not to be found in the earlier book.

ARTICLE I

Observations to determine what force a light must have to make a weaker one disappear.

Let us put at the head of our observations those which show us how strong a light must be to make the effect of a second and very much weaker light truly inappreciable beside it. All our most sensitive organs, like our crudest, are subject to practically these same limitations. Just as a loud noise prevents our hearing another weaker one, so, in the presence of a strong light, we do not see another of far lesser intensity if the two strike each retina in the same spot.

Having placed a candle 1 foot away from a very white surface, I put a ruler of a certain size beside the candle, and then I drew back, now more, now less, a candle of the same size as the first, until I could no longer make out the shadow of the ruler cast by this second candle. The shadow was very appreciable when I had moved this latter candle only 4 or 5 feet away from the surface. The entire space occupied by the shadow was, however, illuminated by the first candle; but next to this space, the light was increased by $\frac{1}{16}$ or $\frac{1}{25}$ owing to the rays from the second candle, and this increase was very appreciable. It was appreciable even when I moved the second candle to 6 or 7 feet; but it finally disappeared—or, in other words, the entire surface appeared to be of an absolutely uniform whiteness—when I placed the candle at a distance of about 8 feet. Thus, the distinction between the two lights ceased to be visible only when the small increment added [by the second candle] had been made about 64 times weaker than the first light. I could have obtained the two lights I was comparing fairly easily from the same source of illumination. I repeated the experiment several times, and I took care to substitute one candle for the other, to see whether they gave equal illumination.

·   ·   ·   ·   ·

Sometimes the same thing happens when the eyes are not directly exposed to light but have retained the impression of it. The retina can be compared to phosphorus, which absorbs, as it were, the light it is exposed to. If the light is very weak, its action must be repeated for the effect to become appreciable. On the other hand, rays striking the back of the eye provoke a disturbance that can last long after we have gone into a dark place; then our eyes remain in the same state as when they were affected by a definite light. And, if the impression is much stronger than that of an object in the dark, we will not discover the object and will begin to perceive it only when the excitation of the retina, having weakened, ceases to exceed by 60 to 80 times the intensity of the object's dim glow. Some places in the dark are better lighted than others; there are often infinitely different gradations of darkness, and it is certainly true that shadows tend to obscure the other parts; but, if the retained impression of outside light is still too strong, then not only do we not see the dark parts of the objects, we do not even distinguish the brightest. Actually, we see only the dominating light that still persists in our eyes, the effect of which, however, keeps weakening, according to the terms of an apparently geometric progression.

16 CHARLES ÉDUARD JOSEPH DELEZENNE (1776–1866) ON THE DIFFERENTIAL THRESHOLD FOR THE PITCH OF TONES, 1827

C. É. J. Delezenne, "Sur les valeurs numériques des notes de la gamme," *Recueil des travaux de la Société des Sciences, de l'Agriculture et des Arts de Lille,* 1827, pp. 4–6. Translated for this book by Mollie D. Boring.

Delezenne was a physicist at the University of Lille who concerned himself with acoustical and electrical problems. Using a sonometer with a taut string, he found that the differential threshold for musical pitch is less than a third of a comma for a musically trained ear, though somewhat greater for the naive ear. A comma is about a fifth of a semitone of equal temperament and is also about the amount of variation in a musical note that is regarded as mistuning rather than as establishing a different note.

Two strings, *absolutely equal in every way,* give two identical sounds: that is absolute unison, and it is self-evident.

Long segments of a single metallic string, cut in equal lengths, have been found to be of equal weight, thus showing the uniformity of their diameter and their density. One of these, mounted on a sonometer, gives the B of the great octave on the fourth string of a cello, according to a steel tuning fork. The length between the fixed bridges is *exactly* 1147 millimeters; thus the string vibrates 120 times per second. At the exact middle of this string, I put a movable bridge which, barely touching the string, does not increase the tension; the string is pressed against the sharp edge of the bridge by another sharp edge. Both sides now being perfectly equal, I make the two halves vibrate, either alternately or simultaneously, by means of a piece of flexible leather inserted in the barrel of a quill. One can then make the two strings vibrate with a slight enough contact to obtain sounds of little intensity; one operates at equal distances from the middle. With these precautions and many others related to the measurements, which I omit for conciseness, one obtains sounds seeming identical to the ear. But if we displace the movable bridge 2 millimeters either to the right or to the left, the difference is discernible to the least trained ear, or so several persons have assured me. If the displacement of the bridge is only 1 millimeter, one must have a fairly sensitive ear to perceive it immediately. The person submitted to this test shuts his eyes so as not to be distracted by the surrounding objects but to remain unaware of actual or feigned displacements of the bridge and to avoid being influenced by any change in direction he could see being made. Under these conditions an extremely fine ear is sensitive to this slight difference. Let us call this the extreme limit of the sensitivity of the human ear, and let us calculate the relation of these two only slightly different sounds. We get:[1]

$$\frac{(1147/2) + 1}{(1147/2) - 1} = \frac{1149}{1145} = \left(\frac{81}{80}\right)^{0.2807}.$$

The finest ear is thus sensitive to a difference of 4 vibrations in 1149!!

To compare this interval with that represented by the known comma 81/80, which we usually take for unity, we shall say that the ear is barely sensitive to [a difference of] a quarter of this comma.

We observed that a displacement of 2 millimeters was discernible by persons who had never tried to compare sounds. We found, for sounds thus compared, the interval[2]

$$\frac{(1147/2) + 2}{(1147/2) - 2} = \frac{1151}{1143} = \left(\frac{81}{80}\right)^{0.561}.$$

These persons are thus sensitive to a difference of 8 vibrations in 1151, to an interval a little larger than half a comma.

We can thus affirm that all ears are sensitive to an interval of a whole comma when they compare two sounds nearly in unison and when they hear them reverberating *alternately*. I say alternately because, in the comparison of simultaneous sounds, the ear tolerates greater differences.

---

[1]Delezenne moved the bridge 1 mm to get two strings of length 1147/2 + 1 and 1147/2 − 1 mm, with a resulting ratio of 1149/1145. The comma is the smallest musical interval, just over a fifth of a semitone, and is represented by the ratio 81/80. The formula says that moving the bridge 1 mm separates the pitches of the two parts of the string by 0.2807 comma, but little more than a quarter of this smallest musical interval. Since rate of vibration is inversely proportional to length of strings, these half strings would vibrate respectively about 239 and 241 times a second, a perceptible difference of only 2 vibrations per second.—Trans.

[2]Moving the bridge 2 mm separates the two pitches by 0.561 comma.—Trans.

The error is appreciable for a displacement of the bridge of 3 millimeters, which corresponds to 0.84 comma. At 4 millimeters it is more than evident and corresponds to 1.12 commas.

A result of these experiments is that an interval of 1 comma between two sounds being compared is most certainly appreciable and cannot be overlooked, at least not of sounds presented as equal.

17   ERNST HEINRICH WEBER (1795–1878) ON WEBER'S LAW, 1834

E. H. Weber, *De pulsu, resorptione, auditu et tactu: annotationes anatomicae et physiologicae* (Leipzig, 1834), pp. 172–175. Translated for this book by Barbara Herrnstein.

Ernst Heinrich Weber, the brother of the distinguished physicist Wilhelm Eduard Weber, was a physiologist and anatomist, and was a professor at the University of Leipzig from the age of twenty-three on. He is best known for his studies of tactual sensibility (see No. 10) and for his demonstration that the just noticeable difference in stimulus intensity is a constant fraction of the total intensity at which it is measured, the relation that came to be called Weber's law. These experiments, described in Latin in 1834, did not attract much attention until Weber brought them together with other facts in his famous chapter on tactual sensibility in *Handwörterbuch der Physiologie*, edited by Rudolph Wagner and published in Brunswick in 1846.

In observing the disparity between things that are compared, we perceive not the difference between the things, but the ratio of this difference to the magnitude of the things compared.

If two weights, one of which is 30 half ounces and the other 29, are compared by handling them, the disparity is not perceived more easily than when two weights of 30 and 29 drams are compared with each other. The difference between the weights is equal to one half ounce in the former instance and to one dram in the latter. Since a half ounce is equivalent to four drams, the weight in which the difference lies is, in the latter instance, four times less than in the former. In fact, since the disparity in the former instance is not perceived more easily than in the latter, it is obvious that what is felt is not the weights of the differences, but their ratios. Indeed, the same appears if we compare weights differing to so small a degree that the disparity is not clearly felt. If two weights, one of 33 and one of 34 half ounces, are compared by handling them, we do not feel the disparity; but it is not therefore because the half ounces differ too little from one weight to the other for us to perceive them by handling (for we perceive clearly, by handling, weights that are a hundred times smaller) but because the relation of the difference between the weights compared is such that it is only $\frac{1}{34}$ of the heavier weight. For experience has taught us that expert and practiced men feel a disparity of weight if it is not less than $\frac{1}{30}$ of the heavier weight, and perceive the disparity to be the same if, in place of half ounces, we put drams.

What I have demonstrated with respect to weights compared by handling holds true of lines compared by sight. For whether you compare longer or shorter lines, you will find that the difference is not felt by most men if one line is shorter than the other by $\frac{1}{100}$. Indeed, the portion by which one line exceeds the other is absolutely larger if the lines you compare are longer, smaller if the lines are shorter. If, for example, you compare a line 100 mm long with another line of 101 mm, the difference will be equal to 1 mm. If you compare a line 50 mm long with a line equal to $50\frac{1}{2}$ mm, the difference will be $\frac{1}{2}$ mm. The length, therefore, in which the disparity lies, even though it is two times less in the former instance, is, however, recognized just as easily, because in both cases the difference between the compared lines is $\frac{1}{100}$ of the longer line.

I have not performed any experiments concerning sounds compared by hearing. It is perfectly clear to me, however, from tests described by Delezenne, that musicians endowed with extremely fine judgment perceive an interval between two tones as small as $\frac{1}{4}$ comma, or $(81/80)^{\frac{1}{4}}$ (that is to say, the difference between two sounds of which the [frequency of] vibrations differ as 645 to 643 or, what is almost the same, as 322 to 321). These musicians, then, have discriminated by ear a sound of $\frac{1}{322}$ of the vibrations. Other men, according to that writer, have recognized a difference of sounds which is at least $\frac{1}{161}$ of the vibrations. Since, however, this author has not said that this difference is more difficult to discern in low tones and easier in higher tones, and also since I have never heard that in higher tones, which consist of a greater number of vibrations [per second], the difference is more easily perceptible than in lower ones, I suspect that in hearing, too, what is discerned is not the absolute difference between the vibrations of two tones, but the relative difference, compared to the number of vibrations of the tones.

Since the observation has been confirmed in most of the senses that men, in observing disparity, perceive not absolute but relative differences between things, I have urged myself again and again to investigate the cause of this phenomenon, and I hope that this cause will sometime be known well enough so that we will be able to judge more correctly concerning the nature of the senses.

I will mention here only that I have, by this observation, disproved certain things which might be suspected as fairly plausible concerning the manner in which we compare two impressions on the senses.

For instance, there is nothing incongruous, in itself, in the idea that we compare the images of two lines pictured on the retina in the same way as we compare two things held in our hand. We compare the length of such things most accurately when we superimpose one upon the other, so that both their congruence and disparity are known. It may be possible to do the same in our eyes. For if a line should produce on the center of the retina an impression persisting throughout some period of time during which the eye was turned elsewhere, and if the eye were then directed on another line so that its image fell on the same place, it might be possible to compare the length of the image of the second line with the impression of the line seen earlier and thus to determine to what degree it was congruent. My observation, however, is inconsistent with this hypothesis. If this were so, we would, of course, see the absolute difference. For if a line which is equal to an ell is superimposed on a line which exceeds an ell [*ulna,* elbow] by one inch [*pollex,* thumb], that inch is seen just as distinctly as if you were to place a line which exceeds a foot [*pes,* foot] by one inch on a line which is equal to a foot. Therefore, if one line is superimposed on another line in the retina, the disparity would be distinguished as easily in the former instance (where it is equal to $\frac{1}{25}$) as in the latter (where it is equal to $\frac{1}{12}$) [sic]. On the contrary, however (as our experience with vision teaches), when objects are placed next to each other we recognize the disparity of length more easily when it is between an object that is 13 inches long and a foot (12 inches long) than when it is between another object 25 inches long and an ell (24 inches long).

Nor indeed would it be clearly absurd for someone to say, through a reasonable analogy, that the lengths of two lines are compared by sight and measured in such a way that the mind, without obvious self-consciousness, counts the nerve endings touched in the retina by each of the two lines. My observations, however, are inconsistent with this hypothesis also, since if this were so, the absolute difference between the lengths of those lines would be seen and, moreover, the difference would not be perceived any better if it were greater when compared to the lengths of the lines.

18 GUSTAV THEODOR FECHNER (1801–1887) ON FECHNER'S LAW, 1860

G. T. Fechner, *Elemente der Psychophysik* (Leipzig, 1860), I, 64–67; II, 9–17. Translated by Herbert S. Langfeld in Benjamin Rand, ed., *Classical Psychologists* (Boston: Houghton Mifflin, 1912), pp. 562–572.

The sixty years of Fechner's productive life were varied: fifteen years as a

Leipzig physicist; ten years in retirement, his eyesight injured by gazing at the sun in order to obtain afterimages of it; starting his war against materialism; ten years at psychophysical experimentation, as part of this war, and ending in the publication of the *Psychophysik;* fifteen years with experimental aesthetics; and then ten more with psychophysics again as criticism of his views forced him back into this battle. His great contribution was his creation of a scale of sensation, a scale that measured sensation as something different from the stimulus so that the relation of the magnitude of the stimulus to the magnitude of the consequent sensation could be determined. That determination came to be known as

Fechner's law (which he called Weber's law), the rule that the magnitude of the sensation is proportional to the logarithm of its stimulus. Theretofore there had been no strictly subjective measurement, and the new scientific psychology, just ready to be founded, hailed this achievement as its warrant and safe-conduct. Now the impalpable mind could actually be measured. Fechner assumed that all just noticeable differences in sensation are equal and therefore provide a subjective unit for stating the magnitude of sensation. It was this assumption that was to create most of the quarrels in psychophysics for the next hundred years.

VII. THE MEASUREMENT OF SENSATION

Weber's law, that equal relative increments of stimuli are proportional to equal increments of sensation, is, in consideration of its generality and the wide limits within which it is absolutely or approximately valid, to be considered fundamental for psychic measurement. There are, however, limits to its validity as well as complications, which we shall have carefully to examine later. Yet even where this law ceases to be valid or absolute, the principle of psychic measurement continues to hold, inasmuch as any other relation between constant increments of sensation and variable increments of stimulus, even though it is arrived at empirically and expressed by an empirical formula, may serve equally well as the fundamental basis for psychic measurement, and indeed must serve as such in those parts of the stimulus scale where Weber's law loses its validity. In fact such a law, as well as Weber's law, will furnish a differential formula from which may be derived an integral formula containing an expression for the measurement of sensation.

This is a fundamental point of view, *in which Weber's law, with its limitations, appears, not as limiting the application of psychic measurement, but as restricted in its own application toward that end and beyond which application the general principle of psychic measurement nevertheless continues to hold.* It is not that the principle depends for its validity upon Weber's law, but merely that the application of the law is involved in the principle.

Accordingly investigation in the interest of the greatest possible generalization of psychic measurement has not essentially to commence with the greatest possible generalization of Weber's law, which might easily produce the questionable inclination to generalize the law beyond its natural limitation, or which might call forth the objection that the law was generalized beyond these limits solely in the interest of psychic measurement; but rather it may quite freely be asked how far Weber's law is applicable, and how far not; for the three methods which are used in psychic measurement are applicable even when Weber's law is not, and where these methods are applicable psychic measurement is possible.

In short, Weber's law forms merely the basis for the most numerous and important applications of psychic measurement, but not the universal and essential one. The most general and more fundamental basis for psychic measurement is rather those methods by which the relation between stimulus increments and sensation increment in general is determined, within, as well as without, the limits of Weber's law; and the development of these methods towards ever greater precision and perfection is the most important consideration in regard to psychic measurement.

And yet a great advantage would be lost, if so simple a law as Weber's law could not be used as an exact or at least sufficiently approximate basis for psychic measurement; just such an advantage as would be lost if we could not use the Kepler law in astronomy, or the laws of simple refraction in the theory of the dioptric instruments. Now there is just the same difficulty with these laws as with Weber's law. In the case of Kepler's law we abstract from deviations. In the case of simple lens refraction we abstract from optical aberration. In fact they may become invalid as soon as the simple hypotheses for which they are true no longer exist. Yet they will always remain decisive for the principle relation with which astronomy and dioptrics are concerned. Weber's law may in like manner, entirely lose its validity, as soon as the average or normal conditions under which the stimulus produces the sensation are unrealized. It will always, however, be decisive for these particular conditions.

Further, just as in physics and astronomy, so can we also in psychic measurement, neglect at first the irregularities and small departures from the law in order to discover and examine the principle relations with which the science has to do. The existence of these exceptions must not, however, be forgotten, inasmuch as the finer development and further progress of the science depends upon the determination and calculation of them, as soon as the possibility of doing so is given.

The determination of psychic measurement is a matter for outer psychophysics and its first applications lie within its boundary; its further applications and consequences, however, extend necessarily into the domain of inner psychophysics and its deeper meaning lies there. It must be remembered that the stimulus does not cause sensation directly, but rather through the assistance of bodily processes with which it stands in more direct connection. The dependence, quantitatively considered of sensation on stimulus, must finally be translated into one of sensation on the bodily processes which directly underlie the sensation—in short the psycho-physical processes; and the sensation, instead of being measured by the amount of the stimulus, will be measured by the intensity of these processes. In order to do this, the relation of the inner process to the stimulus must be known. Inasmuch as this is not a matter of direct experience it must be deduced by some exact method. Indeed it is possible for this entire investigation to proceed along exact lines, and it cannot fail at some time or other to obtain the success of a critical study, if one has not already reached that goal.

Although Weber's law, as applied to the relation of stimulus to sensation, shows only a limited validity in the domain of outer psychophysics, it has, as applied to the relation of sensation to kinetic energy, or as referred to some other function of the psycho-physical process, in all probability an unlimited validity in the domain of inner psychophysics, in that all exceptions to the law which we find in the arousal of sensation by external stimulus, are probably due to the fact that the stimulus only under normal or average conditions engenders a kinetic energy in those inner processes proportional to its own amount. From this it may be foreseen, that this law, after it has been restated as a relation between sensation and the psycho-physical processes, will be as important, general, and fundamental for the relations of mind to body, as is the law of gravity for the field of planetary motion. And it also has that simplicity which we are accustomed to find in fundamental laws of nature.

Although, then, psychic measurement depends upon Weber's law only within certain limitations in the domain of outer psycho-physics, it may well get its unconditional support from this law in the field of inner psychophysics. These are nevertheless for the present merely opinions and expectations, the verification of which lies in the future.

·   ·   ·   ·   ·

XVI. THE FUNDAMENTAL FORMULA AND THE MEASUREMENT FORMULA

Although not as yet having a measurement for sensation, still one can combine in an exact formula the relation expressed in Weber's law,—that the sensation difference remains constant when the relative stimulus difference remains constant,—with the law, established by the mathematical auxiliary principle, that small sensation increments are proportional to stimulus increments. Let us suppose, as has generally been done in the attempts to preserve Weber's law, that the difference between two stimuli, or, what is the same, the increase in one stimulus, is very small in proportion to the stimulus itself. Let the stimulus which is increased be called β, the small increase $d\beta$, where the letter d is to be considered not as a special magnitude, but simply as a sign that $d\beta$ is the small increment of β. This already suggests the differential sign. The relative stimulus increase therefore is $d\beta/\beta$. On the other hand, let the sensation which is dependent upon the stimulus β be called γ, and let the small increment of the sensation which results from the increase of the stimulus by $d\beta$ be called $d\gamma$, where d again simply expresses the small increment. The terms $d\beta$ and $d\gamma$ are each to be considered as referring to an arbitrary unit of their own nature.

According to the empirical Weber's law, $d\gamma$ remains constant when $d\beta/\beta$ remains constant, no matter what absolute values $d\beta$ and β take; and according to the *a priori* mathematical auxiliary principle the changes $d\gamma$ and $d\beta$ remain proportional to one another so long as they remain very small. The two relations may be expressed together in the following equation:

$$d\gamma = Kd\beta/\beta, \tag{1}$$

where κ is a constant (dependent upon the units selected for γ and β). In fact, if one multiplies βd and β by any number, so long as it is the same number for both, the proportion remains constant, and with it also the sensation difference $d\gamma$. This is Weber's law. If one doubles or triples the value of the variation $d\beta$ without changing the initial value β, then the value of the change $d\gamma$ is also doubled or tripled. This is the mathematical principle. The equation $d\gamma = Kd\beta/\beta$ therefore entirely satisfies both Weber's law and this principle; and no other equation satisfies both together. This is to be called the *fundamental formula,* in that the deduction of all consequent formulas will be based upon it.

The fundamental formula does not presuppose the measurement of sensation, nor does it establish any; it simply expresses the relation holding between small relative stimulus increments and sensation increments. In short, it is nothing more than Weber's law and the mathematical auxiliary principle united and expressed in mathematical symbols.

There is, however, another formula connected with this formula by infinitesimal calculus, which expresses a general quantitative relation between the stimulus magnitude as a summation of stimulus increments, and the sensation magnitude as a summation of sensation increments, in such a way, that with the validity of the first formula, together with the assumption of the fact of limen, the validity of this latter formula is also given.

Reserving for the future a more exact deduction, I shall attempt first to make clear in a general way the connection of the two formulas.

One can readily see, that the relation between the increments $d\gamma$ and $d\beta$ in the fundamental formula corresponds to the relation between the increments of a logarithm and the increments of the corresponding number. For as one can easily convince oneself, either from theory or from the table, the logarithm does not increase by equal increments when the corresponding number increases by equal increments, but rather when the latter increases by equal relative amounts; in other words, the increases in the logarithms remain equal, when the relative increases of the numbers remain equal. Thus, for example, the following numbers and logarithms belong together:

Number	Logarithm
10	1.000000
11	1.0413927
100	2.000000
110	2.0413927
1000	3.000000
1100	3.0413927

where an increase of the number 10 by 1 brings with it just as great an increase in the corresponding logarithm, as the increase of the number 100 by 10 or 1000 by 100. In each instance the increase in the logarithm is 0.0413927. Further, as was already shown in explaining the mathematical auxiliary principle, the increases in the logarithms are proportional to the increases of the numbers, so long as they remain very small. Therefore one can say, that Weber's law and the mathematical auxiliary principle are just as valid for the increases of logarithms and numbers in their relation to one another, as they are for the increases of sensation and stimulus.

The fact of the threshold appears just as much in the relation of a logarithm to its number as in the relation of sensation to stimulus. The sensation begins with values above zero, not with zero, but with a finite value of the stimulus—the threshold; and so does the logarithm begin with values above zero, not with a zero value of the number, but with a finite value of the number, the value 1, inasmuch as the logarithm of 1 is equal to zero.

If now, as was shown above, the increase of sensation and stimulus stands in a relation similar to that of the increase of logarithm and number, and, the point at which the sensation begins to assume a noticeable value stands in a relation to the stimulus similar to that which the point at which the logarithm attains positive value stands to the number, then one may also expect that sensation and stimulus themselves stand in a relation to one another similar to that of logarithm to number, which, just as the former (sensation and stimulus) may be regarded as made up of a sum of successive increments.

Accordingly the simplest relation between the two that we can write is $\gamma = \log \beta$.

In fact it will soon be shown that, provided suitable units of sensation and stimulus are chosen, the functional relation between both reduces to this very simple formula. Meanwhile it is not the most general formula that can be derived, but one which is only valid under the supposition of particular units of sensation and stimulus, and we still need a direct and absolute deduction instead of the indirect and approximate one.

The specialist sees at once how this may be attained, namely, by treating the fundamental formula as a differential formula and integrating it. In the following chapter one will find this done. Here it must be supposed already carried out, and those who are not able to follow the simple infinitesimal deduction, must be asked to consider the result as a mathematical fact. This result is the following functional formula between stimulus and sensation, which goes by the name of the measurement formula and which will now be further discussed:

$$\gamma = \kappa \, (\log \beta - \log b). \tag{2}$$

In this formula κ again stands for a constant, dependent upon the unit selected and also the logarithmic system, and b a second constant which stands for the threshold value of the stimulus, at which the sensation γ begins and disappears.

According to the rule, that the logarithm of a quotient of two numbers may be substituted for the difference of their logarithms, . . . one can substitute for the above form of the measurement formula the following, which is more convenient for making deductions:

$$\gamma = \kappa \log (\beta/b). \tag{3}$$

From this equation it follows that the sensation magnitude γ is not to be considered as a simple function of the stimulus value β, but of its relation to the threshold value b, where the sensation begins and disappears. This relative stimulus value, β/b is for the future to be called the fundamental stimulus value, or the fundamental value of the stimulus.

Translated in words, the measurement formula reads:

The magnitude of the sensation (γ) is not proportional to the absolute value of the stimulus (β), but rather to the logarithm of the magnitude of the stimulus, when this last is expressed in terms of its threshold value (b), i.e. that magnitude considered as unit at which the sensation begins and *disappears. In short, it is proportional to the logarithm of the fundamental stimulus value.*

Before we proceed further, let us hasten to show that that relation between stimulus and sensation, from which the measurement formula is derived, may be correctly deduced in turn from it, and that this latter thus finds its verification in so far as these relations are found empirically. We have here at the same time the simplest examples of the application of the measurement formula.

The measurement formula is founded upon Weber's law and the fact of the stimulus threshold; and both must follow in turn from it.

Now as to Weber's law. In the form that equal increments of sensation are proportional to relative stimulus increments, it may be obtained by differentiating the measurement formula, inasmuch as in this way one returns to the fundamental formula, which contains the expression of the law in this form.

In the form, that equal sensation differences correspond to equal relations of stimulus, the law may be deduced in quite an elementary manner as follows.

Let two sensations, whose difference is to be considered, be called γ and γ', and the corresponding stimuli β and β'. Then according to the measurement formula,

$$\gamma = \kappa \, (\log \beta - \log b),$$
$$\gamma' = \kappa \, (\log \beta' - \log b),$$

and likewise for the sensation difference

$$\gamma - \gamma' = \kappa \, (\log \beta - \log \beta'),$$

or, since $\log \beta - \log \beta' = \log \beta/\beta'$,

$$\gamma - \gamma' = \kappa \log (\beta/\beta').$$

From this formula it follows, that the sensation difference $\gamma - \gamma'$ is a function of the stimulus relation β/β', and remains the same no matter what values β, β' may take, so long as the relation remains unchanged, which is the statement of Weber's law.

In a later chapter we shall return to the above formula under the name of the difference formula, as one of the simplest consequences of the measurement formula.

As for the fact of the threshold, which is caused by the sensation having zero value not at zero but at a finite value of the stimulus, from which point it first begins to obtain noticeable values with increasing values of stimulus, it is so far contained in the measurement formula as

γ does not, according to this formula, have the value zero when $\beta = 0$, but when β is equal to a finite value b. This follows as well from equation (2) as (3) of the measurement formula, directly from (2), and from (3) with the additional consideration of the fact, that when β equals b, log (β/b) equals log 1, and log 1 = 0.

Naturally all deduction from Weber's law and the fact of the threshold will also be deductions from our measurement formula.

It follows from the former law, that every given increment of stimulus causes an ever decreasing increment in sensation in proportion as the stimulus grows larger, and that at high values of the stimulus it is no longer sensed, while on the other hand, at low values it may appear exceptionally strong.

In fact the increase of a large number β by a given amount is accompanied by a considerably smaller increase in the corresponding logarithm γ, than the increase of a small number β by the same amount. When the number 10 is increased by 10, (that is, reaches 20), the logarithm corresponding to 10, which is 1, is increased to 1.3010. When, however, the number 1000 is increased by 10, the logarithm corresponding to 1000, namely 3, is only increased to 3.0043. In the first case the logarithm is increased by 1-3 of its amount, in the latter case by about 1-700.

In connection with the fact of the threshold belongs the deduction, that a sensation is further from the perception threshold the more the stimulus sinks under its threshold value. This distance of a sensation from the threshold, is represented in the same manner by the negative values of γ, according to our measurement formula, as the increase above the threshold is represented by the positive values.

In fact one sees directly from equation (2), that when β is smaller than b and with it log β smaller than log b, the sensation takes on negative values, and the same deduction follows in equation (3), in that β/b' becomes a proper fraction when $\beta < b$, and the logarithm of a proper fraction is negative.

In so far as sensations, which are caused by a stimulus which is not sufficient to raise them to consciousness, are called unconscious, and those which affect consciousness are called conscious, we may say that the unconscious sensations are represented in our formula by negative, the conscious by positive values . . .

According to the foregoing our measurement formula corresponds to experience:

1. In the cases of equality, where a sensation difference remains the same when the absolute intensity of the stimulus is altered (Weber's law).

2. In the cases of the thresholds, where the sensation itself ceases, and where its change becomes either imperceptible or barely perceptible.

In the former case, when the sensation reaches its lower threshold; in the latter case, when it becomes so great that a given stimulus increase is barely noticed.

3. In the contrasting cases, between sensations which rise above the threshold of consciousness and those that do not reach it,—in short, conscious and unconscious sensations. From the above the measurement formula may be considered well founded.

In the measurement formula one has a general dependent relation between the size of the fundamental stimulus and the size of the corresponding sensation and not one which is valid only for the cases of equal sensations. This permits the amount of sensation to be calculated from the relative amounts of the fundamental stimulus and thus we have a measurement of sensation.

19 JOSEPH ANTOINE FERDINAND PLATEAU (1801–1883) ON THE MEASUREMENT OF SENSATION, 1872

J. A. F. Plateau, "Sur la mesure des sensations physiques, et sur la loi qui lie l'intensité de ces sensations à l'intensité de la cause excitante," *Bulletins de l'Académie Royale des Sciences, des Lettres et des Beaux-Arts de Belgique* [2] *33,* 376–385 (1872). Translated for this book by Mollie D. Boring.

Plateau was professor of physics at the University of Ghent. He had been a pupil of Adolph Quetelet's, the statistician who invented the word "statistics," Plateau devoted himself to a study of physiological optics and was an inventor of the stroboscope, which is used for observing the nature of rapidly recurring movements. Like Fechner he injured his eyes by gazing at the sun, and fourteen years later (1843) he went blind. Subsequently his visual experiments were conducted, as was the one excerpted here, with the assistance of his friend François J. Duprez, a fellow member of the Académie Royale. This particular experiment was carried out in the 1850's, but Plateau did not publish the results until 1872. Relying on his conviction that equally sensed contrasts represent equal increments of sensations, he showed how a scale of equal supraliminal sense distances can be built up, and for this reason he is known as the inventor of the method of equal sense distances (Titchener's phrase). Because the perception of chiaroscuro in a picture or scene is not altered even by a great change in total illumination, Plateau concluded that a sensed contrast, which he regarded as a sensory ratio, remains constant at all illuminations. He also made the seemingly patent assumption that the stimulus for zero sensation must be zero (Fechner to the contrary notwithstanding), and that a geometric series of stimuli must arouse a geometric series of sensory intensities. In this way Plateau was proposing a power law of sensory intensities as opposed to Fechner's logarithmic law.

E. H. Weber, around 1853, I believe, clearly demonstrated a remarkable fact, already known to a number of natural philosophers, that the least perceptible difference in the intensities of two like exciting causes is a fairly constant fraction of the intensity of one of them. Fechner, in two publications, one in 1859 and the other in 1860, deduced from this a relation between the intensity of sensation and the intensity of its exciting cause. If we call the first S and the second E, the following relation, which expresses the law of the increase in sensation with its cause, is:

$$S = A \log E + C, \tag{1}$$

where A and C are constants.

The idea of measuring physical sensations up to a certain point occurred to me about twenty years ago, and I had started a series of experiments on the subject; but, involved with other research, I did not continue. The aim of this present note is not to claim priority for the idea, since my early attempts have not been published; but, because my method is based on a totally different principle from the one on which Fechner bases his formula and because, further, my results show us to have a particular faculty of judgment, I feel it to be not without interest to make this fact known.

When we experience two like physical sensations of unequal intensity, either simultaneously or successively, we can easily judge which of the two is stronger and can, furthermore, tell whether the difference is slight or great; but the comparison, it would seem, ought to stop there, especially if we restrict ourselves to a direct judgment and thus consider ourselves incapable of determining the numerical ratio of intensity between these sensations. However, in pursuing the question further, we soon realize that our judgment of these relative intensities is not so vague after all. Take, for example, the sensation of light: when we say an object is light gray, we apparently mean a gray more nearly white than black, which is to say that the intensity of our sensation is more than half that of the sensation produced by a white object under the same conditions of illumination. On the other hand, when we say an object is dark gray, we mean a gray more nearly black than white, or, in other words, that the intensity for this sensation is less than half that for the sensation of a white object exposed to the same light. In the end, we can obtain a gray between black and white. Thus we see that the judgment of this last gray can be made with a certain precision and, if it is done well, we shall then have a gray sensation that is just about half the intensity of a white sensation.

I have just said that a gray seeming to be exactly halfway between white and black can be determined with sufficient exactitude. Thus, if you take three paper squares of the same size, one coated with very pure white, a second with gray, and a third with very intense black, and then juxtapose these with the gray square in the middle, you can judge whether the white contrasts more or less with the gray than the gray with the black, and you have only to keep modifying the shade of gray until both contrasts seem equal.

Nevertheless, before taking the precision of such judgments for granted, I wanted to reassure myself by experiment, and, with this aim, I had recourse to the following procedure. I asked a number of persons separately—all of them painters and therefore used to the scrutiny and manipulation of colors—to make me a sample intermediate gray by the process indicated above, using oil-color squares in ordinary daylight. Obviously, if the judgment of equal contrast depends on merely a vague feeling, then the grays furnished by these persons—there were eight of them—would be markedly different, whereas if this is a clear-cut feeling then all the grays would have to be very similar. That last is what happened: the eight samples of gray were practically identical. By placing them in order from the lightest to the darkest, I was able to choose what seemed the mean of them all, and this one was always very close to the gray of sensation just halfway between pure white and pure black.

Actually, the most intense black obtainable by painting still reflects a little bit of light, thus making the above gray sensation a bit more than half that for white; but the difference was truly small, and besides, it would be simple to alter the experiment by substituting, for the square coated with black, a space totally void of light and thus virtually an absolute black.

By the same method of judging, we can get a gray that is exactly halfway between the above one and black, and this second gray will now evoke a sensation equal in intensity to one quarter that for the sensation white. Still using our method, we can next find a third gray, halfway between white and the first gray, and this third gray will give a sensation equal in intensity to three quarters that for white; and thus we see that the intensities for the sensations corresponding to the five shades from black to white can be numbered 0, 1, 2, 3, and 4. In the end, we could increase the intermediate gradations at will and obtain a sort of scale of sensations, the intensities of which will have known ratios.

Thus, though we have no faculty of judging directly the ratio of intensity for two sensations of light, we do have another faculty that enables us indirectly to reach the value of this ratio: namely, the faculty of clearly perceiving equal contrast. By this means, we have seen that it is possible to construct a scale of shades producing sensations in increasing arithmetical progression; and, if we do this so that the first term of the progression is zero, then the succeeding terms will be numbered 1, 2, 3, 4, 5, and so on.

·   ·   ·   ·   ·

We all recall, in another way, that there is little change in the relative intensity of sensations corresponding to different shades with change in the intensity of an illumination common to them: everyone knows that the effect of an engraving stays pretty much the same whether we see it by daylight, by candlelight, by gaslight, or even by sunlight; such different illuminations do not greatly affect the relation of the light to the shadow, not so much so but that the imagination can supply the general effect.

But if the ratio of the intensities of sensations due to two unlike shades is independent of the amount of illumination common to them, then we obtain a formula which does not coincide with Fechner's. Take, for example, very generally, $S = F(E)$; the function F will have to be such that when $E = 0$, then $S = 0$, and S increases with E. This being the case, now imagine that we observe two adjacent squares of different shades, one white and the other gray, illuminated by a given light such as daylight. Each of these squares will send a light to the eye, and the resultant ratio of intensity of these lights we will designate by r. If now we expose these two squares to a light m times stronger, to sunlight, for example, then each of the squares will send a light to the eye m times more intense than under the first illumination, and the ratio of intensity between these lights is still r. Let us now suppose that the ratio of intensity of sensation for any two shades remains constant as we vary the illumination they have in common, and that we give them the values E', E'', E''', E^{iv}, and so on, in such a way that the quotient of any one of these and the one that precedes it remains constant; then the quotient of any one of the corresponding values of S with its preceding value will also be constant. In other words, then, if E increases by a geometric progression, S will also increase by a geometric progression.

From this, if B is the first term of the progression in accordance with which E increases, and λ is the constant factor of the progression, we will have, to represent the nth term, the expression

$$E = B\lambda^{n-1}. \tag{2}$$

In like manner, designating by C and by μ the first term and the constant factor of the progression by means of which S increases, we will have for the corresponding term in this second progression

$$S = C\mu^{n-1}.$$

Now, since μ and λ are constants, we can obtain $\mu = \lambda^p$, the exponent p being likewise constant, and this gives us

$$S = C(\lambda^p)^{n-1} = C(\lambda^{n-1})^p \tag{3}$$

Eliminating λ^{n-1} between equations (2) and (3), we get $S = (B/C^p)E^p$, or, replacing the constant quantity B/C^p by the single letter A,

$$S = AE^p, \tag{4}$$

an equation in which p can be less than unity.

Such, then, is the relation we attain from the hypothesis of the independence of the ratio of sensations of unlike shades and the lighting common to them.

We see that the law expressed by our formula (4) is essentially different from Fechner's, expressed in formula (1). His seems applicable only from a certain value of E on; thus, for $E = 0$, for example, it gives $S =$ negative infinity, which is interpretable only with difficulty. It is true that Fechner partly avoids the difficulty by considering that there is an extremely small, though finite, value of E below which no sensation is perceived and which, therefore, corresponds to $S = 0$. If we call this value E', we obtain the formula

$$S = A \log E/E'.$$

Formula (4) gives $S = 0$ for $E = 0$, but it is founded on a hypothesis perhaps insufficiently justified. Moreover, both this same formula (4) and Fechner's must cease to apply beyond a certain upper limit of E, since, when excitation is too vigorous, it alters the organ perceiving the sensation.

Here we actually have a method for obtaining the relative luminous intensities of different shades, together with the scale of these shades as discussed above. It will now be possible to submit formulas (1) and (4) to experimental proof.

20
JOSEPH RÉMI LÉOPOLD DELBOEUF (1831–1896) ON SENSED CONTRAST AS THE MEASURE OF SENSATION, 1883

J. R. L. Delboeuf, *Examen critique de la loi psychophysique: sa base et sa signification* (Paris, 1883), pp. 140–144. Translated for this book by Mollie D. Boring.

J. R. L. Delboeuf was a professor at the Belgian university of Liége, where he concerned himself with problems of philology, philosophy, and physics. He published four critical monographs on psychophysics (1873–1883) and sought in the one excerpted here to show how the just noticeable sensory differences of Fechner are subjectively equal. In this argument he was borrowing from the thought of his senior Belgian colleague Plateau by insisting that all just noticeable differences show equally sensed contrast. It is only a step from this belief to Titchener's conception that equally sensed contrasts are equally sensed distances. Following Fechner, Delboeuf believed that supraliminal sensed contrasts or distances are equal when each is the sum of the same number of just noticeable contrasts or distances. The argument is confused because Plateau and Delboeuf thought of equal contrast as equal ratios; nevertheless Delboeuf's discussion was the ground for Titchener's development of the concept of the equal sense distance in relation to the scaling of sensations. It should also be noted that at about this time (1903) G. E. Müller, who was responsible for three very influential books on psychophysics, proposed that sensory separation could be indicated by the subjective degree of coherence (*Kohärenzgrad*) of two sensations. Contrast and coherence, being opposites, presumably determine for observation the same continuum.

A NEW THEORY OF SENSIBILITY

Statement of the Theory

The theory, deduced by me from experiments and known observations, can be summed up in three laws: the laws of progression, degradation, and tension. Of course, I do not have absolute faith in them, but up to now I can say that my misgivings have lessened rather than increased. As Fechner does not seem to me to have realized fully the principles involved and since, at this time, these are the only laws which can come into accord with his, it seems to me of use to make this new statement . . .

Imagine an experiment on the sense of temperature. This experiment is purely ideal, for there is, practically speaking, no possibility of executing it as I have here conceived it; but it does make clear my understanding of the phenomenon of sensation.

We presume that our two hands are equally sensitive to heat and that no pain is involved. Two bowls contain water of the same temperature as the hands. On immersing our hands, we have no sensation. Increase the temperature slowly in one of the bowls, and a point will come when we perceive a difference. This will happen, if the temperature of the first bowl is t_0, when the temperature of the second is $t_0 + d_0 = t_1$. We could write $d_0 = mt_0$, m being in general less than 1. Now make the warmth of the two bowls equal to t_1 and be sure that both hands are at the point of not feeling anything. Next, proceed as before and increase the warmth in one bowl. Again we shall feel that there has been an increase, when the difference becomes d_1. We will have $t_1 + d_1 = t_2$. It will be found, we may suppose, that $d_1 = mt_1$ and, likewise, that $d_2 = mt_2$, and so on indefinitely. If I express these results mathematically, I obtain the formula $S = k \log t/t_0$ or, more generally, $S = k \log P/p$, a formula that coincides with one of Fechner's and implies Fechner's law but is interpreted very differently. It expresses what I have called *the law of progression between the quantity of excitation and the quantity of corresponding sensation.*

What does this experiment really show? It shows that sensation is a result of a difference or a contrast, and that successive contrasts, to be felt as equal, must correspond to greater and greater actual differences. And that is the basis of Weber's law.

This experiment used both hands and two bowls. We can omit one hand and one of the bowls, for they both had been introduced only as an example, as a means of comparison to support the judgment of contrast to be felt by the other hand. As we have seen, the feeling of contrast keeps lessening because the hand grows accustomed bit by bit to the increase in warmth of the water in the bowl; and thus the sensation resulting from the inequality of temperature between the hand and the water tends to disappear as equilibrium is re-established. It disappears completely when the hand and the water reach the same temperature. This is the fact of *degradation* of sensation, a phenomenon analogous to the loss of heat a warm body undergoes in a colder medium.

That leaves the third law. We cannot indefinitely increase the temperature of the water in which the hand is immersed. A point comes when the characteristic sensation is succeeded by a sharper and sharper feeling of pain. Actually, we experience a slight discomfort when the temperature in the bowl exceeds normal skin temperature even by very little; and in the degree that it varies from normal skin temperature, sensibility diminishes and judgment is less easily made, so that the ideal proportionality discussed above is not manifested in all its purity, and the difference must, in general, become greater and greater for the effects of contrast to continue to be judged as equal. I say *judged* deliberately, because a point comes when pain overcomes the faculty of judgment. And that is what can be called the phenomenon of *alteration* of sensation; and since it is manifested largely when we submit sensibility to an action that is excessive to the point of destruction, we can name it the phenomenon of *tension,* by way of allusion to the taut spring that is liable to break under undue stress.

21
EDWARD BRADFORD TITCHENER (1867–1927) ON THE SENSE DISTANCE AS THE MEASURE OF SENSATION, 1905

E. B. Titchener, *Experimental Psychology* (New York: Macmillan, 1905), vol. II, pt. 2, "Instructor's Quantitative Manual," pp. xix–xxix.

Oswald Külpe remarked that Titchener's *Experimental Psychology,* or at least the 600 pages of the "Instructor's Quantitative Manual," which is cited above, was the most erudite psychological work then in the English language, and doubtless this psychophysical encyclopedia is of such broad and accurate scholarship as to deserve this accolade. Fechner had based sensory measurement on the assumption of the equality of just noticeable differences, an assumption which many doubted. Plateau had shown that supraliminal differences in sensation can be compared introspectively and had described these differences as sensed contrasts. Delboeuf, in the excerpt already given, described how just noticeable differences can be judged equal in sensed contrast. Titchener based his view on Delboeuf and by implication on Plateau, insisting that distances between sensory magnitudes (the sensed contrasts) are observable as well as the sensations themselves. This bisection of sense distances—and the continued bisection of halves according to Plateau's procedure—results, Titchener argued, in true measurement, and he was right as against Fechner, even though later psychophysics might find a special limitation to the method he proposed.

1. *Measurement.*—Whenever we *measure,* in any department of natural science, we compare a given magnitude with some conventional unit of the same kind, and determine how many times the unit is contained in the magnitude. Let P be the magnitude to be measured, and p the unit in terms of which it is to be expressed. The result of our measurement of P is the discovery of the numerical ratio existing between P and p. We state this result always in terms of an equation: $P = (x/y)\,p$. The object of measurement, then, is the giving of such values to x and y that the equation may be true.

When we say, *e.g.,* that Mt. Vesuvius is 4200 ft. high, we mean that the given linear magnitude $P,$ the distance from sea level to the topmost point of the volcano, contains the conventional unit of linear measurement, 1 ft., four thousand two hundred times: $P = (4200/1)\,p.$ When we say that an operation lasted 40 minutes, we mean that the given temporal magnitude, the time occupied by the operation, contained the conventional time unit, 1 min., forty times: $P = (40/1)\,p.$ When we say that an express package weighs three quarters of a pound, we mean that the package, laid in the scale pan, just balances the sliding weight when this is placed at the twelfth short stroke beyond the first long stroke or zero point of the bar,—the distance between any two long strokes giving the conventional unit 1 lb., and the distance between any two short strokes the conventional sub-unit 1 oz. $= (1/16)\,\text{lb.}$: so that $P = (3/4)\,p$ in terms of lb., or $P = (12/1)\,p$ in terms of oz. In the same way, we might say that the height of Mt. Vesuvius is 4200/5280 mile; or that the time of the operation is 40/60 hr.

These instances show—what must always be borne in mind—that the unit of measurement is conventional. Its choice is simply a matter of practical convenience. Scientific men are now generally agreed that the unit of space shall be the 1 cm., the unit of time the 1 sec., and the unit of mass the 1 gr.; so that the unit of mechanical energy is the amount contained in a body of 1 gr. moving through 1 cm. in 1 sec. There is, however, nothing absolute and nothing sacrosanct about these units. Our measurements would be every bit as valid, just as much true measurements, if we took as units the pace or span or cubit, the average time of a step in walking or of a respiratory movement, the ounce or pound. The metric system makes calculation easy, relates the three fundamental quantities in a very simple manner: but that is its sole, as it is its sufficient claim to acceptance.

The prototype of all measurement is linear measurement in space. We can literally superpose one portion of space upon another, for purposes of measurement: we can hold the compared portions together for as long a time as we like; we can shift the one portion to and fro upon the other. The linear space unit is thus the most easily manipulated of all units of measurement. Hence there is a tendency, in natural science, to reduce all quantitative comparisons to the comparison of spatial magnitudes. We compare masses, with the metric balance, by noting the deflections of the pointer. We measure the intensity of an electric current by noting the deflections of the galvanometer needle. We measure rise and fall of temperature by the rise and fall of mercury in the thermometer. We determine the period of a tuning fork by the graphic method. It is, moreover, with spatial measurements that we are chiefly concerned in everyday life. We are all, to some extent, practised in the estimation of space magnitudes, however helpless we may be when called upon to grade weights or to estimate brightnesses.

Finally, it is to be noticed that every measurement implies three terms, expressed or understood. In measuring the height of Mt. Vesuvius we had the zero point, or sea level; the highest point of the mountain; and the point lying 1 ft. above sea level or 1 ft. below the highest point. In measuring the time of the operation, we had the beginning and end of the period occupied by it, and the time point lying 1 min. (or 1 hr.) distant from the beginning. In weighing our package, we had the zero point upon the scale bar; the limiting point at which the sliding weight was just counterbalanced; and the mark that lay 1 oz. (or 1 lb.) from the zero point. There are various devices—the introduction of submultiples of the unit, the use of the vernier—for increasing the accuracy of measurement; there are other devices for standardising the conditions (temperature, stress) under which a measurement is made; there are mathematical rules for calculating the 'probable error' of a given measurement. These are all refinements of the art of measuring. The essential thing is that we have our three terms: the limiting points of the magnitude to be measured, and a point lying at unit distance from the one or the other limiting point.

The third term, without which measurement is impossible, need not, however, be expressed. Suppose that two black strokes are made upon a sheet of paper, and that you are asked to say how far the one is above the other, or to the right of the other. You reply, without difficulty, "Two inches" or "Five centimetres." But that means that you have mentally introduced a third term: the unit mark, inch or cm., with which you have become familiar in previous measurements. Without this, you could only have said: "The one mark is above the other" or "to the right of the other"; you could not have answered the question "how far." How long is a given stretch of level road? Two hundred and fifty yards? A quarter of a mile? Most people have no mental unit for such a measurement. Either they say "It looks about as long as from so-and-so to so-and-so," —comparing it with a familiar distance; or they make a rough determination by pacing the distance itself. How deep is this well? Very few people can say, even if they can see the water. So a stone is dropped in, and the seconds are counted until the splash is heard. The pace or the familiar distance gives a third term for the measurement of the road; and we know that the distance traversed by the stone in falling is the product of the distance traversed in the first second (about 490 cm.—our third term) into the square of the time. Where there is no such third term, there is no measurement. This rule is universal.

2. *Mental measurement.*—There can be no question but that, in some way or other, mental processes are measurable. It would be strange, indeed, if the processes of the physical universe, which we know only by means of our sense organs or of instruments which refine upon our sense organs, should be capable of measurement, while the sensations of mental science were not: if all measurement in the physical sciences should tend towards spatial, *i.e.,* visual measurement, while yet the visual sensations themselves were unmeasurable. That apart, however, we have only to appeal to introspection to see that our mental processes furnish the raw material of measurement. Sensations differ more or less in quality: a given tone lies higher in the scale than another, a given green is more yellow than another. They differ in intensity: a noise may be louder than another noise, a brightness stronger than another brightness. They differ in duration: the taste of bitter lasts longer than the taste of sweet, the visual after-image lasts longer than the auditory. They differ in extent: one red is larger, one pressure spread more widely than another. These differences are given with the sensations; they obtain whether or not we know anything of the stimuli which arouse the sensations; we have evidence, in the history of science, that they were remarked and utilised long

before the stimuli were known or measured. But if we have differences of more and less, it is only necessary to establish the unit, the third term, in order to convert difference into measured difference.

This establishment of the unit is, however, no small matter. We have said that the units of physical measurement are conventional. On the other hand, the units of modern physics are accurate, objective, universal. It is a far cry to these units—the cm., the sec., the gr.—from such things as the day's journey, the barley-corn, the chaldron. The difficulty of choosing the unit, and of standardising it when chosen, is much greater than might at first thought be supposed, and can be fully appreciated only by one who has followed historically, step by step, the development of scientific theory and practice. What holds in this regard of physics holds also of psychology. Moreover, the psychologist is at a peculiar disadvantage, in that there is no natural unit of mental measurement. The human body affords the natural units of linear measurement: foot, pace, cubit, span. The height of the sun in the heavens, the alternation of day and night, the changes of the moon,—these are all natural units of time measurement. Units of weight are furnished by convenient natural objects (grain, stone) or by the average carrying power of man or animal (pack, load). There are no such obvious points of reference in psychology. Once more: physics is able to relate and combine its units, to reduce one to another, to express one in terms of another; so that the formula for mechanical work, *e.g.,* has the form ML^2/T^2, where M represents mass measured in gr., L length of path measured in cm., and T time measured in sec. This sort of interrelation is forbidden by the very nature of mental processes, every group of which is qualitatively dissimilar to every other group. Hence there can be no single unit of mental measurement, no generalisation of the units employed in special investigations.

Here, then, are difficulties in plenty. And there can be no question but that these, the intrinsic difficulties, are largely responsible for the tardy advent of measurement in psychology, and for the doubts and controversies and confusions that have arisen since the methods of psychological measurement were formulated. The formulation itself dates only from 1860, when Gustav Theodor Fechner (1801–1887), gathering together scattered observations from physics and astronomy and biology, summing up elaborate investigations of his own, putting his physical, mathematical and psychological knowledge at the service of mental measurement, published his *Elemente der Psychophysik.* Fechner is the founder, we might almost say the creator, of quantitative psychology, and the modern student who will understand the principles and methods of mental measurement must still go to school with Fechner.

.   .   .   .   .

There is one mistake, so natural that we might almost call it inevitable, which has sorely delayed the advance of quantitative psychology. It is a mistake with regard to the object of measurement, the mental magnitude. We have seen that every measurement requires three given terms; so that the physical quantity or magnitude is not, so to say, a single term, but rather a distance between terms, a section of some stimulus scale. We are apt to say, carelessly, that we have measured 'the highest point' of Mt. Vesuvius, when we have in reality measured, in terms of our arbitrary unit, the distance between its lowest and highest points. It is not the point that is the magnitude, but the distance between points. So with sensations: we are apt to think of a brightness or a tone of given intensity as a sensation magnitude, as itself measurable. Now the stimulus is measurable: we can measure, in terms of some unit, the amplitude of vibration of the ether or air waves: we have our three terms to measure with. But the sensation, the brightness or the tone, is just a single point upon the sense scale,—no more measurable, of itself, than is 'the highest point' of Mt. Vesuvius. The only thing that we can measure is the distance between two sensations or sense points, and to do this we must have our unit step or unit distance.

Let us take some instances. Suppose that two rooms of equal dimensions are illuminated by two ground glass globes, the one containing five and the other two incandescent lights of the same candle-power. We can say, by eye, that the illumination of the first room is greater than that of the second. How much greater, we cannot possibly say. Even if the globes are removed, so that we can count the lights, we cannot say. The stimuli stand to one another in the ratio 5:2. But the corresponding sensations are simply different as more and less, the one a 'more bright' and the other a 'less bright.' The brightness of the lighter room does not contain within it so and so many of the brightnesses of the darker room. Each brightness is one and indivisible. What we have given is rather this: that on the scale of brightness intensities, which extends from the just noticeable shimmer of light to the most dazzling brilliance that the eye can bear, the illumination of the one room lies higher, that of the other room lower. There is a certain distance between them. If we can establish a sense unit for this distance, we shall be able to say that the greater brightness is, in sensation, so and so many times removed from the lesser brightness; just precisely as the top of the mountain, in terms of the 1 ft. unit, is 4200 times removed from the bottom. Neither of the two brightness sensations is itself a magnitude. The magnitude is the distance which separates them on the intensive brightness scale.

Again: we can say by ear that the roar of a cannon is louder, very much louder, than the crack of a pistol. But the cannon roar, as heard, is not a multiple of the pistol crack, does not contain so and so many pistol cracks within it. What we have given is that, on the scale of noise intensities ranging from the least audible stir to the loudest possible crash, the cannon roar lies very high, the pistol crack a good deal lower. Neither noise, in itself, is a magnitude; both alike are points, positions, upon an intensive scale. The magnitude is the distance between them. With a sense unit of noise distance established, we can measure this given distance, as before.

. . . . .

3. *An analogy.*—The passage . . . from Fechner reads in full as follows: "Our measure of sensation amounts to this: that we divide every sensation into equal parts, *i.e.,* into the equal increments out of which it is built up from the zero point of its existence, and that we regard the number of these equal parts as determined by the number of the corresponding variable increments of stimulus which are able to arouse the equal increments of sensation, just as if the increments of stimulus were the inches upon a yard-stick." Notice that Fechner speaks of the *variable* increments of stimulus which arouse the *equal* increments of sensation. This means, in our own terminology, that equal sense-distances do not correspond always to equal stimulus magnitudes: to obtain equal sense-distances, under different conditions, we must vary the magnitude of the corresponding stimuli. The fact is important: it is also obvious. Go into a small darkened room, and light a candle. There is an immense difference in the illumination of the room. The physical magnitude, the photometric value of the candle, corresponds to a very wide distance upon the scale of subjective brightness intensities. Light a second candle. There is a difference in the illumination, and a marked difference; but it is nothing like so marked as the first difference. The same physical magnitude, then, corresponds now to a lesser sense-distance. Light a third candle, and a fourth, and a fifth. A point will soon come at which the introduction of another candle makes hardly any appreciable difference in the illumination. The same physical magnitude now corresponds to a minimal sense-distance.

Facts of this sort recur in all departments of sense, and it is part of the business of quantitative psychology to take account of them, and to sum them up in a numerical formula. What precisely the programme of quantitative psychology is in this field, what the facts are with which it has to deal, and how these facts are to be grouped under laws, we shall best understand by help of an analogy from physics. The analogy was first suggested by J. R. L. Delbœuf (1831–1896), late professor in the University of Liège,—a psychologist of great originality, to whom we owe the conception of mental measurement set forth in § 2 . . .

If a magnetic needle be suspended at the centre of a circular coil of wire, and an electric current be sent through the wire, the needle is deflected from its position of rest. Suppose that we are seeking to discover the law of this deflection, to find a general expression for the movement of the needle under the influence of currents of different strength. We send a current of so and so many amperes through the coil, and measure the angle of deflection upon a circular scale; then we send through a current of so and so many more amperes, and measure again; and so on. We find that the needle moves farther and farther, as the current is made stronger and stronger. But we find also that the angle of deflection is not simply proportional to the strength of current. If we increase the strength of current by equal amounts, the deflection of the needle becomes progressively smaller and smaller. And however strong we make our current, the needle will never make an excursion of 90°. The mathematical expression of the relation is very simple. If a is the number of amperes in the current, k a constant, and θ the angle of deflection, then $a = k \tan \theta$.

Let us now, instead of taking determinate strengths of current, change the strength of the current continuously, and let us watch the behaviour of the needle. We find the same law in operation; but it is crossed by a second law. If the needle is hanging steady, whether at the zero point of its scale or at any other point at which it is held by the current in the coil, and we increase the current very slowly, we get at first no movement at all. During this period, while the needle remains stationary, our law of correlation is, apparently, not fulfilled; and the greater the increase of the current before movement sets in, the greater, of course, is the apparent deviation from the law. Presently, however, when the current has been increased by a certain amount, the needle goes with a little jump to the position which the law of correlation requires. And as we continue slowly to increase the strength of the current, the phenomenon is repeated, until the limit of the needle's excursion is reached. The law of correlation is not really in abeyance; it is crossed or masked by another law.

We have, then, two things before us. On the one hand, the needle is a magnetic needle, and the amount of its deflection is a continuous function of the current in the coil. On the other hand, the needle does not move without friction; so that we obtain, under the conditions of our second experiment, not a continuous movement of the needle, but a discrete movement, a series of jerks. It is one and the same needle that moves, and one and the same movement that it makes; but the single needle is at once mechanical and magnetic, and the single movement gives evidence of the operation of two distinct laws.

III THE RETINAL IMAGE AND THE ORIENTATION OF PERCEIVED OBJECTS

The text begins with the view of the Greek atomists that perception occurs because all objects give off delicate images of themselves that impinge upon the sense organs and, being conducted to the brain, there represent the objects. Epicurus is quoted as the example. This is the ancient theory of perception which, embedded in the unconscious habits of thought of philosophers and psychologists, has continued to affect thinking on down even to the present day. From Epicurus the exposition passes to Kepler (1604), who concluded that the crystalline body in the eye is a lens that projects an inverted image upon the retina at the back of the eye. This conclusion of Kepler's raised for others the question why we see right side up when the retinal image is upside down, a problem that reveals the whole difficulty of the old Epicurean theory that the mind perceives directly what the nerves bring to it. Molyneux (1692) offered a correct solution to this problem, as did also Johannes Müller much later (1826), but Müller's argument shows how firmly entrenched the ancient theory still was. The problem may be said to have been solved, even if not wholly banished, by Stratton's classical experiment (1897), which showed how the wearing of reversing lenses which right the image on the retina leads to experience's becoming normal as soon as proper new associations have been formed between the visual perception of objects and the behavioral response to them.

22 EPICURUS (341–270 B.C.) ON PERCEPTION OF OBJECTS AS MEDIATED BY THE IMAGES THAT EMANATE FROM THE OBJECTS, ca. 300 B.C.

Epicurus, Letter to Herodotus, from the original text of Diogenes Laertius. Translated by Cyril Bailey in W. J. Oates, ed., *The Stoic and Epicurean Philosophers: Complete Extant Writings of Epicurus, Epictetus, Lucretius, and Marcus Aurelius* (New York: Random House, 1940), pp. 5–7.

The Greek atomist philosophers, Empedocles (ca. 490–ca. 435 B.C.), Democritus (ca. 460–ca. 370 B.C.), and Epicurus, held that the perception of objects occurs because all objects constantly give off representative images of themselves (*simulacra, eidola*), which impinge upon the organs of sense and are conducted to the brain to make perception of the objects possible. This is the view of perception that Johannes Müller was trying to scotch as late as the nineteenth century.

There are images like in shape to the solid bodies, far surpassing perceptible things in their subtlety of texture. For it is not impossible that such emanations should be formed in that which surrounds the objects, nor that there should be opportunities for the formation of such hollow and thin frames, nor that there should be effluences which preserve the respective position and order which they had before in the solid bodies: these images we call idols.

Next, nothing among perceptible things contradicts the belief that the images have unsurpassable fineness of texture. And for this reason they have also unsurpassable speed of motion, since the movement of all their atoms is uniform, and besides nothing or very few things hinder their emission by collisions, whereas a body composed of many or infinite atoms is at once hindered by collisions. Besides this, nothing contradicts the belief that the creation of the idols takes place as quick as thought. For the flow of atoms from the surface of bodies is continuous, yet it cannot be detected by any lessening in the size of the object because of the constant filling up of what is lost. The flow of images preserves for a long time the position and order of the atoms in the solid body, though it is occasionally confused. Moreover, compound idols are quickly formed in the air around, because it is not necessary for their substance to be filled in deep inside: and besides there are certain other methods in which existences of this sort are produced. For not one of these beliefs is contradicted by our sensations, if one looks to see in what way sensation will bring us the clear visions from external objects, and in what way again the corresponding sequences of qualities and movements.

Now we must suppose too that it is when something enters us from external objects that we not only see but think of their shapes. For external objects could not make on us an impression of the nature of their own colour and shape by means of the air which lies between us and them, nor again by means of the rays or effluences of any sort which pass from us to them—nearly so well as if models, similar in colour and shape, leave the objects and enter according to their respective size either into our sight or into our mind; moving along swiftly, and so by this means reproducing the image of a single continuous thing and preserving the corresponding sequence of qualities and movements from the original object as the result of their uniform contact with us, kept up by the vibration of the atoms deep in the interior of the concrete body.

And every image which we obtain by an act of apprehension on the part of the mind or of the sense-organs, whether of shape or of properties, this image is the shape or the properties of the concrete object, and is produced by the constant repetition of the image or the impression it has left. Now falsehood and error always lie in the addition of opinion with regard to what is waiting to be confirmed or not contradicted, and then is not confirmed or is contradicted. For the similarity between the things which exist, which we call real and the images received as a likeness of things and produced either in sleep or through some other acts of apprehension on the part of the mind or the other instruments of judgement, could never be, unless there were some effluences of this nature actually brought into contact with our senses. And error would not exist unless another kind of movement too were produced inside ourselves, closely linked to the apprehension of images, but differing from it; and it is owing to this, supposing it is not confirmed, or is contradicted, that falsehood arises; but if it is confirmed or not contradicted, it is true. Therefore we must do our best to keep this doctrine in mind, in order that on the one hand the standards of judgement dependent on the clear visions may not be undermined, and on the other error may not be as firmly established as truth and so throw all into confusion.

Moreover, hearing, too, results when a current is carried off from the object speaking or sounding or making a noise, or causing in any other way a sensation of hearing. Now this current is split up into particles, each like the whole, which at the same time preserve a correspondence of qualities with one another and a unity of character which stretches right back to the object which emitted the sound: this unity it is which in most cases produces comprehension in the recipient, or, if not, merely makes manifest the presence of the external object. For without the transference from the object of some correspondence of qualities, comprehension of this nature could not result. We must not then suppose that the actual air is moulded into shape by the voice which is emitted or by other similar sounds—for it will be very far from being so acted upon by it—but that the blow which takes place inside us, when we emit our voice, causes at once a squeezing out of certain particles, which produce a stream of breath, of such a character as to afford us the sensation of hearing.

Furthermore, we must suppose that smell too, just like hearing, could never bring about any sensation, unless there were certain particles carried off from the object of suitable size to stir this sense-organ, some of them in a manner disorderly and alien to it, others in a regular manner and akin in nature.

23   JOHANNES KEPLER (1571–1630) ON THE CRYSTALLINE HUMOR AS A LENS AND THE INVERSION OF THE RETINAL IMAGE, 1604

Johannes Kepler, *Ad Vitellionem paralipomena, quibus astronomiae pars optica traditur* (Frankfurt, 1604), chap. 5. Translated by Alistair C. Crombie in I. B. Cohen and René Taton, eds., *Mélanges Alexandre Koyré: L'Aventure de la science* (Paris: Hermann, 1964).

In Kepler's day one of the most important texts about vision was still the *Optics* by Erasmus Vitellio, a Polish physicist and philosopher, who was a contemporary of Roger Bacon and who wrote his book about 1270. More than three hundred years later Kepler undertook significantly to extend Vitellio's account in the volume cited here as "Things Omitted by Vitellio," a book that is one of Kepler's two important works on optics. The other treatise was his *Dioptrice* of 1611.

By 1604 the philosophers and the instrument makers had long known about lenses, and a Dutch manufacturer of spectacles was about to put together the first telescope in 1608. The camera obscura had been known for at least six hundred years. It is a dark chamber with a pinhole opening. Light from a brightly illuminated scene passing in a converging-diverging pencil through the hole forms an inverted image of the scene on the back of the camera. Later a lens was used. Kepler saw the similarity of the eye to the camera obscura. The crystalline body within the eyeball had been supposed by Kepler's distinguished predecessors in the field of optics, Galen (ca. A.D. 150) and Alhazen (ca. A.D. 1000), to be the sensitive organ of vision. It was sometimes called a humor. Kepler recognized it as a lens and argued, from the analogy to the camera obscura, that a tiny replica of the outer scene would be projected by it on the retina at the back of the eye, forming there an inverted image.

As we have just seen, the prevailing theory of perception derived from the Greeks (Democritus, Empedocles, Epicurus), was that perceived objects give off faint copies of themselves (*eidola, simulacra*) that the nerves conduct to the brain, where the soul apprehends them. If this is true and the retinal image is upside down, how then does it come about that we see objects right side up? Many later philosophers were to be troubled by this problem and to undertake an explanation of why we nevertheless do see right side up, but Kepler was not disturbed about the matter. He took pains to show that the picture on the retina corresponds exactly, point for point, with the scene it depicts, and that seemed to be enough for him. He noted that light itself can not enter the optic nerve and that the inverted image on the retina must be the object immediately perceived at the brain, although he also suggested, as had others before him, that binocular vision is not double because the retinal impressions are carried along the optic nerves and united at the optic chiasma.

2. How Vision Takes Place

I say that vision occurs when the image (*idolum*) of the whole hemisphere of the world which is in front of the eye, and a little more, is formed on the reddish white concave surface of the retina (*retina*). I leave it to natural philosophers to discuss the way in which this image or picture (*pictura*) is put together by the spiritual principles of vision residing in the retina and in the nerves, and whether it is made to appear before the soul or tribunal of the faculty of vision by a spirit within the cerebral cavities, or the faculty of vision, like a magistrate sent by the soul, goes out from the council chamber of the brain to meet this image in the optic nerves and retina, as it were descending to a lower court.

For the equipment of opticians does not take them beyond this opaque surface which first presents itself in the eye. I do not think that we should listen to Vitellio . . . who thinks that these images of light (*idola lucis*) go out further through the nerve, until they meet at the junction half-way along each optic nerve [optic chiasma], and then separate again one going to each cerebral cavity. For by the laws of optics . . . what can be said about this hidden motion, which, since it takes place through opaque and hence dark parts and is brought about by spirits which differ in every respect from the humours of the eye and other transparent things immediately puts itself outside the field of optical laws? So, whereas Vitellio argues thus . . . the images (*species*) must be united, therefore . . . refraction must take place at the back of the vitreous humour, and . . . the spirits must be pellucid, I reverse this argument: spirits are not an optical body, and their thin hollow nerve is not optically in a direct line. Even if it were, it would immediately become bent by the movements of the eye and the opaque parts of the nerve would become opposed to the light entering the tiny opening or door of the passage. Hence light neither passes through the posterior surface of the vitreous humour nor is refracted there, but falls upon it. And indeed how could images (*species*) entering perpendicularly be refracted? It is strange that this did not occur to Vitellio . . . Hence . . . he was put into by no means minor difficulties over this union of images (*species*) at the junction of the nerves. Because if this union in the mid-path of the nerves is to be asserted, it must be done in terms of natural philosophy. For it is undeniably certain that no optical image could penetrate to this point. It seems then clear that, if any nerve went to its seat in the brain freely in a straight line, with two eyes we would think we saw two things instead of one. Either this junction takes place so that, when one eye is closed, this hidden seat in the brain should not cease from its function of judging. Or perhaps the actual doubling of the seats is not only on account of the eyes, but is for the purpose of correct judgment of distances, as with the pair of eyes. Therefore, in order that visible things may be judged correctly and a distinction made between what is seen with one and with two eyes, this junction of the passages must take place. Here this one optical conclusion . . . can be stated: the spirits are affected by the qualities of colour and light, and this affection (*passio*) is, so to speak, a colouring and a lighting. For in vision images (*species*) of strong colours remain behind after looking, and these are united with colours printed by a fresh look, and a mixture of both colours is made. This image (*species*) existing separately from the presentation of the thing seen is not present in the humours or coats of the eye . . . hence vision takes place in the spirits and through the impression (*impressio*) of these images (*species*) on the spirit. But really this 'impression' does not belong to optics

but to natural philosophy and the study of the wonderful. But this by
the way. I will return to the explanation of how vision takes place.

Thus vision is brought about by pictures of the thing seen being formed
on the white concave surface of the retina. That which is to the right
outside is depicted on the left on the retina, that to the left on the right,
that above below, and that below above. Green is depicted green, and in
general things are depicted by whatever colour they have. So, if it were
possible for this picture on the retina to persist if taken out into the light
by removing the anterior parts of the eye which form it, and if it were
possible to find someone with sufficiently sharp sight, he would recognize
the exact shape of the hemisphere compressed into the confined space of
the retina. For a proportion is kept, so that if straight lines are drawn
from separate points on the thing seen to some determined point within
the eye, the separate parts are depicted in the eye at almost the same
angle as that at which these lines meet. Thus, not neglecting the smallest
points, the greater the acuity of a given person, the finer will be the
picture formed in his eye.

So that I may go on to treat this process of depiction and prepare for
a demonstration of it, I say that this picture consists of as many cones
[of light] of equal size as there are points in the thing seen, in pairs
always with the same base, namely the width of the lens (*crystallinus*) or
part of it. Thus while one cone of each pair has its vertex at the point
seen and its base on the lens (nothing is altered by refraction through
the cornea), the other has the same base on the lens as the first one and
the vertex at a point in the picture depicted on the retina; this cone [of
light] undergoes refraction on passing out of the lens . . . All the outer
cones meet in the pupil, so that they intersect in that space, and right
becomes left.

.   .   .   .   .

Nature has found an admirable way of preventing the disturbance of
the proportions of the visible hemisphere which would occur if points
outside in the air, mutually opposite to those on the retina on a line
through the centre of the eye, were deflected from an opposite position
through the threefold refraction of the rays, at the cornea and at the
[two] surfaces of the lens, and passed down at an angle into the depths
of the eye and so were focused on to a portion of the retina smaller than
a hemisphere. For Nature has placed the centre of the retina, not at the
junction of the axes of the cones [of light] penetrating the vitreous humour
but a long way inside, and she has advanced the edge of the retina at
the sides, so that the longer cones, which are more widely separated,
intercept perpendicularly placed and therefore narrow sections of the

retina, while the shorter ones, which are less widely separated at the sides of the retina, mark off an acute angle broad sections of the retina obliquely set towards them. Thus the rays coming from opposite points, though after refraction no longer opposite, nevertheless fall on corresponding opposite points of the retina . . . and so there is compensation. And so if, finally, straight lines are drawn from points on the visible hemisphere through the centre of the eye and the vitreous humour, these lines will imprint points forming a picture of the radiating points on the retina opposite . . .

The direct cone [of light] comes to a point at the center in the middle of the retina, the oblique ones do so to the sides. The direct cone falls upright on the retina; the lateral ones fall obliquely because, as already described, the centre of the retina is below the intersection of the axes of the cones in the vitreous humour. Finally, the sensory power or spirit diffused through the nerve is more concentrated and stronger where the retina meets direct cones, because of its source and where it has to go: from that point it is diffused over the sphere of the retina, gets further from the source, and hence becomes weaker. But as in a funnel and in a fishing net with a sack . . . the sides all send the liquid or the fishes into the canal or sack; so the sides of the retina do not usurp for themselves its sensory capacity, but whatever they can they bring into the perfection of direct vision. Thus, when we see a thing perfectly, we see it within the whole surrounding area of the visible hemisphere. For this reason oblique vision satisfies the soul least and only invites the turning of the eyes in that direction so that they may see directly . . .

The colour of the retina is neither dark nor black, in case it should tint the colours of things, nor dazzling white, in case too much brightness should pour into the vitreous humour and the things appearing white and bright above this should be made to seem too coloured . . .

The shape of the retina is larger than a hemisphere. It must in the first place be a hemisphere, thus proportionate to the picture received of things . . . Beyond that, the border extends as far as the ciliary processes so that when the eye-ball is filled with the vitreous humour the retina is kept stretched with the collar narrower than the belly. It cannot be tied on because of the softness and subtlety of the visual spirit, for which indeed there are canals through the nerve, contrary to the nature of the rest of the nerves, so that the substance of the nerve does not impede it. Now if the retina did not occupy more than a hemisphere, it could easily become wrinkled and slip back on to the junction of the nerve. In any case the gap between the hemisphere and the ciliary processes had to be filled, so otherwise this would have had to be done with the choroid or with the vitreous humour. How much better that it should be done with the retina! This extends the visual function into the border. Although none of the cones [of light] formed by the lens reach this border, nevertheless slits appear formed by the ciliary processes, so that some light can enter from the sides through the ciliary processes and be received by this copious border of the retina. For a line drawn from the extreme edge of the cornea through the adjacent edge of the pupil almost falls on to the junction of the lens with the ciliary processes; one drawn through the opposite edge of the pupil almost touches the origin of the ciliary processes from the choroid. By this machinery Nature has brought it about that we see more than a hemisphere with the eyes fixed, or at any rate as much as is admitted through the corner of the eye, with a minimum of movement. Indeed with a little wider vision you would be able to see your ears, especially if they were long with the eyes on the same side. I have often been surprised at seeing the sun and my shadow both appearing as if in front, instead of being opposite each other. It seems to be Nature's precaution to protect the eyes that, when they are not actually turned away, things approaching come immediately into view wherever the eyes are looking. And this preserves the whole living thing, for certainly it takes care to preserve itself and this adaptation helps it to look after its body.

·   ·   ·   ·   ·

3. Demonstration of the Conclusions Stated Concerning How Vision Takes Place Through the Lens

Everything said so far about the lens can be observed in everyday experiments with crystal pestles (*pilum crystallinum*) or glass urinary flasks filled with clear water. For if one stands at the glazed window of a room with a globe of this kind of crystal or water, and arranges a sheet of white paper behind the globe at a distance equal to half the diameter of the globe, the glazed window with the fluted wooden or leaden divisions between the lights will be very clearly depicted on the paper, but inverted. The same effect can be obtained with other things, if the place is darkened a little. Thus, using a globe of water set up in a chamber opposite a small window . . . everything that can reach the globe through the width of the small window or opening will be depicted very clearly and delightfully on the paper opposite. Since the picture is clear at this one distance (namely with the paper a semi-diameter of the globe away from it), it will become indistinct at positions in front of or behind this one. But the direct opposite happens using the eyes. For, if the eye is placed at a distance of a semi-diameter of the globe behind the glass, where before the picture of things represented through the glass had been clearest, it is now most indistinct. For the glass appears either all white, or all red, or all dark, etc. If the eye is moved nearer the globe, it sees things opposite erect and magnified, where they were indistinct on the paper, but if it is moved away from the globe to a distance greater than a semi-diameter of the globe, it sees them as distinct images, inverted, reduced in size, and lying on the surface of the globe itself . . .

DEFINITION

Whereas up to now the Image has been an entity of reason (ens rationale), *the shapes of things really present on the paper, or on any other screen, will now be called pictures.*

·   ·   ·   ·   ·

PROPOSITION XXIII

When a screen with a small window is placed in front of the globe within the limit of the sections of the parallels, and the window is smaller than the globe, a picture of the visible hemisphere is projected on to the paper, formed by most of the rays brought together behind the globe at the limit of the last intersection of the rays from a luminous point. The picture is inverted, but purest and most distinct in the middle. So great is the uncertainty in this matter and indeed such its novelty that unless we take the greatest care, it may easily become confused. Indeed I was held up myself for a long time, until I convinced myself that all the different effects had the same explanation.

·   ·   ·   ·   ·

COROLLARY

Thus is seen the design of Nature concerning the posterior surface of the lens in the eye. She wants all the rays entering the pupil from a visible thing to come together at one point on the retina, both so that each point of the picture will be so much the clearer, and so that the other points of the picture will not be accidentally confused with other, unfocused or focused rays.

It is also seen that the expansion of the pupil has no other purpose than that which I said above, nor does it confuse the picture but only makes it clearer.

24 WILLIAM MOLYNEUX (1656–1698) ON THE INVERTED RETINAL IMAGE, 1692

William Molyneux, *Dioptrica Nova: A Treatise of Dioptrics* (London, 1692), pp. 104–106.

It was not long after Kepler had posed the problem of the inverted retinal image that both Christoph Scheiner (1625) and Descartes (1637) proved Kepler right, at least as far as ocular optics goes, by fitting the fresh eye of a bull into a hole in a window's closed shutter, scraping off the opaque dark coat from the back of the eyeball, and seeing then on the back of the eye the tiny inverted image of the scene outside the window. Yet why does one then see right side up? The correct answer was given by Molyneux, a friend of John Locke's: "The visive faculty takes no notice of its own parts but uses them as an instrument only." The top of the world finds itself represented by the bottom of the retina. With the right answer clearly given in 1692, it is interesting to note that the question was often to be argued again.

'Tis therefore contrived by the *Most Wise and Omnipotent Framer of the Eye,* That it should have a Power of adapting it self in some Measure to *Nigh* and *Distant* Objects. For they require different Conformations of the Eye; Because the Rays proceeding from the Luminous Points of *Nigh* Objects do more Diverge, than those from more Remote Objects.

But whether this variety of Conformation consist in the Crystallines approaching nigher to, or removing farther from the *Retina;* Or in the Crystallines assuming a different Convexity, sometimes greater, sometimes less, according as is requisite, I leave to the scrutiny of others, and particularly of the curious Anatomist. This only I can say, that either of these Methods will serve to explain the various Phænomena of the Eye; And I am apt to believe, that both these may attend each other, *viz.* a Less Convex Crystalline requires an Elongation of the Eye, and a more Convex Crystalline requires a shortning thereof; As a more Flat Convex Object glass or of a Larger Sphere requires a Longer Tube, and one more Protuberant, bulging or of a smaller Sphere requires a shorter Tube.

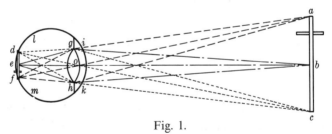

Fig. 1.

(3) [In] the formention'd Scheme [Fig. 1] we perceive[d], the Rays from each Point of the Object are all confused together on the Pupil in *gh,* so that the Eye is placed in the Point of the Greatest Confusion: But by means of the Humors and Coats thereof each Cone of Rays is separated, and brought by it self to determine in its proper Point on the *Retina,* there Painting distinctly the Vivid Representation of the Object; Which Representation is there perceived by the *sensitive Soul* (whatever it be) the manner of whose Actions and Passions, He only knows who Created

and Preserves it, *Whose Ways are Past finding out, and by us unsearchable.* But of this Moral truth we may be assured, *That He that made the Eye shall see.*

(4) We are likewise to observe, that the Representation of the Object *abc* on the Fund of the Eye *fed* is *Inverted.* For so likewise it is on the Paper in a dark Room; there being no other way for the Radious Cones to enter the Eye or the dark Chamber, but by their Axes *ao, bo, co,* crossing in the Pole *o* of the Crystalline or Glass. And here it may be enquired; How then comes it to pass that the Eye sees the Object *Erect?* But this Quæry seems to encroach too nigh the enquiry into the manner of the Visive Faculties *Perception;* For 'tis not properly the Eye that *sees,* it is only the Organ or Instrument, 'tis the *Soul* that *sees* by means of the Eye. To enquire then, how it comes to pass, that the Soul perceives the Object *Erect* by means of an *Inverted* Image, is to enquire into the Souls Faculties; which is not the proper subject of this Discourse. But yet that in this Matter we may offer at somthing, I say, *Erect* and *Inverted* are only Terms of *Relation* to *Up* and *Down,* or *Farther from* and *Nigher to* the Centre of the Earth, in parts of the same thing: And that that is an *Erect* Object, makes an *Inverted* Image in the Eye, and an *Inverted* Object makes an *Erect* Image; That is, that part of the Object which is *farthest from* the Centre of the Earth is Painted on a Part of the Eye *Nigher* the Centre of the Earth, than the other parts of the Image. But the Eye or Visive Faculty takes no Notice of the Internal Posture of its own Parts, but uses them as an Instrument only, contrived by Nature for the Exercise of such a Faculty.

But to come yet a little nigher this difficulty; This enquiry results briefly to no more than this, How comes it to pass, that the Eye receiving the Representation of a Part of an Object on that part of its Fund which is *Lowermost* or nighest the Centre of the Earth, perceives that part of the Object as *Uppermost* or farthest from the Centre of the Earth? And in answer to this, let us imagine, that the Eye in the Point *f* receives an Impulse or Stroke by the Protrusion forwards of the Luminous Axis *aof,* from the Point of the Object *a;* Must not the Visive Faculty be necessarily directed hereby to consider this stroke, as coming from the Top *a,* rather than from the bottom *c,* and consequently should be directed to conclude *f* the Representation of the Top?

Hereof we may be satisfy'd by supposing a Man standing on his Head: For here, tho the Upper Parts of Objects are painted on the Upper Parts of the Eye, yet the Objects are judged to be *Erect.* And from this Posture of a Man, the Reason appears, why we have used the Words *Farthest from,* and *Nighest to the Centre of the Earth,* rather than *Upper* and *Lower.* For in this Posture, because the *Upper* Parts of the

Object are painted on that part of the Eye nighest the Earth, (though really the upper Part of the Eye) they are judged to be farthest removed from the Earth.

What is said of *Erect* and *Reverse* may be understood of *Sinister* and *Dexter:* But of these Physical Conjectures enough.

(5) The Image of an *Erect* Object being Represented on the Fund of the Eye *Inverted,* and yet the sensitive Faculty judging the Object *Erect;* it follows that when the Image of an *Erect* Object is Painted on the Fund of the Eye *Erect,* the sense Judges that Object to be *Inverted.* This is a necessary Conclusion.

25 JOHANNES MÜLLER (1801–1858) ON SUBJECTIVE VISUAL SIZE AND POSITION IN RELATION TO THE RETINAL IMAGE, 1826

Johannes Müller, *Zur vergleichenden Physiologie des Gesichtssinnes* (Leipzig, 1826), pp. 55–66. Translated for this book by Don Cantor.

In order to see how the problem that Kepler raised was getting along 134 years later, we skip to Johannes Müller's little book of 1826 on vision, written when he was only twenty-four years old (perhaps that is why he was so impatient with the stupidities of his elders), the book which really initiates his theory of specific nerve energies (see No. 9 above). By this time the problem of why we see right side up when the retinal image is upside down has been supplemented by the comparable problem of how we can see large objects in so small an eye. It seems preposterous that his problem should have needed Müller's detailed serious consideration in 1826, but the ancient theory of perception had gotten itself so deeply rooted in common sense that it bothered philosophers to consider how the *eidolon* of a large object could properly represent the object's size and yet get into the eye to be perceived by the mind. Müller's doctrine of specific nerve energies asserts, however, that all the mind can ever perceive directly is the states of the nerves as they come to the brain. Thus he comments here that, when sizes are being measured, "the measure of all measures" is physically the size of the eye itself, which is for the moment the sum of the sizes of all objects in the visual field. Such a view makes sense, for it shows that apparent size is relative. On the inversion of the retinal image, Müller appealed to easily demonstrated fact: with the finger press the eyeball from below, thus stimulating the bottoms of both the eyeball and the retina. Tactually the finger is perceived as being below, as is proper, but visually the bright spot that its pressure on the retina creates is seen as coming from above, for the bottom of the retina represents the top of the visually perceived field. Thus two adjacent stimulations produce two localizations remote from each other.

CONCERNING THE ABSOLUTE PHYSIOLOGICAL SIZE OF THE EYE AND ITS
OBJECTS IN RELATION TO THE TRUE AND APPARENT SIZE OF OBJECTS IN
THE PHYSICISTS' SENSE

As soon as one learns through experiment to represent objectively the images at the back of the eye, he is certain to find an apparent confirmation of the optical theory of vision. Nevertheless those who are still impartial can not deny this decisive contradiction: how is it that the images communicated by refractive media to so small an area as the back of the eye lead us, in perceiving objects, to see them not in the size of those projected images but in their so-called natural size? Such a reproach is the voice of men who are impartial and do not deny that their explanation is imperfect, yet who are, through the training of this sense, misled regarding its basic natural conditions, for, no matter how well and clearly the small images may be projected in the leucaethiopic [albino Negro] eye, no one will believe that it is the images that are perceived. If the retina of the eye always perceives only itself, why does it perceive itself in such a size that the image of the entire body—of which the smallest part is, after all, the eye—is but a part of the visual field? This question, it seems, physiology has had to exclude in favor of the optical analysis of vision, docilely allowing the opticians and mathematicians to urge upon physiology not a theory of vision, but rather a discussion of its external circumstances . . . The doctors [for instance] have quietly put up with a statement like the following: Once the small image has been projected through refractive media onto the retina, the retina senses not so much the contact between itself and light in this image as the direction of the light waves to the natural sizes of the objects. Such mystifications of the physiologists as well as the opticians (in which the former are the more often unknowingly betrayed) are frequent enough in the history of optics. In fact, we have recently read in a French periodical a so-called theory of vision in which the contradiction of inverted vision was explained as follows: it is not the contact of light itself that is perceived, but the lines of direction of the light waves which cross one another on their way to the objects [in the visual field]. And thus was inverted seeing taken care of! . . . Let us pass over other irritating examples and occupy ourselves first with a serious solution of the problem of inverted vision, for without this solution all other theories of vision are . . . unsatisfactory and undesirable. The solution to this problem, moreover, will lead us into the inner workings of the physiology of vision.

It is known that the apparent size of things in relation to one another is determined by their distance away; and indeed apparent sizes are related to one another as the tangents of their visual angles. With various distances of the objects from the refractive media, the focal distance of the picture behind the media must change. For the eye alone, this change in the focal distance of the picture from the lens never is greater than one [Paris] line and can be ignored in respect of the law of visual angles.

·   ·   ·   ·   ·

The measure of all measures [and] of all apparent sizes of things is then the invariable true size of the eye and its retina when it directly perceives itself. The apparent sizes of objects appear on the true subjective size of the retina, and the sum of the apparent sizes of all objects present in one and the same visual field remains constant through all changes in objects in every visual image; it is identical with the true size of the eye, the retina itself. The retina, which appears subjectively, and the sum of the images presented simultaneously are one and the same; they *are* the subjective visual field. The image of our corporeal selves constitutes but a part of this subjective true size of the retina or of the visual field, for our own body is then reduced to its apparent size, as something invariable, which we recognize as ourselves bound up with the changing apparent sizes of other objects. The apparent size of our body is, then, much smaller than the true size of the retina, which is identical with the visual field. And the apparent size of the objective eye, as a small part of the apparent size of the objective human body, constitutes but a small part of the true subjective size of the eye. Therefore the poet [Schiller; Arnold-Foster trans.] may well say, in that fine riddle about the eye:

'Tis compassed in the smallest space,
Its framework is the narrowest bound;
Yet all dimensions leave their trace,
And through it everything is found.

·   ·   ·   ·   ·

CONCERNING THE APPARENT POSITION OF SEEN OBJECTS

Men have never lacked ingenuity in eradicating somehow the necessity of [considering] inverted vision . . . To say that the direction of the light waves is perceived is but to explain vision by vision. It is, however, once and for all necessary [to take account of] inverted vision. Moreover, this is not a problem that calls for a physiological solution, for no difficulty at all arises in this matter, no more than [what comes about] through the apparent false relations communicated by the sensory recognition of [external] objects. Furthermore, inverted seeing is not incongruous with perception; rather, the two are perfectly congruent. Because we as perceivers appear in apparent size upon the true size of our eye, the perceiving and touching portions of our bodies will, like that which is felt and touched, be seen inverted. Anyone who doubts [the phenomenon of] inverted vision has never pressed his eye to experience the subjective light phenomena, phenomena which are, here as everywhere else in the physiology of the senses, the single, comprehensive key to physiological truth. If it is true that we see things in an inverted position because of the refraction of external objects, then we should be able to see them in their real, natural position could they be perceived by direct contact, without the mediation of refraction in transparent media. This opportunity is offered by the subjective experiments. The eye sees a finger brought close to it from side *b,* but sees it on the opposite side *a;* yet, if the finger presses the eye on side *b,* it appears in the shining image produced by the pressure on that side *b,* and is not reversed [right for left]. A finger pressing the under [portion of the eye] will be seen above; but the image produced by pressure is underneath. Here is a contradiction of the [principle of] objective and subjective visual phenomena, which is not a deception but one of the most fruitful and beautiful truths about vision. If, then, inverted vision is consistent with perception, how can one expect that persons who, having been totally blind since youth, are given their sight by an operation, will be conscious of the inversion directly after the operation? There is no accredited, exactly reported case in which this [inversion] was clear to the healed person.

26   GEORGE MALCOLM STRATTON (1865–1957) ON VISUAL LOCALIZATION AND THE INVERSION OF THE RETINAL IMAGE, 1897

G. M. Stratton, "Vision without inversion of the retinal image," *Psychological Review 4*, 341–360, 463–481 (1897).

Kepler's demonstration (1604) that the image of the external world as projected on the retina is upside down raised the question why we see right side up, a question answered by Molyneux (1692), who noted, as Kepler himself had implied, that, as long as the pattern of the image remains intact, the sensorium has no way of knowing whether the top of the image is at the top or bottom of the retina. Johannes Müller (1826) faced this same problem in insisting upon the doctrine of the specific energies of nerves: the mind perceives only the state of the nerves as their excitation represents the external world; and he noted evidence that the retinal image is inverted. The *coup de grâce,* however, was given to this persistent doubt by Stratton in his classical experiment, in which, by a set of lenses, he righted his inverted retinal images, turning them right side up on his retinas and reversing them left and right. He arranged a device made of plaster that he wore strapped to his head for eight days during waking hours. It included for each eye a brass tube with the reversing lenses in it. The text gives below Stratton's description of his experience the first day he wore these special glasses, when the world seemed to him upside down and confusion was great, and then of the seventh day, when his adjustment to the use of the glasses was almost complete. It quotes his account of what happened when he again started going without the glasses: the field was seen reversed and normal relations had to be learned again, but were learned quickly. Stratton was writing sixteen years before the school of behaviorism was ready to supply the concepts and language which he might so well have used; nevertheless it is clear that his experiment showed the explanation of the age-old problem to be partly visual and partly behavioral. The top of the world is what is seen as opposite to the bottom of the world; that is visual. But the top of the world is also what elicits behavior in its own direc-

tion, that is to say, when you reach up for your hat or for the jar on the shelf your behavior is a response to stimulation of the part of the retina that represents the top of the world, and that part is indeed the bottom of the retina. Comparable relations hold for the bottom, left, and right of the retina, which by inversion represent respectively the top, right, and left of the visual world outside. The relation between a seen object and the response to it is associative and learned; the relation can be reversed in experience and restored in further experience; the world looks right side up when the viewer reacts correctly and unhesitatingly to the objects in it, whatever the orientation of the images of it on the viewer's retina.

The experience from day to day was as follows:

First Day.—The entire scene appeared upside down. When I moved my head or body so that my sight swept over the scene, the movement was not felt to be solely in the observer, as in normal vision, but was referred both to the observer and to objects beyond. The visual picture seemed to move through the field of view faster than the accompanying movement of my body, although in the same direction. It did not feel as if I were visually ranging over a set of motionless objects, but the whole field of things swept and swung before my eyes.

Almost all movements performed under the direct guidance of sight were laborious and embarrassed. Inappropriate movements were constantly made; for instance, in order to move my hand from a place in the visual field to some other place which I had selected, the muscular contraction which would have accomplished this if the normal visual arrangement had existed, now carried my hand to an entirely different place. The movement was then checked, started off in another direction, and finally, by a series of approximations and corrections, brought to the chosen point. At table the simplest acts of serving myself had to be cautiously worked out. The wrong hand was constantly used to seize anything that lay to one side. In pouring some milk into a glass, I must by careful trial and correction bring the surface of the milk to the spout of the pitcher, and then see to it that the surface of the milk in the glass remained everywhere equally distant from the glass's rim.

The unusual strain of attention in these cases, and the difficulty of finally getting a movement to its goal, made all but the simplest movements extremely fatiguing. The observer was thus tempted to omit all those which required nice guidance, or which included a series of changes or of rapid adaptations to untried visual circumstances. Relief was sometimes sought by shutting out of consideration the actual visual data, and by depending solely on tactual or motor perception and on the older visual representations suggested by these. But for the most part this tendency was resisted, and movements were performed with full attention to what was visually before me. Even then, I was frequently aware that the opposite, the merely represented, arrangement was serving as a secondary guide along with the actual sight perceptions, and that now the one factor and now the other came to the foreground and was put in control. In order to write my notes, the formation of the letters and words had to be left to automatic muscular sequence, using sight only as a guide to the general position and direction on my paper. When hesitation occurred in my writing, as it often did, there was no resort but to picture the next stroke or two in pre-experimental terms, and when the movement was once under way, control it visually as little as possible.

The scene before me was often reconstructed in the form it would have had in normal vision; and yet this translation was not carried to such an extent as at the beginning of the first experiment. The scene was now accepted more as it was immediately presented. Objects of sight had more reality in them—had more the character of 'things,' and less that of phantasms—than when the earlier trial began. Objects were, however, taken more or less isolatedly; so that inappropriateness of place with reference to other objects even in the same visual field was often, in the general upheaval of the experience, passed by unnoticed. I sat for some time watching a blazing open fire, without seeing that one of the logs had rolled far out on the hearth and was filling the room with smoke. Not until I caught the odor of the smoke, and cast about for the cause, did I notice what had occurred.

Similarly, the actual visual field was, for the most part, taken by itself and not supplemented, as in normal vision, by a system of objects gathered and held from the preceding visual experience. Sporadic cases occurred, in which some object out of sight was represented as it had just been seen; but in general all things not actually in view returned to their older arrangement and were represented, if at all, as in normal sight. Usually this was the case also in picturing an unseen movement of some part of my body. At times, however, both the normal and the later representation of the moving part spontaneously arose in the mind, like an object and its mirrored reflection. But such cases occurred only when actual sight had just before revivified the later memory-image.

As regards the parts of the body, their pre-experimental representation often invaded the region directly in sight. Arms and legs in full view were given a double position. Beside the position and relation in which they were actually seen, there was always in the mental background, in intimate connection with muscular and tactual sensations, the older representation of these parts. As soon as my eyes were closed or directed elsewhere, this older representation gathered strength and was the dominant image. But other objects did not usually have this double localization while I looked at them, unless non-visual sensations came from the objects. Touch, temperature, or sounds, brought up a visual image of the source in pre-experimental form.

Anticipations of contact from bodies seen to be approaching, arose as if particular places and directions in the visual field had the same meaning as in normal experience. When one side of my body approached an object in view, the actual feeling of contact came from the side opposite to that from which I had expected it. And likewise in passing under a hanging lamp, the lamp, in moving toward what in normal experience had been the lower part of the visual field, produced a distinct anticipatory shrinking in the region of the chin and neck, although the light really hung several inches above the top of my head.

Whether as a result of the embarassment under which nearly all visually guided movements, were performed, or as a consequence of the swinging of the scene, described above, there were signs of nervous disturbance, of which perhaps the most marked was a feeling of depression in the upper abdominal region, akin to mild nausea. This disappeared, however, toward evening; so that by half-past seven it was no longer perceptible.

·   ·   ·   ·   ·

Seventh Day.—In the morning the flow of ideas while I was blindfolded was like that described for the evening before. But I noticed in bathing that the old representation of those parts of my body which I had so frequently seen (at least in their clothing) during the experiment, was decidedly less vivid, the outline more blurred, the color paler, grayer, more 'washed out,' than of the parts which had never come within the limits of the visual field.

Later, with my lenses on, it seemed at first as if the experience was in all respects the same as on the previous day. But when I began to pace rapidly up and down the room, I felt that I was more at home in the scene than ever before. There was perfect reality in my visual surroundings, and I gave myself up to them without reserve and without being conscious of a single note of discord with what I saw. This feeling of complete harmony throughout, lasted as long as I kept my legs either within or near the borders of my field of view. Otherwise the older, inappropriate representation of my body arose at times, but faded, while the new representation revived, as soon as some passing object was seen to enter the region into which the older image of my body extended. The absence of any tactual experiences such as a real body in that position would imply, cast, for the moment, an illusory character over the older form of representation.

To what extent objects in view suggested the idea of other things in harmonious relation with the seen things is best shown by the following cases: As I walked into my bedroom and saw the bedstead, I involuntarily thought of the windows, representing them in the appropriate direction fixed by the position of the bed. The general outlines of the room, and the more important points of reference, arose in harmony with the new sight-perceptions. But the detailed filling of this outline was far less complete than is usual in my case in normal sight. A large number of important things in the room simply did not arise in my mind until their relation to the field of seen things had been brought home afresh by perception. During the first days of the experiment ideas of objects frequently arose in opposition to the new sight-perception; now they either did not arise at all, or came in the newer form. The idea of the sofa or chair on which I *passively* sat did still come up in discord with the general experience, together with the dim feeling of my shoulders and of the uppper parts of my back. But these were now a comparatively isolated group, and not a vigorous *Apperceptions-masse* to call up a host of surrounding things in orderly relation to itself.

In regard to movements, the most striking fact was that the *extent* of the movement now was inappropriate, movements in the wrong *direction* being comparatively rare in the case of the hands, and even still rarer in the case of the feet. My hands frequently moved too far or not far enough, especially when coming from beyond the visual field to something in sight. In trying to take a friend's hand, extended into the (new) lower portion of my visual field, I put my hand too high. In brushing a speck from my paper in the (new) upper portion of the field I did not move my hand far enough. And in striking with my index finger the outstretched fingers of my other hand the movement was much less accurate when I looked at my hands than when I closed my eyes and depended on motor guidance. The actual distance that my hand moved, in such cases, would, under the normal conditions of sight, doubtless have been appropriate to bring my hand to the desired spot. But an object in what had before been the upper part of the field was now at a shorter distance from my hands than formerly; the movement, under the influence of the habitual interpretation of the visual position, would therefore go too far. And, *vice versa,* a movement to an object in what had formerly been the lower part of the visual field would now fall short of its destination. For the visual position would now require a more extended movement of the arm than formerly, in order to reach it.

When I watched one of my limbs in motion, no involuntary suggestion arose that it was in any other place or moved in any other direction than as sight actually reported it, except that in moving my arm a slightly discordant group of sensations came from my unseen shoulder. If, while looking at the member, I summoned an image of it in its old position, then I could feel the limb there too. But this latter was a relatively weak affair, and cost effort. When I looked away from it, however, I involuntarily felt it in its pre-experimental position, although at the same time conscious of a solicitation to feel it in its new position. This representation of the moving part in terms of the new vision waxed and waned in strength, so that it was sometimes more vivid than the old, and sometimes even completely overshadowed it.

The conflict between the old and the new localization of the parts of my body was shown in several instances. The mistaken visual localization of a contact in the palm of one of my hands, and the sudden reversal of even the touch-localization when I detected by sight the true source of the sensations, occurred as on the preceding day. Somewhat similarly, when I moved a heated iron with my right hand to that border of the visual field just beyond which, according to pre-experimental localization, my left hand would have been lying, I involuntarily felt an anticipatory shrinking in my unseen left hand, as if it were on the point of being burnt; although the iron in my right hand was actually several feet from my left, and was moving away from it. When I put my left hand in sight, or looked at it afresh to make sure where it was, the hot iron caused no premonitory feeling whatever on approaching the visual locality which had before been so suggestive of danger.

Seated by the open fire, I happened to rest my head on my hands in such a way that the fire shone directly on the top of my head. I closed my eyes, and the image of the fire remained true to the recent perception. But soon I noticed that I was representing the fire in pre-experimental terms, and I finally discovered that the change was caused by the growing sensations of warmth on the top of my head. My hair and scalp were persistently felt in their older position, no doubt because I never directly saw them in any other. And the old localization of the fire was the only one consistent with this old localization of the hair and scalp. But by passing my hands rapidly back and forth before my open eyes, ending the movement each time with a touch upon the top of my head, it was not difficult to produce a vivid localization of my scalp in harmony with the new sight-perceptions. And with this change the old localization of the fire was suppressed. During the walk in the evening, I enjoyed the beauty of the evening scene, for the first time since the experiment began. Evidently the strangeness and inconvenience of the new relations no longer kept me at such a tension as hitherto.

On removing the glasses, my visual images relapsed into their older form, with a constant interplay and accompaniment, however, of the new.

.   .   .   .   .

When the time came for removing the glasses at the close of the experiment, I thought it best to preserve as nearly as possible the size of visual field to which I had now grown accustomed; so that any results observed might be clearly due solely to the reversion of my visual objects and not to a sudden widening of the visual field. Instead, therefore, of removing the plastercast from my face, I closed my eyes and had an assistant slip out the brass tube which held the lenses, and insert in its place an empty black-lined paper tube that gave about the same range of vision. On opening my eyes, the scene had a strange familiarity. The visual arrangement was immediately recognized as the old one of pre-experimental days; yet the reversal of everything from the order to which I had grown accustomed during the past week, gave the scene a surprising, bewildering air which lasted for several hours. It was hardly the feeling, though, that things were upside down.

When I turned my body or my head, objects seemed to sweep before me as if they themselves were suddenly in motion. The 'swinging of the scene,' observed so continously during the first days of the experiment, had thus returned with great vividness. It rapidly lost this force, however, so that at the end of an hour the motion was decidedly less marked. But it was noticeable the rest of the day, and in a slight degree even the next morning.

Movements which would have been appropriate to the visual arrangement during the experiment, were now repeatedly performed after this arrangement had been reversed. In walking toward some obstacle on the floor of the room—a chair, for instance—I turned the wrong way in trying to avoid it; so that I frequently either ran into things in the very effort to go around them, or else hesitated, for the moment, bewildered what I should do. I found myself more than once at a loss which hand I ought to use to grasp the door-handle at my side. And of two doors, side by side, leading to different rooms, I was on the point of opening the wrong one, when a difference in the metal work of the locks made me aware of my mistake. On approaching the stairs, I stepped up when I was nearly a foot too far away. And in writing my notes at this time, I continually made the wrong movement of my head in attempting to keep the centre of my visual field somewhere near the point where I was writing. I moved my head upward when it should have gone downward; I moved it to the left when it should have gone to the right. And this to such a degree as to be a serious disturbance. While walking, there were distinct signs of vertigo and also the depression in the upper abdominal region, noticed during the earlier days of the experiment. The feeling that the floor and other visual objects were swaying, in addition to the symptoms just mentioned, made my walking seem giddy and uncontrollable. No distinct errors in localizing parts of my body occurred; I was more than once surprised, however, to see my hands enter the visual field from the old lower side.

Objects in the room, at a distance of ten or twelve feet from me, seemed to have lost their old levels and to be much higher than they were either during the experiment or before the experiment. The floor no longer seemed level, but appeared to slope up and away from me, at an angle of perhaps five degrees. The windows and other prominent objects seemed also too high. This strange aspect of things lasted (as did also the swinging of the scene, the feeling of giddiness, and certain inappropriate movements) after the plaster cast had been removed and the normal compass of the visual field was restored. In the dim light of the next morning, the upward slope of the floor and the unusual position of the windows were distinctly noticeable.

It is clear, from the foregoing narrative, that our total system of visual objects is a comparatively stable structure, not to be set aside or transformed by some few experiences which do not accord with its general plan of arrangement. It might perhaps have been supposed beforehand that if one's visual perceptions were changed, as in the present experiment, the visual ideas of things would without resistance conform to the new visual experiences. The results show, however, that the harmony comes only after a tedious course of adjustment to the new conditions, and that the visual system has to be built anew, growing from an isolated group of perceptions. The older visual representations for the most part have to be suppressed rather than reformed.

Why then do the old visual ideas persist in their old form, and not come immediately into accord with the new perceptions? If their position were merely relative to the sight-perceptions they undoubtedly would come into harmony with these perceptions, at least after the first moments of dismay were past. But the fact that the ideas can for some time refuse spatially to conform to the new experience, shows that their position and direction is fixed with reference to something other than the immediate perceptions of sight. What is it which caused the older visual images to preserve a spatial arrangement whose lines of direction were opposed to those of the actual field of view?

To say that the older visual directions persisted because the older tactual directions remained in force, is certainly no sufficient answer unless we can show that visual direction is dependent on tactual direction. But the preceding narrative furnishes strong evidence against such a view. If there is any dependence either way (which I doubt), the evidence seems to favor the primacy of sight.

However that may be, the facts in the present case are more accurately described when we say that the discord was not between tactual directions and visual directions, but between the visual directions suggested by touch and the visual directions given in the actual sight. The real question then is: Why did touch-perceptions so persistently suggest visual images whose positions and directions were in discord with the actual scene? The answer is found, I think, in the familiar doctrine of 'local signs' in touch and in sight, and in the farther assumption that a system of correspondence exists whereby a sign in one sense comes to be connected with and to suggest a particular sign in the other sense.

In the organized experience, a perception in one sensory field not only has in it that peculiar qualitative or intensive character which is its own 'local sign,' but, through this local sign, suggests in the other sensory field the local sign which is most intimately associated with the first. A perception in one sensory field suggests, therefore, in terms of the other sense an image in that place whose local sign is most strongly associated with the local sign of the original perception. According to this view, the local signs of sight correspond to the signs of touch, and *vice versa;* so that each member in this system of *corresponding signs* has its particular correlate in the other sensory field. The correspondence here indicated, does not, however, consist in any spatial or qualitative identity or even similarity of the particular signs which correspond, but only in the fact that both have come to mean the same thing. They have occurred in connection with disparate sensory perceptions whose times of appearing and whose 'curve' of change have been so continuously and repeatedly identical that the perceptions themselves come, in time, to be referred to the same source, or, in other words, give the perception to the same object. The perceptions of the two senses are thus identified; and, at the same time, the disparate local signs (in the different senses) which are simultaneously aroused in the perception of the one object come to have the same spatial meaning.

· · · · ·

We are now enabled also to see what the harmony between touch and sight really is. The experiment clearly shows that an object need not appear in any particular position in the visual field in order to admit of a union or identification of the tactual and visual perceptions of the object. The visual position which any tactual experience suggests—the visual place in which we 'feel' that an object is—is determined, not by some fundamental and immutable relation of tactual and visual 'spaces,' but by the mere fact that we have constantly seen the object there when we have had that particular touch-experience. If this particular touch-experience were the uniform and exclusive accompaniment of a visual object in some different visual position, the two sensory reports would mean the same thing, and the places of their object would be identical. Of course, the harmony of touch the sight also implies that visual appearances have the same relations to one another as tactual appearances have to one another; so that a given object in sight must have the same spatial relation to the rest of my visual world as the accompanying touch-object has with respect to the rest of my tactual world. But this harmony does not require that the visual manifestation of a tactual object should be just here and not there, or in this direction and not in that.

The inverted position of the retinal image is, therefore, not essential to 'upright vision,' for it is not essential to a harmony between touch and sight, which, in the final analysis, is the real meaning of upright vision. For some visual objects may be inverted with respect to other visual objects, but the *whole system* of visual objects can never by itself be either inverted or upright. It could be inverted or upright only with respect to certain non-visual experiences with which I might compare my visual system—in other words, with respect to my tactual or motor perceptions.

·   ·   ·   ·   ·

To return to the more significant features of the experiment. These are, without doubt, found in the results bearing on the relation between touch and sight, and through them on the interrelation of the senses generally. The experiment makes it clear that the harmony between sight and touch does not depend on the inversion of the retinal image. The spatial identity of tactual and visual objects evidently does not require that there should be a visual transposition of objects or that they should be given some special direction in the visual field. The chief reason for the existence of the projection theory is therefore taken away.

IV THE VISUAL PERCEPTION OF SIZE AND DISTANCE

Descartes (1638) discussed the psychological cues to the visual perception of distance, size, and shape, and Berkeley later (1709) gave a more sophisticated account of the same phenomena. At this stage the cues to the perception of distance were to be found in the convergence of the two eyes on near and far objects, in the optical accommodation of each eye to near and far vision, and in the many secondary criteria that are effective in monocular vision and in paintings. It was not until Wheatstone's paper (1838) that stereoscopic vision was understood, and it was realized that the perception of depth and solidity may emerge from the disparity of the two binocular retinal images that occurs because of the parallax of the two eyes in viewing near objects. The relative importance and the interaction of these same cues to the perception of size and distance still constitute fundamental problems in the visual perception of space, as, for example, in the moon illusion and in the constancy of perceived size when distance to the stimulus object varies.

27 RENÉ DESCARTES (1596–1650) ON THE VISUAL PERCEPTION OF SIZE, SHAPE, AND DISTANCE, 1638

René Descartes, *La Dioptrique* (Leiden, 1638), discours 6. Translated for this book by Mollie D. Boring from Victor Cousin, ed., *Oeuvres de Descartes*, V (Paris, 1824), 59–66.

We may start with Descartes. This difficult scientific work of his, usually appended to his *Discours de la méthode* but never before translated into English, includes a very early systematic account of the various cues to the visual perception of distance, size, and shape. For distance it discusses the primary criteria of the equivalents of visual accommodation and convergence (but not of retinal disparity, which was not discovered as a cue until almost two centuries later) and the secondary criteria of aerial perspective and relative size. It asserts that the peripheral field is projected upon the brain, where the soul takes account of sizes and shapes, correcting the distorted shapes of the projected pattern in accordance with its past experience with the shapes of objects. Since the soul can make contact with the outer world only through the brain, illusion occurs when the pattern at the brain is distorted and no means of correction is available.

As to the way in which each part of an object is placed with respect to our body, we perceive it no differently through our eyes than with our hands. Moreover, our knowledge of it does not depend on an image or on any action by the object, but only on the arrangement of the small parts of the brain where the nerves originate—for this arrangement, which

changes so little with each shift in disposition of the members reached by
the nerves, is of such a nature as not only to make known to the soul
where each part of the body it animates is in relation to the other parts but
also to enable the soul to shift its attention to the entire area that lies
within such straight lines as we might imagine drawn from the end of
each of these parts and extended to infinity. Similarly, when the blind
man . . . turns his hand *A* toward *E* (Fig. 18), or his hand *C* toward *E,*
the nerves leading from the hand effect a certain change in his brain, and
that gives his soul a means of knowing the locus not only of *A* or *C* but
also of all the other points along the straight lines *AE* and *CE,* so that
the soul can then turn its attention to the objects *B* and *D* to determine
their locations, without needing to know or to think of the location of the
two hands.

Likewise, when our eye or our head is turned toward one side, our soul
is alerted by the change that the nerves leading from the muscles used for
these movements cause in our brain. In the case of the eye *RST* (Fig. 16),
we must realize that the disposition of the small fibers of the optic nerve,
at *R* or *S* or *T,* is followed by a certain other disposition of part 7 or 8 or 9
in the brain, and this fact enables the soul to know all points along the
line *RV* or *SX* or *TY.* Thus we should not find it strange that objects can
be seen in their true relation, in spite of the fact that the picture they
imprint upon the eye is a complete reversal.

Likewise, our blind man (Fig. 18) can simultaneously feel object *B* at
his right by means of his left hand, and *D* at his left by means of his
right hand; and, just as the blind man does not believe that one body is
two when he is touching it with two hands, so, when both our eyes are
turned in the direction proper for bringing our attention to a single point,
we see only one object, in spite of the fact that a picture is formed in each
eye.

Now does the perception of distance depend only on the disposition of
images sent out from the objects but, in the first place, on the shape of
the eyeball, for . . . the shape must be a little different for seeing what is
near our eyes than for seeing what is farther away; and, in so far as we
change the shape in proportion to the distance of the objects, we are
changing also a certain part of our brain in such a way that our soul can
perceive the distance. Usually this happens without our thinking of it,
just as, when we grasp a body with our hand, we adapt our hand to the
size and shape of the body, in order to feel it, without having to think of
these movements.

Then, secondly, we know about the distance by the relation of the eyes
to each other, for just as our blind man, who holds the two sticks *AE* and
CE (of whose length I assume him to be ignorant) and who knows only

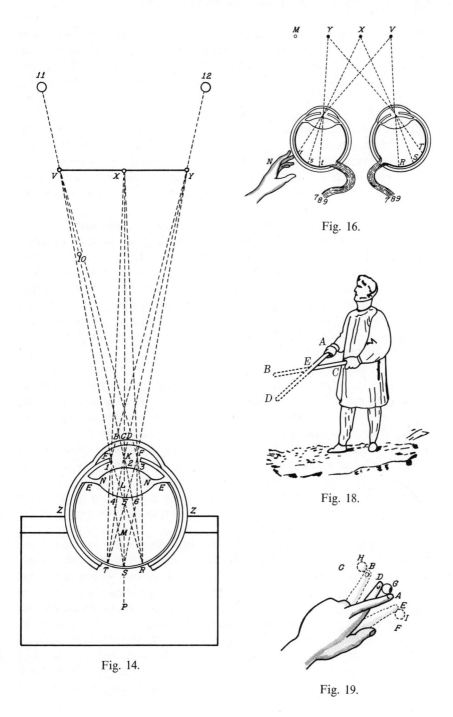

Fig. 14.

Fig. 16.

Fig. 18.

Fig. 19.

the distance AC between his hands and the size of the angles ACE and CAE, can deduce, by a sort of natural geometry, the location of point E, so we ourselves, when our eyes RST and rst are turned toward X (Fig. 16), can deduce the location of X from the length of the line Ss and the size of the angles XSs and XsS. We can even do so with only one eye if we move the eye; for example, keeping it turned toward X, we place it first at point S and then immediately afterward at point s, which enables us to know in imagination the length of the line Ss, as well as the size of the angles XSs and XsS, and thereby to perceive the distance of point X. All this occurs by an act of thinking which, being simple imagination, does not entail [explicit] reasoning like that of surveyors' measuring inaccessible places from two separate stations.

There is yet a [third] way of perceiving distance, and that is by the distinctness or vagueness of shape, together with the strength or weakness of light. For instance, when we look steadily at X (Fig. 14), the rays emanating from objects 10 and 12 do not meet as exactly at R and at T at the back of our eye as they would for objects at V and at Y, and thus we see them as nearer or farther than X. Since the light to our eye from object 10 is stronger than if it were at V, we judge it to be closer; and since the light from object 12 is weaker than if it were at Y, we judge it to be farther away.

Finally, in imagining the size of an object, or its pattern, or the distinction of its shape and colors, or merely the strength of light emanating from it, we not only aid ourselves to see it but also to imagine its distance. Likewise, in looking from afar at a body that we have been accustomed to seeing close up, we can judge its distance much better than if its size were less familiar. When we look at a mountain exposed to sunshine beyond a forest in shadow, only the position of the forest makes us judge it to be nearer. In looking at two ships at sea, one of which is smaller yet proportionately nearer so that they seem equal in size, we can judge which is farther away by the difference in their shapes and the colors of the light they send toward us.

For the rest, concerning the way in which we see the size and shape of objects, there is no need for me to say anything in particular, other than that these characteristics are presumed from what we see of the distance and arrangements of the parts. In other words, their size is estimated by the knowledge or opinion we have of their distance as compared with the size of image they cast at the back of the eye, and not absolutely by the size of these images, as is fairly evident from the fact that, whereas [the images of] objects [in the eye] may be a hundred times bigger than when they were [for objects] ten times farther away, they still do not seem one hundred times bigger, but practically equal, as long as we are not deceived as to their distances. It is also evident that shape is judged by our knowledge or opinion of the disposition of the diverse parts of the objects and not by the shape of the pictures formed in the eye, for these pictures usually consist only of ovals and diamonds, when what we see are [really] circles and squares.

That you may have no further doubts of that vision is accomplished as I have explained it, let us here consider how vision can sometimes deceive us. In the first place, it can deceive us because the soul, and not the eye, does the seeing, and the soul sees directly only by the mediation of the brain, and that is why distracted persons or people in their sleep often see, or believe they see, various objects actually not before their eyes. In other words, vapors, stirring the brain, rearrange the parts ordinarily used in vision exactly as the objects, had they actually been present, would have arranged them. Since impressions from outside reach the common sense [Aristotle's *sensus communis*, the general seat of all sensations] by the mediation of the nerves, then if the arrangement of these nerves is restrained by some unusual cause, objects will be seen in places other than where they are. Thus, if the eye *rst* (Fig. 16), while looking toward *X*, is constrained by the finger *N* to turn toward *M*, the parts of the brain reached by its nerves will not be arranged in the manner they would be if the muscles themselves had turned the eye toward *M*, nor even as they would be in actually looking at *X*, but in a manner somewhere be-tween these two, as if the eye were looking at *Y;* and thus *M* will seem to be in the place of *Y*, by the mediation of this eye, and *Y* in the place of *X*, and *X* in the place of *V*. Since these objects appear to be at the same time in their actual locations by the mediation of the other eye *RST*, they will now seem to be double.

In the same way, a person touching the small ball *G* (Fig. 19), with his fingers *A* and *D* crossed, will believe he is touching two balls, and that is because, when the fingers are crossed this way, the muscles tend to spread the fingers apart—*A* toward *C*, and *D* toward *F*—so that the parts of the brain from which nerves lead into these muscles become arranged in the requisite fashion to make *A* seem to be at *B*, and *D* seem to be at *E*, as if they were touching two separate balls, *H* and *I*.

28   GEORGE BERKELEY (1685–1753) ON THE VISUAL PERCEPTION OF DISTANCE AND MAGNITUDE, 1709

George Berkeley, *An Essay towards a New Theory of Vision* (Dublin, 1709), sects. 1–28, 52–58.

Visual space perception is tridimen-sional, whereas the perceiving organ, the retina, is bidimensional. How do you get three dimensions out of two?

Berkeley faced this problem in his *New Theory of Vision,* advancing thinking about the matter considerably further than had Descartes. The perception of distance is tied in with the perception of size, and vice versa, he said. This book also contains his anticipation of the associative context theory of objective reference (No. 36).

I. My Design is to shew the manner, wherein we perceive by Sight the Distance, Magnitude, and Situation of *Objects.* Also to consider the Difference there is betwixt the *Ideas* of Sight and Touch, and whether there be any *Idea* common to both Senses. In treating of all which, it seems to me, the Writers of *Optics* have proceeded on wrong Principles.

II. It is, I think, agreed by all that *Distance* of it self, and immediately cannot be seen. For *Distance* being a Line directed end-wise to the Eye, it projects only one Point in the Fund of the Eye. Which Point remains invariably the same, whether the Distance be longer or shorter.

III. I find it also acknowledg'd, that the Estimate we make of the Distance of *Objects* considerably remote, is rather an Act of Judgment grounded on *Experience,* than of *Sense.* For Example, When I perceive a great number of intermediate *Objects,* such as Houses, Fields, Rivers, and the like, which I have experienced to take up a considerable Space; I thence form a Judgment or Conclusion, that the *Object* I see beyond them is at a great Distance. Again, when an *Object* appears Faint and Small, which at a near Distance I have experienced to make a vigorous and large Appearance; I instantly conclude it to be far off. And this, 'tis evident, is the result of *Experience;* without which, from the Faintness and Littleness, I should not have infer'd any thing concerning the Distance of *Objects.*

IV. But when an *Object* is placed at so near a Distance, as that the Interval between the Eyes bears any sensible Proportion to it. It is the receiv'd Opinion that the two *Optic Axes* (the Fancy that we see only with one Eye at once being exploded) concurring at the *Object* do there make an *Angle,* by means of which, according as it is Greater or Lesser, the *Object* is perceiv'd to be nearer or farther off.

V. Betwixt which, and the foregoing manner of Estimating Distance, there is this remarkable Difference. That whereas, there was no apparent, necessary Connexion between small Distance and a large and strong Appearance, or between great Distance, and a little and faint Appearance. Yet there appears a very necessary Connexion between an obtuse Angle and near Distance, and an acute Angle and farther Distance. It does not in the least depend upon Experience, but may be evidently known by any one before he had experienc'd it, that the nearer the Concurrence of the *Optic Axes,* the greater the *Angle,* and the remoter their Concurrence is, the lesser will be the *Angle* comprehended by them.

VI. There is another way, mention'd by the *Optic Writers,* whereby they will have us judge of those Distances, in respect of which, the breadth of the *Pupil* hath any sensible bigness. And that is the greater or lesser Divergency of the Rays, which issuing from the visible Point, do fall on the *Pupil:* That Point being judged nearest, which is seen by most diverging Rays; and that remoter, which is seen by less diverging Rays. And so on, the apparent Distance still increasing, as the Divergency of the Rays decreases, till at length it becomes infinite, when the Rays that fall on the *Pupil* are to Sense Parallel. And after this manner it is said we perceive Distances when we look only with one Eye.

VII. In this Case also, 'tis plain we are not beholding to Experience: It being a certain, necessary Truth, that the nearer the direct Rays falling on the Eye approach to a *Parallelism,* the farther off is the Point of their Intersection, or the visible Point from whence they flow.

VIII. I have here set down the common, current Accounts that are given of our perceiving near Distances by Sight, which, tho' they are unquestionably receiv'd for true by *Mathematicians,* and accordingly made use of by them in determining the apparent Places of *Objects,* do nevertheless seem to me very unsatisfactory: And that for these following Reasons.

IX. *First,* It is evident that when the Mind perceives any *Idea,* not immediately and of it self, it must be by the means of some other *Idea.* Thus, for Instance, the Passions which are in the Mind of another, are of themselves, to me invisible. I may nevertheless perceive them by Sight, tho' not immediately yet, by means of the Colours they produce in the Countenance. We do often see Shame or Fear in the Looks of a Man, by perceiving the Changes of his Countenance to Red or Pale.

X. Moreover it is evident, that no *Idea* which is not it self perceiv'd, can be to me the means of perceiving any other *Idea.* If I do not perceive the Redness or Paleness of a Man's Face themselves, it is impossible I shou'd perceive by them the Passions which are in his Mind.

XI. Now from *Sect.* II. 'Tis plain that Distance is in it's own nature imperceivable, and yet it is perceiv'd by Sight. It remains therefore, that it be brought into view by means of some other *Idea,* that is it self immediately perceiv'd in the Act of *Vision.*

XII. But those *Lines* and *Angles,* by means whereof *Mathematicians* pretend to explain the Perception of Distance, are themselves not at all perceiv'd, nor are they in Truth, ever thought of by those unskilful in *Optics.* I appeal to any ones Experience, whether upon Sight of an *Object,* he compute it's Distance by the bigness of the *Angle,* made by the meeting of the two *Optic Axes?* Or whether he ever think of the greater or lesser Divergency of the Rays, which arrive from any Point to his *Pupil.*

Nay, whether it be not perfectly impossible for him to perceive by Sense, the various Angles wherewith the Rays according to their greater, or lesser Divergence do fall on his Eye. Every one is himself the best Judge of what he perceives, and what not. In vain shall all the *Mathematicians* in the World tell me, that I perceive certain *Lines* and *Angles* which introduce into my Mind the various *Ideas* of *Distance;* so long as I my self am conscious of no such thing.

XIII. Since therefore those *Angles* and *Lines* are not themselves perceiv'd by Sight, it follows from *Sect.* X. that the Mind does not by them judge of the Distance of *Objects.*

XIV. *Secondly,* The Truth of this Assertion will be, yet, farther evident to any one that considers those *Lines* and *Angles* have no real Existence in Nature, being only an *Hypothesis* fram'd by *Mathematicians,* and by them introduc'd into *Optics,* that they might treat of that Science in a *Geometrical* way.

XV. The *Third* and Last Reason I shall give for my Rejecting that Doctrine, is, that tho' we should grant the real Existence of those *Optic Angles,* &c. and that it was possible for the Mind to perceive them; yet these Principles wou'd not be found sufficient to explain the *Phænomena* of *Distance.* As shall be shewn hereafter.

XVI. Now, It being already shewn that Distance is suggested to the Mind, by the Mediation of some other *Idea* which is it self perceiv'd in the Act of Seeing. It remains that we enquire what *Ideas,* or *Sensations* there be that attend *Vision,* unto which we may suppose the *Ideas* of Distance are connected, and by which they are introduced into the Mind. And *First,* It is certain by Experience, that when we look at a near *Object* with both Eyes, according as it approaches, or recedes from us, we alter the Disposition of our Eyes, by lessening or widening the Interval between the *Pupils.* This Disposition or Turn of the Eyes is attended with a Sensation, which seems to me, to be that which in this Case brings the *Idea* of greater, or lesser Distance into the Mind.

XVII. Not, that their is any natural or necessary Connexion between the Sensation we perceive by the Turn of the Eyes, and greater or lesser Distance. But because the Mind has by constant *Experience,* found the different Sensations corresponding to the different Dispositions of the Eyes, to be attended each, with a Different Degree of Distance in the *Object:* There has grown an Habitual or Customary Connexion, between those two sorts of Ideas. So that the Mind no sooner perceives the Sensation arising from the different Turn it gives the Eyes, in order to bring the *Pupils* nearer, or farther asunder; but it withal perceives the different *Idea* of Distance which was wont to be connected with that Sensation. Just as upon hearing a certain Sound, the *Idea* is immediately suggested to the Understanding, which Custom had united with it.

XVIII. Nor do I see, how I can easily be mistaken in this Matter. I know evidently that Distance is not perceived of it self. That by consequence, it must be perceived by means of some other *Idea* which is immediately perceiv'd, and varies with the different Degrees of Distance. I know also that the Sensation arising from the Turn of the Eyes is of it self, immediately perceiv'd, and various Degrees thereof are connected with different Distances; which never fail to accompany them into my Mind, when I view an *Object* distinctly with both Eyes, whose Distance is so small that in respect of it, the Interval between the Eyes has any considerable Magnitude.

XIX. I know it is a receiv'd Opinion, that by altering the disposition of the Eyes, the Mind perceives whether the Angle of the *Optic Axes* is made greater or lesser. And that accordingly by a kind of *Natural Geometry,* it judges the Point of their Intersection to be nearer, or farther off. But that this is not true, I am convinc'd by my own Experience. Since I am not conscious, that I make any such use of the Perception I have by the Turn of my Eyes. And for me to make those Judgments, and draw those Conclusions from it, without knowing that I do so, seems altogether incomprehensible.

XX. From all which it plainly follows, that the Judgment we make of the Distance of an *Object,* view'd with both Eyes, is entirely the *Result of Experience.* If we had not constantly found certain Sensations arising from the various Disposition of the Eyes, attended with certain degrees of Distance. We shou'd never make those sudden Judgments from them, concerning the Distance of *Objects;* no more than we wou'd pretend to judge of a Man's Thoughts, by his pronouncing Words we had never heard before.

XXI. *Secondly,* An *Object* placed at a certain Distance from the Eye, to which the breadth of the *Pupil* bears a considerable Proportion, being made to approach, is seen more confusedly. And the nearer it is brought, the more confused Appearance it makes. And this being found constantly to be so, there arises in the Mind an *Habitual* Connexion between the several Degrees of Confusion and Distance. The greater Confusion still implying the lesser Distance, and the lesser Confusion, the greater Distance of the *Object.*

XXII. This confused Appearance of the *Object,* doth therefore seem to me to be the *Medium,* whereby the Mind judges of Distance in those Cases, wherein the most approv'd Writers of *Optics* will have it judge, by the different Divergency, with which the Rays flowing from the Radiating Point fall on the *Pupil.* No Man, I believe, will pretend to see or feel those imaginary Angles, that the Rays are supposed to form according to their various Inclinations on his Eye. But he cannot choose seeing whether the *Object* appear more, or less confused. It is therefore a manifest Consequence from what has been Demonstrated, that instead of the greater, or lesser Divergency of the Rays, the Mind makes use of the greater or lesser Confusedness of the Appearance, thereby to determine the apparent Place of an *Object*.

XXIII. Nor doth it avail to say, there is not any necessary Connexion between confused *Vision,* and Distance great, or small. For I ask any Man, What necessary Connexion he Sees, between the Redness of a Blush and Shame? And yet no sooner shall he behold that Colour to arise in the Face of another. But it brings into his Mind the *Idea* of that Passion which has been observ'd to accompany it.

XXIV. What seems to have misled the Writers of *Optics* in this Matter is, that they imagine Men judge of Distance, as they do of a Conclusion in *Mathematics;* betwixt which and the Premises, it is indeed absolutely requisite there be an apparent, necessary Connexion. But it is far otherwise, in the sudden Judgments Men make of Distance. We are not to think, that Brutes and Children, or even grown reasonable Men, whenever they perceive an *Object* to approach, or depart from them, do it by vertue of *Geometry* and *Demonstration.*

XXV. That one *Idea* may suggest another to the Mind, it will suffice that they have been observ'd to go together; without any demonstration of the necessity of their Coexistence, or without so much as knowing what it is that makes them so to Coexist. Of this there are innumerable Instances, of which no one can be Ignorant.

XXVI. Thus greater Confusion having been constantly attended with nearer Distance, no sooner is the former *Idea* perceiv'd, but it suggests the latter to our Thoughts. And if it had been the ordinary Course of Nature, that the farther off an *Object* were placed, the more Confused it shou'd appear. It is certain, the very same Perception that now makes us think an *Object* approaches, would then have made us to imagine it went farther off. That Perception, abstracting from *Custom* and *Experience,* being equally fitted to produce the *Idea* of great Distance; or small Distance, or no Distance at all.

XXVII. *Thirdly,* an *Object* being placed at the Distance above specified, and brought nearer to the Eye, we may nevertheless prevent, at least for some time, the Appearance's growing more confus'd, by straining the Eye. In which Case, that Sensation supplys the place of confused *Vision,* in aiding the Mind to judge of the Distance of the *Object.* It being esteemed so much the nearer, by how much the effort, or straining of the Eye in order to [obtain] distinct *Vision,* is greater.

XXVIII. I have here set down those Sensations or *Ideas,* that seem to me to be the constant and general Occasions of introducing into the Mind, the different *Ideas* of near Distance. 'Tis true in most Cases, that divers other Circumstances contribute to frame our *Idea* of Distance, *viz.* the particular Number, Size, Kind, *&c.* of the things seen. Concerning which as well as all other the forementioned Occasions which suggest Distance, I shall only observe, they have none of them, in their own Nature, any Relation or Connexion with it. Nor is it possible, they shou'd ever signifie the various Degrees thereof, otherwise than as by *Experience* they have been found to be connected with them.

$$\cdot \qquad \cdot \qquad \cdot \qquad \cdot \qquad \cdot$$

LII. I have now done with Distance, and proceed to shew, how it is, that we perceive by Sight, the Magnitude of Objects. It is the Opinion of some, that we do it by Angles, or by Angles in Conjunction with Distance. But neither Angles, nor Distance being perceivable by Sight: And the things we see being, in truth, at no Distance from us; it follows, that as we have demonstrated Lines and Angles not to be the *Medium,* the Mind makes use of in apprehending the Apparent Place, so neither are they, the *Medium* whereby it apprehends the Apparent Magnitude of Objects.

LIII. It is well known that the same Extension, at a near Distance, shall subtend a greater Angle, and at a farther Distance, a lesser Angle. And by this Principle (we are told) the Mind estimates the Magnitude of an *Object,* comparing the Angle under which it is seen, with its Distance, and thence infering the Magnitude thereof. What inclines Men to this Mistake (beside the Humour of making one see by *Geometry*) is, that the same Perceptions or *Ideas* which suggest Distance, do also suggest Magnitude. But if we examine it, we shall find they suggest the latter, as immediately as the former. I say, they do not first suggest Distance, and then leave it to the Judgment to use that as a *Medium,* whereby to collect the Magnitude; but they have as close, and immediate a Connexion with the Magnitude, as with the Distance; and suggest Magnitude as independently of Distance, as they do Distance independently of Magnitude. All which will be evident, to whoever considers what has been already said, and what follows.

LIV. It has been shewn, there are two sorts of *Objects* apprehended by Sight; each whereof hath its distinct Magnitude, or Extension. The one, properly Tangible, *i.e.* to be perceiv'd and measur'd by Touch, and not immediately falling under the Sense of Seeing. The other, properly and immediately Visible, by Mediation of which, the former is brought in View. Each of these Magnitudes are greater or lesser, according as they contain in them more or fewer Points; they being made up of Points or *Minimums.* For, whatever may be said of Extension in *Abstract,* it is certain sensible Extension is not infinitely Divisible. There is a *Minimum*

Tangibile, and a *Minimum Visibile,* beyond which Sense cannot perceive. This, every ones Experience will inform him.

LV. The Magnitude of the *Object* which exists without the Mind, and is at a Distance, continues always invariably the same. But the Visible *Object* still changing as you approach to, or recede from the Tangible *Object,* it hath no fixed and determinate Greatness. Whenever therefore, we speak of the Magnitude of any thing, for Instance a *Tree* or a *House,* we must mean the Tangible Magnitude, otherwise there can be nothing steddy, and free from Ambiguity spoken of it. Now, tho' the Tangible and Visible Magnitude do, in truth, belong to two distinct Objects: I shall nevertheless (especially since those Objects are called by the same Name, and are observ'd to coexist) to avoid tediousness and singularity of Speech, sometimes speak of them, as belonging to one and the same thing.

LVI. Now in order to discover by what means, the Magnitude of Tangible Objects is perceived by Sight; I need only reflect on what Passes in my own Mind. And observe what those things be, which introduce the *Ideas* of *greater* or *lesser* into my Thoughts, when I look on any *Object.* And these I find to be, First, the Magnitude or Extension of the Visible *Object,* which being immediately perceived by Sight, is connected with that other which is Tangible, and placed at a Distance. Secondly, The Confusion or Distinctness. And, Thirdly, The Vigorousness or Faintness of the aforesaid Visible Appearance. *Cæteris paribus,* by how much the greater or lesser, the Visible *Object* is, by so much the greater or lesser, do I conclude the Tangible *Object* to be. But, be the *Idea* immediately perceived by Sight never so large, yet if it be withal Confused, I judge the Magnitude of the thing to be but small. If it be Distinct and Clear, I judge it greater. And if it be Faint, I apprehend it to be yet greater . . .

LVII. Moreover, the Judgments we make of Greatness do, in like manner as those of Distance, depend on the Disposition of the Eyes, also on the Figure, Number of intermediate Objects, and other Circumstances that have been observ'd to attend great, or small Tangible Magnitudes. Thus, for Instance, The very same Quantity of Visible Extension, which in the Figure of a Tower, doth suggest the *Idea* of great Magnitude, shall, in the Figure of a Man, suggest the *Idea* of much smaller Magnitude. That this is owing to the Experience we have had, of the usual Bigness of a Tower and a Man, no one, I suppose, need be told.

LVIII. It is also evident, that Confusion, Faintness, &c. have no more a necessary Connexion with little or great Magnitude, than they have with little or great Distance. As they suggest the latter, so they suggest the former to our Minds. And, by Consequence, if it were not for *Experience,* we shou'd no more judge a faint or confused Appearance to be connected, with great or little Magnitude, than we shou'd that it was connected with great or little Distance.

29   CHARLES WHEATSTONE (1802–1875) ON BINOCULAR PARALLAX AND THE STEREOSCOPIC PERCEPTION OF DEPTH, 1838

Charles Wheatstone, "Contributions to the physiology of vision: on some remarkable and hitherto unobserved phenomena of binocular vision," *Philosophical Transactions of the Royal Society of London*, 1838, pt. 1, pp. 371–394.

We have already seen how Descartes (1638) and Berkeley (1709) accounted for the visual perception of distance in terms of what are now called secondary cues, but it was left for Wheatstone to demonstrate how the disparity of the binocular retinal images produced by the parallax of binocular vision results in what since his discovery has been called stereoscopic depth. Others had touched on the fact of binocular disparity for the perception of near objects (Leonardo da Vinci, before 1519), but Wheatstone was the first to exploit the full significance of this primary cue to the perception of depth in 1838, after his discovery had already been described by H. Mayo in 1833. The stereoscope was Wheatstone's invention.

1

When an object is viewed at so great a distance that the optic axes of both eyes are sensibly parallel when directed towards it, the perspective projections of it, seen by each eye separately, are similar, and the appearance of the two eyes is precisely the same as when the object is seen by one eye only. There is, in such case, no difference between the visual appearance of an object in relief and its perspective projection on a plane surface; and hence pictorial representations of distant objects, when those circumstances which would prevent or disturb the illusion are carefully excluded, may be rendered such perfect resemblances of the objects they are intended to represent as to be mistaken for them; the Diorama is an instance of this. But this similarity no longer exists when the object is placed so near the eyes that to view it the optic axes must converge; under these conditions a different perspective projection of it is seen by each eye, and these perspectives are more dissimilar as the convergence of the optic axes becomes greater. This fact may be easily verified by placing any figure of three dimensions, an outline cube for instance, at a moderate distance before the eyes, and while the head is kept perfectly steady, viewing it with each eye successively while the other is closed. Fig. 13 [below] represents the two perspective projections of a cube; *b* is that seen by the right eye, and *a* that presented to the left eye; the figure being supposed to be placed about seven inches immediately before the spectator.

The appearances, which are by this simple experiment rendered so obvious, may be easily inferred from the established laws of perspective; for the same object in relief is, when viewed by a different eye, seen from two points of sight at a distance from each other equal to the line joining the two eyes. Yet they seem to have escaped the attention of every philosopher and artist who has treated of the subjects of vision and perspective. I can ascribe this inattention to a phenomenon leading to the important and curious consequences, which will form the subject of the present communication, only to this circumstance; that the results being contrary to a principle which was very generally maintained by optical writers, viz. that objects can be seen single only when their images fall on corresponding points of the two retinæ, an hypothesis which will be hereafter discussed, if the consideration ever arose in their minds, it was hastily discarded under the conviction, that if the pictures presented to the two eyes are under certain circumstances dissimilar, their differences must be so small that they need not be taken into account.

It will now be obvious why it is impossible for the artist to give a faithful representation of any near solid object, that is, to produce a painting which shall not be distinguished in the mind from the object itself. When the painting and the object are seen with both eyes, in the case of the painting two *similar* pictures are projected on the retinæ, in the case of the solid object the pictures are *dissimilar;* there is therefore an essential difference between the impressions on the organs of sensation in the two cases, and consequently between the perceptions formed in the mind; the painting therefore cannot be confounded with the solid object.

After looking over the works of many authors who might be expected to have made some remarks relating to this subject, I have been able to find but one, which is in the *Trattato della pittura* of Leonardo da Vinci. This great artist and ingenious philosopher observes, "that a painting, though conducted with the greatest art and finished to the last perfection, both with regard to its contours, its lights, its shadows and its colours, can never show a relievo equal to that of the natural objects, unless these be viewed at a distance and with a single eye. For," says he, "if an object . . . be viewed by a single eye . . . all objects in the space behind it . . . are invisible to the eye . . . but when the other eye . . . is opened, part of these objects become visible to it . . . Thus the object . . . seen with both eyes becomes, as it were, transparent, according to the usual definition of a transparent thing; namely, that which hides nothing beyond it. But this cannot happen when an object, whose breadth is bigger than that of the pupil, is viewed by a single eye. The truth of this observation is therefore evident, because a painted figure intercepts all the space behind its apparent place, so as to preclude the eyes from the sight of every part of the imaginary ground behind it."

Had Leonardo da Vinci taken, instead of a sphere, a less simple figure for the purpose of his illustration, a cube for instance, he would not only have observed that the object obscured from each eye a different part of the more distant field of view, but the fact would also perhaps have forced itself upon his attention, that the object itself presented a different appearance to each eye. He failed to do this, and no subsequent writer within my knowledge has supplied the omission; the projection of two obviously dissimilar pictures on the two retinæ when a single object is viewed, while the optic axes converge, must therefore be regarded as a new fact in the theory of vision.

2

It being thus established that the mind perceives an object of three dimensions by means of the two dissimilar pictures projected by it on the two retinæ, the following question occurs: What would be the visual effect of simultaneously presenting to each eye, instead of the object itself, its projection on a plane surface as it appears to that eye? To pursue this inquiry it is necessary that means should be contrived to make the two pictures, which must necessarily occupy different places, fall on similar parts of both retinæ. Under the ordinary circumstances of vision the object is seen at the concourse of the optic axes, and its images consequently are projected on similar parts of the two retinæ; but it is also evident that two exactly similar objects may be made to fall on similar parts of the two retinæ, if they are placed one in the direction of each optic axis, at equal distances before or beyond their intersection.

.   .   .   .   .

Now if . . . the two perspective projections of the same solid object be so disposed, the mind will still perceive the object to be single, but instead of a representation on a plane surface, as each drawing appears to be when separately viewed by that eye which is directed towards it, the observer will perceive a figure of three dimensions, the exact counterpart of the object from which the drawings were made.

.   .   .   .   .

The inconveniences [of free stereoscopy] are removed by the instrument I am about to describe; the two pictures (or rather their reflected images) are placed in it at the true concourse of the optic axes, the focal adaptation of the eye preserves its usual adjustment, the appearance of lateral images is entirely avoided, and a large field of view for each eye is obtained. The frequent reference I shall have occasion to make to this instrument, will render it convenient to give it a specific name, I therefore propose that it be called a Stereoscope, to indicate its property of representing solid figures.

3

The stereoscope is represented by fig. 8 . . . *A A'* are two plane mirrors, about four inches square, inserted in frames, and so adjusted that their

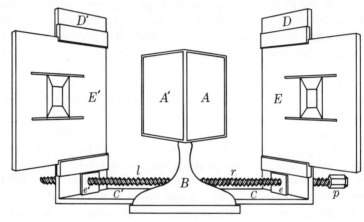

Fig. 8.

backs form an angle of 90° with each other; these mirrors are fixed by their common edge against an upright *B,* or which was less easy to represent in the drawing, against the middle line of a vertical board, cut away in such manner as to allow the eyes to be placed before the two mirrors. *C C'* are two sliding boards, to which are attached the upright boards *D D',* which may thus be removed to different distances from the mirrors. In most of the experiments hereafter to be detailed, it is necessary that each upright board shall be at the same distance from the mirror which is opposite to it. To facilitate this double adjustment, I employ a right and a left-handed wooden screw, *rl;* the two ends of this compound screw pass through the nuts *e e',* which are fixed to the lower parts of the upright boards *D D',* so that by turning the screw pin *p* one way the two boards will approach, and by turning it the other they will recede from each other, one always preserving the same distance as the other from the middle line *f. E E'* are pannels, to which the pictures are fixed in such manner that their corresponding horizontal lines shall be on the same level: these pannels are capable of sliding backwards and forwards in grooves on the upright boards D D'. The apparatus having been described, it now remains to explain the manner of using it. The observer must place his eyes as near as possible to the mirrors, the right eye before the right hand mirror, and the left eye before the left hand mirror, and he must move the sliding pannels E E' to or from him until the two reflected images coincide at the intersection of the optic axes, and form an image of the same apparent magnitude as each of the component pictures. The pictures will indeed coincide when the sliding pannels are in a variety of different positions, and consequently when viewed under different inclinations of the optic axes; but there is only one position in which the binocular image will be immediately seen single, of its proper magnitude, and without fatigue to the eyes, because in this position only the ordinary relations between the magnitude of the pictures on the retina, the inclination of the optic axes, and the adaptation of the eye to distinct vision at different distances are preserved . . .

4

A few pairs of outline figures, calculated to give rise to the perception of objects of three dimensions when placed in the stereoscope in the manner described, are represented [in the figures] . . . As the drawings are reversed by reflection in the mirrors, I will suppose these figures to be the reflected images to which the eyes are directed in the apparatus; those marked b being seen by the right eye, and those marked a by the left eye. The drawings, it has been already explained, are two different projections of the same object seen from two points of sight, the distance between which is equal to the interval between the eyes of the observer; this interval is generally about $2\frac{1}{2}$ inches . . .

Fig. 11. A series of points all in the same horizontal plane, but each towards the right hand successively nearer the observer . . .

Fig. 13. A cube.

Fig. 14. A cone, having its axis perpendicular to the referent plane, and its vertex towards the observer . . .

The other figures require no observation.

For the purposes of illustration I have employed only outline figures, for had either shading or colouring been introduced it might be supposed that the effect was wholly or in part due to these circumstances, whereas by leaving them out of consideration no room is left to doubt that the entire effect of relief is owing to the simultaneous perception of the two monocular projections, one on each retina. But if it be required to obtain the most faithful resemblances of real objects, shadowing and colouring may properly be employed to heighten the effects. Careful attention would enable an artist to draw and paint the two component

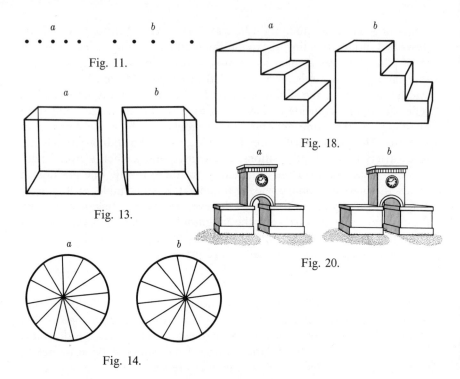

Fig. 11.

Fig. 13.

Fig. 18.

Fig. 20.

Fig. 14.

pictures, so as to present to the mind of the observer, in the resultant perception, perfect identity with the object represented. Flowers, crystals, busts, vases, instruments of various kinds, &c., might thus be represented so as not to be distinguished by sight from the real objects themselves. [Photographs were not yet available in 1838.]

.   .   .   .   .

5

A very singular effect is produced when the drawing originally intended to be seen by the right eye is placed at the left hand side of the stereoscope, and that designed to be seen by the left eye is placed on its right hand side. A figure of three dimensions, as bold in relief as before, is perceived, but it has a different form from that which is seen when the drawings are in their proper places. There is a certain relation between the proper figure and this, which I shall call its *converse* figure. Those points which are nearest the observer in the proper figure are the most remote from him in the converse figure, and *vice versa,* so that the figure is, as it were, inverted; but it is not an exact inversion, for the near parts of the converse figure appear smaller, and the remote parts larger than the same parts before the inversion. Hence the drawings which, properly placed, occasion a cube to be perceived, when changed in the manner described, represent the frustum of a square pyramid with its base remote from the eye: the cause of this is easy to understand.

This conversion of relief may be shown by all the pairs of drawings . . . In the case of simple figures like these the converse figure is as readily apprehended as the original one, because it is generally a figure of as frequent occurrence; but in the case of a more complicated figure, an architectural design, for instance, the mind, unaccustomed to perceive its converse, because it never occurs in nature, can find no meaning in it.

NATIVISTIC AND EMPIRISTIC THEORIES
OF SPACE PERCEPTION

Kant (1781) was the forerunner of the nativistic theories of space, and Lotze (1852), with his special theory of local signs, became the specific promoter of the opposing empiristic theories. E. H. Weber (1852) described the new experimental measurement of cutaneous spatial sensitivity and also contributed his theory of sensory circles to the physiological account of spatial localization. Presently Hering (1864), influenced by Kant and Johannes Müller, emerged as the leading proponent of nativism, holding that spatial relations are given innately with sensation, even though these raw data have to undergo an elaborate reorganization before the normal perception of space is fully formed. Helmholtz (1866), Hering's chief opponent in respect of more than one psychological theory, took his stand with the empiricist, Lotze. His doctrine of unconscious inference is included in his general theory. Hering's nativistic position reappears in the phenomenology of Gestalt psychology. Here Wertheimer (1912) is cited, because the Phi phenomenon of seen movement constitutes an excellent example of a spatial perception that is quite unlearned and independent of past experience.

30 IMMANUEL KANT (1724–1804) ON THE A PRIORI NATURE OF SPACE, 1781

Immanuel Kant, *Kritik der reinen Vernunft* (Riga, 1781), pt. I, bk. I, "Tranzendentale Aesthetik," chap. 2, sect. 1. Translated by John Watson in *The Philosophy of Kant as Contained in Extracts of His Own Writings* (New York, 1888), pp. 22–29.

Kant, generally regarded as the greatest of modern philosophers, was the founder of German idealism, which was opposed to the British empiricism of Locke, Berkeley, and Hume. It was against Hume's skeptical empiricism that Kant primarily oriented his philosophy. Thus the lines were drawn between empiricism and idealism, and in nineteenth-century psychology between empiricism and nativism. J. S. Mill, Lotze, Helmholtz, Wundt, and the later Wundtians stood for empiricism in the perception of space: man's conception of space is built up in experience, which is a genetic theory. Others, like Johannes Müller and later Ewald Hering and Carl Stumpf, either influenced by Kant and the idealistic movement or else resenting empiricism and glad to accept Kant's sponsoring intellect, settled for nativism, which the empiricist Lotze thought was not a theory at all but a disavowal of the possibility of theory. This excerpt shows Kant arguing in his famous *Critique of Pure Reason* for the a priori nature of space (and time), given as subjective patterns of pure perception and as the mold into which experience fits rather than the structure into which it is fabricated. (See also No. 105 for the paragraphs introducing this discussion of space by Kant.)

2. METAPHYSICAL EXPOSITION OF SPACE

In external sense we are conscious of objects as outside of ourselves, and as all without exception in space. In space their shape, size, and relative position are marked out, or are capable of being marked out. Inner sense, in which we are conscious of ourselves, or rather of our own state, gives us, it is true, no direct perception of the soul itself as an object; but it nevertheless is the one single form in which our own state comes before us as a definite object of perception; and hence all inner determinations appear to us as related to one another in time. We cannot be conscious of time as external, any more than we can be conscious of space as something within us. What, then, are space and time? Are they in themselves real things? Are they only determinations, or perhaps merely relations of things, which yet would belong to things in themselves even if those things were not perceived by us? Or, finally, have space and time no meaning except as forms of perception, belonging to the subjective constitution of our own mind, apart from which they cannot be predicated of anything whatever? To answer these questions I shall begin with a metaphysical exposition of space. An *exposition* I call it, because it gives a distinct, although not a detailed, statement of what is implied in the idea of space; and the exposition is *metaphysical,* because it brings forward the reasons we have for regarding space as given *a priori.*

(1) Space is not an empirical conception, which has been derived from external experiences. For I could not be conscious that certain of my sensations are relative to something outside of me, that is, to something in a different part of space from that in which I myself am; nor could I be conscious of them as outside of and beside one another, were I not at the same time conscious that they not only are different in content, but are in different places. The consciousness of space is, therefore, necessarily presupposed in external perception. No experience of the external relations of sensible things could yield the idea of space, because without the consciousness of space there would be no external experience whatever.

(2) Space is a necessary *a priori* idea, which is presupposed in all external perceptions. By no effort can we think space to be away, although we can quite readily think of space as empty of objects. Space we therefore regard as a condition of the possibility of phenomena, and not as a determination dependent on phenomena. It is thus *a priori,* and is necessarily presupposed in external phenomena.

(3) Space is not a discursive or general conception of the relation of things, but a pure perception. For we can be conscious only of a single

space. It is true that we speak as if there were many spaces, but we really mean only parts of one and the same identical space. Nor can we say that these parts exist *before* the one all-embracing space, and are put together to form a whole; but we can think of them only as *in* it. Space is essentially single; by the plurality of spaces, we merely mean that because space can be limited in many ways, the general conception of spaces presupposes such limitations as its foundation. From this it follows, that an *a priori* perception, and not an empirical perception, underlies all conceptions of pure space. Accordingly, no geometrical proposition, as, for instance, that any two sides of a triangle are greater than the third side, can ever be derived from the general conceptions of line and triangle, but only from perception. From the perception, however, it can be derived *a priori,* and with demonstrative certainty.

(4) Space is *presented* before our consciousness as an infinite magnitude. Now, in every conception we certainly think of a certain attribute as common to an infinite number of possible objects, which are subsumed *under* the conception; but, from its very nature, no conception can possibly be supposed to contain an infinite number of determinations *within* it. But it is just in this way that space is thought of, all its parts being conceived to co-exist *ad infinitum.* Hence the original consciousness of space is an *a priori* perception, not a conception.

3. TRANSCENDENTAL EXPOSITION OF SPACE

A transcendental exposition seeks to show how, from a certain principle, the possibility of other *a priori* synthetic knowledge may be explained. To be successful, it must prove (1) that there really are synthetic propositions which can be derived from the principle in question, (2) that they can be so derived only if a certain explanation of that principle is adopted.

Now, geometry is a science that determines the properties of space synthetically, and yet *a priori.* What, then, must be the nature of space, in order that such knowledge of it may be possible? Our original consciousness of it must be perception, for no new truth, such as we have in the propositions of geometry, can be obtained from the mere analysis of a given conception. And this perception must be *a priori,* or, in other words, must be found in us before we actually observe an object, and hence it must be pure, not empirical perception. For all geometrical propositions, as, for instance, that space has but three dimensions, are of demonstrative certainty, or present themselves in consciousness as necessary; and such propositions cannot be empirical, nor can they be derived from judgments of experience.

How, then, can there be in the mind an external perception, which is antecedent to objects themselves, and in which the conception of those objects may be determined *a priori?* Manifestly, only if that perception has its seat in the subject, that is, if it belongs to the formal constitution of the subject, in virtue of which it is so affected by objects as to have a direct consciousness or perception of them; therefore, only if perception is the universal *form* of outer sense.

.   .   .   .   .

INFERENCES

(*a*) Space is in no sense a property of things in themselves, nor is it a relation of things in themselves to one another. It is not a determination that still belongs to objects even when abstraction has been made from all the subjective conditions of perception. For we never could perceive *a priori* any determination of things, whether belonging to them individually or in relation to one another, antecedently to our perception of those things themselves.

(*b*) Space is nothing but the form of all the phenomena of outer sense. It is the subjective condition without which no external perception is possible for us. The receptivity of the subject, or its capability of being affected by objects, necessarily exists before there is any perception of objects. Hence it is easy to understand, how the form of all phenomena may exist in the mind *a priori,* antecedently to actual observation, and how, as a pure perception in which all objects must be determined, it may contain the principles that determine beforehand the relations of objects when they are met with in experience . . .

Our exposition, therefore, establishes the *reality,* or objective truth of space, as a determination of every object that can possibly come before us as external; but, at the same time, it proves the *ideality* of space, when space is considered by reason relatively to things in themselves, that is, without regard to the constitution of our sensibility. We, therefore, affirm the *empirical reality* of space, as regards all possible external experience; but we also maintain its *transcendental ideality,* or, in other words, we hold that space is nothing at all, if its limitation to possible experience is ignored, and it is treated as a necessary condition of things in themselves.

31   RUDOLF HERMANN LOTZE (1817–1881) ON LOCAL SIGNS IN THEIR RELATION TO THE PERCEPTION OF SPACE, 1852

R. H. Lotze, *Medicinische Psychologie, oder Physiologie der Seele* (Leipzig, 1852), bk. II, chap. 4, sect. 28, paras. 289, 290, 292–294. Translated for this book by Don Cantor.

Lotze, a philosopher best known for his metaphysical writings, played an important role in the founding of the new experimental physiological psychology by his enthusiastic support of the movement. G. E. Müller and Carl Stumpf, respectively in command of the Göttingen and Berlin institutes, the two more important laboratories next to Wundt's at Leipzig, were Lotze's pupils, and Franz Brentano, in his professional difficulties brought on by his Catholicism, was Lotze's protegé. Lotze had succeeded Herbart at Göttingen and G. E. Müller succeeded Lotze there, the three successive incumbencies of the same chair spanning 87 years (1833-1920). By intent Lotze was writing in 1852 a physiological psychology, for *medicinische* was hardly to be distinguished from *physiologische* at that time, and the subtitle was "Physiology of the Mind." Lotze was not a very good physiologist, but he became important because his theory of local signs stood clearly at the head of the movement for empiricism in the

perception of space. The local sign is a distinctive tag that every visual or tactual sensation carries, distinctive as to its physical locus in bidimensional visual and tactual space, but without psychological indications of spatial order or arrangement. Actually, as the excerpt shows, Lotze was half a nativist, for he gave that degree of homage to Kant which was implicit in his faith that the mind has in it certain inherent propensities for spatial ordering and is thus able to use the local signs in constructing a single total space. Most of the empiricists of that day were partly nativists, and most of the nativists found themselves turning to experience to put the finishing touches on a spatial system that they thought had been given a priori at the start. The two vigorously opposed schools were closer together than either of them realized.

289. A general maxim for the understanding of the anatomical structure of the sensory organs follows: . . . wherever we find arrangements prepared so as to allow a multiplicity of external stimuli to act in ordered geometrical relations upon the nervous system, these arrangements become very important to us as hints that nature intends to build up something for consciousness out of those spatial relations. Since the arrangements in themselves clarify nothing, it becomes necessary to search within the sensory organs for other means by which the loci of the stimulated points, as well as the qualities of the stimulus, are communicated to the mind. Since the spatial specification of a sensory element is independent of its qualitative content (so that at different moments very different sensations can fill the same places in our picture of space), each stimulation must have a characteristic peculiarity, given it by the point in the nervous system at which it occurs. We shall call this its *local sign*. Presently we shall speak of the nature of these local signs in greater detail. Here we can only characterize them generally as physical nervous processes, one for every place in the nervous system, consistently associated with the variable nervous process that is the basis for the qualitative content of the changing sensations that arise at the same place. The two processes do not interfere with each other at all, or only to a very small degree. While the mind continues to form its usual qualitative sen-

sations under the influence of the second process, each sensation is [also] accompanied by another stimulus that governs its later disposition in a spot in imagined space and is dependent upon the local sign.

290. One can think of two uses for this supposition. First of all, [by it] we draw attention to the fact that the mind itself cannot have the least reason to apprehend as manifold a quality of sensation aroused in it by many nerves in exactly the same way and at the same time. It is much more likely that those things that are alike would inevitably merge. Matters would, on the other hand, be different were each individual sensation to be accompanied by a local sign characteristic of it alone, which it received at the place of its entry into the body or in the central organs. These secondary traits would hinder the unification of what is sensed without changing its quality, and they would make it necessary for the mind to allow the sensory content to become separate and to be perceived as often as it is thrust into consciousness . . .

That all perception of space touches upon this is now easy to see; however, we have here only a necessary condition, and not the sufficient foundation of this perception. For as long as we regard the local signs only as different (as would be suggested by the letters α, β, γ), but not as members of an ordered series—the units of which could for comparison be designated by numbers—a separation of the sensations they accompany might be expected to take place, although without the development of any sort of clear spatial perception . . .

We see something similar in hearing. If one and the same tone is sounded simultaneously by instruments having different characteristics, whose varying timbres would be comparable to local signs, we clearly feel, to be sure, a certain breadth of the tones along with an increase in their volume, but the single tones do not seem separated in space. This phenomenon, however, which of course has still other characteristics, is valid not as an example but only as an analogy.

On the other hand, this comment applies to the sensations of temperature and pressure in a true sense. If we thought of all cutaneous nerves as functionally equivalent in organization, then the same degree of warmth applied to many of them would be perceived by us just as though the total heat came through one nerve; for the many entirely equal sensations that would be awakened by the single nerves would have no conceivable barrier between them in consciousness to hinder their full merging. On the other hand, were we to suppose that each single nerve gives to its warm stimulus a local sign, characteristic of that nerve alone and different from that of all the others, then the conception of a frequent repetition of the same warm stimulus would have to appear in the consciousness. And these manifold sensations would not only be separable in a general way; the mind's custom of creating images of space would cause them to separate spatially . . .

The difficulty of giving the sensations of warmth and pressure definite spatial forms in spite of their perceived extension comes from the fact that, although the cutaneous nerves contribute local signs, these signs are not units of an accurate graduated system of comparable elements. Therefore, although we easily perceive the stimuli on the various points of the skin as generally different in place, their disposition on definite points in space on the body very often involves associations between them and the images of vision.

. . . . . .

292. There is, however, still another point of general interest to be mentioned. Admitting that simultaneously with every qualitative sensation its local sign will be brought to the mind, we leave the question what that sign consists of in order to ask another: is it not dubious to suppose that the mind is enabled through that phenomenon alone—and forced—not only to separate its sensations but also to differentiate them spatially? To at least part of this question we must say "yes." Surely, we may regard it as a certainty that the consciousness must in general differentiate those sensations which have associated themselves with different local signs; and surely the differentiation would take place as thoroughly as that which occurs when we perceive the same tones in different timbres . . .

It is, however, in no sense our intention to deduce from the local signs the mind's general ability to perceive space or its need to assimilate that which is sensed into this spatial perception. We suppose, rather, that there are features to the nature of the mind which make it not only capable of a type of spatial perception but force it to exercise this capability on the content of that which it senses. We do not try to explain either this capability or this necessity from the assumed physiological relations of the local signs. If we suppose and grant as an accepted fact that the mind can form images of space and is willing to do so, there still arises the other question: according to what principles of selection does it assign place to one sensation here, to another there; or to what principles does it adjust in perceiving sensations a and b as neighbors, but a and c as distant from one another in this general image of space which it has formed? If all geometrical relations that exist between the parts of the external stimuli and also between their effects on the nerves merge in the simple intensive presence which belongs to the images in the mind, and if these relations are to be reconstructed from this presence, then definite individual characteristics must be present in the single sensations that represent the place in the corresponding stimuli, [thus] allowing the mind to establish a spatial order. It is to be supposed that the local signs serve this purpose alone—not that they instill in the mind the desire or ability (both of which it lacks) for spatial perception, but rather that they are the means by which the mind, driven by its nature to the spatial unfolding of its innermost contents, uses its general manner of perceiving in accordance with the nature and the mutual relations of the [perceived] objects.

293. We can with an easy conscience leave to philosophical psychology the question of whether what we here only suppose to be a fact is capable of further extension and clarification. The idea that spatial perception is an original, a priori property of the nature of the mind, provoked to certain uses but not created by external impressions, is sufficient for all of our physiological observations. We do not mean that space, extending infinitely in three dimensions, is in itself a perpetual property of our consciousness, [something] which we have perhaps observed in our thoughts since birth, eager to fill it with images. We mean only that the original nature of our mind requires us to arrange what we sense in positions in space, and that the infinite quantity of such arrangements, performed unconsciously by us, brings to our consciousness the more or less vivid total concept of all-embracing infinite space. At the same time, we believe that a spatial trait must exist in the nature of the mind, one that makes it capable of conceiving [space] in this inevitable form, and that an intelligent mind, similar to ours in all other respects but lacking this trait, would not be able to perceive space as we do by the finely adjusted relations between impressions that impel us to use spatial forms; rather, such a mind would possess only a variety of impressions that were thoroughly mixed together. We must therefore disapprove of all attempts to explain our perception of space through only two elements: the mind's ability to form images, on the one hand, and the simultaneous or successive ordering of the stimuli, on the other . . .

294. Finally, let us add a word about the relation of the use of local signs to consciousness. If one wished to suppose that this local characteristic always came to be consciously perceived with the sensation, and that the mind, reflecting on it, assigned each element in the sensation its place in space, then one would be confusing a situation that arises only in the last applications of our perception of space, confusing it with the simple processes that are the basis for the possibility of such applications. If we use trigonometry to determine the place of distant external points, we make a fully conscious use, of course, of the angles enclosed by our lines of sight to them, employing these angles as local signs, and

determining the place of the points by deliberate calculation based on them. Furthermore, when in ordinary life we relate a stimulus on the surface of the body to a particular point of that surface, the relation that we make is based mostly on an association borrowed from [other] experience. This association couples the quality of the sensation with an image of the place on the skin and its locale known from some [earlier experience]. And this process is generally not clear to us, although, if we consider it carefully, we find that through such a fast and unconscious repetition we do, in fact, revive the earlier experience of the stimulus on this spot. When, however, we finally locate the simultaneously perceived points of color in a visual field on definite spots, the reason remains entirely unconscious, and the local signs, which we suppose must exist also here, are as consistently denied to consciousness as the stimuli of other sensory nerves when they elicit reflex movements that are conscious only in that they are known to occur . . . Spatial localization belongs therefore to those things that the mind can do unconsciously and what it can do because of the mechanics of its inner workings.

32   ERNST HEINRICH WEBER (1795–1878) ON SENSORY CIRCLES AND CUTANEOUS SPACE PERCEPTION, 1852

E. H. Weber, "Ueber den Raumsinn und die Empfindungskreise in der Haut und im Auge," *Berichte der königlich-sächsischen Gesellschaft der Wissenschaften zu Leipzig, mathematisch-physische Classe 4,* 87–105 (1852). Translated for this book by Don Cantor.

Weber, best known for the law of sensory intensity that bears his name (see No. 17), as the pioneer in the investigation of tactual sensibility (see No. 10) included in his research studies of the cutaneous perception of space. He devised a means of determining the error of tactual localization and also the compass test, which measures the two-point threshold (the minimal separation of two tactual stimuli at which the two can be perceived as separate). Both of these measures of cutaneous sensitivity vary with bodily region and were supposed by Weber to depend on the closeness of separate nerve endings in the skin. He suggested that the skin be considered as divided into tiny areas ("sensory circles"), with one fiber to each area. Two adjacent stimuli would be felt as two, he thought, only if they stimulate different sensory circles, with at least one unstimulated circle lying in between. The theory implies that the peripheral distribution of the fibers is projected upon the brain. That opinion was being advanced, of course, before the belief in brain centers had been generally accepted; even Descartes had expressed a similar belief.

EXPERIMENTS FOR DETERMINING THE ACCURACY OR PRECISION OF THE SPATIAL SENSIBILITY OF VARIOUS PARTS OF THE SKIN

1. A subject points out the place on his skin that is being or has just been touched.

If one blackens the rounded end of a small rod (for example, a thick knitting needle) with fine carbon powder, fastens a weight to its upper end, and places it perpendicularly upon various parts of a person's skin so that its weight is felt, one finds that the subject cannot, without looking, point out exactly the place of contact in every region of the skin, but only in those places provided with very accurate spatial sensibility.

If the subject uses a short probe to indicate the place where he believes he has just been touched, he can often himself feel that the probe is rather distant from the point of contact and that he could approach that point by moving the probe in a specific direction. Often, having succeeded in doing so, he moves farther away in his effort to come still closer. If one uses a compass and ruler to measure by how much the subject misses the sought-for point (which is marked by the black smudge) when he believes that he is closest to it, and if one takes the average of many such measurements, one finds that the greater the error, the more imperfect is spatial sensibility in the part of the skin where the experiment has been performed. When the knitting needle was taken away from my skin just as I tried to determine its place of contact, the distances between it and the probe were approximately as follows:

up to 7.0 Paris lines[1]	on the middle of the anterior aspect of the thigh
3.8	on the middle of the volar aspect of the upper arm
2.9	on the middle of the back of the hand
1.9	on the middle of the hollow of the hand
0.5	on the volar side of the fingertips
2.8	on the forehead
2.4	on the chin
0.5	on the lips

On the lips and the volar side of the fingertips one often succeeds in hitting the black spot that the knitting needle made as it pressed the skin. To be sure, when one makes such experiments using various subjects, one finds spatial sensibility more accurate in some persons than in others; but for everybody it is always most accurate or inaccurate in the same parts.

In the method just described, the pressure exerted on the skin does not persist while we try to determine the point of contact. We are directed only by the memory or continuing feeling of the impression we have had. One might, therefore, believe that, were the pressure to continue, we would be able to indicate exactly on all parts of the skin the

[1]Paris line = 0.2256 cm.—Trans.

place where the pressure was exerted, but such is not the case. Even when the knitting needle remains on the skin, the subject mistakes the place of contact as long as he is prevented from striking the knitting needle with his probe. (This is done by removing the needle when the probe is about to strike it, and then immediately replacing it on the black mark it has made.) When this last method is used, the place is, to be sure, determined somewhat more accurately, but only a little.

2. Experiments that determine the accuracy of spatial sensibility for various parts of the skin, in which one observes how far apart the points of an open compass seem to be when various areas of the skin are touched with it, and in which a subject tries to indicate the direction of the line which he thinks would connect the two points.

When, without our looking, the skin is touched with equal pressure simultaneously at two places that are not too close, we perceive two points of contact and a space between them. When the distance between the compass points remains constant, it [nevertheless] seems to us very different on different parts of the skin: very large where spatial sensitivity is very accurate, and small or even imperceptible where this sensibility is dull; that is to say, large on the face, the palm of the hand, and the sole of the foot, [but] small on the torso, the thigh, and the upper arm.

Spatial sensibility is more accurate on the head than on the torso, and on the hands and feet than on the other parts of the arm and leg. The face has more accurate sensibility than the hairy part of the head. The palm of the hand and the sole of the foot are much more sensitive than the backs of the hand or foot; on the torso, too, spatial sensibility on the stomach is somewhat more accurate than on the back. The tip of the tongue has the greatest accuracy of all the parts of the body. From there in all directions the fineness of discrimination declines more and more and rather rapidly, so that it is very inaccurate in the gums, the palate, and the root of the tongue. The lips are the most sensitive part of the face; from them sensitivity declines slowly in all directions, so that surprisingly soon it becomes dull toward the protuberances of the lower jawbone and, in general, toward the neck. The volar sides of the last joint of the fingers and toes—in short, the tips of the fingers and toes—are the most sensitive regions of the arms and legs; in each limb, the sensibility becomes less accurate as one approaches the metacarpus, and still less so as one goes farther up the arm to the torso. On the edge of the palm of the hand, where the sensory spots cease to be ordered in ridges, there is a very rapid decline in the accuracy of spatial sensibility. The hand and foot are alike in this respect, but the sensibility in the hand and arm is considerably more accurate than in the foot.

Individual experiments performed on set parts of the skin constitute the best way to discover how distant the points of a compass must be in order that the two impressions produced by their touching the skin simultaneously and with equal pressure can be perceived as separate (rather than merging into a single impression) and divided by an interval of space. For this one needs a reliable and practiced observer, and [only] a very few experiments should be performed at any one time lest he tire—for tiredness produces false results. [Moreover,] to be fully reliable the experiments must be repeated often and the repetitions separated by long intervals of time.

In the following tabulation I have brought together the results of many observations made at various times on my own body. The distance, given in Paris lines, is that which was necessary between compass points, touching my skin simultaneously and with equal pressure, for me to perceive and distinguish two contacts, whether the points lay along or were perpendicular to the length of my body and limbs.

Part of the skin[2]	Distance between the compass points (Paris lines)
Tip of the tongue	0.5
Volar side of the last member of the finger	1
Red part of the lips	2
Tip of the nose	3
Back of the tongue	4
Cheeks	5
Middle of the hard palate	6
Skin on the anterior aspect of the malar bone	7
Skin on the posterior aspect of the malar bone	10
Hairy part of the rear of the head	12
Crown of the head	15
Small of the back	18
Instep near the toes	18
Skin on the backbone in the area of the neck, chest, and hips	21–30
On many places in the middle of the upper arm and thigh	16–30

·   ·   ·   ·   ·

If one traces two parallel lines on the skin of a subject, using the points of an opened compass and taking care that they exert equal pressure, the subject will believe that he feels the lines approach one another on some parts of the skin and grow farther apart on other parts. They seem to diverge where the compass points move from less to more sen-

[2]The tabulation includes only 15 of Weber's 37 entries.—Trans.

sitive parts of the skin and to converge where they move from more to less sensitive parts. When they are a certain very small distance apart, the compass points bring forth the feeling that a single line only is being traced on the skin. For example, if the compass is opened so that the points are separated by 7–10 Paris lines and one touches the forearm with it at the elbow, holding the compass perpendicular to the arm, many people believe that they feel a single point, [although] many who know that the points of the compass are really separated imagine that they feel two contacts . . .

If one moves the compass in constant contact with the skin over the forearm, hand, and fingers (see Fig. 1), the two points trace the parallel

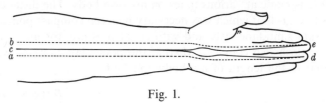

Fig. 1.

lines *ad* and *be*. The observer, on the other hand, believes that he feels the compass points describing the path *ced;* that is to say, a path that begins as a single line, but separates itself near the wrist into two lines that move quickly apart at the beginning of the palm of the hand, converge [again] somewhat in the middle of the palm, and diverge again on the fingers.

The paths that the points of the compass seem to describe show us how far apart they would seem were they placed perpendicular to and touching the skin at the various points. The paths demonstrate that, though the points are separated by 7–10 Paris lines, they seem for a long stretch of the forearm as though they touched at only a single spot, and seem to be considerably separated for the first time on that part of the surface of the palm of the hand that is distinguished by volar ridges sensitive to touch.

If the compass is opened wider, to 18 Paris lines, say, one perceives a distance between the points even on those parts of the forearm where before only the impression of a single point was perceived, and where before the distance seemed minute, it now seems greater.

The same phenomenon may be perceived in the face if one first touches the skin on the lower jawbone near the earlobe with the points of a compass, opened to 6–9 Paris lines, so that the spots touched are vertically aligned. The interval between them seems very small to the observer; to many observers it is imperceptible, being like a single point. If the compass is then moved transversely over the face so that its points describe two parallel lines, the observer believes that he feels the paths diverge and that they separate further as they approach the middle of the lips. When the one point touches the upper lip and the other the lower one, they seem more distant from each other than they really are. If one continues to move them in constant contact with the skin over the other half of the face, the observer then believes that he feels the paths traced by the points of the compass converge as they approach the rear border of the other ridge of the jawbone. (See Fig. 2, in which

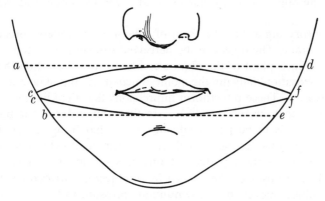

Fig. 2.

the two dotted parallel lines *ad* and *be* represent the true paths of the compass points, and the curved lines *cc* to *ff* the paths we feel.) . . .

One's perception of the distance between the compass points, then, does not correspond at all to the size of the piece of skin that they span, but rather to the abundance of nerve stems in that piece. It is clear from the above table that a distance of 24 [Paris] lines between the points of a compass when it touches the back may seem smaller than a distance of 0.5 Paris line touching the tip of the tongue. It is because of the arrangement of the organs of touch that the same distance between points of a compass may in some parts of the skin seem to us almost imperceptibly small, in others very great, and in still others of middle size; and this phenomenon opens to us a means for discovering the causes of the perception of space as it is awakened by an object touching our skin.

·   ·   ·   ·   ·

THE ARRANGEMENTS WITHIN THE NERVOUS SYSTEM FOR EFFECTING
SPATIAL SENSIBILITY IN THE SKIN, AND ESPECIALLY THE
SENSORY CIRCLES

In connection with all sensory systems we must distinguish between the part located in the brain (the central part, about which we know nothing detailed), the peripheral part, and the nerve fibers that connect the two parts.

.   .   .   .   .

The peripheral sensory organ for the sense of touch—that is, the organ through which we receive sensations of pressure and temperature and also of space—is so made that the patterns, distances, and movements of the objects that are perceived are reflected upon it at the same time.

This sensory organ has the form of a large, very widely spread, sensitive membrane. The objects to be perceived can come into contact with it, and then each point of the objects communicates the action that is to call forth sensation only to the point touched on the skin and to the nearby area . . . These punctiform impressions that the [different] points of an object make on the skin lie . . . in the same order as the points of the object that is to be perceived, the object that brings forth the impressions. If a cross section of a three- or four-sided cylindrical tube is pressed onto the skin, a pattern of this cross-section is formed on the skin . . . In just this way, the contact of an object is mirrored on the skin when the object moves, brings continuous pressure to bear on the skin, or affects the skin by giving or drawing away heat. If the skin is simultaneously touched with two compass points, the distance between them is mirrored on it.

But the mirrored patterns, movements, and distances will not be perceived if impressions of the same kind call forth sensations that are entirely alike and therefore cannot be distinguished one from another. The main characteristics of the organization of the sense of space consist therefore of the following:

(1) The skin of this sensory organ is a mosaic of sensitive parts (sensory circles), each of which has its characteristic capacity to receive sensation; and, because of this, two actions upon two [different] parts of this mosaic always bring forth two different sensations, which do not merge into one even when the two actions are otherwise entirely alike.

(2) The difference between sensations on neighboring sensory circles is, to be sure, extremely small, but it becomes more distinct as the number of specific sensory circles that lie between the parts of the skin touched increases.

(3) Experience shows that these sensory circles are neither of equal size nor of equivalent form in different areas of the skin; and this difference explains the fact that the sense of space in many areas of the skin is sharp, in others dull, whereas on the arms and legs it is more accurate

in a transverse direction and duller in the direction of the length of the limb. For the smaller the sensory circles are, the more of them can exist next to each other on [a given] area on the skin.

The greater the number of such sensory circles lying next to each other on equal areas of skin, the more points of the object to be sensed can act to bring forth disparate impressions upon the skin and the more accurate and detailed is the sensing of the spatial relations of the object and its parts. If the sensory circles are oblong and their larger diameters always lie in the same direction, the skin will have duller spatial sensibility in that direction than in the direction of the smaller diameters.

Everyone can see easily that a compass in Fig. 3 will span more sensory circles than in Fig. 4, where they are larger, and that it will span more sensory circles in a horizontal line in Fig. 5, where they are placed transversely, than in Fig. 6. So, when the distance between the compass

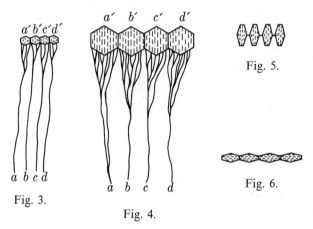

Fig. 5.

Fig. 3.

Fig. 4.

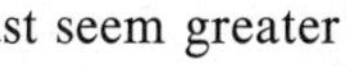

Fig. 6.

points seems greater, more sensory circles must lie within the parts of the skin which it touches, and this distance must seem greater in an area where their size is like that in Fig. 3, rather than in an area like that in Fig. 4. In the former case one will, given a certain opening of the compass, be able to distinguish two sensations and therefore a double contact, whereas in the latter case the same circumstances would lead us to believe we were perceiving a single contact only. If the sensory circles are oblong and all of their long diameters lie in one direction, as in Fig. 5, then a compass opened to the same width will span more sensory circles when it spans the short diameters than when it spans the long ones; and this fact will have the effect that the compass points will seem, as they touch us, to be farther apart in the first case than in the second, and that in the first case one will be able to distinguish two contacts for

147

a sufficient distance between the compass points, while in the second case he would believe he had been touched but once. And this is really the way it is on the arms and legs . . . There, given a certain opening of the compass, one distinguishes two contacts when the compass is placed in a transverse direction and only a single one when it is placed along the length of the limb.

33 EWALD HERING (1834–1918) ON THE NATIVISTIC THEORY OF VISUAL SPACE PERCEPTION, 1864

Ewald Hering, *Beiträge zur Physiologie: Zur Lehre vom Ortsinn der Netzhaut* (Leipzig, 1861–1864), pt. 5, "Vom binocularen Tiefsehen" (1864), sect. 124, pp. 323–328. Translated for this book by Don Cantor.

Hering, the consistent opponent of Helmholtz the empiricist, fits into the middle of the course of nativistic thinking about space: Kant, Johannes Müller, Hering, Carl Stumpf, and the Gestalt psychologists Wertheimer and Köhler. The empiricists were Lotze, Helmholtz, and Wundt. Lotze held that every point on the retina or the skin has a qualitatively unique local sign, and that the local signs come through experience to be related in a spatial bidimensional order. The propensity for finding spatial order is given in the nature of mind, but spatial order is built up in experience. Hering held that every retinal point has a simple spatial feeling that is composed of three simpler spatial qualities of height, breadth, and depth. Spatiality is thus given a priori, yet real visual space, including the perception of direction, has to be built up in experience. There is really not such a great difference between the nativist and the empiricist, although these bitter antagonists of the 1860's thought there was.

FUNDAMENTALS OF THE GENERAL THEORY OF VISUAL SPACE

Visual space—that is to say, the sum of everything we see at any given moment—is the creation of our sensorium and is produced by the combined effects of two factors: (1) the sensations of light and space that are induced directly through the binocular retinal images and are based on an inborn mechanism, and (2) the constantly changing state of the sensorium, which depends upon the infinite number of experiences, opinions, and thoughts by means of which, in the course of our lives, the sensorium is, so to speak, continuously recreated.

Let us deal first of all with the first factor, assuming a primitive and absolutely inexperienced sensorium—one that is, as it were, capable neither of opinions nor of thought.

Such a sensorium, consisting only of pure sensation, is possessed of consciousness but lacks self-consciousness. It perceives space and light because of its given retinal images, but it does not orient itself in contrast to these perceptions as an "I." The images that it receives of parts of its own body seem to it equal in importance to the images of other things. It perceives a space and distinguishes separate, variously colored forms within this space. There may be, for example, hands and feet among these images, but the sensorium has no reason to set them apart from the other forms as belonging especially to the narrow "I." The sensorium does not reflect but only perceives, and every moment it enlarges itself in its perceptions.

Such a sensorium does not see things as lying in this or that direction, for [the concept of] direction presupposes a point from which all directions begin, that is to say, an "I" as the center of all spatial relations. Therefore, at this juncture, one cannot yet speak of visual direction. One can, however, deal with the spatial relations of forms perceived in visual space and with the laws governing the dependence of these relations on the relations of the single parts of the retinal image itself . . .

The most appropriate starting point for the spatial determinations of the perceived images is the central point of visual space. It has no set place. Rather, it can be fixed in space only in so far as its position is related to other simultaneously perceived parts of visual space. In this manner it is related to every other point in visual space. Thus we cannot at first consider absolute determinations of place in visual space—that is, those [localizations] that correspond to real space or to real things with a fixed size. Rather, we can deal only with the relations of the single points themselves in visual space. For this reason no other measurement may be applied to the perceived images than that of their relations among themselves, for we deal [only] with proportions, never with absolute sizes (sizes corresponding to reality).

In binocular vision, the centers of the two retinas correspond to the center of visual space. The images elicited at all the other points of the retina group themselves around this central point in the [total] perception of space. Their [perceived] spatial relations are dependent on the spatial values of the single points on the retina or on the spatial feelings that are aroused by the [respective] points.

There are only three simple spatial feelings and, correspondingly, three systems of spatial relations for the doubled retina. The first simple spatial feeling corresponds to the ability to perceive height, the second to the ability to perceive breadth, and the third to the ability to perceive depth. They are all elicited by every retinal point.

Every retinal point has a particular value for height, breadth, and depth. Its magnitude is, in general, dependent on the direct distance of the retinal point from the vertical or horizontal dividing line, that is to say, the middle longitudinal or latitudinal cross sections. Every retinal point thus elicits a spatial feeling composed of these three simple qualities of feeling combined in particular proportions.

Every simple spatial feeling can be decomposed, as it were, into a positive and a negative quality . . . For the feeling of depth, there appears, on the one side, a feeling for closeness, and, on the other, a feeling for distance. Thus [it comes about that] one can speak of six simple qualities of spatial perception.

The vertical dividing line separates the places on the retina with positive depth and breadth values from those with the same [corresponding] negative values; the depth and breadth values of the dividing line itself are therefore zero. The corresponding values of single longitudinal sections increase in proportion to their distance from the central longitudinal division. Analogously, the height value for the horizontal dividing line is zero, whereas the positive or negative values of latitudinal cross sections grow in proportion to their distance from the middle latitudinal cross section. All points of one and the same latitudinal cross section have then one and the same depth and breadth value and [one and] the same height value. The intersection of the middle longitudinal and latitudinal dividing lines—that is to say, the middle of the retina—is the zero point for the spatial values in [all] three dimensions.

[Thus] each retinal point elicits a spatial feeling that is made up of three simple spatial feelings and varies with the [relative] strengths of the individual simple spatial feelings. In this way every point of the image in the visual space is allotted a definite location relative to the central point—for, as in geometry the position of a point is determined by its spatial relation to three coordinates, so likewise the location of a point in a [retinal] image is determined by the three simple spatial feelings, simultaneously united in the mixed feeling that it produces in the sensorium. On the other hand, if one thinks of the three spatial feelings elicited by stimulating a certain point on the retina as forces that come together at the central point of visual space at right angles to one another and computes the resultant sum, the resultant shows the point at which the image of the retinal point in question will appear in the visual space.

According to our theory, we have only three qualitatively different spatial feelings instead of an infinite number of them or of local signs, and, through the mixture of these three in various relations of intensity, it becomes possible to characterize spatially every point on the retina.

Two corresponding [binocular] points have identical values for height and breadth, and it is therefore of no importance for localization of height or breadth whether a point of the image is delineated on the right retina or at the corresponding place on the left one. Two [symmetrically] opposite points have identical value as to depth; therefore it is of no importance for the perception of depth whether a point is delineated on the right retina or at the symmetrical place on the left one.

.   .   .   .   .

It is clear that in this most primitive seeing of space no relation to far and near exists, for [distance] appears first when the "I" contrasts itself with the perceived images, and the imaginary picture of the body is [seen as a part of] visual space. For this reason apparent size does not grow originally with apparent distance; rather, the images retain the same apparent size, being determined by their values of breadth and height, without regard to whether they are localized (according to their positive or negative distance values) on one or the other side of the central plane.

If, however, the spatial "I" detaches itself from the entire mass of that which has been perceived as space and contrasts itself with the remainder of that mass, the relation of the apparent position of the perceived images to the place taken by the spatial "I" immediately becomes important, and then may one speak also of this other [type of] direction . . . Because the "I" has to a certain extent made itself the midpoint of visual space, it becomes at the same time the starting point for the directions of vision, which are consequently now to be regarded as diverging [from the eyes] in visual space.

34   HERMANN LUDWIG FERDINAND VON HELMHOLTZ (1821–1894) ON EMPIRICISM IN PERCEPTION, 1866

H. L. F. von Helmholtz, *Handbuch der physiologischen Optik,* vol. III (Leipzig, 1866), sect. 26. Translated by J. P. C. Southall as *Helmholtz' Treatise on Physiological Optics,* vol. III ([Rochester, N. Y.]: Optical Society of America, 1925).

This is Helmholtz's argument for the empirical theory of perception, which was the theory of Lotze before and of Wundt after him. He is against the nativism, or, as he calls it, the intuitionism, of Kant and Hering. He undertakes to marshal his many instances in which perception is built up in experience so that it becomes something more than the bare sensory core on which it is founded. Whatever in a perception can be changed by experience, he held, is not sensation. The empirical theory is associationistic. In developing it Helmholtz had also to ask for consideration of the principle of unconscious inference (unconscious conclusion), which to a large extent affects the fabrication of perception. Unconscious inference comes under further consideration in the next chapter (No. 40).

The general rule determining the ideas of vision that are formed whenever an impression is made on the eye, with or without the aid of optical instruments, is that *such objects are always imagined as being present in the field of vision as would have to be there in order to produce the same impression on the nervous mechanism, the eyes being used under ordinary normal conditions* . . . Suppose that the eyeball is mechanically stimulated at the outer corner of the eye. Then we imagine that we see an appearance of light in front of us somewhere in the direction of the bridge of the nose. Under ordinary conditions of vision, when our eyes are stimulated by light coming from outside, if the region of the retina in the outer corner of the eye is to be stimulated, the light actually has to enter the eye from the direction of the bridge of the nose. Thus, in accordance with the above rule, in a case of this kind we substitute a luminous object at the place mentioned in the field of view, although as a matter of fact the mechanical stimulus does not act on the eye from in front of the field of view nor from the nasal side of the eye, but, on the contrary, is exerted on the outer surface of the eyeball and more from behind. The general validity of the above rule will be shown by many other instances that will appear in the following pages.

In the statement of this rule mention is made of the ordinary conditions of vision, when the visual organ is stimulated by light from outside; this outside light comes from the opaque objects in its path that were the last to be encountered, and reaches the eye along rectilinear paths through an uninterrupted layer of air. This is what is meant here by the normal use of the organ of vision, and the justification for using this term is that this mode of stimulation occurs in such an enormous majority of cases that all other instances where the paths of the rays of light are altered by reflections or refractions, or in which the stimulations are not produced by external light, may be regarded as rare exceptions. This is because the retina in the fundus of the firm eyeball is almost completely protected from the actions of all other stimuli and is not easily accessible to anything but external light. When a person is in the habit of using an optical instrument and has become accustomed to it (for example, if he is used to wearing spectacles) to a certain extent he learns to interpret the visual images under these changed conditions.

Incidentally, the rule given above corresponds to a general characteristic of all sense-perceptions and not simply to the sense of sight alone. For example, the stimulation of the tactile nerves in the enormous majority of cases is the result of influences that affect the terminal extensions of these nerves in the surface of the skin. It is only under exceptional circumstances that the nerve-stems can be stimulated by more powerful agencies. In accordance with the above rule, therefore, all stimulations of cutaneous nerves, even when they affect the stem or the nerve-centre itself, are perceived as occurring in the corresponding peripheral surface of the skin. The most remarkable and astonishing cases of illusions of this sort are those in which the peripheral area of this particular portion of the skin is actually no longer in existence, as, for example, in case of a person whose leg has been amputated. For a long time after the operation the patient frequently imagines he has vivid sensations in the foot that has been severed. He feels exactly the places that ache on one toe or the other. Of course, in a case of this sort the stimulation can affect only what is left of the stem of the nerve whose fibres formerly terminated in the amputated toes. Usually, it is the end of the nerve in the scar that is stimulated by external pressure or by contraction of the scar tissue. Sometimes at night the sensations in the missing extremity get to be so vivid that the patient has to feel the place to be sure that his limb is actually gone.

Thus it happens, that when the modes of stimulation of the organs of sense are unusual, incorrect ideas of objects are apt to be formed— which used to be described, therefore, as *illusions of the senses.* Obviously, in these cases there is nothing wrong with the activity of the organ of sense and its corresponding nervous mechanism which produces the illusion. Both of them have to act according to the laws that govern their activity once for all. It is rather simply an illusion in the judgment of the material presented to the senses, resulting in a false idea of it.

The psychic activities that lead us to infer that in front of us at a certain place there is a certain object of a certain character are generally not conscious activities, but unconscious ones. In their result they are equivalent to a *conclusion* [or *inference*] to the extent that the observed action on our senses enables us to form an idea as to the possible cause of this action; although, as a matter of fact, it is invariably simply the nervous stimulations that are perceived directly, that is, the actions, but never the external objects themselves. But what seems to differentiate them from a conclusion, in the ordinary sense of that word, is that a conclusion is an act of conscious thought. An astronomer, for example, comes to real conscious conclusions of this sort, when he computes the positions of the stars in space, their distances, etc., from the perspective images he has had of them at various times and as they are seen from different parts of the orbit of the earth. His conclusions are based on a conscious knowledge of the laws of optics. In the ordinary acts of vision this knowledge of optics is lacking. Still it may be permissible to speak of the psychic acts of ordinary perception as *unconscious conclusions,* thereby making a distinction of some sort between them and the common so-called conscious conclusions. And while it is true that there has been, and probably always will be, a measure of doubt as to the similarity of the psychic activity in the two cases, there can be no doubt as to the similarity between the results of such unconscious conclusions and those of conscious conclusions.

These unconscious conclusions derived from sensation are equivalent in their consequences to the so-called *conclusions from analogy*. Inasmuch as in an overwhelming majority of cases, whenever the parts of the retina in the outer corner of the eye are stimulated, it has been found to be due to external light coming into the eye from the direction of the bridge of the nose, the inference we make is that it is so in every new case whenever this part of the retina is stimulated; just so we assert that every single individual now living will die, because all previous experience has shown that all men who were formerly alive have died.

But, moreover, just because they are not free acts of conscious thought, these unconscious conclusions from analogy are irresistible, and the effect of them cannot be overcome by a better understanding of the real relations. It may be ever so clear how we get an idea of a luminous phenomenon in the field of vision when pressure is exerted on the eye; and yet we cannot get rid of the conviction that this appearance of light is actually there at the given place in the visual field; and we cannot seem to comprehend that there is a luminous phenomenon at the place where the retina is stimulated. It is the same way in case of all the images that we see in optical instruments.

On the other hand, there are numerous illustrations of fixed and inevitable associations of ideas due to frequent repetition, even when they have no natural connection, but are dependent merely on some conventional arrangement, as, for example, the connection between the written letters of a word and its sound and meaning. Still to many physiologists and psychologists the connection between the sensation and the conception of the object usually appears to be so rigid and obligatory that they are not much disposed to admit that, to a considerable extent at least, it depends on acquired experience, that is, on psychic activity. On the contrary, they have endeavoured to find some mechanical mode of origin for this connection through the agency of imaginary organic structures. With regard to this question, all those experiences are of much significance which show how the judgment of the senses may be modified by experience and by training derived under various circumstances, and may be adapted to the new conditions. Thus, persons may learn in some measure to utilize details of the sensation which otherwise would escape notice and not contribute to obtaining any idea of the object. On the other hand, too, this new habit may acquire such a hold that when the individual in question is back again in the old original normal state, he may be liable to illusions of the senses.

.   .   .   .   .

Another general characteristic property of our sense-perceptions is, that *we are not in the habit of observing our sensations accurately, except as they are useful in enabling us to recognize external objects. On the contrary, we are wont to disregard all those parts of the sensations that are of no importance so far as external objects are concerned.* Thus in most cases some special assistance and training are needed in order to observe these latter subjective sensations. It might seem that nothing could be easier than to be conscious of one's own sensations; and yet experience shows that for the discovery of subjective sensations some special talent is needed, such as Purkinje manifested in the highest degree; or else it is the result of accident or of theoretical speculation. For instance, the phenomena of the blind spot were discovered by Mariotte from theoretical considerations. Similarly, in the domain of hearing, I discovered the existence of those combination tones which I have called summation tones. In the great majority of cases, doubtless it was accident that revealed this or that subjective phenomenon to observers who happened to be particularly interested in such matters. It is only when subjective phenomena are so prominent as to interfere with the perception of things, that they attract everybody's attention. Once the phenomena have been discovered, it is generally easier for others to perceive them also, provided the proper precautions are taken for observing them, and the attention is concentrated on them. In many cases, however—for example, in the phenomena of the blind spot, or in the separation of the overtones and combination tones from the fundamental tones of musical sounds, etc.—such an intense concentration of attention is required that, even with the help of convenient external appliances, many persons are unable to perform the experiments. Even the after-images of bright objects are not perceived by most persons at first except under particularly favourable external conditions. It takes much more practice to see the fainter kinds of after-images. A common experience, illustrative of this sort of thing, is for a person who has some ocular trouble that impairs his vision to become suddenly aware of the so-called *mouches volantes* in his visual field, although the causes of this phenomenon have been there in the vitreous humor all his life. Yet now he will be firmly persuaded that these corpuscles have developed as the result of his ocular ailment, although the truth simply is that, owing to his ailment, the patient has been paying more attention to visual phenomena. No doubt, also, there are cases where one eye has gradually become blind, and yet the patient has continued to go about for an indefinite time without noticing it, until he happened one day to close the good eye without closing the other, and so noticed the blindness of that eye.

When a person's attention is directed for the first time to the double images in binocular vision, he is usually greatly astonished to think that he had never noticed them before, especially when he reflects that the only objects he has ever seen single were those few that happened at the moment to be about as far from his eyes as the point of fixation. The great majority of objects, comprising all those that were farther or nearer than this point, were all seen double.

Accordingly, the first thing we have to learn is to pay heed to our individual sensations. Ordinarily we do so merely in case of those sensations that enable us to find out about the world around us. In the ordinary affairs of life the sensations have no other importance for us. Subjective sensations are of interest chiefly for scientific investigations only. If they happen to be noticed in the ordinary activity of the senses, they merely distract the attention. Thus while we may attain an extraordinary degree of delicacy and precision in objective observation, we not only fail to do so in subjective observations, but indeed we acquire the faculty in large measure of overlooking them and of forming our opinions of objects independently of them, even when they are so pronounced that they might easily be noticed.

. . . . . .

The same difficulty that we have in observing subjective sensations, that is, sensations aroused by internal causes, occurs also in trying to analyze the compound sensations, invariably excited in the same connection by any simple object, and to resolve them into their separate components. In such cases experience shows us how to recognize a compound aggregate of sensations as being the sign of a simple object. Accustomed to consider the sensation-complex as a connected whole, generally we are not able to perceive the separate parts of it without external help and support . . . For instance the perception of the apparent direction of an object from the eye depends on the combination of those sensations by which we estimate the adjustment of the eye, and on being able to distinguish those parts of the retina where light falls from those parts where it does not fall. The perception of the solid form of an object of three dimensions is the result of the combination of two different perspective views in the two eyes. The gloss of a surface, which is apparently a simple effect, is due to differences of colouring or brightness in the images of it in the two eyes. These facts were ascertained by theory and may be verified by suitable experiments. But usually it is very difficult, if not impossible, to discover them by direct observation and analysis of the sensations alone. Even with sensations that are much more involved and always associated with frequently recurring complex objects, the oftener the same combination recurs, and the more used we have become to regarding the sensation as the normal sign of the real nature of the object, the more difficult it will be to analyze the sensation by observation alone. By way of illustration, it is a familiar experience that the colours of a landscape come out much more brilliantly and definitely by looking at them with the head on one side or upside down than they do when the head is in the ordinary upright position. In the usual mode of observation all we try to do is to judge correctly the objects as such. We know that at a certain distance green surfaces appear a little different in hue. We get in the habit of overlooking this difference, and learn to identify the altered green of distant meadows and trees with the corresponding colour of nearer objects. In the case of very distant objects like distant ranges of mountains, little of the colour of the body is left to be seen, because it is mainly shrouded in the colour of the illuminated air. This vague blue-grey colour, bordered above by the clear blue of the sky or the red-yellow of the sunset glow, and below by the vivid green of meadows and forests, is very subject to variations by contrast. To us it is the vague and variable colour of distance. The difference in it may, perhaps, be more noticeable sometimes and with some illuminations than at other times. But we do not determine its true nature, because it is not ascribed to any definite object. We are simply aware of its variable nature. But the instant we take an unusual position, and look at the landscape with the head under one arm, let us say, or between the legs, it all appears like a flat picture; partly on account of the strange position of the image in the eye, and partly because, as we shall see presently, the binocular judgment of distance becomes less accurate. It may even happen that with the head upside down the clouds have the correct perspective, whereas the objects on the earth appear like a painting on a vertical surface, as the clouds in the sky usually do. At the same time the colours lose their associations also with near or far objects, and confront us now purely in their own peculiar differences. Then we have no difficulty in recognizing that the vague blue-grey of the far distance may indeed be a fairly saturated violet, and that the green of the vegetation blends imperceptibly through blue-green and blue into this violet, etc. This whole difference seems to me to be due to the fact that the colours have ceased to be distinctive signs of objects for us, and are considered merely as being different sensations. Consequently, we take in better their peculiar distinctions without being distracted by other considerations.

The connection between the sensations and external objects may interfere very much with the perception of their simplest relations. A good illustration of this is the difficulty about perceiving the double images of binocular vision when they can be regarded as being images of one and the same external object.

.   .   .   .   .

It is likewise true with respect to the perception of space-relations. For example, the spectacle of a person in the act of walking is a familiar sight. We think of this motion as a connected whole, possibly taking note of some of its most conspicuous singularities. But it requires minute attention and a special choice of the point of view to distinguish the upward and lateral movements of the body in a person's gait. We have to pick out points or lines of reference in the background with which we can compare the position of his head. But look through an astronomical telescope at a crowd of people in motion far away. Their images are upside down, but what a curious jerking and swaying of the body is produced by those who are walking about! Then there is no trouble whatever in noticing the peculiar motions of the body and many other singularities of gait; and especially differences between individuals and the reasons for them, simply because this is not the everyday sight to which we are accustomed. On the other hand, when the image is inverted in this way, it is not so easy to tell whether the gait is light or awkward, dignified or graceful, as it was when the image was erect.

Consequently, it may often be rather hard to say how much of our apperceptions (*Anschauungen*) as derived by the sense of sight is due directly to sensation, and how much of them, on the other hand, is due to experience and training. The main point of controversy between various investigators in this territory is connected also with this difficulty. Some are disposed to concede to the influence of experience as much scope as possible, and to derive from it especially all notion of space. This view may be called the *empirical theory* (*empiristische Theorie*). Others, of course, are obliged to admit the influence of experience in the case of certain classes of perceptions; still with respect to certain elementary apperceptions that occur uniformly in the case of all observers, they believe it is necessary to assume a system of innate apperceptions that are not based on experience, especially with respect to space-relations. In contradistinction to the former view, this may perhaps be called the *intuition theory* (*nativistische Theorie*) of the sense-perceptions.

In my opinion the following fundamental principles should be kept in mind in this discussion.

Let us restrict the word *idea* (*Vorstellung*) to mean the image of visual objects as retained in the memory, without being accompanied by any present sense-impressions; and use the term *apperception* (*Anschauung*) to mean a perception (*Wahrnehmung*) when it is accompanied by the sense-impressions in question. The term *immediate perception* (*Perzeption*) may then be employed to denote an apperception of this nature in which there is no element whatever that is not the result of direct sensations, that is, an apperception such as might be derived without any recollection of previous experience. Obviously, therefore, one and the same apperception may be accompanied by the corresponding sensations in very different measure. Thus idea and immediate perception may be combined in the apperception in the most different proportions.

A person in a familiar room which is brightly lighted by the sun gets an apperception that is abundantly accompanied by very vivid sensations. In the same room in the evening twilight he will not be able to recognize any objects except the brighter ones, especially the windows. But whatever he does actually recognize will be so intermingled with his recollections of the furniture that he can still move about in the room with safety and locate articles he is trying to find, even when they are only dimly visible. These images would be utterly insufficient to enable him to recognize the objects without some previous acquaintance with them. Finally, he may be in the same room in complete darkness, and still be able to find his way about in it without making mistakes, by virtue of the visual impressions formerly obtained. Thus, by continually reducing the material that appeals to the senses, the perceptual-image (*Anschauungsbild*) can ultimately be traced back to the pure memory-image (*Vorstellungsbild*) and may gradually pass into it. In proportion as there is less and less material appeal to the senses, a person's movements will, of course, become more and more uncertain, and his apperception less and less accurate. Still there will be no peculiar abrupt transition, but sensation and memory will continually supplement each other, only in varying degrees.

But even when we look around a room of this sort flooded with sunshine, a little reflection shows us that under these conditions too a large part of our perceptual-image may be due to factors of memory and experience. The fact that we are accustomed to the perspective distortions of pictures of parallelopipeds and to the form of the shadows they cast has much to do with the estimation of the shape and dimensions of the room, as will be seen hereafter. Looking at the room with one eye shut, we think we see it just as distinctly and definitely as with both eyes. And yet we should get exactly the same view in case every point in the room were shifted arbitrarily to a different distance from the eye, provided they all remained on the same lines of sight.

Thus in a case like this we are really considering an extremely multiplex phenomenon of sense; but still we ascribe a perfectly definite explanation to it, and it is by no means easy to realize that the monocular image of such a familiar object necessarily means a much more meagre perception than would be obtained with both eyes. Thus too it is often hard to tell whether or not untrained observers inspecting stereoscopic views really notice the peculiar illusion produced by the instrument.

We see, therefore, how in a case of this kind reminiscences of previous experiences act in conjunction with present sensations to produce a perceptual image (*Anschauungsbild*) which imposes itself on our faculty of perception with overwhelming power, without our being conscious of how much of it is due to memory and how much to present perception.

Still more remarkable is the influence of the comprehension of the sensations in certain cases, especially with dim illumination, in which a visual impression may be misunderstood at first, by not knowing how to attribute the correct depth-dimensions; as when a distant light, for example, is taken for a near one, or *vice versa*. Suddenly it dawns on us what it is, and immediately, under the influence of the correct comprehension, the correct perceptual image also is developed in its full intensity. Then we are unable to revert to the previous imperfect apperception.

· · · · ·

Hence, at all events it must be conceded that, even in what appears to the adult as being direct apperception of the senses, possibly a number of single factors may be involved which are really the product of experience, although at the time it is difficult to draw the line between them.

Now in my opinion we are justified by our previous experiences in stating that no indubitable present sensation can be abolished and overcome by an act of the intellect; and no matter how clearly we recognize that it has been produced in some anomalous way, still the illusion does not disappear by comprehending the process. The attention may be diverted from sensations, particularly if they are feeble and habitual; but in noting those relations in the external world, that are associated with these sensations, we are obliged to observe the sensations themselves. Thus we may be unmindful of the temperature-sensation of our skin when it is not very keen, or of the contact-sensations produced by our clothing, as long as we are occupied with entirely different matters. But just as soon as we stop to think whether it is warm or cold, we are not in the position to convert the feeling of warmth into that of coldness; maybe because we know that it is due to strenuous exertion and not to the temperature of the surrounding air. In the same way the apparition of light when pressure is exerted on the eyeball cannot be made to vanish simply by comprehending better the nature of the process, supposing the attention is directed to the field of vision and not, say, to the ear or the skin.

On the other hand, it may also be that we are not in the position to isolate an impression of sensation, because it involves the composite sense-symbol of an external object. However, in this case the correct comprehension of the object shows that the sensation in question has been perceived and used by the consciousness.

My conclusion is, that *nothing in our sense-perceptions can be recognized as sensation which can be overcome in the perceptual image and converted into its opposite by factors that are demonstrably due to experience.*

Whatever, therefore, can be overcome by factors of experience, we must consider as being itself the product of experience and training. By observing this rule, we shall find that it is merely the qualities of the sensation that are to be considered as real, pure sensation; the great majority of space-apperceptions, however, being the product of experience and training.

· · · · ·

Heretofore practically nothing has been ascertained as to the nature of psychic processes. We have simply an array of facts. Therefore, it is not strange that no real explanation can be given of the origin of sense-perceptions. The *empirical theory* attempts to prove that at least no other forces are necessary for their origin beyond the known faculties of the mind, although these forces themselves may remain entirely unexplained. Now generally it is a useful rule in scientific investigation not to make any new hypothesis so long as known facts seem adequate for the explanation, and the necessity of new assumptions has not been demonstrated. That is why I have thought it incumbent to prefer the empirical view essentially. Still less does the *intuition theory* attempt to give any explanation of the origin of our perceptual images; for it simply plunges right into the midst of the matter by assuming that certain perceptual images of space would be produced directly by an innate mechanism, provided certain nerve fibres were stimulated. The earlier forms of this theory implied some sort of self-observation of the retina; inasmuch as we were supposed to know by intuition about the form of this membrane and the positions of the separate nerve terminals in it. In its more recent development, especially as formulated by E. Hering, there is an hypothetical subjective visual space, wherein the sensations of the separate nerve fibres are supposed to be registered according to certain intuitive laws. Thus in this theory not only is Kant's assertion adopted, that the general apperception of space is an original form of our imagination, but certain special apperceptions of space are assumed to be intuitive.

The naturalistic view has been called also a special *theory of identity,* because in it the perfect fusion of the impressions on the corresponding places of the two retinas has to be postulated. On the other hand, the *empirical* theory is spoken of as a *theory of projection,* because according to it the perceptual images of objects are projected in space by means of psychic processes. I should like to avoid this term, because both supporters and opponents of this view have often attached undue importance to the idea that this projection must take place parallel to the lines of direction; which was certainly not the correct description of the psychic process. And, even if this construction were admitted as being valid simply with respect to the physiological description of the process, the idea would be incorrect in very many instances.

I am aware that in the present state of knowledge it is impossible to refute the intuition theory. The reasons why I prefer the opposite view are because in my opinion:

1. The intuition theory is an unnecessary hypothesis.

2. Its consequences thus far invariably apply to perceptual images of space which only in the fewest cases are in accordance with reality and with the correct visual images that are undoubtedly present; as will be shown in detail later. The adherents of this theory are, therefore, obliged to make the very questionable assumption, that the *space sensations,* which according to them are present originally, are continually being improved and overruled by knowledge which we have accumulated by experience. By analogy with all other experiences, however, we should have to expect that the sensations which have been overruled continued to be present in the apperception as a conscious illusion, if nothing else. But this is not the case.

3. It is not clear how the assumption of these original "*space sensations*" can help the explanation of our visual perceptions, when the adherents of this theory ultimately have to assume in by far the great majority of cases that these sensations must be overruled by the better understanding which we get by experience. In that case it would seem to me much easier and simpler to grasp, that all apperceptions of space were obtained simply by experience, instead of supposing that the latter

have to contend against intuitive perceptual images that are generally false.

This is by way of justifying my point of view. A choice had to be made simply for the sake of getting at least some sort of superficial order amid the chaos of phenomena; and so I believed I had to adopt the view I have chosen. However, I trust it has not affected the correct observation and description of the facts.

35   MAX WERTHEIMER (1880–1943) ON THE PHI PHENOMENON AS AN EXAMPLE OF NATIVISM IN PERCEPTION, 1912

Max Wertheimer, "Experimentelle Studien über das Sehen von Bewegung," *Zeitschrift für Psychologie 61*, 162–163, 221–227 (1912). Translated for this book by Don Cantor.

Hering in his defense of nativism and Helmholtz in his argument for empiricism represent in psychological thinking two poles of thought that are found in many different contexts. Helmholtz believed in rigor, in causes, and in precise thinking, and was guided by mathematics whenever possible. The empiristic theory of space is a genetic theory and is thus a causal theory. It was natural for Helmholtz to accept Lotze in these matters and so to lead on to Wundt's rigorous and rigid elementism and to other associationists, like Titchener, who followed Wundt. If Helmholtz were alive today he would be a positivist. Hering, on the other hand, believed in description and observation and the consequent insights into relations. He was a phenomenologist in the tradition of Goethe, and he stimulated the later nativists like Stumpf and still later the Gestalt psychologists, who resisted Helmholtzian thinking and believed in Hering's approach to psychology's problems. Thus today Gestalt psychologists denigrate modern positivism, whereas Wertheimer's initiation of Gestalt theory was sometimes known as experimental phenomenology. For this reason it is natural for us in concluding this chapter to pass to Wert-heimer, the first Gestalt psychologist, who never called himself a nativist, but who carried on the tradition of Hering, exhibiting nativism in its nineteenth-century form.

Wertheimer was off and on at Prague, where Hering had been, although there seems to have been no close personal influence of the older man on the younger. Phenomenology was nevertheless in the air, the phenomenology of Husserl; and Külpe knew of Husserl along about the time Wertheimer was taking his degree with Külpe. Stumpf at Berlin was adopting Husserl's kind of phenomenology and Wertheimer was for a time at Berlin. Wertheimer's new idea that movement is perceived merely as itself, even when the displacement of its stimulus is discrete and not continuous, is the beginning of his phenomenology and a modern equivalent of nativism, which is content with observation and does not look for causes. In this excerpt he describes movement as immediately seen, movement that has no color or pattern, that is not the movement of an object, but movement that does, nevertheless, have both direction and location: it is given and observers insist upon its actuality, and that is all there ever is to say about a Phi phenomenon.

One sees a movement: an object moves from one position to another. One describes the physical circumstances: until the point of time t_1, the object was in position p_1 (in the region r_1); at the moment t_n, it was in position p_n (in the region r_n); in the time between t_1 and t_n, the object was in the positions between p_1 and p_n successively and continuously in time and space until it reached p_n.

One sees this movement. It is not that one merely sees that the object was earlier somewhere else and knows therefore that it has moved (as is the case with the slow clock hand), but one [actually] sees the movement. What is physically presented here?

One might almost say, in simple analogy to the physical state, that the seeing of movement depends on the movement of the thing seen—the physical visualized object—from the seen position p_1, through the intermediate positions in space, continuously until it reaches p_n, and that the fact that we are given such a series of positions allows us to see movement.

If this seeing of movement were attained as an "illusion" (that is, if there were physically in truth merely a resting position and afterward a second resting position clearly separated from the first one), then, in addition to the two perceptions of resting objects and because of them, a subjective completion would somehow be [thought to be] present: the going through, the taking in of the in-between positions would somehow be filled up subjectively.

The following investigation will deal with impressions of movement attained by presenting two such successive positions, even where there is considerable distance in space between them.

.   .   .   .   .

What is physically given in the field of movement? The hypothesis cited [above] says that the positions taken by the object between p_1 and p_n are filled in subjectively. (One could also bring in the a priori proposition that movement is not conceivable unless an object, article, seen thing moves.)

If optimal movement occurred only when an object clearly moved or rotated from its original position through the field to its final position, this statement would be at once feasible.

It can, however, be shown that the essential character of straight movement or rotation has nothing to do with subjective intermediate positions. There are cases where Phi, the straight movement or the rotation, is clearly present without the [stimulus] strip's appearing in any way in the field of movement; the original and final positions were there as well as the movement between them, but there was no optical filling in, seeing, or conceiving of the intermediate positions of the strip. This [result] occurred spontaneously for all observers. In the experimental arrangements discussed below, this "pure" Phi without the filling in of intermediate points was easily demonstrated.

In impressions of uniform complete movement, careful observation of what was presented in the field of movement often revealed the following: if, for example, in an experiment with an angular or a straight movement, *a* and *b* were white strips, the field was black, and uniform movement from *a* to *b* occurred, then this movement was seen, although the strips in the field of movement certainly did not in any sense move through the intermediate positions and, [moreover], the color of the moving strips was missing except in positions *a* and *b* themselves and at the start and finish of the movement.

For example, [the following comment was made] when the angular pattern was being used with *a* horizontal, *b* vertical, and red strips on a black background: "Very clear, uniform rotation, easy to describe physically. One sees the horizontal strip turn, moving a bit upwards. It moves a bit in its final vertical position, but the entire thing is a unity. One sees clearly not a broken movement, but a complete rotation from *a* to *b*. For the rest, it is to be said that there is no trace of strips or of red in the middle."

Here is a similar example for white strips: "It is remarkable that I do not see the white strip anywhere during the movement. Wait! At the end of the movement, that is, at about 15°, when the white is already there, there is white in the closing portion of the movement, but before that it is not there. I never see the white strip at all in, let's say, an area of 45°."

And again, with *a* vertical: "There is a kind of clear, forceful, certain movement. There is a rotation of about 90°. It cannot be thought of as a succession. The white vertical [strip] does not move, but there is some sort of process or going across. One sees the horizontal line 'lay itself down.' Earlier positions of the strip, as white, in an area of 45°, say, are certainly not there—nothing of the sort. Although nothing white rotates (that is, the object itself does not turn), it is [nevertheless] clear that the movement is there and that the end of the movement is set off by the 'laying down' of the [final] horizontal line."

And so it was, quite spontaneously, in many cases . . .

There were also cases in which the thought that an object had moved across was absent. Such objects as were there were present in two positions. Neither one nor the other of them, nor a similar one, was concerned in the movement; instead, there was movement between them, and this was not the movement of an object. The subject did not think that the object had moved across without his having seen it. Rather, movement was simply there, without being related to an object.

What was it that was physically there, in this field where, except for the bare background, nothing of the remaining optical qualities were to be seen? [Can it be that] no image of the moving strip on the intermediate places was formed, and it was not thought that the strip went across?

When the [subject's] attention was concentrated there, the impression was still there, even stronger, that there was nothing else of the optical —in no sense any trace of the strip's passing through intermediate points; but at the same time "a strong, uniform movement here in this field, a specific, forceful 'going across' or turning."

I gave this arrangement as a right triangle under optimal conditions for identical rotation, but adding a short strip, *c,* which had the same color, to the exposure field, *B.* (See Figure XXa.) The short piece did

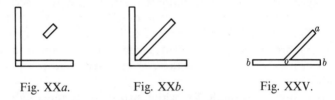

Fig. XX*a*.   Fig. XX*b*.   Fig. XXV.

not reach the vertex, and was placed in the area of 45°—in an area, then, over which the strip to be added in the intermediate stages would have to pass. If, for example, *a* and *b* were white strips $\frac{1}{2}$ or 1 cm by 6 or 8 cm, the *c* was just as wide but shorter—1.2 cm. ([It was], so to speak, [like] a part of the form of the intermediate space.) Attention was focused on *c*, its inner end, or on the space between *c* and the vertex. The movement *ab,* with rotation of 90°, remained optimal. Was the white strip *c* supplemented in any way? Does *c* seem for a moment to be somehow lengthened by being passed over by the filled-in, interpolated moving strip, or does any faint glimmer of light hurry over the place between *c* and the vertex? Here follow characteristic [remarks that were made] again and again in repeated observations: "Clear forceful movement of 90° was there. The specific 'going across' was clearly to be seen. Nothing white hurried over the place between *c* and the vertex. The background there remained quietly black." [So] at no moment was anything added, but the "going across" was there in the [intermediate] space—not as a going across of the strip, but simply as a "going across," a [seen] "turning."

.   .   .   .   .

In this detachment of the phenomenon from the visual objects *a* and *b*, there also occurred cases in which two pure Phi movements or turnings appeared from a single *a*, but in no sense did it seem as though *a* split itself in two directions. With the arrangement shown in Figure XXV, with a smaller vertical [strip], *a*, in the middle of the longer horizontal [one], *b*, the phenomenon of rotation was often seen clearly and simultaneously both to the right and to the left, as long as no one direction was favored . . . It never seemed as though the vertical [strip] turned itself or as though there were two lines [going] toward opposite sides; rather, the vertical and horizontal strips were seen along with the two rotations . . .

Several times this Phi phenomenon, this "movement across," this "turning," was so strong in tachistoscopic experiments, especially where there was something new in the arrangements (like a reversal of *ab* to *ba*), that the observer could report nothing about the objects themselves . . . [He would say:] "It cannot be said what objects were there. I saw [only] a strong movement." And then he would indicate the correct direction. "I know nothing about objects and know nothing about having seen objects!" There was a similar case . . . with the arrangements of Figure XXb. Here the subject did not know the procedure, or that *b* was exposed as a right triangle and *a* as a strip at 45° to the angle of *b*. He reported: "A movement was there. At the end there stood a right angle. There was in the lower part of the right angle a turning movement around the vertex and downward toward the horizontal. I don't know what it was that moved; the horizontal [line] lay quiet, as did the vertical [one], and it did not seem that the horizontal [one] had turned in its position."

·   ·   ·   ·   ·

These experiments are more than a theoretical means of proving that Phi occurs [in perception] without the addition of intermediate positions for the object. The essence of the Phi process is not touched on by [a discussion of] the absence of this addition. Rather, we are dealing here with an experimental demonstration in the exact sense, one in which the pure Phi process appears.

Except for the color of the background, none of the usual optical qualities of the field of movement is given. There is nothing about color or contour [to Phi]. In the usual optical sense, nothing changes on the surface of the intermediate field, the background. The observer does not say here that the strip moves across, nor does he believe that it moves across, *a* toward *b*, or that it seems to move across. Rather, he says, "I see *a*; I see *b*; I see movement between the two; I see the going across, the turning—not a turning of the strip and not the strip itself, for both are in their places, *a* and *b*. I see a strong or a weak 'moving across,' just that." "I see movement, thus [he indicates it]; not a moving across by something." And this [experience] occurs with the most intense concentration of attention on the field and with [the most] critical observation.

One might think that where there is no addition, no thought of an object's moving across, and no intermediate points of the object, the "illusion" of movement would disappear. To the contrary, the movement is there, forceful and characteristic in its own way. It is clearly present and always observable.

VI OBJECTIVE REFERENCE

The fundamental problem of psychology has been the one of empiricism: how does man come to know about the external world in which he lives in order that he may act in accordance with it and adjust to it? The problem of knowing came first with the British empiricists and the question of adjustment later with functional psychology and the behaviorists. Locke (1690), getting empiricism off to a good start, said that knowledge of objects comes about with experience, which writes, as if on a piece of white paper, on the impressionable and originally blank mind. An object is known as it is imprinted on the mind in the idea that represents it. Berkeley (1709) inverted this question, assuming that conscious experience is all that there is initially and that the problem lies, therefore, in finding out how the mind creates objects out of experience. It does so by association, as the excerpt from Berkeley explains. With Thomas Reid (1785) the issue became more explicit. He distinguished between sensation—the raw data of experience—and perception, which, based on sensation, also refers to an object. Reid held that the belief in the existence of external objects is divinely implanted in man and is inexorable, but Thomas Brown (1820), complementing Reid, explained the phenomenon as being association or suggestion, in manner resembling Berkeley a century earlier and Titchener a century later. John Stuart Mill (1865) dealt with the same problem by arguing that the mind becomes aware of the corporeal causes of its sensations as the Permanent Possibilities of Sensation. Helmholtz (1866) discussed the general problem of perception in the setting of the new experimental physiological psychology, introducing the conception of unconscious inference to explain how sensory experience provides the knower with information about stimulus objects without the grounds of the implication necessarily being clear to him—as when one perceives fierce anger in a human face. The perception of the anger is immediate and irresistible, whereas the inferential process by which the visual pattern of a facial expression has come to imply anger remains inscrutable. It was Titchener (1910) who extended the associative theory of Berkeley and Brown to include the unconscious objective references of familiar perceptions, thus also extending Helmholtz's view and anticipating the behavioral theory of cognition propounded by Holt (1915). In fact, this problem presently moved on to more adequate treatment by Tolman and other later behaviorists. On the other hand, the Gestalt psychologists felt no need to explain the objective reference inherent in perception, for they held, as did Wertheimer (1923), that objects are given as such, whole and immediately in experience, and that they do not come into existence by associative synthesis. That is a nativistic theory of objective reference, or, if not a theory, it is a fundamental nativistic fact, at least for those who see experience as being constituted of Gestalten.

36 GEORGE BERKELEY (1685–1753) ON THE ROLE OF ASSOCIATION IN THE OBJECTIVE REFERENCE OF PERCEPTION, 1709

George Berkeley, *An Essay towards a New Theory of Vision* (Dublin, 1709), sects. 45–47, 50, 51. Compare sects. 9, 10, and 16–18 in Chapter IV, No. 28.

In giving this account of how perception comes by association to mean the existence of particular external objects, Bishop Berkeley is really presenting a conscious context theory of meaning, not unlike Thomas Brown's (1820) and the first part of Titchener's (1910). It was Titchener who applied the term *context theory* to this use of the associative principle.

XLV. . . . Having of a long time, experienced certain *Ideas,* perceivable by Touch, as Distance, Tangible Figure, and Solidity to have been connected with certain *Ideas* of Sight, I do upon perceiving these *Ideas* of Sight, forthwith conclude what Tangible *Ideas* are, by the wonted, ordinary course of Nature like to follow. Looking at an *Object* I perceive a certain Visible Figure, and Colour with some degree of Faintness and other Circumstances; which, from what I have formerly observ'd, determin me to think, that if I advance forward so many Paces, Miles, &c. I shall be affected with such, and such *Ideas* of Touch. So that in truth, and strictness of Speech, I neither see Distance it self, nor any thing that I take to be at a Distance. I say, neither Distance, nor things placed at a Distance are themselves, or their *Ideas,* truly perceiv'd by Sight. This I am perswaded of, as to what concerns my self. And I believe whoever will look narrowly into his own Thoughts, and examin what he means by saying, he sees this, or that thing at a Distance, will agree with me that, what he sees only suggests to his Understanding, that after having passed a certain Distance, to be measur'd by the Motion of his Body, which is perceivable by Touch, he shall come to perceive such, and such Tangible *Ideas* which have been usually connected with such and such Visible *Ideas.* But that one might be deceived by these suggestions of Sense, and that there is no necessary Connexion, between Visible, and Tangible *Ideas* suggested by them, we need go no farther than the next Looking-Glass or Picture, to be convinced. Note, that when I speak of Tangible *Ideas,* I take the Word *Idea* for any the immediate *Object* of Sense, or Understanding, in which large Signification it is commonly used by the Moderns.

XLVI. From what we have shewn it is a manifest Consequence, that the *Ideas* of Space, Outness, and things placed at a Distance are not, strictly speaking, the *Object* of Sight. They are no otherwise perceived by the Eye, than by the Ear. Sitting in my Study I hear a Coach drive along the Streets. I look through the Casement and see it. I walk out and enter into it. Thus, common Speech wou'd incline one to think, I heard, saw, and touch'd the same Thing, *viz.* the Coach. It is, nevertheless, certain, the *Ideas* intromitted by each Sense are widely different, and distinct from each other; but having been observed constantly to go together, they are spoken of as one and the same thing. By the variation of the Noise, I perceive the different Distances of the Coach, and know that it approaches before I look out. Thus by the Ear I perceive Distance, just after the same manner, as I do by the Eye.

XLVII. I do not, nevertheless, say I hear Distance, in like manner as I say that I see it, the *Ideas* perceiv'd by Hearing not being so apt to be confounded with the *Ideas* of Touch, as those of Sight are. So likewise, a Man is easily convinced that Bodies, and external Things are not properly the *Object* of Hearing, but only Sounds, by the Mediation whereof the *Idea* of this or that Body, or Distance is suggested to his Thoughts. But then one is with more difficulty, brought to discern the difference there is, betwixt the *Ideas* of Sight and Touch: Tho' it be certain, a Man no more Sees and Feels the same Thing, than he Hears and Feels the same Thing.

·　　·　　·　　·　　·

L. In order therefore to treat accurately, and unconfusedly of *Vision,* we must bear in Mind, that there are two sorts of *Objects* apprehended by the Eye: The one, primarily and immediately, the other, secondarily and by intervention of the former. Those of the first sort neither are, nor appear to be without the Mind, or at any distance off. They may, indeed, grow Greater, or Smaller, more Confused, or more Clear, or more Faint. But, they do not, cannot Approach, or even seem to Approach, or Recede from us. Whenever we say an *Object* is at a Distance, whenever we say, it draws near, or goes farther off; we must always mean it of the latter sort, which properly belong to the *Touch,* and are not so truly perceived, as suggested, by the Eye, in like manner as Thoughts by the Ear.

LI. No sooner do we hear the Words of a familiar Language pronounced in our Ears, but the *Ideas* corresponding thereto present themselves to our Minds. In the very same instant, the Sound and the Meaning enter the Understanding. So closely are they United, that 'tis not in our Power to keep out the one, except we exclude the other also. We even act in all respects, as tho' we heard the very Thoughts themselves. So likewise, the Secondary *Objects,* or those which are only suggested by Sight, do often more strongly affect us, and are more regarded than the proper *Objects* of that Sense; along with which they enter into the Mind, and with which they have a far more strict and near Connexion, than

Ideas have with Words. Hence it is, we find it so difficult to discriminate, between the immediate and mediate *Objects* of Sight, and are so prone to attribute to the former, what belongs only to the latter. They are, as it were, most closely twisted, blended, and incorporated together. And the Prejudice is confirm'd, and riveted in our Thoughts, by a long tract of Time, by the use of Language, and want of Reflexion. However, I doubt not, but any one that shall attentively consider what we have already said, and shall say upon this Subject before we have done, (especially if he pursue it in his own Thoughts) may be able to deliver himself from that Prejudice. Sure I am, 'tis worth some Attention, to whoever wou'd understand the true nature of *Vision*.

37   THOMAS REID (1710–1796) ON THE DISTINCTION BETWEEN SENSATION AND PERCEPTION, 1785

Thomas Reid, *Essays on the Intellectual Powers of Man* (Edinburgh, 1785), essay 2, chaps. 5, 16.

The Scottish school of psychological philosophy comes after Locke, Berkeley, and Hume and before the two Mills and Spencer in the evolution of British empiricism and associationism. Thomas Reid was the originator of the Scottish school and the creator of the distinction between sensation, as the raw stuff of experience, and perception, as sensation that carries with itself an inevitable reference to an external object. This is a distinction that persisted for over 150 years, although later, after Wundt, it became confused with the difference between the qualitative and intensive attributes (sensation) and the spatial and temporal ones (perception). Faced with the inescapable fact of objective reference in perception, Reid, like many others, fell back on nativism: reference to the object is a fact of man's divinely given nature.

CHAPTER V

In speaking of the impressions made on our organs in perception, we build upon facts borrowed from anatomy and physiology, for which we have the testimony of our senses. But, being now to speak of perception itself, which is solely an act of the mind, we must appeal to another authority. The operations of our minds are known, not by sense, but by consciousness, the authority of which is as certain and as irresistible as that of sense.

In order, however, to our having a distinct notion of any of the operations of our own minds, it is not enough that we be conscious of them; for all men have this consciousness. It is farther necessary that we attend to them while they are exerted, and reflect upon them with care, while they are recent and fresh in our memory. It is necessary that, by employing ourselves frequently in this way, we get the habit of this attention and reflection; and, therefore, for the proof of facts which I shall have occasion to mention upon this subject, I can only appeal to the reader's own thoughts, whether such facts are not agreeable to what he is conscious of in his own mind.

If, therefore, we attend to that act of our mind which we call the perception of an external object of sense, we shall find in it these three things:—*First,* Some conception or notion of the object perceived; *Secondly,* A strong and irresistible conviction and belief of its present existence; and, *Thirdly,* That this conviction and belief are immediate, and not the effect of reasoning.

First, It is impossible to perceive an object without having some notion or conception of that which we perceive. We may, indeed, conceive an object which we do not perceive; but, when we perceive the object, we must have some conception of it at the same time; and we have commonly a more clear and steady notion of the object while we perceive it, than we have from memory or imagination when it is not perceived. Yet, even in perception, the notion which our senses give of the object may be more or less clear, more or less distinct, in all possible degrees.

Thus we see more distinctly an object at a small than at a great distance. An object at a great distance is seen more distinctly in a clear than in a foggy day. An object seen indistinctly with the naked eye, on account of its smallness, may be seen distinctly with a microscope. The objects in this room will be seen by a person in the room less and less distinctly as the light of the day fails; they pass through all the various degrees of distinctness according to the degrees of the light, and, at last, in total darkness they are not seen at all. What has been said of the objects of sight is so easily applied to the objects of the other senses, that the application may be left to the reader.

In a matter so obvious to every person capable of reflection, it is necessary only farther to observe, that the notion which we get of an object, merely by our external sense, ought not to be confounded with that more scientific notion which a man, come to the years of understanding, may have of the same object, by attending to its various attributes, or to its various parts, and their relation to each other, and to the whole. Thus, the notion which a child has of a jack for roasting meat, will be acknowledged to be very different from that of a man who understands its construction, and perceives the relation of the parts to one another, and to the whole. The child sees the jack and every part of it as well as the man. The child, therefore, has all the notion of it which sight gives; whatever there is more in the notion which the man forms of it, must be derived from other powers of the mind, which may afterwards be explained. This observation is made here only that we may not confound the operations of different powers of the mind, which by being always conjoined after we grow up to understanding, are apt to pass for one and the same.

Secondly, In perception we not only have a notion more or less distinct of the object perceived, but also an irresistible conviction and belief of its existence. This is always the case when we are certain that we perceive it. There may be a perception so faint and indistinct as to leave us in doubt whether we perceive the object or not. Thus, when a star begins to twinkle as the light of the sun withdraws, one may, for a short time, think he sees it without being certain, until the perception acquire some strength and steadiness. When a ship just begins to appear in the utmost verge of the horizon, we may at first be dubious whether we perceive it or not; but when the perception is in any degree clear and steady, there remains no doubt of its reality; and when the reality of the perception is ascertained, the existence of the object perceived can no longer be doubted.

By the laws of all nations, in the most solemn judicial trials, wherein men's fortunes and lives are at stake, the sentence passes according to the testimony of eye or ear witnesses of good credit. An upright judge will give a fair hearing to every objection that can be made to the integrity of a witness, and allow it to be possible that he may be corrupted; but no judge will ever suppose that witnesses may be imposed upon by trusting to their eyes and ears. And if a sceptical counsel should plead against the testimony of the witnesses, that they had no other evidence for what they declared but the testimony of their eyes and ears, and that we ought not to put so much faith in our senses as to deprive men of life or fortune upon their testimony, surely no upright judge would admit a plea of this kind. I believe no counsel, however sceptical, ever dared to offer such an argument; and, if it was offered, it would be rejected with disdain.

Can any stronger proof be given that it is the universal judgment of mankind that the evidence of sense is a kind of evidence which we may securely rest upon in the most momentous concerns of mankind; that it is a kind of evidence against which we ought not to admit any reasoning; and, therefore that to reason either for or against it is an insult to common sense?

The whole conduct of mankind in the daily occurrences of life, as well as the solemn procedure of judicatories in the trial of causes civil and criminal, demonstrates this. I know only of two exceptions that may be offered against this being the universal belief of mankind.

· · · · ·

It appears, therefore, that the clear and distinct testimony of our senses carries irresistible conviction along with it to every man in his right judgment.

I observed, *Thirdly,* That this conviction is not only irresistible, but it is immediate; that is, it is not by a train of reasoning and argumentation that we come to be convinced of the existence of what we perceive; we ask no argument for the existence of the object, but that we perceive it; perception commands our belief upon its own authority, and disdains to rest its authority upon any reasoning whatsoever.

The conviction of a truth may be irresistible, and yet not immediate. Thus, my conviction that the three angles of every plain triangle are equal to two right angles, is irresistible, but it is not immediate; I am convinced of it by demonstrative reasoning. There are other truths in mathematics of which we have not only an irresistible but an immediate conviction. Such are the axioms. Our belief of the axioms in mathematics is not grounded upon argument—arguments are grounded upon them; but their evidence is discerned immediately by the human understanding.

It is, no doubt, one thing to have an immediate conviction of a self-evident axiom; it is another thing to have an immediate conviction of the existence of what we see; but the conviction is equally immediate and equally irresistible in both cases. No man thinks of seeking a reason to believe what he sees; and, before we are capable of reasoning, we put no less confidence in our senses than after. The rudest savage is as fully convinced of what he sees, and hears, and feels, as the most expert logician. The constitution of our understanding determines us to hold the truth of a mathematical axiom as a first principle, from which other truths may be deduced, but it is deduced from none; and the constitution of our power of perception determines us to hold the existence of what we distinctly perceive as a first principle, from which other truths may be deduced; but it is deduced from none. What has been said of the irresistible and immediate belief of the existence of objects distinctly perceived, I mean only to affirm with regard to persons so far advanced in understanding as to distinguish objects of mere imagination from things which have a real existence. Every man knows that he may have a notion of Don Quixote, or of Garagantua, without any belief that such persons ever existed; and that of Julius Cæsar and Oliver Cromwell, he has not only a notion, but a belief that they did really exist. But whether children, from the time that they begin to use their senses, make a distinction between things which are only conceived or imagined, and things which really exist, may be doubted. Until we are able to make this distinction, we cannot properly be said to believe or to disbelieve the existence of anything. The belief of the existence of anything seems to suppose a notion of existence—a notion too abstract, perhaps, to enter into the mind of an infant. I speak of the power of perception in those that are adult and of a sound mind, who believe that there are some things which do really exist; and that there are many things conceived by themselves, and by others, which have no existence. That such persons do invariably ascribe existence to everything which they distinctly perceive, without seeking reasons or arguments for doing so, is perfectly evident from the whole tenor of human life.

The account I have given of our perception of external objects, is intended as a faithful delineation of what every man, come to years of understanding, and capable of giving attention to what passes in his own mind, may feel in himself. In what manner the notion of external objects, and the immediate belief of their existence, is produced by means of our senses, I am not able to shew, and I do not pretend to shew. If the power of perceiving external objects in certain circumstances, be a part of the original constitution of the human mind, all attempts to account for it will be vain. No other account can be given of the constitution of things, but the will of Him that made them. As we can give no reason why matter is extended and inert, why the mind thinks and is conscious of its thoughts, but the will of Him who made both; so I suspect we can give no other reason why, in certain circumstances, we perceive external objects, and in others do not.

The Supreme Being intended that we should have such knowledge of the material objects that surround us, as is necessary in order to our supplying the wants of nature, and avoiding the dangers to which we are constantly exposed; and he has admirably fitted our powers of perception to this purpose. If the intelligence we have of external objects were to be got by reasoning only, the greatest part of men would be destitute of it; for the greatest part of men hardly ever learn to reason; and in infancy and childhood no man can reason: Therefore, as this intelligence of the objects that surround us, and from which we may receive so much benefit or harm, is equally necessary to children and to men, to the ignorant and to the learned, God in his wisdom conveys it to us in a way that puts all upon a level. The information of the senses is as perfect, and gives as full conviction to the most ignorant as to the most learned.

CHAPTER XVI

Having finished what I intend, with regard to that act of mind which we call the preception of an external object, I proceed to consider another, which, by our constitution, is conjoined with perception, and not with perception only, but with many other acts of our minds; and that is sensation.

Almost all our perceptions have corresponding sensations which constantly accompany them, and, on that account, are very apt to be confounded with them. Neither ought we to expect that the sensation, and its corresponding perception, should be distinguished in common language, because the purposes of common life do not require it. Language is made to serve the purposes of ordinary conversation; and we have no reason to expect that it should make distinctions that are not of common use. Hence it happens, that a quality perceived, and the sensation corresponding to that perception, often go under the same name.

This makes the names of most of our sensations ambiguous, and this ambiguity hath very much perplexed philosophers. It will be necessary to give some instances, to illustrate the distinction between our sensations and the objects of perception.

When I smell a rose, there is in this operation both sensation and perception. The agreeable odour I feel, considered by itself, without relation to any external object, is merely a sensation. It affects the mind in a certain way; and this affection of the mind may be conceived, without a thought of the rose, or any other object. This sensation can be nothing else than it is felt to be. Its very essence consists in being felt; and, when it is not felt, it is not. There is no difference between the sensation and the feeling of it—they are one and the same thing. It is for this reason that we before observed that, in sensation, there is no object distinct from that act of the mind by which it is felt—and this holds true with regard to all sensations.

Let us next attend to the perception which we have in smelling a rose. Perception has always an external object; and the object of my perception, in this case, is that quality in the rose which I discern by the sense of smell. Observing that the agreeable sensation is raised when the rose is near, and ceases when it is removed, I am led, by my nature, to conclude some quality to be in the rose, which is the cause of this sensation. This quality in the rose is the object perceived; and that act of my mind by which I have the conviction and belief of this quality, is what in this case I call perception.

But it is here to be observed, that the sensation I feel, and the quality in the rose which I perceive, are both called by the same name. The smell of a rose is the name given to both: so that this name hath two meanings; and the distinguishing its different meanings removes all perplexity, and enables us to give clear and distinct answers to questions about which philosophers have held much dispute.

Thus, if it is asked, whether the smell be in the rose, or in the mind that feels it, the answer is obvious: That there are two different things signified by the smell of a rose; one of which is in the mind, and can be in nothing but in a sentient being; the other is truly and properly in the rose. The sensation which I feel is in my mind. The mind is the sentient being; and, as the rose is insentient, there can be no sensation, nor anything resembling sensation in it. But this sensation in my mind is occasioned by a certain quality in the rose, which is called by the same name with the sensation, not on account of any similitude, but because of their constant concomitancy.

All the names we have for smells, tastes, sounds, and for the various degrees of heat and cold, have a like ambiguity; and what has been said of the smell of a rose may be applied to them. They signify both a sensation, and a quality perceived by means of that sensation. The first is the sign, the last the thing signified. As both are conjoined by nature, and as the purposes of common life do not require them to be disjoined in our thoughts, they are both expressed by the same name: and this ambiguity is to be found in all languages, because the reason of it extends to all.

38   THOMAS BROWN (1778–1820) ON SENSATION, PERCEPTION, AND THE ASSOCIATIVE EXPLANATION OF OBJECTIVE REFERENCE, 1820

Thomas Brown, *Lectures on the Philosophy of the Human Mind* (Edinburgh, 1820), vol. II, lecture 25.

Thomas Brown, a not wholly docile disciple of the Scottish school, undertakes here to comment constructively on Reid's distinction between sensation and perception and to appeal to association, after the manner of Berkeley, to show how objective reference comes to be added to the sensory core of perception. In this passage he uses the concept of association quite freely to effect his purpose, although in general he preferred the term *suggestion* as more discreet in the Presbyterian atmosphere in which he wrote, for the word *association*, being identified with the skeptic Hume, was not looked upon favorably in the Edinburgh vicinage.

My last Lecture, Gentlemen, was chiefly employed in considering the nature of that complex process which takes place in the mind, when we ascribe the various classes of our sensations to their various external objects,—to the analysis of which process we were led, by the importance which Dr Reid has attached to the distinction of *sensation* and *perception;*—a sensation, as understood by him, being the simple feeling that immediately follows the action of an external body on any of our organs of sense, considered merely as a feeling of the mind; the corresponding perception being the reference of this feeling to the external body as its cause.

The distinction I allowed to be a convenient one, if the nature of the complex process which it expresses be rightly understood. The only question that seemed, philosophically, of importance, with respect to it, was, whether the *perception* in this sense,—the reference of the sensation to its external *corporeal* cause,—imply, as Dr Reid contends, a peculiar mental power, coextensive with sensation, to be distinguished by a peculiar name in the catalogue of our faculties, or be not merely one of the results of a more general power, which is afterwards to be considered by us,—the *power of association,*—by which one feeling suggests, or induces, other feelings that have formerly coexisted with it.

It would be needless to recapitulate the argument minutely, in its relation to *all* the senses. That of smell, which Dr Reid has himself chosen as an example, will be sufficient for our retrospect.

Certain particles of odorous matter act on my nostrils,—a peculiar *sensation* of fragrance arises,—I refer this sensation to a rose. This reference, which is unquestionably something superadded to the original sensation itself, is what Dr Reid terms the *perception* of the *fragrant body*. But what is the reference itself, and to what source is it to be ascribed? That we should have supposed our sensations to have had a cause of some sort, as we suppose a cause of all our feelings *internal* as well as external, may indeed be admitted. But if I had had *no other* sense than that of smell,—if *I had never seen a rose,*—or, rather, since the knowledge which vision affords is chiefly of a secondary kind, if I had *no mode of becoming acquainted* with the compound of *extension* and *resistance,* which the mere sensations of smell, it is evident, are incapable of affording,—could I have made this reference of my sensation to a quality of a fragrant *body?* Could I, in short, have had more than the mere sensation itself, with that general belief of a cause of *some sort,* which is not confined to our sensations, but is common to them with all our other feelings?

By *mere smell,* as it appears to me, I could not have become acquainted with the existence of *corporeal substances,*—in the sense in which we now understand the term *corporeal,*—nor, consequently, with the *qualities* of corporeal substances; and, if so, how could I have had that *perception* of which Dr Reid speaks,—that reference to a fragrant *body,* of which, as a body, I was before in absolute ignorance? I should, indeed, have ascribed the sensation to some cause or antecedent, like every other feeling; but I could as little have ascribed it to a bodily cause, as any feeling of joy or sorrow. I refer it now *to a rose;* because, being endowed with *other* sensitive capacities, I have previously learned, from another source, the existence of *causes without,* extended and resisting,—because I have previously touched or seen a rose, when the sensation of fragrance

179

coexisted with my visual or tactual sensation; and all which distinguishes the *perception* from the mere *sensation,* is this suggestion of former experience, which reminds me now of other feelings, with the continuance or cessation of which, in innumerable former instances, the fragrance itself also continued or ceased. The perception, in short, in smell, taste, hearing, is a *sensation suggesting, by association,* the notion of some extended and resisting substance, *fragrant, vapid, vibratory,*—a notion which smell alone, taste alone, hearing alone, never could have afforded; but which, when once received from any other source, may be suggested by these as readily as any other associate feeling that has frequently coexisted with them. To the simple primary sensations of *vision* the same remark may be applied. A mere sensation of colour could not have made me acquainted with the existence of bodies, that would resist my effort to grasp them. It is only in one sense, therefore,—that which affords us the knowledge of *resistance,*—that any thing like original *perception* can be found; and even in *this,* the process of perception, as I formerly explained to you, implies no *peculiar* power, but only common *sensations,* with *associations* and *inferences* of precisely the same kind, as those which are continually taking place in *all* our reasonings and trains of thought.

Extension and *resistance,* I need scarcely repeat, are the complex elements of what we term matter; and nothing is *matter* to our conception, or a *body,* to use the simpler synonymous term, which does not involve these elements. If we had no other sense than that of smell, and, therefore, could not have referred the sensations to any fragrant *body,* what, in Dr Reid's meaning of this term, would the supposed power of perception, in these circumstances, have been? What would it have been, in like manner, if we had had only the sense of taste in sweetness and bitterness,—or of hearing in melody,—or of vision in colour,—without the capacity of knowing light as a material substance, or the *bodies* that *vibrated,* or the *bodies* of another kind that were *sweet* or *bitter?* It is only by the sense of *touch,* or, at least, by that class of perceptions which Dr Reid ascribes to touch,—and which, therefore, though traced by us, in part, to another source, I, for brevity's sake, comprehend under that term in our present discussion,—it is only by *touch* that we become acquainted with those elements which are essential to our very notion of a *body;* and to *touch,* therefore, in his own view of it, we must be indebted, directly or indirectly, as often as we refer the sensations of any other class to a *corporeal* cause. Even in the supposed perceptions of touch itself, however, as we have seen, the reference of our feelings to an external cause is not demonstrative of any peculiar power of the mind, to be classed separately from its other faculties. But when a body is first grasped, in infancy, by fingers that have been accustomed to *contract* without being impeded, we learn to consider the sensation as the result of a cause that is *different from our own mind,* because it breaks an accustomed series of feelings, in which all the antecedents, felt by us at the time, were such as were before uniformly followed by a different consequent, and were expected, therefore, to have again their usual consequent. The *cause* of the new sensation, which is thus believed to be something different from our sentient self, is regarded by us as something which *has parts,* and which *resists our effort,* that is to say, as an external *body;*—because the muscular feeling, excited by the object grasped, is, in the *first* place, the very feeling of that which we term resistance; and, *secondly,* because, by uniformly supplying the place of a definite portion of a progressive series of feelings, it becomes ultimately representative of that particular length of series, or number of parts, of which it thus uniformly supplies the place. Perception, then, even in that class of feelings by which we learn to consider ourselves as surrounded by substances extended and resisting, is only another name, as I have said, for the result of certain *associations* and *inferences* that flow from other more general principles of the mind; and with respect to all our other sensations, it is only another name for the *suggestion* of these *very perceptions of touch,* or at least of the feelings, tactual and muscular, which are, by Dr Reid, ascribed to that single sense. If we had been unsusceptible of these tactual and muscular feelings, and, consequently, had never conceived the existence of any thing extended and resisting *till* the sensation of fragrance, colour, sweetness, or sound had arisen, we should, after any one or all of these sensations, have still known as little of *bodies* without, as if no sensation whatever had been excited.

The distinction, then, on which Dr Reid has founded so much, involves, in his view of it, and in the view that is generally taken of it, a false conception of the nature of the process which he describes. The two words *sensation* and *perception,* are, indeed, as I have already remarked, very convenient for expressing, in one case, the mere existence of an external feeling,—in the other case, the reference which the percipient mind has made of this feeling to an external cause. But this reference is *all,* which the *perception* superadds to the *sensation;*—and the source of the reference itself we are still left to seek, in the other principles of our intellectual nature. We have no need, however, to invent a peculiar power of the mind for producing it; since there are other principles of our nature, from which it may readily be supposed to flow,—the principle by which we are led to believe, that every new consequent, in a train of changes, must have had a new antecedent of some sort in the train,—and the principle of association, by which feelings, that have usually coexisted, suggest or become representative of each other. With these principles, it certainly is not wonderful, that when the fragrance of a rose has uniformly affected our sense of smell, as often as the flower itself was presented to us, we should ascribe the fragrance to the flower which we have seen and handled;—but though it would not be wonderful, that we should make it, it *would indeed* be wonderful, if, with these principles, we *did not* make that very reference, for which Dr Reid thinks it necessary to have recourse to a peculiar faculty of perception.

Such, then, is the view, which I would take of that distinction of *sensation* and *perception,* which Dr Reid, and the philosophers who have followed him, and many of the philosophers, too, that preceded him,—for the distinction, as I have said, is far from being an original one,—have understood in a *different* sense; in consequence, as I cannot but think, of a defective analysis of the mental process, which constitutes the *reference* of our feelings of this class to *causes* that are *without*.

39   JOHN STUART MILL (1806–1873) ON THE PERMANENT POSSIBILITIES OF SENSATION, 1865

J. S. Mill, *An Examination of Sir William Hamilton's Philosophy* (London, 1865), chap. 11.

J. S. Mill, a British empiricist and associationist of great sophistication, undertook to resolve Berkeley's problem by assuming that the mind is capable of being aware of possibilities and that the corporeal objects that give rise to sensation and thus to matter itself exist in the mind as the Permanent Possibilities of Sensation. The mind has the capacity for Expectation, he said, and here the careful reader will note that he subtly anticipates Titchener's context theory of meaning (see No. 41), in that he is making perceptual objective reference depend on the predisposition of the perceiver, just as Titchener made it depend on a predisposing set or determining tendency.

We have seen Sir W. Hamilton at work on the question of the reality of Matter, by the introspective method, and, as it seems, with little result. Let us now approach the same subject by the psychological. I proceed, therefore, to state the case of those who hold that the belief in an external world is not intuitive, but an acquired product.

This theory postulates the following psychological truths, all of which are proved by experience, and are not contested, though their force is seldom adequately felt, by Sir W. Hamilton and the other thinkers of the introspective school.

It postulates, first, that the human mind is capable of Expectation.

In other words, that after having had actual sensations, we are capable of forming the conception of Possible sensations; sensations which we are not feeling at the present moment, but which we might feel, and should feel if certain conditions were present, the nature of which conditions we have, in many cases, learnt by experience.

It postulates, secondly, the laws of the Association of Ideas. So far as we are here concerned, these laws are the following: 1st. Similar phænomena tend to be thought of together. 2nd. Phænomena which have either been experienced or conceived in close contiguity to one another, tend to be thought of together. The contiguity is of two kinds; simultaneity, and immediate succession. Facts which have been experienced or thought of simultaneously, recall the thought of one another. Of facts which have been experienced or thought of in immediate succession, the antecedent, or the thought of it, recalls the thought of the consequent, but not conversely. 3rd. Associations produced by contiguity become more certain and rapid by repetition. When two phænomena have been very often experienced in conjunction, and have not, in any single instance, occurred separately either in experience or in thought, there is produced between them what has been called Inseparable, or less correctly, Indissoluble Association: by which is not meant that the association must inevitably last to the end of life—that no subsequent experience or process of thought can possibly avail to dissolve it; but only that as long as no such experience or process of thought has taken place, the association is irresistible; it is impossible for us to think the one thing disjoined from the other. 4th. When an association has acquired this character of inseparability—when the bond between the two ideas has been thus firmly riveted, not only does the idea called up by association become, in our consciousness, inseparable from the idea which suggested it, but the facts or phænomena answering to those ideas come at last to seem inseparable in existence: things which we are unable to conceive apart, appear incapable of existing apart; and the belief we have in their co-existence, though really a product of experience, seems intuitive. Innumerable examples might be given of this law. One of the most familiar, as well as the most striking, is that of our acquired perceptions of sight. Even those who, with Mr. Bailey, consider the perception of distance by the eye as not acquired, but intuitive, admit that there are many perceptions of sight which, though instantaneous and unhesitating, are not intuitive. What we see is a very minute fragment of what we think we see. We see artificially that one thing is hard, another soft. We see artificially that one thing is hot, another cold. We see artificially that what we see is a book, or a stone, each of these being not merely an inference, but a heap of inferences, from the signs which we see, to things not visible. We see, and cannot help seeing, what we have learnt to infer, even when we know that the inference is erroneous, and that the apparent perception is deceptive. We cannot help seeing the moon larger when near the horizon, though we know that she is of precisely her usual size. We cannot help seeing a mountain as nearer to us and of less height, when we see it through a more than ordinarily transparent atmosphere.

Setting out from these premises, the Psychological Theory maintains, that there are associations naturally and even necessarily generated by the order of our sensations and of our reminiscences of sensation, which, supposing no intuition of an external world to have existed in consciousness, would inevitably generate the belief, and would cause it to be regarded as an intuition.

What is it we mean, or what is it which leads us to say, that the objects we perceive are external to us, and not a part of our own thoughts? We mean, that there is concerned in our perceptions something which exists when we are not thinking of it; which existed before we had ever thought of it, and would exist if we were annihilated; and further, that there exist things which we never saw, touched, or otherwise perceived, and things which never have been perceived by man. This idea of something which is distinguished from our fleeting impressions by what, in Kantian language, is called Perdurability; something which is fixed and the same, while our impressions vary; something which exists whether we are aware of it or not, and which is always square (or of some other given figure) whether it appears to us square or round—constitutes altogether our idea of external substance. Whoever can assign an origin to this complex conception, has accounted for what we mean by the belief in matter. Now all this, according to the Psychological Theory, is but the form impressed by the known laws of association, upon the conception or notion, obtained by experience, of Contingent Sensations; by which are meant, sensations that are not in our present consciousness, and individually never were in our consciousness at all, but which in virtue of the laws to which we have learnt by experience that our sensations are subject, we know that we should have felt under given supposable circumstances, and under these same circumstances, might still feel.

I see a piece of white paper on a table. I go into another room. If the phænomenon always followed me, or if, when it did not follow me, I believed it to disappear *e rerum natura,* I should not believe it to be an external object. I should consider it as a phantom—a mere affection of my senses: I should not believe that there had been any Body there. But, though I have ceased to see it, I am persuaded that the paper is still there. I no longer have the sensations which it gave me; but I believe that when I again place myself in the circumstances in which I had those sensations, that is, when I go again into the room, I shall again have them; and further, that there has been no intervening moment at which this would not have been the case. Owing to this property of my mind, my conception of the world at any given instant consists, in only a small proportion, of present sensations. Of these I may at the time have none at all, and they are in any case a most insignificant portion of the whole which I apprehend. The conception I form of the world existing at any moment, comprises, along with the sensations I am feeling, a countless variety of possibilities of sensation: namely, the whole of those which past observation tells me that I could, under any supposable circumstances, experience at this moment, together with an indefinite and illimitable multitude of others which though I do not know that I could, yet it is possible that I might, experience in circumstances not known to me. These various possibilities are the important thing to me in the world. My present sensations are generally of little importance, and are moreover fugitive: the possibilities, on the contrary, are permanent, which is the character that mainly distinguishes our idea of Substance or Matter from our notion of sensation. These possibilities, which are conditional certainties, need a special name to distinguish them from mere vague possibilities, which experience gives no warrant for reckoning upon. Now, as soon as a distinguishing name is given, though it be only to the same thing regarded in a different aspect, one of the most familiar experiences of our mental nature teaches us, that the different name comes to be considered as the name of a different thing.

There is another important peculiarity of these certified or guaranteed possibilities of sensation; namely, that they have reference, not to single sensations, but to sensations joined together in groups. When we think of anything as a material substance, or body, we either have had, or we think that on some given supposition we should have, not some *one* sensation, but a great and even an indefinite number and variety of sensations; generally belonging to different senses, but so linked together, that the presence of one announces the possible presence at the very same instant of any or all of the rest. In our mind, therefore, not only is this particular Possibility of sensation invested with the quality of permanence when we are not actually feeling any of the sensations at all; but when we are feeling some of them, the remaining sensations of the group are conceived by us in the form of Present Possibilities, which might be realized at the very moment. And as this happens in turn to all of them, the group as a whole presents itself to the mind as permanent, in contrast not solely with the temporariness of my bodily presence, but also with the temporary character of each of the sensations composing the group; in other words, as a kind of permanent substratum, under a set of passing experiences or manifestations: which is another leading character of our idea of substance or matter, as distinguished from sensation.

Let us now take into consideration another of the general characters of our experience, namely, that in addition to fixed groups, we also recognise a fixed Order in our sensations; an Order of succession, which, when ascertained by observation, gives rise to the ideas of Cause and Effect, according to what I hold to be the true theory of that relation, and is on any theory the source of all our knowledge what causes produce what effects. Now, of what nature is this fixed order among our sensations? It is a constancy of antecedence and sequence. But the constant antecedence and sequence do not generally exist between one actual sensation and another. Very few such sequences are presented to us by experience. In almost all the constant sequences which occur in Nature, the antecedence and consequence do not obtain between sensations, but between the groups we have been speaking about, of which a very small portion is actual sensation, the greater part being permanent possibilities of sensation, evidenced to us by a small and variable number of sensations actually present: Hence, our ideas of causation, power, activity, do not become connected in thought with our sensations as *actual* at all, save in the few physiological cases where these figure by themselves as the antecedents in some uniform sequence. Those ideas become connected, not with sensations, but with groups of possibilities of sensation. The sensations conceived do not, to our habitual thoughts, present themselves as sensations actually experienced, inasmuch as not only any one or any number of them may be supposed absent, but none of them need be present. We find that the modifications which are taking place more or less regularly in our possibilities of sensation, are mostly quite independent of our consciousness, and of our presence or absence. Whether we are asleep or awake the fire goes out, and puts an end to one particular possibility of warmth and light. Whether we are present or absent the corn ripens, and brings a new possibility of food. Hence we speedily learn to think of Nature as made up solely of these groups of possibilities, and the active force in Nature as manifested in the modification of some of these by others. The sensations, though the original foundation of the whole, come to be looked upon as a sort of accident depending on us, and the possibilities as much more real than the actual sensations, nay, as the very realities of which these are only the representations, appearances, or effects. When this state of mind has been arrived at, then, and from that time forward, we are never conscious of a present sensation without instantaneously referring it to some one of the groups of possibilities into which a sensation of that particular description enters; and if we do not yet know to what group to refer it, we at least feel an irresistible conviction that it must belong to some group or other; *i.e.* that its presence proves the existence, here and now, of a great number and variety of possibilities of sensation, without which it would not have been. The whole set of sensations as possible, form a permanent back-ground to any one or more of them that are, at a given moment, actual; and the possibilities are conceived as standing to the actual sensations in the relation of a cause to its effects, or of canvas to the figures painted on it, or of a root to the trunk, leaves, and flowers, or of a substratum to that which is spread over it, or, in transcendental language, of Matter to Form.

When this point has been reached, the Permanent Possibilities in question have assumed such unlikeness of aspect, and such difference of apparent relation to us, from any sensations, that it would be contrary to all we know of the constitution of human nature that they should not be conceived as, and believed to be, at least as different from sensations as sensations are from one another. Their groundwork in sensation is forgotten, and they are supposed to be something intrinsically distinct from it. We can withdraw ourselves from any of our (external) sensations, or we can be withdrawn from them by some other agency. But though the sensations cease, the possibilities remain in existence; they are independent of our will, our presence, and everything which belongs to us. We find, too, that they belong as much to other human or sentient beings as to ourselves. We find other people grounding their expectations and conduct upon the same permanent possibilities on which we ground ours. But we do not find them experiencing the same actual sensations. Other people do not have our sensations exactly when and as we have them: but they have our possibilities of sensation; whatever indicates a present possibility of sensations to ourselves, indicates a present possibility of similar sensations to them except so far as their organs of sensation may vary from the type of ours. This puts the final seal to our conception of the groups of possibilities as the fundamental reality in Nature. The permanent possibilities are common to us and to our fellow-creatures; the actual sensations are not. That which other people become aware of when, and on the same grounds, as I do, seems more real to me than that which they do not know of unless I tell them. The world of Possible Sensations succeeding one another according to laws, is as much in other beings as it is in me; it has therefore an existence outside me; it is an External World.

If this explanation of the origin and growth of the idea of Matter, or External Nature, contains nothing at variance with natural laws, it is at least an admissible supposition, that the element of Non-ego which Sir W. Hamilton regards as an original datum of consciousness, and which we certainly do find in our present consciousness, may not be one of its primitive elements—may not have existed at all in its first manifestations. But if this supposition be admissible, it ought, on Sir W. Hamilton's principles, to be received as true. The first of the laws laid down by him for the interpretation of Consciousness, the law (as he terms it) of Parcimony, forbids to suppose an original principle of our nature in order to account for phænomena which admit of possible explanation from known causes. If the supposed ingredient of consciousness be one which might grow up (though we cannot prove that it did grow up) through later experience; and if, when it had so grown up, it would, by known laws of our nature, appear as completely intuitive as our sensations themselves; we are bound, according to Sir W. Hamilton's and all sound philosophy, to assign to it that origin. Where there is a known cause adequate to account for a phænomenon, there is no justification for ascribing it to an unknown one. And what evidence does Consciousness furnish of the intuitiveness of an impression, except instantaneousness, apparent simplicity, and unconsciousness on our part of how the impression came into our minds? These features can only prove the impression to be intuitive, on the hypothesis that there are no means of accounting for them otherwise. If they not only might, but naturally would, exist, even on the supposition that it is not intuitive, we must accept the conclusion to which we are led by the Psychological Method, and which the Introspective Method furnishes absolutely nothing to contradict.

Matter, then, may be defined, a Permanent Possibility of Sensation. If I am asked, whether I believe in matter, I ask whether the questioner accepts this definition of it. If he does, I believe in matter: and so do all Berkeleians. In any other sense than this, I do not. But I affirm with confidence, that this conception of Matter includes the whole meaning attached to it by the common world, apart from philosophical, and sometimes from theological, theories. The reliance of mankind on the real existence of visible and tangible objects, means reliance on the reality and permanence of Possibilities of visual and tactual sensations, when no such sensations are actually experienced. We are warranted in believing that this is the meaning of Matter in the minds of many of its most esteemed metaphysical champions, though they themselves would not admit as much.

40

HERMANN LUDWIG FERDINAND VON HELMHOLTZ (1821–1894) ON PERCEPTION AND THE UNCONSCIOUS CONCLUSION, 1866

H. L. F. von Helmholtz, *Handbuch der physiologischen Optik,* vol. III (Leipzig, 1866), sect. 26. Translated by J. P. C. Southall as *Helmholtz' Treatise on Physiological Optics,* vol. III ([Rochester, N.Y.]: Optical Society of America, 1925).

Helmholtz's doctrine of perception holds that it may gain its meaning by being founded on cues that are not immediately available as such to consciousness, and that specific objective reference may come by way of an unconscious conclusion (*unbewusster Schluss*) derived by unconscious inferences. (*Conclusion* is the correct translation for *Schluss,* but the more common English phrase is *unconscious inference.*) Had Helmholtz adopted this theory before he reached the third volume of his monumental work in 1866, he could have provided better instances of unconscious inference, such as the way in which the retinal disparity of binocular images is transformed into a perception of the relative distances of perceived objects—as in the case of stereoscopy, where the cues which are adequate to the geometric conclusion remain wholly unavailable to consciousness. (On Helmholtz and unconscious inference, see also the preceding chapter, No. 34.)

We must speak now of the manner in which our ideas and perceptions are formed by inductive conclusions. The best analysis of the nature of our conclusions I find in J. S. Mill's Logic. As long as the premise of the conclusion is not an injunction imposed by outside authority for our conduct and belief, but a statement related to reality, which can therefore be only the result of experience, the conclusion, as a matter of fact, does not tell us anything new or something that we did not know already before we made the statement. Thus, for example:

Major: All men are mortal.
Minor: Caius is a man.
Conclusion: Caius is mortal.

The major premise, that all men are mortal, which is a statement of experience, we should scarcely venture to assert without knowing beforehand whether the conclusion is correct, namely, that Caius, who is a man, either is dead or will die. Thus we must be sure of the conclusion before we can state the major premise by which we intend to prove it. That seems to be proceeding in a circle. The real relation evidently is, that, in common with other folks, we have observed heretofore without exception that no person has ever survived beyond a certain age. Observers have learned by experience that Lucius, Flavius and other individuals of their acquaintance, no matter what their names are, have all died; and they have embraced this experience in the general statement, that *all* men die. Inasmuch as this final result occurred regularly in all the instances they observed, they have felt justified in explaining this general law as being valid also for all those cases which might come up for observation hereafter. Thus we preserve in our memory the store of experiences heretofore accumulated on this subject by ourselves and others in the form of the general statement which constitutes the major premise of the above conclusion.

However, the conviction that Caius would die might obviously have been reached directly also without formulating the general statement in our consciousness, by having compared his case with all those which we knew previously. Indeed, this is the more usual and original method of reasoning by induction. Conclusions of this sort are reached without conscious reflection, because in our memory the same sort of thing in cases previously observed unites and reinforces them; as is shown especially in those cases of inductive reasoning where we cannot succeed in deducing from previous experiences a rule with precisely defined limits to its validity and without any exceptions. This is the case in all complicated processes. For instance, from analogy with previous similar cases, we can sometimes predict with tolerable certainty what one of our acquaintances will do, if under certain circumstances he decides to go into business; because we know his character and that he is, let us say, ambitious or timid. We may not be able to say exactly how we have estimated the extent of his ambition or timidity, or why this ambition or timidity of his will be enough to decide that his business will turn out as we expect.

In the case of conclusions properly so-called, which are reached consciously, supposing they are not based on injunctions but on facts of experience, what we do, therefore, is really nothing more than deliberately and carefully to retrace those steps in the inductive generalizations of our experiences which were previously traversed more rapidly and without conscious reflection, either by ourselves or by other observers in whom we have confidence. But although nothing essentially new is added to our previous knowledge by formulating a general principle from our previous experiences, still it is useful in many respects. A definitely stated general principle is much easier to preserve in the memory and to be imparted to others than to have to do this same thing with every individual case as it arises. In formulating it we are led to test accurately every new case that occurs, with reference to the correctness of the generalization. In this way every exception will be impressed on us twice as forcibly. The limits of its validity will be recalled much sooner when we have the principle before us in its general form, instead of having to go over each separate case. By this sort of conscious formulation of inductive reasoning, there is much gain in the convenience and certainty of the process; but nothing essentially new is added that did not exist already in the conclusions which were reached by analogy without reflection. It is by means of these latter that we judge the character of a person from his countenance and movements, or predict what he will do in a given situation from a knowledge of his character.

Now we have exactly the same case in our sense-perceptions. When those nervous mechanisms whose terminals lie on the right-hand portions of the retinas of the two eyes have been stimulated, our usual experience, repeated a million times all through life, has been that a luminous object was over there in front of us on our left. We had to lift the hand toward the left to hide the light or to grasp the luminous object; or we had to move toward the left to get closer to it. Thus while in these cases no particular conscious conclusion may be present, yet the essential and original office of such a conclusion has been performed, and the result of it has been attained; simply, of course, by the unconscious processes of association of ideas going on in the dark background of our memory. Thus too its results are urged on our consciousness, so to speak, as if an external power had constrained us, over which our will has no control.

These inductive conclusions leading to the formation of our sense-perceptions certainly do lack the purifying and scrutinizing work of conscious thinking. Nevertheless, in my opinion, by their peculiar nature they may be classed as *conclusions,* inductive conclusions unconsciously formed.

There is one circumstance quite characteristic of these conclusions which operates against their being admitted in the realm of conscious thinking and against their being formulated in the normal form of logical conclusions. This is that we are not able to specify more closely what has taken place in us when we have experienced a sensation in a definite nerve fibre, and how it differs from corresponding sensations in other nerve fibres. Thus, suppose we have had a sensation of light in certain fibres of the nervous mechanism of vision. All we know is that we have had a sensation of a peculiar sort which is different from all other sensations, and also from all other visual sensations, and that whenever it occurred, we invariably noticed a luminous object on the left. Naturally, without ever having studied physiology, this is all we can say about the sensation, and even for our own imagination we cannot localize or grasp the sensation except by specifying it in terms of the conditions of its occurrence. I have to say, "I see something bright there on my left." That is the only way I can describe the sensation. After we have pursued scientific studies, we begin to learn that we have nerves, that these nerves have been stimulated, and that their terminals in fact lie on the right-hand side of the retina. Then for the first time we are in a position to define this mode of sensation independently of the mode in which it is ordinarily produced.

It is the same way with most sensations. The sensations of taste and smell usually cannot be described even as to their quality except in terms of the bodies responsible for them; although we do have a few rather vague and more general expressions like "sweet," "sour," "bitter" and "sharp."

These judgments, in which our sensations in our ordinary state of consciousness are connected with the existence of an external cause, can never once be elevated to the plane of conscious judgments. The inference that there is a luminous object on my left, because the nerve terminals on the right-hand side of my retina are in a state of stimulation, can only be expressed by one who knows nothing about the inner structure of the eye by saying, "There is something bright over there on my left, because I see it there." And accordingly from the standpoint of everyday experience, the only way of expressing the experience I have when the nerve terminals on the right-hand side of my eyeball are stimulated by exerting pressure there, is by saying, "When I press my eye on the right-hand side, I see a bright glow on the left." There is no other way of describing the sensation and of identifying it with other previous sensations except by designating the place where the corresponding external object appears to be. Hence, therefore, these cases of experience have the peculiarity that the connection between the sensation and an external object can never be expressed without anticipating it already in the designation of the sensation, and without presupposing the very thing we are trying to describe.

Even when we have learned to understand the physiological origin and connection of the illusions of the senses, it is impossible to get rid of the illusion in spite of our better knowledge. This is because inductive reasoning is the result of an unconscious and involuntary activity of the memory; and for this very reason it strikes our consciousness as a foreign and overpowering force of nature. Incidentally, manifold analogies for it are to be found in all other possible modes of *apparition.* We might say that all apparition originates in premature, unmeditated inductions, where from previous cases conclusions are deduced as to new ones, and where the tendency to abide by the false conclusions persists in spite of the better insight into the matter based on conscious deliberation. Every evening apparently before our eyes the sun goes down behind the stationary horizon, although we are well aware that the sun is fixed and the horizon moves. An actor who cleverly portrays an old man is for us an old man there on the stage, so long as we let the immediate impression sway us, and do not forcibly recall that the programme states that the person moving about there is the young actor with whom we are acquainted. We consider him as being angry or in pain according as he shows us one or the other mode of countenance and demeanour. He arouses fright or sympathy in us, we tremble for the moment, which we see approaching, when he will perform or suffer something dreadful; and the deep-seated conviction that all this is only show and play does not hinder our emotions at all, provided the actor does not cease to play his part. On the contrary, a fictitious tale of this sort, which we seem to enter into ourselves, grips and tortures us more than a similar true story would do when we read it in a dry documentary report.

The experiences we have that certain aspects, demeanours and modes of speech are indicative of fierce anger, are generally experiences concerning the external signs of certain emotions and peculiarities of character which the actor can portray for us. But they are not nearly so numerous and regular in recurrence as those experiences by which we have ascertained that certain sensations correspond with certain external objects. And so we need not be surprised if the idea of an object which is ordinarily associated with a sensation does not vanish, even when we know that in this particular instance there is no such object.

41   EDWARD BRADFORD TITCHENER (1867–1927) ON THE CONTEXT THEORY OF MEANING, 1910

E. B. Titchener, *A Text-Book of Psychology* (New York: Macmillan, 1910), pp. 367–371, 274–275. The theory was first formulated in 1909 in *Lectures on the Experimental Psychology of the Thought-Processes* (New York: Macmillan, 1909), pp. 174–194, but the reference of 1910 is clearer and more concise, and quotes the earlier one repeatedly.

Titchener's theory that the meaning of a perception or idea is carried by a context that consists of the sensations or images that accrue to the conscious core of sensations in a perception or the conscious core of images in an idea is a purely associative theory, like Berkeley's modernized and clarified, though Titchener does not use the word *association* in developing his view. Once when he was charged with merely repeating in 1909 what Berkeley had said in 1709, he replied that the context theory of meaning claims originality because it includes the principle that in familiar perceptions the conscious context lapses and the meaning is carried unconsciously, an addition that is new as far as its anticipations by J. S. Mill and Helmholtz go, and one that was explicitly brought over by Titchener from the work of Narziss Ach (1905) of the Würzburg school on the determining tendency. Much more interesting is the fact that Titchener, in this provision, was himself anticipating the behavioral theory of meaning soon to be presented by Holt, Tolman, and later behaviorists. If the meaning of a familiar perception

is "there" without conscious represen- lie in the specificity of the behavior
tation (like the meaning of the word of the organism. Titchener would not
but when one is reading easy material have wished to be classified as a be-
rapidly and adequately, or the meaning haviorist, but the history of thinking
of *B-is-flatted* when one is playing eas- is, though deliberately slow, continuous
ily in the key of F), then the evidence in just this manner.
for the meaning's being "there" must

103. *Meaning.*—Perceptions are selected groups of sensations, in which
images are incorporated as an integral part of the whole process. But
that is not all: the essential thing about them has still to be named: and
it is this,—that perceptions have meaning. No sensation means; a sensation simply goes on in various attributive ways, intensively, clearly, spatially, and so forth. All perceptions mean; they go on, also, in various
attributive ways; but they go on meaningly. What then, psychologically,
is meaning?

Meaning, psychologically, is always context; one mental process is the
meaning of another mental process if it is that other's context. And
context, in this sense, is simply the mental process which accrues to the
given process through the situation in which the organism finds itself.
Originally, the situation is physical, external; and, originally, meaning
is kinaesthesis; the organism faces the situation by some bodily attitude,
and the characteristic sensations which the attitude arouses give meaning
to the process which stands at the conscious focus, are psychologically
the meaning of that process. For ourselves, the situation may be either
external or internal, either physical or mental, either a group of adequate stimuli or a constellation of ideas; image has now supervened
upon sensation, and meaning can be carried in imaginal terms. For us,
therefore, meaning may be mainly a matter of sensations of the special
senses, or of images, or of kinaesthetic or other organic sensations, as
the nature of the situation demands.

Of all its possible forms, however, two appear to be of especial importance: kinaesthesis and verbal images. We are locomotor organisms, and
change of bodily attitude is of constant occurrence in our experience;
so that typical kinaesthetic patterns become, so to say, ingrained in our
consciousness. And words themselves, let us remember, were at first bodily attitudes, gestures, kinaesthetic contexts: complicated, of course, by
sound, but still essentially akin to the gross bodily attitudes of which
we have been speaking. The fact that words are thus originally con-
textual, and the fact that they nevertheless as sound, and later as sight,
possess and acquire a content-character,—these facts render language
preeminently available as the vehicle of meaning. The words that we
read are both perception and context of perception; the auditory-kin-

aesthetic idea is the meaning of the visual symbols. And it is obvious that all sorts of sensory and imaginal complexes receive their meaning from some mode of verbal representation: we understand a thing, place a thing, as soon as we have named it.

Hence, in minds of a certain constitution, it may well be that all conscious meaning is carried by total kinaesthetic attitude or by words. As a matter of fact, however, mental constitution is widely varied, and meaning is carried by all sorts of sensory and imaginal processes.

The gist of this account is that it takes at least two sensations to make a meaning. If an animal has a sensation of light, and nothing more, there is no meaning in consciousness. If the sensation of light is accompanied by a strain, it becomes forthwith a perception of light, with meaning; it is now 'that bright something'; and it owes the 'that something' to its strain-context. Simple enough!—only be clear that the account is not genetic, but analytic. We have no reason to believe that mind began with meaningless sensations, and progressed to meaningful perceptions. On the contrary, we must suppose that mind was meaningful from the very outset. We find, by our analysis, that sensation does not mean; and we find, in synthesis, that the context which accrues from the situation, however simple or however complex the context may be, makes it mean, is its meaning.

What, then, precisely, is a situation? The physical or external situation is the whole external world as an organism, at any given moment, takes it; it consists of those stimuli to which the organism, by virtue of its inherited organisation and its present disposition, is responsive,—which it selects, unifies, focalises, supplements, and, if need be, acts upon. The mental or internal situation is, in like manner, some imaginative or memorial complex which is fitted, under the conditions obtaining in the nervous system, to dominate consciousness, to maintain itself in the focus of attention, to serve as the starting-point for further ideas or for action. To put the definition in a word, a situation is the meaningful experience of a conscious present.

But is meaning always conscious meaning? Surely not: meaning may be carried in purely physiological terms. In rapid reading, the skimming of pages in quick succession; in the rendering of a musical composition, without hesitation or reflection, in a particular key; in shifting from one language to another as you turn to your right- or left-hand neighbour at a dinner-table: in these and similar cases meaning has, time and time again, no discoverable representation in consciousness. The course and connection of ideas may be determined beforehand and from without; a word, an expression of face, an inflection of the voice, a bodily attitude, presses the nervous button, and consciousness is switched, automatically, into new channels. We find here an illustration of an universal law of mind, of which we shall have more to say when we come to deal with Action: the law that all conscious formations, as the life of the organism proceeds, show like phenomena of rise and fall, increase and decrease in complexity, expansion and reduction; so that, in the extreme case, what was originally a focal experience may presently lapse altogether. We learned our French and German with pains and labour; the conscious context that gave meaning to words and sentences was elaborate; but now all this context has disappeared, and a certain set of the nervous system, itself not accompanied by consciousness, gives the sounds that fall upon our ears a French-meaning, or changes us into German-speakers.

This predetermination of consciousness by influences that, during the course of consciousness, are not themselves conscious, is a fact of extreme psychological importance, and the reader should verify it from his own experience. It has a threefold bearing upon the psychological system. First, it reminds us that consciousness is a temporal affair, to be studied in longitudinal as well as in transverse section. It is part of the direct business of psychology to trace the fate of meaning from its full and complete conscious representation, through all the stages of its degeneration, to its final disappearance. Secondly, our psychology is to be explanatory, and our explanations are to be physiological. To explain the way in which consciousness runs, the definite line that it takes, we must have recourse to physiological organisation; and the tracing of the stages of mental decay helps us to follow and understand the organising process. Thirdly, if we lose sight of nervous predisposition, we shall make grave mistakes in our psychological analysis; we shall read into mental processes characters that, in fact, they do not possess . . . A simple instance is given [below]. Here we must either say that the meaning of the experiment, after the week's work, is carried for the observer in purely physiological, non-conscious terms; or we must say that his observation is untrustworthy, that there is a mental context which he has overlooked. But if we take this latter alternative, we shall be constructing mind as the naturalist in the story constructed the camel; we shall be inventing, not describing.

· · · · ·

The guiding influence of nervous bias is not a matter of inference, still less a matter of speculation; it can be demonstrated in the psychological laboratory. Suppose that we are measuring the time required to reply to a spoken word by another word of the same class or kind: to associate dog to cat, table to chair, and so on. The experimenter prepares a long list of words: cat, chair, and so forth. Then he explains to

the observer the precise nature of the experiment: I shall call out certain words, he says, and you are to reply, as quickly as you can, with
words of the same class; if I say horse, you will mention some other
animal, and if I say pen, you will mention something else that has to do
with writing. The observer understands, and the experiment begins. Suppose, further, that the experiments have been continued for some days.
The experimenter has no need to repeat his explanation at every sitting;
the observer takes it for granted that he is still to reply with a coordinate
word. And suppose, finally, that some day, after a week's work, the experimenter interrupts the series, and asks: Are you thinking about what
I told you to do? The observer, fearing that he has made some error,
and feeling very repentant, will say: No! to tell the truth I had absolutely forgotten all about it; it had gone altogether out of my mind; have
I done anything wrong? He had not done anything wrong; but his answer shows that a certain tendency, impressed upon his nervous system
by the experimenter's original explanation, has been effective to direct
the course of his ideas long after its conscious correlate has disappeared.
And what happens here, in the laboratory, happens every day of our
lives in the wider experience outside the laboratory.

42   EDWIN BISSELL HOLT (1873–1946) ON RESPONSE AS THE ESSENCE OF COGNITION, 1915

E. B. Holt, *The Freudian Wish and Its Place in Ethics* (New York: Henry Holt, 1915),
"Supplement: Response and cognition," pp. 154–160, 171–174. This supplement is
reprinted from the *Journal of Philosophy, Psychology and Scientific Methods 12*,
365–373, 393–409 (1915).

The evidence for what man knows lies in what he does, as we have just seen even Titchener arguing in the instance of familiar perceptions. Holt, writing only two years after John B. Watson had inaugurated behaviorism, was the first psychologist to argue that the quiddity of cognition is response—and by response he meant not a mere movement or secretion but the whole pattern of stimulus and response, a reaction to a situation. Holt argued against the "bead theory," which states that causally connected events, strung out along a line, can adequately describe behavior, just as John Dewey had 19 years earlier in his attack on the reflex-arc theory (see No. 64). In this respect Holt was, like Dewey, an-ticipating Gestalt psychology, which later was also objecting to the elementism of both Titchenerian introspective psychology and Watsonian behaviorism. (Cf. Wertheimer, in the next selection). The influence of Holt in establishing the behavioral theory of meaning appeared a little later on, most clearly in the work of Edward C. Tolman (*Purposive Behavior in Animals and Men*, New York: Century, 1932), who was a student of Holt's at Harvard but who flourished so well along in the twentieth century as to be beyond the limits of this book. Nowadays in American psychology awareness and discrimination are pretty well identified, an evidence of the midcentury culmination of positivism.

Most of us believe that the appearance of life was . . . a critical moment in the evolution of the universe: that life came into existence when some, perhaps a specific, sort of chemical process was set up under such conditions as maintained it around a general point of equilibrium. The result was undoubtedly novel; and more novelties were to come. Living substance was to acquire a protective envelope, to become irritable, conductive, and contractile, to develop specific irritability to many stimuli, to get the power of locomotion, and much else. Now in the course of this further evolution, there is a critical point which is worthy of notice. It is the point where the irritable, contractile, and conductive tissues develop systematic relations which enable them to function as an integral whole. Here, too, novelty ensues.

How 'critical' this point is, how sudden and well-defined, or, on the other hand, how gradual, cannot as yet be told. The integrative process in the nervous system, as Sherrington so well calls it, has not, even yet, been observed in sufficient detail. But this is of secondary importance; and the result of the process we do know definitely. This is, that the phenomena evinced by the integrated organism are no longer merely the excitation of nerve or the twitching of muscle, nor yet the play merely of reflexes touched off by stimuli. These are all present and essential to the phenomena in question, but they are merely components now, for they have been integrated. And this integration of reflex arcs, with all that they involve, into a state of systematic interdependence has produced something that is not merely reflex action. The biological sciences have long recognized this new and further thing, and called it 'behavior.'

Of recent years, many of the workers in animal psychology have been coming to call this the science of behavior, and have been dwelling less and less on the subject of animal 'consciousness.' They do not doubt, any of them, that at least the higher animals are 'conscious'; but they find that nothing but the behavior of animals is susceptible of scientific observation. Furthermore, several students in the human field have come to the same conclusion—that not the 'consciousness,' but the behavior of one's fellow-men, and that alone, is open to investigation. Several volumes have been put forth which even undertake to construe human psychology entirely in terms of behavior. It is obvious that this is an unstable condition in which the science now finds itself. We cannot continue thus, each man proclaiming his own unquestionable gift of 'consciousness,' but denying that either his fellow-men or the animals evince the slightest indication of such a faculty. Now I believe that a somewhat closer definition of 'behavior' will show it to involve a hitherto unnoticed feature of novelty, which will throw light on this matter.

Precisely how does this new thing, 'behavior,' differ, after all, from mere reflex action? Cannot each least quiver of each least muscle fiber be wholly explained as a result of a stimulus impinging on some sense-organ, and setting up an impulse which travels along definite nerves with definite connections, and comes out finally at a definite muscle having a certain tonus, etc., all of which is merely reflex action? Yes, exactly *each least* component can be so explained, for that is just what, and all that, it is. But it is the coördinated totality of these least components which *cannot* be described in such terms, nor indeed in terms resembling these. For such neural and reflex terms fail to seize that integration factor which has now transformed reflex action into something else, i.e., behavior. We require, then, an exact definition of behavior.

But before proceeding to this definition we shall probably find useful an illustration from another science, which was once in the same unstable state of transition as psychology is now. In physics a theory of causation once prevailed, which tried to describe causal process in terms of successive 'states,' the 'state' of a body at one moment being the *cause* of its 'state' and position at the next. Thus the course of a falling body was described as a series of states (*a, b, c, d,* etc.), each one of which was the effect of the state preceding, and cause of the one next following. This may be designated as the 'bead theory' of causation. Inasmuch, however, as nothing could be observed about one of these 'states' which would show why the next 'state' must necessarily follow, or, in other words, since the closest inspection of 'states' gave no clew toward explaining the course or even the continuance of the process, an unobservable impetus (*vis viva, Anstoss,* 'force') was postulated. This hidden impetus was said to be the ultimate secret of physical causation. But, alas, a secret! For it remained, just as the 'consciousness' of one's fellow-man remains to-day in psychology, utterly refractory to further investigation. Now 'myth' is the accepted term to apply to an entity which is believed in, but which eludes empirical inquiry. This mythical *vis viva* has now, in good part owing to the efforts of Kirchhoff and Hertz, been rejected, and, what is more important, with it has gone the bead theory itself. It is not the 'previous state' of the falling body which causes it to fall, but the earth's mass. And it is not in the 'previous state' but in laws that explanation resides, and no laws for falling bodies or for any other process could, on the terms of the bead theory, be extracted from the phenomena. But laws were easily found for physical processes, if the observer persuaded himself to make the simple inquiry, *What* are the objects *doing?* Now the falling body is not merely moving downwards past the successive divisions of a meter-stick which I have placed beside it (which is all that the bead theory would have us consider), nor is it essentially moving toward the floor which, since a floor happens to be there, it will presently strike. The body is *essentially* moving toward the center of the earth, and these other objects could be removed without altering the influence of gravity. In short, the fall of a body is adequately described as a function of its mass, of the earth's mass, and of the distance between the centers of the two. And the *function* is *constant,* is that which in change remains unchanged (in the case cited it is a constant acceleration). The physical sciences, of course, have now explicitly adopted this function theory of causation. Every physical law is in the last analysis the statement of a constant function between one process or thing and some other process or thing. This abandonment of the bead theory in favor of the function theory requires, at the first, some breadth and some bravery of vision.

.　.　.　.　.

II. COGNITION AS RESPONSE

We have now a compact and, as I believe, a rather precise definition of behavior or, as it might be called, the relation of specific response. And we are in a position to compare it with the cognitive relation, the relation between the 'psychological subject and its object of consciousness.' Our aim would be to see how far those phenomena which we ordinarily attribute to 'consciousness' may be intrinsically involved by this strictly objective and scientifically observable behavior.

Firstly, as to the object cognized, the 'content of consciousness.' It is obvious that the object of which an organism's behavior is a constant function corresponds with singular closeness to the object of which an organism is aware, or of which it is conscious. When one is conscious of a thing, one's movements are adjusted to it, and to precisely those features of it of which one is conscious. The two domains are conterminous. It is certain, too, that it is not generally the stimulus to which one is adjusted, or of which one is conscious: as such classic discussions as those about the inverted retinal image and single vision (from binocular stimulation) have shown us. Even when one is conscious of things that are not there, as in hallucination, one's body is adjusted to them as if they were there; and it behaves accordingly. In some sense or other they are there; as in some sense there are objects in mirrored space. Of course the objects of one's consciousness, and of one's motor adjustments, may be past, present, or future: and similar temporally forward and backward functional relations are seen in many inorganic mechanisms. If it be thought that there can be consciousness without behavior, I would say that the doctrine of dynamogenesis, and indeed the doctrine of psycho-physical parallelism itself, assert just the contrary. Of course muscle tonus and 'motor set' are as much behavior as is the more extensive play of limb. In short, I know not what distinction can be drawn between the object of consciousness and the object of behavior.

Again, if the object of which behavior is a constant function is the object of consciousness, *that function of it which behavior is* presents a close parallel to volition. Psychological theory has never quite succeeded in making will a content of knowledge in the same sense as sensation, perception, and thought; the heterogeneous (motor-image) theory being manifestly untrue to rather the larger part of will acts. Indeed, in the strict sense the theory of innervation feelings is the only one which ever allowed will to be, in its own right, a content. All other views, including the heterogeneous, show one's knowledge of one's own will acts to be gained by a combination of memory and the direct observation of what one's own *body is doing.* And this is quite in harmony with the idea that what one wills is that which one's body does (in attitude or overt act) toward the environment. In a larger sense, however, and with less deference to the tendencies of bead theorizing, one's volitions are obviously identical with that which one's body in the capacity of released mechanism *does.* If a man avoids draughts, that is both the behavior and the volition at once, and any motor-image, 'fiat,' or other account of it merely substitutes some subordinate aspect for that which is the immediate volition.

43   MAX WERTHEIMER (1880–1943) ON OBJECTS AS IMMEDIATELY GIVEN TO CONSCIOUSNESS, 1923

Max Wertheimer, "Untersuchungen zur Lehre von der Gestalt," *Psychologische Forschung* 4, 301–303 (1923). Translated for this book by Don Cantor.

So far all the theories of perceptual objective reference except Holt's have been empiricistic and, for the most part, associationistic, and also—perhaps for those very reasons—written in English. An empiricistic theory is genetic: one tells how objective reference or meaning is established for a perception. Germany has, however, been the proper home of nativism and of the immediately given conscious datum. When both behaviorism and Gestalt psychology, new movements in the second decade of the present century, complained about nineteenth-century introspection, behaviorism was found objecting to the older dualism, and Gestalt psychology to the older elementism. This excerpt from the writing of Max Wertheimer, who was recognized as the leader of the Gestalt movement, is a piece of his early protest against the sensory elementism of the introspectionists—Wundt, Titchener, and the others. He never would have said, as Titchener did, that it takes at least two sensations to make a meaning, but he would have eagerly supported Holt's objections to the bead theory. Gestalt psychology has insisted on dealing with wholes because it finds wholes given in conscious experience. That they are not compounded like Wundt's perceptions, or prefabricated like Helmholtz's, but given in their various individual entireties, is the argument that Wertheimer is making here. Perhaps nativism is not a theory, but we have to take nature as it comes. Later Wolfgang Köhler was to make this argument more effectively (for example, in his *Gestalt Psychology,* New York: Liveright, 1929), but like Tolman's his very considerable contribution lies beyond the limit of this book.

I stand at the window and see a house, trees, sky.

And I could, then, on theoretical grounds, try to sum up: there are 327 brightnesses (and tones of color).

(Have I "327"? No: sky, house, trees; and no one can realize the having of the "327" as such.)

Let there be, in this particular calculation, about 120 for the house, 90 for the trees, and 117 for the sky; in any case I have *this* togetherness and *this* distribution, not 127 and 100 and 100, or 150 and 177.

I *see* it in specific togetherness and specific distribution; and the manner of unity and separation [togetherness and distribution] in which I see it is not determined simply by my whimsy. It is quite certain that I cannot actualize another desired kind of coherence according to the dictates of my whimsy.

(And what a remarkable process, when suddenly something of that sort is attained! What astonishment when, after staring at a window for a long time, making many vain attempts, and in a bemused mood, I discover that the pieces of a dark frame and a bare branch combine to form a [Roman] N.)

Or take two faces cheek to cheek.

I see one (with, if one wishes, "57" brightnesses) and the other (with its "49"), but not in the division 66 plus 40 or 6 plus 100.

Theories that would suggest that I see "106" there remain on paper; it is two faces that I see. But first of all let the important point here be *only* the manner of the unification and separation, which is, in any case, *so* certain. To begin with, only this modest but theoretically important set of circumstances will be dealt with.

Or: I hear a melody (17 tones!) with its accompaniment (32 tones!). I hear melody and accompaniment, not simply "49"; or, at least, certainly not usually 20 plus 29 or entirely according to my whimsy.

Thus it is when no stimulus continua come under consideration—as, for example, when the melody and its accompaniment are played in short single bell tones by one of the old toy clocks, or, as in the case of vision, when forms consisting of discontinuous parts (for example, points) contrast with one another on an otherwise homogeneous background. It may be that establishing different kinds of unity is here easier than in the earlier cases; nevertheless, for the most part it is also true here that a spontaneous, "natural," normally predictable kind of unity and separation are present. Furthermore, it is only sometimes, seldom under definite conditions, that another configuration can be obtained or produced through special artificial means—and it is harder to accomplish.

In general:

Where a number of stimuli are simultaneously effective, a corresponding (just as large) number of single experiences is generally not present for human beings; there is not the one, the other, and the third experience, and so on. Rather there are experiences in larger areas with definite contrasts, definite unities, definite separations. Nor does it matter how the theoretical structure may be constituted; whether it is far from what is simply seen on theoretical grounds or whether it does or does not establish the sum of "327 . . . sensations," there still remains for every configuration a problem in reality.

Are there principles for [forming each] kind of "unity" and "separation"? *What principles?*

If the stimuli *a, b, c, d, e* are effective together, what are the principles governing the fact that, given these stimuli or this configuration, experiences arise that [belong] typically to the unity and to the separation, as *a, b, c/d, e,* and not *a, b/c, d, e?* And this whether the first is the inevitable result (and any other unobtainable) or only the spontaneous "natural" one, normally expected.

VII CEREBRAL LOCALIZATION

The belief that the brain is the chief organ of the mind is quite old. Pythagoras and Plato held this view, although the great Aristotle assigned the mental functions primarily to the heart. Galen (ca. A.D. 129–199), whose medical pronouncements were accepted as unimpeachable for many centuries after they were laid down, firmly established the belief that the mind has its seat in the brain, yet it was not until Descartes that the identification of the soul with consciousness became so clear and explicit that the question of how mind and body interact became substantial. For this reason the text begins with Descartes (1650), a dualist as to the distinction between mind and body, who localized the interaction between the soul and the body at the pineal gland in the brain.

From Descartes the chapter skips to Gall (1825) the phrenologist, who localized more than thirty mental faculties at specific regions or spots in the brain and on the skull, and to Flourens (1824), an experimental physiologist who opposed Gall's views and interpreted his own findings in a less presumptuous manner. After Flourens we pass to Broca (1861), who located what he believed to be the speech center in the brain, and then to Fritsch and Hitzig (1870), who used electric shock to locate the motor centers in the cerebral cortex. These early specific localizations were followed by a period of active research and controversy during which the belief in the localization of motor and sensory centers in the cerebral cortex increased. The positive view in favor of exact localization of function may be said to have been most explicitly supported by David Ferrier,

but, on the other side, there was the important, influential, and yet not easily understood Hughlings Jackson, who, stimulated by Herbert Spencer and guided by his clinical observations, held an evolutionary view that there are higher, intermediate, and lower levels of the nervous system in nervous disease and that, as the magnitude of the disturbance increases, there is a continuous regression from the higher levels to the lower.

Franz (1902, 1915) began in the tradition of Ferrier, but his experiments with extirpation of cerebral tissue led him over to a position somewhat resembling Flourens's and thus away from exact localization. Lashley (1917, 1929), with his conceptions of equipotentiality and mass action, moved still further away from the belief in exact localization.

Meanwhile, Sigmund Freud in clinical studies of aphasia in 1891 found himself supporting Hughlings Jackson and formulated an evolutionary view in which the degree of aphasia is thought of as regression in respect of evolutionary advance. Later Henry Head (1915, 1926), on the basis of clinical observation, adopted a similar counterevolutionary theory of aphasia, being ignorant of what Freud had said on the subject but ready to found his conceptions on Hughlings Jackson's contentions. In general it may be said that the exact localizationists lost out, although there is, of course, a general correlation between behavioral and sensory function on the one hand and region or level of the nervous system on the other, as indeed Flourens had argued, although without sufficient specification.

44   RENÉ DESCARTES (1596–1650) ON THE INTERACTION OF MIND AND BRAIN, 1650

René Descartes, *Les Passions de l'âme* (Amsterdam, 1650), pt. 1, arts. 2–4, 16, 17, 30–32, 34, 35, 41–44. Translated by E. S. Haldane and G. R. T. Ross in *The Philosophical Works of Descartes* (Cambridge, Eng.: University Press, 1931), I, 332–351. The translation of article 44 has been added by the editors.

Descartes was the father both of French psychological materialism and of later reflexology, since he held that animals, having no rational souls, are machines. He was also the father of mind-body dualism, an interactionist as contrasted with the later psychophysical parallelists. Convinced that the brain is the organ of the mind, he chose the pineal gland as the means of interaction between the soul and the brain because it is the only organ in the brain not duplicated bilaterally, and thus is proper as the agent of a soul that is necessarily undivided.

ARTICLE II

That in order to understand the passions of the soul its functions must be distinguished from those of body.

Next I note also that we do not observe the existence of any subject which more immediately acts upon our soul than the body to which it is joined, and that we must consequently consider that what in the soul is a passion is in the body commonly speaking an action; so that there is no better means of arriving at a knowledge of our passions than to examine the difference which exists between soul and body in order to know to which of the two we must attribute each one of the functions which are within us.

ARTICLE III

What rule we must follow to bring about this result.

As to this we shall not find much difficulty if we realise that all that we experience as being in us, and that to observation may exist in wholly inanimate bodies, must be attributed to our body alone; and, on the other hand, that all that which is in us and which we cannot in any way conceive as possibly pertaining to a body, must be attributed to our soul.

ARTICLE IV

That the heat and movement of the members proceed from the body, the thoughts from the soul.

Thus because we have no conception of the body as thinking in any way, we have reason to believe that every kind of thought which exists in us belongs to the soul; and because we do not doubt there being inanimate bodies which can move in as many as or in more diverse modes than can ours, and which have as much heat or more (experience demonstrates this to us in flame, which of itself has much more heat and movement than any of our members), we must believe that all the heat and all the movements which are in us pertain only to body, inasmuch as they do not depend on thought at all.

· · · · ·

ARTICLE XVI

How all the members may be moved by the objects of the senses and by the animal spirits without the aid of the soul.

We must finally remark that the machine of our body is so formed that all the changes undergone by the movement of the spirits may cause them to open certain pores in the brain more than others, and reciprocally that when some one of the pores is opened more or less than usual (to however small a degree it may be) by the action of the nerves which are employed by the senses, that changes something in the movement of the spirits and causes them to be conducted into the muscles which serve to move the body in the way in which it is usually moved when such an action takes place. In this way all the movements which we make without our will contributing thereto (as frequently happens when we breathe, walk, eat, and in fact perform all those actions which are common to us and to the brutes), only depend on the conformation of our members, and on the course which the spirits, excited by the heat of the heart, follow naturally in the brain, nerves, and muscles, just as the movements of a watch are produced simply by the strength of the springs and the form of the wheels.

ARTICLE XVII

What the functions of the soul are.

After having thus considered all the functions which pertain to the body alone, it is easy to recognise that there is nothing in us which we ought to attribute to our soul excepting our thoughts, which are mainly of two sorts, the one being the actions of the soul, and the other its passions. Those which I call its actions are all our desires, because we find by experience that they proceed directly from our soul, and appear to depend on it alone: while, on the other hand, we may usually term one's passions all those kinds of perception or forms of knowledge which are found in us, because it is often not our soul which makes them what they are, and because it always receives them from the things which are represented by them.

· · · · ·

ARTICLE XXX

That the soul is united to all the portions of the body conjointly.

But in order to understand all these things more perfectly, we must know that the soul is really joined to the whole body, and that we cannot, properly speaking, say that it exists in any one of its parts to the exclusion of the others, because it is one and in some manner indivisible, owing to the disposition of its organs, which are so related to one another that when any one of them is removed, that renders the whole body defective; and because it is of a nature which has no relation to extension, nor dimensions, nor other properties of the matter of which the body is composed, but only to the whole conglomerate of its organs, as appears from the fact that we could not in any way conceive of the half or the third of a soul, nor of the space it occupies, and because it does not become smaller owing to the cutting off of some portion of the body, but separates itself from it entirely when the union of its assembled organs is dissolved.

ARTICLE XXXI

That there is a small gland in the brain in which the soul exercises its functions more particularly than in the other parts.

It is likewise necessary to know that although the soul is joined to the whole body, there is yet in that a certain part in which it exercises its functions more particularly than in all the others; and it is usually believed that this part is the brain, or possibly the heart: the brain, because it is with it that the organs of sense are connected, and the heart because it is apparently in it that we experience the passions. But, in examining the matter with care, it seems as though I had clearly ascertained that the part of the body in which the soul exercises its functions immediately is in nowise the heart, nor the whole of the brain, but merely the most inward of all its parts, to wit, a certain very small gland which is situated in the middle of its substance and so suspended above the duct whereby the animal spirits in its anterior cavities have communication with those in the posterior, that the slightest movements which take place in it may alter very greatly the course of these spirits; and reciprocally that the smallest changes which occur in the course of the spirits may do much to change the movements of this gland.

ARTICLE XXXII

How we know that this gland is the main seat of the soul.

The reason which persuades me that the soul cannot have any other seat in all the body than this gland wherein to exercise its functions immediately is that I reflect that the other parts of our brain are all of them double, just as we have two eyes, two hands, two ears, and finally all the organs of our outside senses are double; and inasmuch as we have but one solitary and simple thought of one particular thing at one and the same moment, it must necessarily be the case that there must somewhere be a place where the two images which come to us by the two eyes, where the two other impressions which proceed from a single object by means of the double organs of the other senses, can unite before arriving at the soul, in order that they may not represent to it two objects instead of one. And it is easy to apprehend how these images or other impressions might unite in this gland by the intermission of the spirits which fill the cavities of the brain; but there is no other place in the body where they can be thus united unless they are so in this gland.

$$. \quad \bullet \quad \bullet \quad \bullet \quad \bullet \quad \bullet$$

ARTICLE XXXIV

How the soul and the body act on one another.

Let us then conceive here that the soul has its principal seat in the little gland which exists in the middle of the brain, from whence it radiates forth through all the remainder of the body by means of the animal spirits, nerves, and even the blood, which, participating in the impressions of the spirits, can carry them by the arteries into all the members. And, recollecting what has been said above about the machine of our body, i.e., that the little filaments of our nerves are so distributed in all its parts, that on the occasion of the diverse movements which are there excited by sensible objects, they open in diverse ways the pores of the brain, which causes the animal spirits contained in these cavities to enter in diverse ways into the muscles, by which means they can move the members in all the different ways in which they are capable of being moved; and also that all the other causes which are capable of moving the spirits in diverse ways suffice to conduct them into diverse muscles; let us here add that the small gland which is the main seat of the soul is so suspended between the cavities which contain the spirits that it can be moved by them in as many different ways as [it] may also be moved in diverse ways by the soul, whose nature is such that it receives in itself as many diverse impressions, that is to say, that it possesses as many diverse perceptions as there are diverse movements in this gland. Reciprocally, likewise, the machine of the body is so formed that from the simple fact that this gland is diversely moved by the soul, or by such other cause, whatever it is, it thrusts the spirits which surround it towards the pores of the brain, which conduct them by the nerves into the muscles, by which means it causes them to move the limbs.

ARTICLE XXXV

Example of the mode in which the impressions of the objects unite in the gland which is in the middle of the brain.

Thus, for example, if we see some animal approach us, the light reflected from its body depicts two images of it, one in each of our eyes, and these two images form two others, by means of the optic nerves, in the interior surface of the brain which faces its cavities; then from there, by means of the animal spirits with which its cavities are filled, these images so radiate towards the little gland which is surrounded by these spirits, that the movement which forms each point of one of the images tends towards the same point of the gland towards which tends the movement which forms the point of the other image, which represents the same part of this animal. By this means the two images which are in the brain form but one upon the gland, which, acting immediately upon the soul, causes it to see the form of this animal.

* * * * *

ARTICLE XLI

The power of the soul in regard to the body.

But the will is so free in its nature, that it can never be constrained; and of the two sorts of thoughts which I have distinguished in the soul (of which the first are its action, i.e., its desires, the others its passions, taking this word in its most general significance, which comprises all kinds of perceptions), the former are absolutely in its power, and can only be indirectly changed by the body, while on the other hand the latter depend absolutely on the actions which govern and direct them, and they can only indirectly be altered by the soul, excepting when it is itself their cause. And the whole action of the soul consists in this, that solely because it desires something, it causes the little gland to which it is closely united to move in the way requisite to produce the effect which relates to this desire.

ARTICLE XLII

How we find in the memory the things which we desire to remember.

Thus when the soul desires to recollect something, this desire causes the gland, by inclining successively to different sides, to thrust the spirits towards different parts of the brain until they come across that part where the traces left there by the object which we wish to recollect are found; for these traces are none other than the fact that the pores of the brain, by which the spirits have formerly followed their course because of the presence of this object, have by that means acquired a greater facility than the others in being once more opened by the animal spirits which come towards them in the same way. Thus these spirits in coming in contact with these pores, enter into them more easily than into the others, by which means they excite a special movement in the gland which represents the same object to the soul, and causes it to know that it is this which it desired to remember.

ARTICLE XLIII

How the soul can imagine, be attentive, and move the body.

Thus when we desire to imagine something we have never seen, this desire has the power of causing the gland to move in the manner requisite to drive the spirits towards the pores of the brain by the opening of which pores this particular thing may be represented; thus when we wish to apply our attention for some time to the consideration of one particular object, this desire holds the gland for the time being inclined to the same side. Thus, finally, when we desire to walk or to move our body in some special way, this desire causes the gland to thrust the spirits towards the muscles which serve to bring about this result.

ARTICLE XLIV

That each desire is naturally connected with some motion of the gland, but that, by intention or habit, the will may be connected with others.

Nevertheless, it is not always the desire to excite a certain motion or other effect which causes it to be excited, for this relation varies according as nature or habit has variously united each motion of the gland to each thought. Thus, for example, if we wish to adjust our eyes to look at a very distant object, this desire causes the pupil to expand, and, if we wish to adjust our eyes so as to see an object very near, this volition makes it contract; but if we simply think of expanding the pupil, we will in vain—the pupil will not expand for that, for nature has not connected the motion of the gland, which serves to impel the [animal] spirits towards the optic nerve in the manner required for expanding or contracting the pupil, with the desire to expand or contract but instead with the desire for looking at objects distant or near. And when, in talking, we think only of the meaning of what we wish to say, that makes us move the tongue and lips much more rapidly and better than if we thought to move them in all the ways required for the utterance of the same words, inasmuch as the habit we have acquired in learning to talk has made us join the action of the mind—which, through the medium of the gland, can move the tongue and lips—with the meaning of the words that follow these motions rather than with the motions themselves.

45   FRANZ JOSEPH GALL (1758–1828) ON PHRENOLOGY, THE LOCALIZATION OF THE FUNCTIONS OF THE BRAIN, 1825

F. J. Gall, *Sur les fonctions du cerveau et sur celles de chacune de ses parties* . . . (Paris, 1825), IV, 128–151, VI, 307–310. Translated by Winslow Lewis as *Gall's Works: On the Functions of the Brain and Each of Its Parts, with Observations on the Possibility of Determining the Instincts, Propensities and Talents, and the Moral and Intellectual Dispositions of Men and Animals by the Configuration of the Brain and Head* (Boston, 1835), 6 vols. Volumes IV and VI, from which these excerpts are taken, are entitled respectively *Organology, or an Exposition of the Instincts, Propensities, Sentiments and Talents of the Moral Qualities and the Fundamental Intellectual Faculties in Man and Animals and the Seat of their Organs,* and *Critical Review of Anatomico-Physiological Works with an Explanation of a New Philosophy of Moral Qualities and Intellectual Faculties.*

Franz Joseph Gall, a good neuro-anatomist, was carried away by his enthusiasm for his anatomical personology, which he named phrenology. Phrenology had great vogue, but most of Gall's scientific contemporaries set their faces against it. It is hard to make a selection from the 600,000 words of these six volumes, but the text illustrates Gall's versatility and ingenuity and the nature of his evidence by quoting his argument for the location of the sentiment of property and especially of the associated propensity to theft. It also adds Gall's summary of his enterprise.

[EXCERPT 1]

VII. SENTIMENT OF PROPERTY; INSTINCT OF PROVIDING; COVETOUSNESS; PROPENSITY TO THEFT (EIGENTHUMSSINN, HANG ZU STEHLEN)

History

The errand-boys, and others of that class of people, whom I used to assemble in my house in great numbers, would frequently charge each other with petty larcenies, or, as they called them, *chiperies,* and took great pleasure in pointing out those who excelled in these practices; while the *chipeurs* themselves would come forward, proud of their superior address. What particularly struck me was, that some of these people showed the utmost abhorrence of thieving, and preferred starving to accepting any part of the bread and fruits their companions had stolen; while the *chipeurs* would ridicule such conduct and think it very silly.

When I could assemble a considerable number, I would often divide them into three classes. The first would include the *chipeurs;* the second, those who had an abhorrence of theft; and the third, those who regard it with indifference. On examining their heads, I was astonished to find that the most inveterate *chipeurs* had a long prominence, extending from the organ of cunning, almost as far as the external angle of the superciliary ridge; and that this region was *flat* in all those who showed a horror of theft; while, in those who were indifferent about it, the part

211

was sometimes more and sometimes less developed, but never so much as in the professed thieves.

These observations were not long in impressing me with the idea, that the propensity to steal might also be the result of organization. All the subjects of my observations were mere children of nature, left exclusively to themselves. None had received the least education, and their conduct, therefore, might well be regarded as the result of organization. Those who detested stealing, were often the very ones whose education had been completely neglected. The wants and circumstances of all of them were nearly the same, and the examples set before them were the same. To what cause, then, could the difference be attributed?

At this time I was physician to the Deaf and Dumb Institution, where pupils were received, from six to fourteen years of age, without any preliminary education. M. May, a distinguished physiologist, then director of the establishment; M. Venus, the teacher; and myself, had it in our power to make the most exact observations on the primitive moral condition of these children. Some of them were remarkable for a decided propensity for stealing; while others did not show the least inclination to it. The most of those who had stolen at first, were corrected of the vice in six weeks; while there were others, with whom we had more trouble, and some were quite incorrigible. On one of them, were several times inflicted the severest chastisements, and he was put into the house of correction; but it was all in vain. As he felt incapable of resisting temptation, he wished to learn the trade of a *tailor;* because, as he said, he might then indulge his inclination with impunity.

The more active and incorrigible the propensity to theft in these young people, the more indubitably was my first observation confirmed. Here, too, education could not come into the account; for, from the moment they came into the institution, their wants, their instruction, and the examples before them, were all the same. I was therefore obliged to conclude, that the propensity to steal is not an artificial product, but *is natural* to some people, and inherent in their organization. I took casts of the heads of all the confirmed thieves, to increase my means of comparison.

At this time, there was in the house of correction, a lad, fifteen years old, who had been a thief from his infancy, notwithstanding all the punishment he had received, and was finally condemned to confinement for life, as absolutely incorrigible . . . His head was small and unsymmetrical . . . and the forehead very retreating. His intellectual powers were so far below mediocrity, that I was astonished the incorrigible nature of his thieving propensity was not always attributed to this circumstance. In him, the region just mentioned, is very prominent, and the corresponding cerebral part was the only one very active; and as its activity was not balanced by the action of other parts, and the subject was incapable of higher motives, it became predominant. This case was conclusive proof to me, that the propensity to theft, is produced by a particular cerebral part, or has its proper organ.

Two citizens of Vienna, who had always lived irreproachable lives, became insane, and from that time they were distinguished in the hospital for an extraordinary propensity to steal. They wandered over the hospital, from morning till night, picking up whatever they could lay their hands on,—straw, rags, clothes, wood, &c.,—which they carefully concealed in the apartment which they inhabited in common; and though lodged in the same chamber, they stole from each other. In both, the cerebral part in question, was very much developed, and the corresponding region of the skull very prominent. The case of these two persons proves, that a man, whose intellect is not quite too feeble, may, in health, overcome the unfortunate impulses of certain organs. It proves, also, that the propensity to steal, proceeds from a particular cerebral part; for a quality which, independently of all others, may be carried to such a pitch of activity, as to deprive the individual of all control over the actions that result from it, must be referred to a cerebral part, independent of all the rest.

These facts were enough to induce me to pursue the natural history of the propensity to theft. My readers being probably acquainted with all that remains to be said on this subject, it will not be difficult to convince them, that the propensity to theft is innate, and has its proper organ.

Natural history of the propensity to steal

The following cases . . . conclusively prove, that the propensity to steal is not the result of moral depravement, nor of a defective education, but is an inherent quality in human nature.

Victor Amadeus I., King of Sardinia, was in the constant habit of stealing trifles. Saurin, pastor at Geneva, though possessing the strongest principles of reason and religion, frequently yielded to the propensity to steal. Another individual was, from early youth, a victim to this inclination. He entered the military service, on purpose that he might be restrained by the severity of the discipline; but having continued his practices, he was on the point of being condemned to be hanged. Ever seeking to combat his ruling passion, he studied theology and became a capuchin. But his propensity followed him even to the cloister. Here, however, as he found only trifles to tempt him, he indulged himself in his strange fancy with less scruple. He seized scissors, candlesticks, snuffers, cups, goblets, and conveyed them to his cell. An agent of the government at Vienna had the singular mania for stealing nothing but kitchen utensils. He hired two rooms as a place of deposit; he did not sell, and made no use of them. The wife of the famous physician Gaubius, had such a propensity to pilfer, that when she made a purchase, she always sought to take something. Countesses M., at Wesel, and P., at Frankfort, also had this propensity. Madame de W. had been educated with peculiar care. Her wit and talents secured her a distinguished place in society. But neither her education nor her fortune saved her from the most decided propensity to theft. Lavater speaks of a physician, who never left the room of his patients without robbing them of something, and who never thought of the matter afterward. In the evening his wife used to examine his pockets; she there found keys, scissors, thimbles, knives, spoons, buckles, cases, and sent them to their respective owners. Moritz, in his experimental treatise on the soul, relates with the greatest minuteness the history of a robber, who had the propensity to theft in such a degree, that, being "in articulo mortis," at the point of death, he stole the snuff-box of his confessor. Doctor Bernard, physician of his majesty the king of Bavaria, speaks of an Alsatian of his acquaintance, who was always committing thefts, though he had every thing in abundance, and was not avaricious. He had been educated with care, and his vicious propensity had repeatedly exposed him to punishment. His father had him enlisted as a soldier, but even this measure failed to correct him. He committed some considerable thefts, and was condemned to be hanged. The son of a distinguished literary man offers us a similar example. He was distinguished among all his comrades for his talents; but, from his early infancy, he robbed his parents, sister, domestics, comrades, and professors. He stole the most valuable books from his father's library. Every kind of means was tried to correct him: he was sent into the service, and underwent several times the most rigorous punishments: but all was useless. The conduct of this unhappy young man was regular in all other respects; he did not justify his thefts; but, if they addressed to him on this subject the most earnest and the most amicable representations, he remained indifferent; he seemed not to understand them. The almoner of a regiment of Prussian cuirassiers, a man otherwise well educated and endowed with moral qualities, had so decided a propensity to theft, that on the parade he frequently robbed the officers of their handkerchiefs. His general esteemed him highly; but as soon as he appeared, every thing was shut up with the greatest care; for he had often carried away handkerchiefs, shirts, and even stockings belonging to the women. For the rest, when he was asked for what he had taken, he always returned it cheerfully. M. Kneisler, director of the prison at Prague, once spoke to us of the wife of a rich shopkeeper, who continually robbed her husband in the most ingenious manner. It was found necessary to confine her in gaol; but she had no sooner escaped than she robbed again, and was shut up for the second time. Being set at liberty, new thefts caused her to be condemned to a third detention longer than the preceding. She even robbed in the prison. She had contrived, with great skill, an opening in a stove which warmed the room, where the money-box of the establishment was placed. The repeated depredations she committed on it were observed. They attached bells to the doors and windows to discover her, but in vain; at length, by the discharge of pistols which went off the moment she touched the box, she was so much terrified that she had not time to escape by the stove. In a prison at Copenhagen, we have seen an incorrigible robber, who sometimes distributed his gains to the poor. In another place, a robber shut up for the seventh time, assured us with sorrow, that it did not seem possible to him to conduct otherwise. He eagerly begged to be retained in prison, and be furnished with the means of gaining his living.

.   .   .   .   .

On the innate sentiment of Property

While embarrassed by the revolting idea of an innate propensity to theft, I thought of the following objection: Theft supposes property, but in nature there is no such thing as property; it is only the result of certain social conventions; therefore, there can exist no innate propensity to theft, nor organ of such propensity.

In all my public courses of lectures, I have noticed this objection, and refuted it. The opponents of organology have universally received it, as an unanswerable argument against the existence of the propensity to steal, and they have been busy in making it known. Although my reply may be found in the numerous works of my pupils; yet all my opponents have been dishonest enough to pass it over in silence, and have made the public acquainted with the objection only, for they calculated upon it for a certain victory. Let us see, therefore, if property does not really exist in nature, and whether property has produced the laws, or the laws, property.

Property is an institution of nature in brutes

Brutes have none of those laws and social conventions, from which property is said to result in man; yet property does exist with them, and they have a strong sense of it, too. They have their fixed abode, and the ardor with which they defend it against all usurpation, proves well enough that they consider it their property. When there is any fear lest the soil they occupy, should prove insufficient for their support, they are careful to drive away immediately every intruder. A certain number of chamois will inhabit a certain mountain, upon which they will suffer no other whatever to come. The wolf, fox, hare, marten, &c., occupy a district of a size proportioned to the quantity of nourishment, from which they instantly drive away all intruders. They who imagine that wild beasts wander at hazard through the woods, are deceived; for each of these animals has, in fact, a chosen abode, which it never abandons, unless forcibly driven away from it. When they are obliged to leave it, in the rutting season, or on account of inundations or the chase, they return as soon as circumstances will permit. The same pairs of storks, swallows, nightingales and redbreasts, return in the spring or in autumn, to the same country in which they had passed the season the preceding year, and establish themselves; the storks on the same steeple, the swallows under the same roof, and the nightingales in the same bushes. If another pair of birds attempt to seize the place already appropriated, war is immediately waged against them, and the intruders are forced to depart. These facts are known to every hunter and naturalist, and my own observations have confirmed them. He who would repeat them, must mark the old and not the young ones; for, with the lower animals, as with our own species, the parents remain in their establishment, and the young people go out.

Cows, returning from pasture, are observed, not only to enter their usual stable, but each one to take its own place and suffer no other to occupy it. We see the same thing in respect to geese and pigs. Each one of the thousand bees that come home loaded with honey, enters its own hive, and wo to the pilfering bees that undertake to lay a foreign hive under contribution! With what courage do all creatures defend their nest, their female, their young! What rash intrepidity will not the dog display in his master's house! Warmed, by the sense of property, how boldly he defends his bone against a stronger dog than himself! The stag leading along his harem, with a proud step and a firm look, seems to threaten every one that would encroach on his rights. Among the lower animals, a leader never yields the prerogatives obtained by his strength and address, and sanctioned by all the members of the republic. The dog and cat, in hiding food to be used when hunger returns; and the squirrel, hamster and jackdaw, which collect provisions for the future, undoubtedly have the notion of property in the stores they accumulate. If they have no such notion, why this ardor in collecting provisions, this anxiety to conceal them? Where do we see in nature such a contradiction between the instincts of animals and the end of those instincts? The manners of brutes, therefore, prove that the sense of property is inherent in their nature.

Property is an institution of nature in man

The opinion is constantly thrown out, even now, that the idea of property is unknown to the savage. "The idea of property," says Cuvier, "does not exist in savages, and they cannot have the same notion of theft as civilized nations have." But, let us see what travellers, such as Lafitau, Charlevoix, and the history of the Caribs teach us on this subject.

· · · · ·

Every uncivilized nation is a band of robbers, that pillage their neighbors without measure or remorse. The cattle in the fields are always fair game, and, in the spirit of such jurisprudence, the coasts of the Egean Sea are ravaged by Homer's heroes, for no other reason than that these same heroes were fond of seizing on the brass, iron, cattle, slaves and women of the surrounding people.

A Tartar mounted on a horse, is a real beast of prey, who knows only where cattle are to be found, and how far he must go to seize them. The same spirit reigned in all the uncivilized nations of Europe, Asia, and Africa. The antiquities of Greece and Italy, and the fables of all the ancient poets, are full of the examples of its influence. It was this spirit that first impelled our ancestors to enter the Roman empire; and in later times this spirit, rather than respect for the cross, led them into the East, to divide with the Tartars the spoils of the Saracenic empire. Even the lower animals steal, such as the cat, dog, magpie. I know a dog, that will eat only what he has stolen. When these creatures commit a larceny so skilfully as not to be detected, they are extremely delighted. A tame magpie cares nothing for a piece of money given to it, but the moment you hide it away, and pretend to look after it or not to think of it, he will take all the trouble in the world to get possession of it . . . These considerations, in relation to the sense of property, naturally lead us to the solution of the question, What is the fundamental quality, to which the propensity to steal, belongs?

The fundamental quality, to which the propensity to steal belongs, is the sentiment of property, or the propensity to make provisions

The sentiment of property and the propensity to provide for the future, are not only useful, but really indispensable, both in man and in brute. It was impossible to discover the organ of this propensity, confined to its primitive destination; this could be done, only when the organ was excessively developed. But, when the organ has obtained this degree of development and of corresponding activity, the legitimate sentiment of property, the rational propensity to make provision, of acquiring a competence, grows into a cupidity, which engenders a desire for appropriating the goods of another; and, finally, when the organ is developed in the very highest degree, unless prevented by internal and external motives, it degenerates into an irresistible impulse to theft. All these different vitiations are so many degrees of activity in a fundamental propensity, which is essential to the sentiment of property and the propensity to provide.

•   •   •   •   •

Seat and external appearance of the organ of Property and the propensity to make provision. Modifications of the manifestation of this organ

This organ is formed by [certain] convolutions . . . When these cerebral parts are very much developed, they produce a prominence on the head and skull, extending in a longitudinal direction . . . from the organ of cunning . . . nearly to the outer angle of the superior superciliary arch.

I have constantly found this prominence, in all inveterate thieves confined in prison, in all idiots with an irresistible propensity to steal, and in all those who, otherwise well endowed with intellect, take an inconceivable pleasure in stealing, and even feel incapable of resisting the passion which forces them to theft. One of my friends, a man of the finest talents, a good husband and father, and remarkably inclined to religious fanaticism, has this organ very large. When he sees scissors, knives, and other similar trifles, he feels a certain uneasiness (*malaise*) until he has put those objects in his pockets. He often has a store of tools of this kind in his house. If they are found there by the owner, he restores them with a hearty laugh; if not, he often presents them to his friends. He appears to be greatly delighted because two of his children have the same propensities, and manifests not the least concern about the influence they may have on their lot. These two children have the same organization as their father.

•   •   •   •   •

[EXCERPT 2]

SUMMARY REVIEW

In all my researches, my object has been to find out the laws of organization, and the functions of the nervous system in general, and of the brain in particular. The exposition of the nervous systems of the chest and abdomen, and of the spinal column, or of voluntary motion, has shown us the same laws, both as to their organization and purposes. The nervous fibres invariably originate from the gray substance, as their matrix, and finally expand on the surface. Whenever an essentially different function appears, there is invariably a particular nervous organization, or system, independent of the rest. I have demonstrated the same laws of organization in the brain. The cerebral nervous fibres all originate in the gray substance, and are successively reinforced by new masses of the gray substance. Many bundles of nervous fibres exist independently of one another, and all finally expand in a nervous membrane, either spread out, or rolled up in the form of convolutions. This uniformity of the laws of organization of all the nervous systems leaves no doubt as to the correctness of the anatomical discoveries of the nervous systems in general and of the brain in particular.

Having determined the functions of the nervous systems of the chest, abdomen, vertebral column, and the five senses, there still remains the great difficulty of determining the functions of the brain and its different parts . . . I have established the fact by a great many proofs, negative and positive, and by refuting the most important objections, that the brain alone has the great prerogative of being the organ of the mind. From further researches, on the degree of intelligence possessed by both man and brutes, we draw the conclusion, that the complexity of the brain of brutes is in proportion to that of their propensities and faculties; that the different regions of the brain are devoted to different classes of functions; and that finally, the brain of each species of animals, man included, is formed by the union of as many particular organs, as there are essentially distinct moral qualities and intellectual faculties.

The moral and intellectual dispositions are innate; their manifestation depends on organization; the brain is exclusively the organ of the mind; the brain is composed of as many particular and independent organs, as there are fundamental powers of the mind;—these four incontestable principles form the basis of the whole physiology of the brain.

These principles having been thoroughly established, it was necessary to inquire, how far the inspection of the form of the head, or cranium, presents a means of ascertaining the existence or absence, and the degree of development, of certain cerebral parts; and consequently the presence or absence, the weakness or energy of certain functions. It was necessary to indicate the means for ascertaining the functions of particular cerebral parts, or the seat of the organs, and finally, it was indispensable, to distinguish the primitive fundamental qualities and faculties, from their general attributes.

After that I was enabled to introduce my readers into the sanctuary of the soul and the brain, and give the history of the discovery of each primitive moral and intellectual power, its natural history in a state of health and of disease, and numerous observations in support of the seat of its organ.

An examination of the forms of heads of different nations, a demonstration of the futility of physiognomy, a theory of natural language, or pathognomy, added new weight to preceeding truths.

The thorough development of the physiology of the brain, has unveiled the defects of the theories of philosophers, on the moral and intellectual powers of man, and has given rise to a philosophy of man founded on his organization, and consequently, the only one in harmony with nature.

. . . . .

The physiology of the brain is entirely founded on observations, experiments, and researches for the thousandth time repeated, on man and brute animals. Here, reasoning has had nothing more to do with it, than to seize the results, and deduce the principles that flow from the facts; and therefore it is, that the numerous propositions, though often subversive of commonly received notions, are never opposed to or inconsistent with one another. All is connected and harmonious; every thing is mutually illustrated and confirmed. The explanation of the most abstruse phenomena of the moral and intellectual life of man and brutes, is no longer the sport of baseless theories; the most secret causes of the difference in the character of species, nations, sexes, and ages, from birth to decrepitude, are unfolded; mental derangement is no longer connected with a spiritualism that nothing can reach; man, finally, that inextricable being, is made known; organology composes and decomposes, piece by piece, his propensities and talents; it has fixed our ideas of his destiny, and the sphere of his activity; and it has become a fruitful source of the most important applications to medicine, philosophy, jurisprudence, education, history, &c. Surely, these are so many guarantees of the truth of the physiology of the brain—so many titles of gratitude to HIM, who has made them known to me!

46 PIERRE JEAN MARIE FLOURENS (1794–1867) ON THE FUNCTIONS OF THE BRAIN, 1824

P. J. M. Flourens, *Recherches expérimentales sur les propriétés et les fonctions du système nerveux dans les animaux vertébrés* (Paris, 1824), pp. 236–241. Translated for this book by Mollie D. Boring.

Pierre Flourens was a distinguished French neurophysiologist who suc- ceeded Georges Cuvier as perpetual secretary of the Académie des Sciences and was later elected to the Académie de France over Victor Hugo. He was a staunch opponent of Gall's phrenology. This excerpt cites the results of his experiments that led him to believe in the common action (*action commune*) and the specific action (*action propre*) of the different parts of the brain, a theory that combines equipotentiality of response with particular localization of function in a manner that foreshadows the findings of K. S. Lashley a century later. *Action commune* also furnished a basis for the neurological views of cerebral action held by those who opposed exact localization, such as Hughlings Jackson in 1884 and Henry Head in 1915.

THE UNITY OF THE NERVOUS SYSTEM

1. Each essentially distinct part of the nervous system has, as we have seen, a proper and fixed function.

The function of the cerebral lobes is to *will,* to *judge,* to *remember,* to *see,* to *hear,* or—in a word—to *feel.* The cerebellum *directs* and *coordinates* the movements of locomotion and grasping and the medulla oblongata those of conservation. The spinal cord *links* muscular contractions immediately excited by the nerves into whole movements.

2. Yet, independently of this proper and exclusive action of each part, each part has its common action, that is to say, an action of each upon the others and of the others upon each.

Thus, the cerebral lobes *wish* and *feel;* that is their proper action. The suppression of these lobes weakens the activity of the entire nervous system; that is their *common action.* The *proper action* of the cerebellum *coordinates* the movements of locomotion; its common action is to affect the activity of the entire system, and so forth and so on.

Each part of the nervous system—the cerebral lobes, the corpora quadrigemina, the medulla oblongata, the nerves—thus has a proper function; and that is what makes it a *distinct part:* but the activity of each of these parts affects the activities of all the others; and that is what makes them *parts* of a particular system.

3. If these facts be granted, the whole question of the *Unity of the Nervous System* is clearly reduced to an experimental evaluation of the way each distinct part of this system contributes to the common activity.

4. We have seen that the removal of the cerebral lobes leads to the weakening of movements, and the removal of the cerebellum to even further weakening, whereas the excision of the spinal cord, the medulla oblongata, or the nerves abolishes movements altogether. For, as we have also seen, the cerebral lobes determine the *willing* of movement, the cerebellum its *coordination,* whereas the medulla oblongata, the spinal cord, and the nerves *produce* movement.

5. In general we give the name *paralysis* indifferently to loss or weak-

ness of movement, no matter what nervous part is the source of this loss or weakness.

All this shows, then, that the word *paralysis,* applied to the destruction of the cerebral lobes or the cerebellum, can mean only *weakening,* as far as the locomotive faculties are concerned, whereas, applied to the destruction of the spinal cord or medulla oblongata, it means *complete loss* of those faculties.

6. We have seen further that, among the different parts of the nervous system involved in movements, some are involved in the movements of locomotion and the others in the movements of conservation. The destruction of these latter would therefore be more immediately fatal than the destruction of the former, since life depends directly on the movements of conservation and only in a roundabout and indirect manner on the movements of locomotion.

7. But there should, moreover, properly be brought to light another order of phenomena that includes both this efficacious *Unity* of the nervous system that joins all the parts of the system together, in spite of their diversity of action, and also the degree of influence that each of these parts contributes to the common activity.

8. When, by transverse section, we sever the spinal cord at some fixed point in its length, the posterior portion dies, while the anterior lives.

When, on the other hand, we cut the cerebral lobes by a similar transverse section, the posterior portion lives, while the anterior dies.

9. In ascending from the caudal end of the spinal cord toward a given point on the brain, the *portion* separated from the brain always dies.

In descending, on the other hand, from the cerebral lobes toward this point, the *portions* detached from the spinal cord always die.

Whether or not these cut-off *portions* live or die thus depends on whether or not they are connected with this point.

10. My experiments locate this important point in the *medulla oblongata,* that is to say, in the whole region of cord from the base of the corpora quadrigemina to the eighth pair of nerves.

This point is remarkable in many respects: impressions must pass through it to be perceived; the orders of the will must pass through it to be executed; the parts where feeling *resides* must terminate in it; the parts which *excite* movement must originate in it: life is maintained if the other parts of the nervous system are connected to it; life is lost if they are cut off from it: thus it is the central focus and the common bond of all these parts.

11. From the foregoing facts it follows that:

(1). Despite the diversity of action of each of its parts, the whole nervous system is still a particular system;

(2). Independently of the *proper action* of each part, each part has a *common action* with all the others, as have all the others with it;

(3). The word *paralysis,* applied to the destruction of the parts that *will* or *coordinate* movement, signifies only *weakness;* but, applied to the parts that *excite* or *produce* movement, it signifies *complete loss;*

(4). The influence of each part of the nervous system on life as a whole depends particularly on what sort of movement (whether of conservation or locomotion) is derived from it;

(5). Finally, there is, in the nervous system, a certain point between the parts of feeling and the parts of movement, very much like the *collar* between the stem and the root of vegetables: impressions must come to this point to be perceived; the orders of the will must go through it to be executed; the attachment of the parts to it makes them live; their detachment from it causes them to die. This point, therefore, is the central focus, the common bond, and—as Monsieur de Lamarck so aptly said of the vegetable *collar*—the *vital knot* of this system.

47 PAUL BROCA (1824–1880) ON THE SPEECH CENTER, 1861

Paul Broca, "Remarques sur le siège de la faculté du langage articulé, suivies d'une observation d'aphémie," *Bulletin de la Société Anatomique de Paris* [2] 6, 343–357 (1861). Translated for this book by Mollie D. Boring.

In spite of the scientific opposition to Gall, there was considerable resistance to acceptance of Flourens's doctrine of common action for the large portions of the brain. It seemed as if there must be specific functions for the various small regions. As early as 1825 Jean Bouillaud, a French physician and physiologist of some importance and influence, had assigned the speech function to the anterior portion of the cerebral lobes, and Paul Broca, a much younger man, had accepted this view and defended Bouillaud. Presently Broca came upon the case of the man with loss of speech or, as Broca put it, loss of the memory for words (aphemia). The man died of an infection shortly after Broca had examined him thoroughly, and Broca took the brain at once to the Société d'Anthropologie to demonstrate to them the deterioration in the third convolution of the left frontal lobe. This evidence made a great impression on scientists who were ready to believe in exact localization. Broca also argued that the convolutions of the cerebral cortex can be used to establish exact localization. Previously there had been uncertainty as to how to distinguish among the small regions of the brain.

A TWENTY-ONE-YEAR CASE OF APHEMIA PRODUCED BY THE CHRONIC AND PROGRESSIVE SOFTENING OF THE SECOND AND THIRD CONVOLUTIONS OF THE SUPERIOR PORTION OF THE LEFT FRONTAL LOBE

On 11 April 1861 there was brought to the surgery of the general infirmary of the hospice at Bicêtre a man named Leborgne, fifty-one years old, suffering from a diffused gangrenous cellulitis of his whole right leg, extending from the foot to the buttocks. When questioned the next day as to the origin of his disease, he replied only with the monosyllable *tan,* repeated twice in succession and accompanied by a gesture of his left hand. I tried to find out more about the antecedents of this man, who had been at Bicêtre for twenty-one years. I questioned his attendants, his comrades on the ward, and those of his relatives who came to see him, and here is the result of this inquiry.

Since youth he had been subject to epileptic attacks, yet he was able to become a maker of lasts, a trade at which he worked until he was thirty years old. It was then that he lost his ability to speak and that is why he was admitted to the hospice at Bicêtre. It was not possible to discover whether his loss of speech came on slowly or rapidly or whether some other symptom accompanied the onset of this affliction.

When he arrived at Bicêtre he had already been unable to speak for two or three months. He was then quite healthy and intelligent and differed from a normal person only in his loss of articulate language. He came and went in the hospice, where he was known by the name of "Tan." He understood all that was said to him. His hearing was actually very good, but whenever one questioned him he always answered, "Tan, tan," accompanying his utterance with varied gestures by which he succeeded in expressing most of his ideas. If one did not understand his gestures, he was apt to get irate and added to his vocabulary a gross oath ["Sacré nom de Dieu!"] . . . Tan was considered an egoist, vindictive and objectionable, and his associates, who detested him, even accused him of stealing. These defects could have been due largely to his cerebral lesion. They were not pronounced enough to be considered pathological, and, although this patient was at Bicêtre, no one ever thought of transferring him to the insane ward. On the contrary, he was considered to be completely responsible for his acts.

Ten years after he lost his speech a new symptom appeared. The muscles of his right arm began to get weak, and in the end they became completely paralyzed. Tan continued to walk without difficulty, but the paralysis gradually extended to his right leg; after having dragged the leg for some time, he resigned himself to staying in bed. About four years had elapsed from the beginning of the paralysis of the arm to the time when paralysis of the leg was sufficiently advanced to make standing absolutely impossible. Before he was brought to the infirmary, Tan had been in bed for almost seven years. This last period of his life is the one for which we have the least information. Since he was incapable of doing harm, his associates had nothing to do with him any more, except to amuse themselves at his expense. This made him angry, and he had by now lost the little celebrity which the peculiarity of his disease had given him at the hospice. It was also noticed that his vision had become notably weaker during the last two years. Because he kept to his bed this was the only aggravation one could notice. As he was not incontinent, they changed his linen only once a week; thus the diffused cellulitis for which he was brought to the hospital on 11 April 1861 was not recognized by the attendants until it had made considerable progress and had infected the whole leg . . .

The study of this unfortunate person, who could not speak and who, being paralyzed in his right hand, could not write, offered some difficulty. His general state, moreover, was so grave that it would have been cruel to torment him by long interviews.

I found, in any case, that general sensitivity was present everywhere, although it was unequal. The right half of his body was less sensitive than the left, and that undoubtedly contributed to the diminished pain at the site of the diffuse cellulitis. As long as one did not touch him, the patient did not suffer much, but palpation was painful and the incisions that I had to make provoked agitation and cries.

The two right limbs were completely paralyzed. The left ones could be moved voluntarily and, though weak, could without hesitation execute all movements. Emission of urine and fecal matter was normal, but swallowing was difficult. Mastication, on the other hand, was executed very well. The face did not deviate from normal. When he whistled, however, his left cheek appeared a little less inflated than his right, indicating that the muscles on this side of the face were a little weak. There was no tendency to strabismus. The tongue was completely free and normal; the patient could move it anywhere and stretch it out of his mouth. Both of its sides were of the same thickness. The difficulty in swallowing . . . was due to incipient paralysis of the pharynx and not to a paralysis of the tongue, for it was only the third stage of swallowing that appeared labored. The muscles of the larynx did not seem to be altered. The timbre of the voice was natural, and the sounds that the patient uttered to produce his monosyllable were quite pure.

Tan's hearing remained acute. He heard well the ticking of a watch but his vision was weak. When he wanted to see the time, he had to take the watch in his left hand and place it in a peculiar position about twenty centimeters from his right eye, which seemed better than his left.

The state of Tan's intelligence could not be exactly determined. Certainly he understood almost all that was said to him, but, since he could express his ideas or desires only by movements of his left hand, this moribund patient could not make himself understood as well as he understood others. His numerical responses, made by opening or closing his fingers, were best. Several times I asked him for how many days had he been ill. Sometimes he answered five, sometimes six days. How many years had he been in Bicêtre? He opened his hand four times and then added one finger. That made 21 years, the correct answer. The next day I repeated the question and received the same answer, but, when I tried to come back to the question a third time, Tan realized that I wanted to make an exercise out of the questioning. He became irate and uttered the oath, which only this one time did I hear from him. Two days in succession I showed him my watch. Since the second hand did not move, he could distinguish the three hands only by their shape and length. Still, after having looked at the watch for a few seconds, he could each time indicate the hour correctly. It cannot be doubted, therefore, that the man was intelligent, that he could think, that he had to a certain extent retained the memory of old habits. He could understand even quite complicated ideas. For instance, I asked him about the order in which his paralyses had developed. First he made a short horizontal gesture with his left index finger, meaning that he had understood; then he showed successively his tongue, his right arm, and his right leg. That was perfectly correct, for quite naturally he attributed his loss of language to paralysis of his tongue.

Nevertheless there were several questions to which he did not respond, questions that a man of ordinary intelligence would have managed to answer even with only one hand. At other times he seemed quite annoyed when the sense of his answers was not understood. Sometimes his answer was clear but wrong—as when he pretended to have children when actually he had none. Doubtless the intelligence of this man was seriously impaired as an effect either of his cerebral lesion or of his devouring fever, but obviously he had much more intelligence than was necessary for him to talk.

From the anamnesis and from the state of the patient it was clear that he had a cerebral lesion that was progressive, had at the start and for the first ten years remained limited to a fairly well-circumscribed region, and during this first period had attacked neither the organs of motility nor of sensitivity; that after ten years the lesion had spread to one or more organs of motion, still leaving unaffected the organs of sensitivity; and that still more recently sensitivity had become dulled as well as vision, particularly the vision of the left eye. Complete paralysis affected the two right limbs; moreover, the sensitivity of these two limbs was slightly less than normal. Therefore, the principal cerebral lesion should lie in the left hemisphere. This opinion was reinforced by the incomplete paralysis of the left cheek and of the left retina, for, needless to say, paralyses of cerebral origin are crossed for the trunk and the extremities but are direct for the face.

.   .   .   .   .

The patient died on 17 April [1861]. The autopsy was performed as soon as possible—that is to say, after 24 hours. The weather was warm but the cadaver showed no signs of putrefaction. The brain was shown a few hours later to the Société d'Anthropologie and was then put immediately into alcohol. It was so altered that great care was necessary to preserve it. It was only after two months and several changes of the fluid that it began to harden. Today it is in perfect condition and has been deposited in the Musée Depuytren . . .

.   .   .   .   .

The organs destroyed are the following: the small inferior marginal convolution of the temporal lobe, the small convolutions of the insula, and the underlying part of the striate body, and, finally, in the frontal lobe, the inferior part of the transverse frontal convolution and the posterior part of those two great convolutions designated as the second and third frontal convolutions. Of the four convolutions that form the superior part of the frontal lobe, only one, the superior and most medial one, has been preserved, although not in its entirety, for it is softened and atrophied, but nevertheless indicates its continuity, for, if one puts back in imagination all that has been lost, one finds that at least three quarters of the cavity has been hollowed out at the expense of the frontal lobe.

Now we have to decide where the lesion started. An examination of the cavity caused by the lack of substance shows at once that the center of the focus corresponds to the frontal lobe. It follows that, if the softening spread out uniformly in all directions, it would have been this lobe in which the disease began. Still we should not be guided solely by a study of the cavity, for we should also keep an eye on the parts that surround it. These parts are very unequally softened and over an especially variable extent. Thus the second temporal convolution, which bounds the lesion from below, exhibits a smooth surface of firm consistency; yet it is without doubt softened, though not much and only in its superficial parts. On the opposite side on the frontal lobe, the softened material is almost fluid near the focus; still, as one goes away from the focus, the substance of the brain becomes gradually firmer, although the softening extends in reality for a considerable distance and involves almost the whole frontal lobe. It is here that the softening mainly progressed and it is almost certain that the other parts were affected only later.

If one wished to be more precise, he could remark that the third frontal convolution is the one that shows the greatest loss of substance, that not only is it cut transversely at the level of the anterior end of the Sylvian fissure but it is also completely destroyed in its posterior half, and that it alone has undergone a loss of substance equal to about one-half of its total. The second or middle frontal convolution, although deeply affected, still preserves its continuity in its innermost parts; consequently it is most likely that the disease began in the third convolution.

. . . . .

Anatomical inspection shows us that the lesion was still progressing when the patient died. The lesion was therefore progressive, but it progressed very slowly, taking twenty-one years to destroy a quite limited part of the brain. Thus it is reasonable to believe that at the beginning there was a considerable time during which degeneration did not go past the limits of the organ where it started. We have seen that the original focus of the disease was situated in the frontal lobe and very likely in its third frontal convolution. Thus we are compelled to say, from the point of view of pathological anatomy, that there were two periods, one in which only one frontal convolution, probably the third one, was attacked, and another period in which the disease gradually spread toward other convolutions, to the insula, or to the extraventicular nucleus of the corpus striatum.

When we now examine the succession of the symptoms, we also find two periods, the first of which lasted ten years, during which the faculty of speech was destroyed while all other functions of the brain remained intact, and a second period of eleven years, during which paralysis of movement, at first partial and then complete, successively involved the arm and the leg of the right side.

With this in mind it is impossible not to see that there was a correspondence between the anatomical and the symptomological periods. Everyone knows that the cerebral convolutions are not motor organs. Of all the organs attacked, the corpus striatum of the left hemisphere is the only one where one could look for the cause of the paralysis of the two right extremities. The second clinical period, in which the motility changed, corresponds to the second anatomical period, when the softening passed beyond the limit of the frontal lobe and invaded the insula and the corpus striatum.

It follows that the first period of ten years, clinically characterized only by the symptom of aphemia, must correspond to the period during which the lesion was still limited to the frontal lobe.

48   GUSTAV FRITSCH (1838–1927) AND EDUARD HITZIG (1838–1907) ON CEREBRAL MOTOR CENTERS, 1870

Gustav Fritsch and Eduard Hitzig, "Ueber die elektrische Erregbarkeit des Grosshirns," *Archiv für Anatomie, Physiologie, und wissenschaftliche Medicin,* 1870, pp. 308–314. Translated for this book by Don Cantor.

Fritsch and Hitzig were contemporaries, both young lecturers at the University of Berlin when this research was done. Hitzig was the principal investigator, or at least he remained in brain physiology and wrote about his work with Fritsch later, whereas Fritsch moved over into physical anthropology, with an especial interest in South Africans. In 1870 it had long been believed that the tissues of the brain could not be excited by direct stimulation, but Hitzig had noted eye movements in a patient whose cortex was stimulated electrically. This experiment with Fritsch, which established a series of motor centers in the precentral region of the cerebral cortex, seemed also to validate the general fact of brain centers and thus led to the many investigations of the late nineteenth century that sought to discover motor and sensory centers and to localize them.

These experiments began with observations that one of us had occasion to make concerning the first recorded movements of voluntary muscles as elicited by direct stimulation of the central organ in man. It was found that, by leading continuous galvanic current through the back of the head, one easily obtains eye movements of a kind that can be brought about only through direct stimulation of the cerebral centers. In so far as these movements appear only when the quadrigeminal body is galvanized, one could consider them to depend on stimulation of that area of the head or its neighboring parts (and there is much to suggest this relation). Since, however, the movements were also found when certain devices that heighten excitability were used in galvanizing the temporal region, the question arose whether (1) the current that penetrated to the base caused the eye movements . . . or (2) the cerebral hemispheres do after all possess the capacity to be [directly] stimulated by electricity, contrary to the general view.

After a preliminary experiment in which one of us, using a rabbit, obtained a generally positive result, we proposed definitively to resolve the latter question in the following way.

The skulls of dogs were opened by a trephine at as flat a place as possible. The animals were given narcotics in the later experiments, but in the earlier ones they were not. Either an entire half of the skullcap or only that part of it covering the frontal lobe was removed with bone forceps rounded in front. In most cases, after finishing with the first hemisphere, we treated the remaining one in exactly the same way. Once a dog bled to death because of a slight lesion of the sagittal sinus, and thereafter we left intact one of these medial bone bridges, which shield the blood vessel. Next the dura, which so far had not been injured, was slightly incised, grasped with the forceps, and completely removed up to the margin of the bone. At this point the dogs showed acute pain by yelping and by characteristic reflex movements. Later, moreover, when the irritation from the air had been effective for a long time, the remnants of the dura became much more sensitive, a fact that had to be kept very carefully in mind in planning [subsequent] stimulation. We could, however, injure the pia mechanically or in any other way without eliciting any sign of sensitivity from the animals.

The apparatus for electrical stimulation was arranged as follows. The poles of a series of ten Daniell's cells were connected through a commutator to the terminals of a Pohl's switch from which the cross had been removed. The opposite terminals were connected with the secondary of an induction coil, and the middle terminals in parallel with a rheostat that had a resistance of 0.2100 Siemens unit [0.2226 ohm]. The main line continued by way of a Dubois key to two small insulated cylindrical screws that carried the electrodes, very fine platinum wires ending in tiny heads. These platinum wires passed through two small pieces of cork arranged so that one could change the distance between the two heads quickly and easily. This separation was usually 2–3 mm. It was necessary that the wires should have a very small mechanical resistance and be furnished with these heads; otherwise any unsteadiness of the hand and even the respiratory movements of the brain would lead immediately to injuries of the soft mass of the central organ. Generally the resistance in parallel was low, about 30–40 Siemens units [31.8–42.4 ohms]. [Thus] the intensity of the current was so low that, when the switch was closed and the tongue was touched by the two heads, a sensation of mere feeling was produced. Much higher intensities of current and cutting off of the parallel circuit were used only in control experiments. For most of the experiments we used an intensity that just barely evoked a sensation of feeling on the tongue.

Using this method, we arrived at the following results, which we present as the outcome of a very large number of the experiments on the brain of the dog. We do this without describing all the experiments separately because most of them agree even in the smallest details. With the exact description of our method just given, and with the following facts at hand, it is so easy to repeat our experiments that confirmations will soon appear.

One part of the convexity of the cerebrum of the dog is motor and another is not . . .

The motor area is generally more toward the front; the nonmotor area is more toward the back. In stimulating the motor area electrically, one elicits combined muscular contractions in the opposite half of the body. By using very weak current, one can localize these contractions exactly in narrowly delimited groups of muscle. If more intense current is used, other muscles—and indeed, also corresponding muscles in the [same] half of the body—immediately take part [in the reaction] as a result of the stimulation of the same or neighboring places. The possibility of isolated stimulation of a limited group of muscles is therefore limited to weak current on very small areas. For brevity's sake, we call these areas "centers." Minute shifting of the electrodes, to be sure, generally sets the same extremities in movement. But if, for example, extension is the first reaction, shifting the electrodes may produce flexion or rotation. We found that the part of the surface of the brain that lies between the centers lacked the capacity to be stimulated by the method described or by using minimal electrical intensity. If we increased either the distance between the electrodes or the intensity of the current, then convulsive movements could be brought forth; but these muscular contractions took place throughout the body, so that we could not once distinguish whether they were on one or both sides.

In dogs, the location of the centers (which will soon be described) is very consistent. The exact demonstration of this fact at first met with several difficulties, which we avoided by first seeking the place where the weakest electrical stimulus elicited the strongest contraction of the appropriate muscular group. Then we stuck a pin into the brain of the still-living animal between the two electrodes. After the brain had been removed we compared the points thus marked with those from earlier experiments that had been preserved in alcohol. How consistently the same centers are placed may best be seen in the fact that we repeatedly succeeded in finding the desired center in the middle of a single trephined hole without opening the skull further. When the dura was removed, the muscles dependent on that area contracted with the same regularity as if the entire hemisphere had been opened.

In the beginning we had great difficulties, even when the field of

operations was entirely free, for, as is known, although the single gyri
are quite uniform, developments in their individual parts and their
relation to one another show very significant differences. It is, in fact,
the rule rather than the exception that the corresponding gyri in the
hemispheres of the same animal are different in individual parts—sometimes in the middle part of the convexity, which is more pronounced
and sometimes in the portions to the front or rear. If one adds to this
[variation] the necessity of leaving much of the brain in its sheath and
the obscuring of the picture by the distribution of blood vessels (which
is for each case different but which generally makes the gyri indistinct),
then he will hardly be surprised at our initial difficulties . . .

The center for the neck muscles (see 1 in the illustration) is the
middle of the prefrontal gyrus, where its surface falls off steeply. The
outermost end of the postfrontal gyrus encloses the center for the ex-

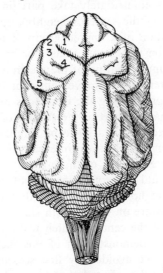

Dorsal view of dog's brain. The numbers indicate areas referred to in the text.

tensors and abductors of the anterior leg and lies at the end of the
frontal fissure (see 2). The areas of the centers for the flexion and
rotation of this member lie somewhat in back of this and nearer the
coronal fissure (see 3). The place for the posterior leg (see 4) is likewise
in the postfrontal gyrus, more toward the middle and somewhat more
to the rear than that for the anterior leg. The facial nerve (see 5) is
controlled from the middle portion of the supersylvian gyrus. This area
is often larger than 0.5 cm and stretches down and forward from the
main bend of the Sylvian fissure.

It is necessary to add that we were not always able to set the neck muscles in motion from the first-named center. We often elicited contractions in the back, tail, and stomach muscles from areas between the marked points, but we were unable to establish with certainty fixed points from which they alone could be stimulated. We could not stimulate the entire portion of the convexity behind the center for the facial nerve, even with excessively high intensities; no muscular contractions were obtained even when the parallel circuit was switched off and a current of ten Daniell's cells was used.

49   JOHN HUGHLINGS JACKSON (1835–1911) ON DISSOLUTION OF THE NERVOUS SYSTEM, 1884

J. Hughlings Jackson, *The Croonian Lectures on the Evolution and Dissolution of the Nervous System* (London, 1884), pp. 3–5.

The investigations stimulated by Fritsch and Hitzig's establishment of the motor centers belong mostly to the period between 1870 and 1900. Three important investigators were David Ferrier (1843–1928), who published *The Functions of the Brain* in 1876, Hermann Munk (1839–1912), who tended to support Ferrier in finding exact localizations, and Friedrich Leopold Goltz (1834–1902), who held more closely to Flourens's view that function tends to be common and localization only gross. Unfortunately their works are not easily excerpted, for the discussion is controversial and the experiments are very specific and are enmeshed in detail.

Meanwhile Hughlings Jackson, a man with a keen mind but a literary style that was not readily comprehensible, had been publishing widely during the period 1864–1884 scattered interpretative reports of his clinical cases of nervous disease. His brilliance was perceived at the time, for he was invited to give three important sets of lectures (1869, 1884, 1890), yet his general meaning was not clearly understood, so that his full significance had to be discovered later, first by Sigmund Freud in 1891 and then again by Henry Head in 1915, who had not discovered Freud's discovery.

Quite early Jackson confirmed Broca's finding for the speech center without knowing about Broca's discovery; he readily admitted Broca's priority when he learned of it. Nevertheless, in general Jackson's observations led him to reject exact localization and to develop an evolutionary view of higher and lower levels in the nervous system that functioned respectively for more complex and less complex functions. Herbert Spencer (1820–1903) was still living, and Jackson was his ardent disciple, knowing little about any other kind of psychology. It was Jackson more than anyone else who worked out the view of the evolutionary hierarchy of the nervous system, a conception that enabled him to consider nervous disease as the dissolution—that was Spencer's term for the opposite of evolution—of the nervous system. Disease attacks the higher levels first and proceeds as it increases to the lower; the lower are never disrupted without the higher having been affected. The following excerpt from the Croonian Lectures of 1884 gives this view.

The doctrine of Evolution daily gains new adherents. It is not simply synonymous with Darwinism. Herbert Spencer applies it to all orders of phenomena. His application of it to the nervous system is most important for medical men. I have long thought that we shall be very much helped in our investigations of diseases of the nervous system by considering them as reversals of Evolution, that is, as Dissolutions. Dissolution is a term I take from Spencer as a name for the reverse of the process of evolution.

.   .   .   .   .

Beginning with evolution, and dealing only with the most conspicuous parts of the process, I say of it that it is an ascending development in a particular order. I make three statements which, although from different standpoints, are about the very same thing. 1. Evolution is a passage from the most to the least organised; that is to say, from the lowest, well organised, centres up to the highest, least organised, centres; putting this otherwise, the progress is from centres comparatively well organised at birth up to those, the highest centres, which are continually organising through life. 2. Evolution is a passage from the most simple to the most complex; again, from the lowest to the highest centres. There is no inconsistency whatever in speaking of centres being at the same time most complex and least organised. Suppose a centre to consist of but two sensory and two motor elements; if the sensory and motor elements be well joined, so that "currents flow" easily from the sensory into the motor elements, then that centre, although a very simple one, is highly organised. On the other hand, we can conceive a centre consisting of four sensory and four motor elements, in which, however, the junctions between the sensory and motor elements are so imperfect that the nerve-currents meet with much resistance. Here is a centre twice as complex as the one previously spoken of, but of which we may say that it is only half as well organised. 3. Evolution is a passage from the most automatic to the most voluntary.

The triple conclusion come to is that the highest centres, which are the climax of nervous evolution, and which make up the "organ of mind" (or physical basis of consciousness) are the least organised, the most complex, and the most voluntary. So much for the positive process by which the nervous system is "put together"—Evolution. Now for the negative process, the "taking to pieces"—Dissolution.

Dissolution being the reverse of the process of evolution just spoken of, little need be said about it here. It is a process of undevelopment; it is a "taking to pieces" in the order from the least organised, from the most complex and most voluntary, towards the most organised, most simple, and most automatic. I have used the word "towards," for, if dissolution were up to and inclusive of the most organised, etc., if, in other words, dissolution were total, the result would be death. I say nothing of total dissolution in these lectures. Dissolution being partial, the condition in every case of it is duplex. The symptomatology of nervous diseases is a double condition; there is a negative and there is a positive element in every case. Evolution not being entirely reversed, some level of evolution is left. Hence the statement, "to undergo dissolution" is rigidly the equivalent of the statement, "to be reduced to a lower level of evolution." In more detail: loss of the least organised, most complex, and most voluntary, implies the retention of the more organised, the less complex, and the more automatic. This is not a mere truism, or, if it be, it is one that is often neglected. Disease is said to "cause" the symptoms of insanity. I submit that disease only produces negative mental symptoms answering to the dissolution, and that all elaborate positive mental symptoms (illusions, hallucinations, delusions, and extravagant conduct) are the outcome of activity of nervous elements untouched by any pathological process; that they arise during activity on the lower level of evolution remaining. The principle may be illustrated in another way, without undue recapitulation. Starting this time with health, the assertion is that each person's normal thought and conduct are, or signify, survivals of the fittest states of what we may call the topmost "layer" of his highest centres: the normal highest level of evolution. Now, suppose that from disease the normal highest level of evolution (the topmost layer) is rendered functionless. This is the dissolution, to which answer the negative symptoms of the patent's insanity. I contend that his positive mental symptoms are still the survivals of his fittest states, are survivals on the lower, but *then* highest, level of evolution. The most absurd mentation, and most extravagant actions in insane people are the survivals of their fittest states. I say "fittest," not "best;" in this connection the evolutionist has nothing to do with good or bad. We need not wonder that an insane man believes in what we call his illusions; they are his perceptions. His illusions, etc., are not caused by disease, but are the outcome of activity of what is left of him (of what disease has spared), of all there then is of him; his illusions, etc., are his mind.

After this brief sketch, I mention what may appear to be a drawback. Scarcely ever, if ever, do we meet with a case of dissolution which we can suppose to be the exact opposite of evolution. Often enough, however, do we meet with its near opposites. I will try to dissipate any difficulties that may arise. We make two broad divisions of cases of dissolution, Uniform and Local.

In Uniform Dissolution the whole nervous system is under the same conditions or evil influence; the evolution of the whole nervous system is comparatively evenly reversed. In these cases the whole nervous system is "reduced," but the different centres are not equally affected. An injurious agency, such as alcohol, taken into the system, flows to all parts of it; but the highest centres, being least organised, "give out" first and most; the middle centres, being more organised, resist longer; and the lowest centres, being most organised, resist longest. Did not the lowest centres for respiration and circulation resist much more than the highest do, death by alcohol would be a very common thing. Another way of stating the foregoing is to say that increasing uniform dissolution follows a "compound order;" these stages may be rudely symbolised thus, using the initial letters of, highest, middle, and lowest centres. First stage, or depth, of dissolution, h; second stage, $h^2 + m$; third stage, $h^3 + m^2 + l$; etc. Although I shall say very little, later on, of involvement of middle and lowest centres in cases of uniform dissolution, it is most important, especially with regard to clear notions on localisation, to recognise that the order of dissolution is a Compound Order.

The next division is Local Dissolution. Obviously, disease of a part of the nervous system could not be a reversal of the evolution of the whole; all that we can expect is a local reversal of evolution, that there should be loss in the order from voluntary towards automatic in what the part diseased represents. Repeating, in effect, what was said on uniform dissolution, it is only when dissolution occurs in all divisions of the highest centres that we can expect a reduction, from the most voluntary of all, towards the most automatic of all. Dissolution may be local in several senses. Disease may occur on any evolutionary level, on one side or on both sides; it may affect the sensory elements chiefly, or the motor elements chiefly. It must be particularly mentioned that there are local dissolutions of the highest centres. It will be granted that, in every case of insanity, the highest centres are morbidly affected. Since there are different kinds, as well as degrees, of insanity (for examples, general paralysis and melancholia), it follows of necessity that different divisions of the highest centres are morbidly affected in the two cases. Different kinds of insanity are different local dissolutions of the highest centres.

50   SHEPHERD IVORY FRANZ (1874–1933) ON THE VARIABILITY OF THE MOTOR CENTERS, 1915

S. I. Franz, "Variations in the distribution of the motor centers," *Psychological Monographs 19*, no. 81, 147–160 (1915).

Franz began work in the tradition of Broca, Fritsch and Hitzig, and David Ferrier, extirpating brain tissue to see what loss in behavior would result. He knew about Friedrich Goltz's failure to find exact localization but presumably not about Hughlings Jackson's ideas. The frontal lobes presented a great mystery at the turn of the century because, although they were considered to be "association areas," their exact function had not been determined empirically. In 1902 Franz published his monograph *On the Functions of the Cerebrum: The Frontal Lobes in Relation to the Production and Retention of Simple Sensory Habits.* He found that frontal lobectomy on one side only has very little effect on learning and memory but that on both sides it may cause the loss of recently formed habits, while old habits remain intact. He also found that habits lost by the destruction of brain tissue can be relearned, although the tissue destroyed does not regenerate. Thus under Franz's experimentation the pendulum was beginning to swing back toward Flourens and, although no one seems fully to have realized it at the time, toward Jackson. Franz poked fun at the "new phrenology" of Ferrier and the others. The following text excerpts his report of experiments that disparage Fritsch and Hitzig's finding for the motor centers, presumably until then the most generally accepted fact of cortical localization.

The results of the present research, in conjunction with the data of others which have been recorded above, indicate that the connections which are made by way of the cortical motor cells are not definite in the sense, for example, that there is a passage of an impulse from a Betz cell in the anatomically defined cerebral motor region to another particular efferent cell in the spinal cord, but that the connection is, in special senses of the terms, promiscuous or irregular. By these last terms I mean only that the connections which one particular efferent or afferent cell makes are connections with a great number of neurones, and that the impulses resulting from the activity of a cell body may affect many other cells. Or, in other terms, an impulse arising in one cell may activate or influence only one, or any number, of the cells which are anatomically associated with the particular cell with which we deal. It is quite generally admitted that a certain cell has the possibility of sending its impulses along the main neuraxon and this is the view which is implicitly apparent in most discussions of cerebral function. But it is also obvious that since this neuraxon gives off, as it passes to its final goal, certain collaterals it is quite as reasonable and quite as logical to conclude that it has also the possibility of sending impulses along any one of these, or along the main neuraxon and any number of the collaterals, or along one or more of the collaterals to the exclusion of the main trunk. It is this later method of looking at the activities of the cerebral cells which appears to me to solve some of the great difficulties of the exclusive neuraxon activity hypothesis.

. . . . .

This neurological conception may be applied to the understanding of the behavior differences of individuals, and also of the same individual from time to time. It appears probable that the variations in behavior of different animals and of the same animal at different times to the same form of stimulation are dependent upon the great number of connections and upon the variations in activity which the connection variations make possible. On the hypothesis that the connection between cortical cells is definite, in the sense that one cell acts solely or principally upon one other cell, we shall have great difficulty in explaining the phenomena in man or in animals which are grouped under the general heading of habit formation. To show this, let us briefly consider the facts regarding the formation of habit in several animals. We shall then realize how the same stimulus may result in different reactions in different animals, and how in one animal at different times different reactions may result from the same stimulus. Conversely also we shall get some neurological insight into the possible reason for similar reactions in different animals from different stimuli. On the assumption of definite connections and definite paths of discharge such facts are neurologically almost unexplainable.

Let us take for consideration a young cat, four to six months old, since an animal of this kind is readily "educable." If the animal is hungry it will be better, since the formation of the habit is then more readily obtained if the habit has one of its elements concerned with the obtaining of food. We prepare for our experiment a box with narrow slats in the front and a small door which is closed with a bolt. The knob of the bolt is attached to a cord which runs along, but an inch under, the top of the box and which the animal can reach either with its claws or by arching its back, or by biting with its teeth. When the cord is pulled downwards or pushed upwards or moved sideways the bolt is also moved. Any one of these actions (there may be others and also combinations of two or more of these actions) will, if sufficiently strong, result in the loosening of the bolt which keeps the door closed, and when the door is thus opened the cat is enabled to escape and to get a particle of food which is placed outside.

When we place a cat in an enclosed space of this character there is a very decided change in the behavior of the animal. It usually becomes very active. This activity we may describe, in terms which are not directly scientific in their psychological aspect, as being due to the desire on the cat's part to escape from the uncomfortable situation of being in an enclosed place of such small compass, and perhaps partly to the desire for the food which in some experiments it may see outside. The actions of the particular cat under these conditions are about the same as those of other animals of the same species which are placed in such a situation. The animal begins to scratch at the front of the cage, at the door, at the sides, at the top. It turns here and there, it takes hold of everything or anything which it can reach. These movements are not performed in any apparently logical order or in any apparently intelligent manner since the animal may at first try one corner, then the top, perhaps next the door. If these movements do not result in the escape of the animal from the "unpleasant" situation the cat may remain quiet for a time and begin all over again scratching at a front corner or a back corner, trying the top, the door, the slats at the front. Even though the special movements do not result in the release which is sought the movements are continued, and if the cat tries one thing and does not escape by so doing it may return to the first which it had previously found unsuccessful. The random movements, if they are continued for a sufficient length of time, eventually result in the animal's moving, either by clawing or by arching its back or by biting, the cord which holds the bolt. When the bolt has thus been lifted the activities of the animal may be continued for some seconds or minutes before it realizes or recognizes that the door is open and there is a possibility of escape. When the animal escapes from the situation it finds the food or it is given a small piece of food. When it is returned to the box which is again bolted it goes through the same kinds of activities, clawing here, biting there, resting, performing movements which are apparently purposeless since they are not directed to the part of the box by which escape becomes possible, or towards the mechanism whereby the door can be opened. In its random movements it again scratches the cord, and again escapes and gets food. At the next trial the animal goes through the same sort of movements. Finally it claws the cord, gets out, and in succeeding tests it is found that this animal which at first escaped because of biting the cord and then later by arching its back against it, and again by clawing at the cord eventually acquires the habit of escape by utilizing only one of these types of movement, namely the scratching or clawing at the cord. Furthermore it is found that when an animal is placed in this situation it eventually acquires the habit to such a degree, or the reaction is facilitated to such an extent, that immediately the animal is dropped into the box it goes to the particular location, claws at the cord thus opening the door, escapes and obtains the food.

Another animal goes through the same general kinds of activities in its escape or its attempts at escape, but instead of acquiring the habit of escaping by means of clawing at or by pulling the cord, it acquires the habit of arching its back and rubbing against the cord, thus putting the cord on a stretch and raising the bolt. A third animal learns to escape from the box by biting and pulling upon the cord.

It will be observed that as far as we can determine all three animals have been stimulated by exactly the same primary forms of stimuli. They have been stimulated by the sight of the box, by the appearance of the slats in the front, by the closed door, by other ill-defined sensations which are obtained from the confinement, perhaps from the stimulation of a variety of organs which go to make up, in human perceptual terms, the general feeling of being enclosed in the box. The sensory elements which are present in these three cases we most likely have the right to conclude are the same. The emotional elements or concomitants we do not know, if any exist, and we have at present no means of determining the similarity or variety of these mental conditions if they exist. It is to be noted however that although the sensory stimuli are the same the behavior to which the stimuli lead differs in the three animals. The reactions, it will be observed, have one thing in common, namely that they result in the escape of the animal. The actual means, however, of producing this desired situation differs for the three animals. Neurologically it is not only likely but it is almost certain that the impulses from the sense areas, those so-called associational impulses which start from the cells in the sensory regions of the brain, eventually concentrate in these three animals in different motor areas, or to put the matter in more probable terms, that the impulses originating in similar sensory cells in all three animals reach (a) the same or (b) a different frontal lobe cell or group of cells in all animals, and that (a) this similar frontal cell or group of cells discharges into different cells in the precentral area, or that (b) the different frontal cells influence motor cells.

Now it will furthermore be found that if an animal which has acquired the habit of escape from a box of this character, either by clawing or biting or rubbing against the cord, be placed in the same box and the movement which it has been accustomed to make results in no food or in no release, this movement is gradually given up. The situation becomes different, although the sensory stimuli remain the same. By holding the bolt or by making some external change in the mechanism (which is not seen by the animal) to prevent the escape by any movement of the cord, but to permit the escape whenever the animal sits quietly and licks itself, or washes its face by the characteristic series of paw movements, or scratches itself, the animal soon gives up the first habit which it had formed and replaces it by behavior which in itself has not apparently any direct bearing upon the desired result. We then have a similar primary series of stimuli which at one time results in a particular mode of activity (clawing the cord), and at another time in a different mode of activity (licking itself) in the same animal. Both lead to what may be considered the desired result, namely the escape from the enclosed box.

It should be understood that the sensory stimuli in two experiments of this character are not the same in their totality. The initial or primary sensory stimuli are, however, the same. When after the receipt of the primary sensory stimulation a reaction is produced the reaction results in an additional sensory stimulation, and this secondary stimulation, or the combination of the secondary with the primary, may give rise to another reaction. The animal which claws first at the front of the cage after the receipt of the primary stimulation has thereby a character or combination of stimulation different from that of the animal which first reaches for the top of the cage and tries to climb out in that way. Each animal however does have the same primary stimulation, or at least the same general primary stimulation, visual, tactile, organic, etc. To go back to the original stimulation we may even wonder why such similar primary stimuli have produced such diverse methods of behavior as that of clawing at the slats at the front of the cage and that of trying to bite the slats at the top. In either case, whether we consider the primary stimulus or the collection of stimulations which make up the whole experience of the animal in the box the sensory stimulations are sufficiently alike to presuppose (on the basis of exactness of neurological connections) an approximate similarity in the activity of the cerebral sensory areas, and to suggest (on the same hypothesis) that the efferent cerebral activity should be the same. This is, of course, on the very generally accepted belief that the impulses from corresponding sensory cells will always go to corresponding efferent cells.

On the hypothesis that there are definite connections established by means of certain cerebral neurones, and the hypothesis that when the stimulation reaches a particular sensory center it flows into other areas, eventually reaching the motor area and resulting in a particular type of movement, the varying activities of these animals are not understandable. It is not an explanation to say that one animal has certain sensory stimuli like those of another, but that there are different activities. Neurologically, there must be a basis for the different kinds of behavior. When we consider the possibility that the discharge from a certain cell may pass not only along the main neuraxon but also along any one or all of the collaterals and that in this manner we have the neural activity diffused, we have a possible explanation of the variety of the

241

actions of the same animal under similar conditions. If the receiving cell were definitely and solely (anatomically and physiologically) connected with a special cell or group of cells, the same sensory stimulus should result in the same kind of reaction in different animals and in the same animal at different times. But we find that at first the cat makes many random movements. In other words, neurologically we are led to conceive that the discharge of the sensory or receptive element is not only along the main neuraxon but is along all of the collaterals as well, and each in turn acts upon its cells or group of cells, producing impulses which eventually result in movements. These movements are random, i.e., not directly correlated with the stimuli nor with the desired result, but as the experience is repeated the animal gives up all but a certain amount of the reaction. Its behavior has changed. It is not only believable but probable that in the development of a particular type of activity or in the production of a particular association or habit, such as that of scratching or of biting or of arching the back, we may have two different neurological conditions . . . The variation in behavior of two animals may then be due to the primary stimulation of corresponding cells, but in one case the habitual reaction is determined by the flow of the impulses from these cells along the course of the main neuraxon and in the other case the habitual reaction is determined by the passage of the impulse along a collateral. These impulses reaching different efferent elements produce the varieties of behavior.

.   .   .   .   .

Thus far we have been considering what is doubtless the most simple neurological system, a system much simpler by far than that which is active in the production of any form of behavior higher than that of a reflex. When we deal with a system containing more than the two elements, afferent and efferent, or receptor and effector, the complexities of connections and the possibilities of variation in the physiological connections become apparent.

In this respect the cerebral cortex, or the cerebrum as a whole, may be looked at as a very labile organ because of the numerous possibilities of connections which may be made. One cell, let us say, may have close connections with a half-dozen or a dozen other cells, and the activity of the primary cell need not always be through all the branches. There is a possibility of a change in the direction of the impulse within the neurone. Thus at one time the main effect may be due to the influence exerted through a certain collateral, and at another time the effect may be due to the impulse passing through the main axon or through a second collateral. If this be true, it helps to understand why there is a possibility of change in reaction and a variability of reaction in the same individual from time to time. At one time the individual may have a discharge from a cortical motor cell along the main neuraxon acting upon a definite cell located in a definite region of the spinal cord. At another time the discharge may take place not only along the main neuraxon, but along one or more of the collateral branches, the actions resulting from the impulses passing through the collaterals being added to that due to the impulse along the main fiber, and the actions along these collaterals producing effects on other cells which either inhibit or alter in character the actions which were formerly produced, or new reactions may entirely replace the original activity by an activity of a very different character.

Nor does it appear necessary to believe that once a path, by way of the main trunk or by one of the collaterals, has been fixed that this fixity is a permanency. There may be a greater tendency to use this particular path after it has been used a number of times, but it may be said with certainty that the impulse may under suitable conditions traverse any one or all of the other collateral paths. In a state of "mental panic" a man acts very differently to a particular stimulus than at other times. His actions may be more diffuse or they may be the opposite of those which he habitually performs at normal times. Thus, the sounds of a rifle-shot heard at two different times although both be of equal intensity may give rise to varying reactions . . . Neurologically, however, it is not satisfying to say that the emotional condition gives the "set" to the discharge of a particular cell, or that it directs the character of the discharge, for we know nothing of the neurological conditions which give rise to or accompany affective states. But, should we admit that the emotional state can alter the character of the motor response due to such a simple stimulus as that of the sound of a rifle-shot, we are admitting at the same time that the impulse from a sensory cell, or group of cells, may pass through certain paths at one time and through other paths at other times . . . An illustration of this is that of the differences of speech, which are special reactions or forms of behavior, when the same picture of an object is shown at different times. At one time such a stimulus (the picture of an apple) may bring forth the reaction *"Apple,"* at another time *"Apfel,"* and at a third time *"Pomme."*

It seems most likely that these variations in activity are due to physiological variations in the traversing of the axon or the collaterals. It is not unlikely that as conductors the axon and the collaterals are physiologically equal, that they may be utilized equally well or equally often if occasion demands it, and that the definiteness of response to any particular stimulus is only a relative definiteness.

In considering the functions of the cerebrum, therefore, we must rid ourselves of any preconceived notions regarding the fixity or definiteness of connections. Fixity or definiteness of an anatomical nature there undoubtedly is, but this fixity or definiteness is on the physiological side a multiplicity of fixities and definitenesses. One cell undoubtedly communicates with many others, and while this is an anatomical fixity it does not result in a physiological definiteness since at one time such a cell may be conceived to discharge in one direction along one collateral and at another time in another direction along another collateral. At present we may not have sufficient information to guide us in determining the reasons for the discharge in this or that direction but the facts at hand indicate that discharges do take place in this manner.

51   KARL SPENCER LASHLEY (1890–1958) ON CEREBRAL
       EQUIPOTENTIALITY AND MASS ACTION, 1929

K. S. Lashley, *Brain Mechanisms and Intelligence: A Quantitative Study of Injuries to the Brain* (Chicago: University of Chicago Press, 1929), pp. 23–25, 74, 175–176.

Franz's contribution seemed largely to be negative: cerebral localization is not precise. Lashley worked with Franz when Lashley was at Johns Hopkins and Franz was at St. Elizabeth's Hospital in Washington, and they published a joint research paper in 1917, the last of Franz's contributions in this field and the first of Lashley's. Lashley advanced the understanding of the problem of cerebral localization and, in a certain sense, reinforced the regression toward Flourens by his development of the principles of cerebral equipotentiality and of mass action, which reached maturity in his thinking in this monograph of 1929.

DIFFICULTIES OF INTERPRETATION IN STUDIES OF BRAIN INJURIES

In analysis of the symptoms of brain injury it seems that we must take into account a number of variables which, because of practical difficulties of technique, are almost impossible of independent control. They enormously complicate the problem; yet, until we have some means of evaluating them severally, we can form no true conception of the cerebral mechanisms. The variables which may be clearly recognized in a series of cases seem to be:

1. *Individual variation in localization.*—Anatomical studies of the area striata have shown that the cortical fields delimited by cell structure vary considerably from one individual to another. Adequate data of this sort are available only for this area, and even here give only the fact of variation without determination of the limits of the range or the distribution of variates. By physiological methods also, indication of this variability is obtained. Observations here are unambiguous only for the

motor area, but the results of Franz [*Psychol. Monogr. 19,* No. 81, 80–161 (1915)] show clearly that even in the two hemispheres of the same animal the arrangement of excitable points differs greatly. Whether this variability is primarily the result of anatomical differences or whether it indicates that functional organization is in some measure independent of structure is uncertain. My observations on temporal variation in the function of the motor area [*Amer. J. Physiol. 65,* 585–602 (1923)] suggest that both anatomical variation and changes in physiological organization may be effective agents in producing the appearance of functional variability.

2. *Specific shock or diaschisis effects.*—Monakow [*Die Lokalisation im Grosshirn,* 1914] has emphasized the rôle played in the production of recoverable symptoms by temporary loss of function in one center as a result of destruction of another. The conception is doubtless a valuable one for the understanding of many cases of spontaneous recovery, but its practical application is complicated by the frequent difficulty in distinguishing between spontaneous recovery and recovery as a result of re-education. We have as yet no understanding of the manner in which the diaschisis effect is produced or any way of predicting the most probable shock effects from injury to any particular locus.

3. *Vicarious function.*—Improvement through re-education has been interpreted as the assumption of the functions of injured parts by others which have escaped injury. There is much incorrect speculation in the older literature concerning the parts functioning vicariously, as the assumption that the precentral gyrus of one side can assume the functions of that of the other; but there is no certain evidence that the reacquired functions are carried out vicariously by any specific loci. Attempts to discover such loci have been in almost all cases fruitless [Lashley, *Amer. J. Physiol. 59,* 44–71 (1922)] and point rather to a reorganization of the entire neural mass than to an action of specific areas. The spontaneous and re-educative improvements after cerebral lesions make it exceedingly difficult to draw final conclusions from any syndrome concerning cerebral function, since a gradual improvement may be ascribed to recovery from shock, even though it occurs during a post-operative retraining.

4. *Equipotentiality of parts.*—The term "equipotentiality" I have used to designate the apparent capacity of any intact part of a functional area to carry out, with or without reduction in efficiency, the functions which are lost by destruction of the whole. This capacity varies from one area to another and with the character of the functions involved. It probably holds only for the association areas and for functions more complex than simple sensitivity or motor co-ordination.

5. *Mass function.*—I have already given evidence [*J. comp. Neurol.*

41, 1–58 (1926)], which is augmented in the present study, that the equipotentiality is not absolute but is subject to a law of mass action whereby the efficiency of performance of an entire complex function may be reduced in proportion to the extent of brain injury within an area whose parts are not more specialized for one component of the function than for another.

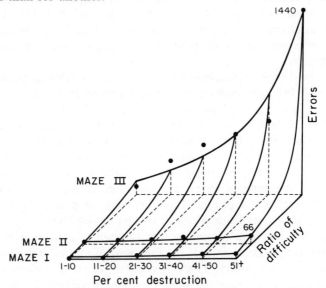

The relation between the extent of cerebral lesion, difficulty of the problem to be learned, and degree of retardation. The separation of the curves represents the relative difficulty of the problems for normal animals; the abscissae of the curves, the percentage destruction; and the ordinates, the number of errors made during training.

6. *Disturbances of the equilibrium within functional systems.*—There is a considerable mass of evidence which suggests that some symptoms, particularly in the class of motor inco-ordinations, may result from disturbances in the functional equilibrium between centers, although no tissue essential to the performance of the disturbed activities is directly involved in the lesion. Thus unilateral lesions to the corpus striatum or to the cerebellum may produce marked disturbances of co-ordination although bilaterally symmetrical lesions involving the same structures produce but slight effects.

. . . . .

SUMMARY

The influence of the extent of cerebral destruction in the rat was tested for a variety of functions, including retention of maze habits formed before cerebral insult, and learning and retention of several habits after the insult. The results may be summarized as follows:

1. The capacity to form maze habits is reduced by destruction of cerebral tissue.

2. The reduction is roughly proportional to the amount of destruction.

3. The same retardation in learning is produced by equal amounts of destruction in any of the cyto-architectural fields. Hence the capacity to learn the maze is dependent upon the amount of functional cortical tissue and not upon its anatomical specialization.

4. Additional evidence is presented to show that the interruption of association or projection paths produces little disturbance of behavior, so long as cortical areas supplied by them remain in some functional connection with the rest of the nervous system.

5. The more complex the problem to be learned, the greater the retardation produced by any given extent of lesion.

6. The capacity to form simple habits of sensory discrimination is not significantly reduced by cerebral lesions, even when the entire sensory field is destroyed.

7. This immunity is probably due to the relative simplicity of such habits.

8. The capacity to retain is reduced, as is the capacity to learn.

9. The maze habit, formed before cerebral insult, is disturbed by lesions in any part of the cortex. The amount of reduction in efficiency of performance is proportional to the extent of injury and is independent of locus.

10. Reduction in ability to learn the maze is accompanied by many other disturbances of behavior, which cannot be stated quantitatively but which give a picture of general inadequacy in adaptive behavior.

11. No difference in behavior in maze situations could be detected after lesions in different cerebral areas, and the retardation in learning is not referable to any sensory defects.

12. A review of the literature on cerebral function in other mammals, including man, indicates that, in spite of the greater specialization of cerebral areas in the higher forms, the problems of cerebral function are not greatly different from those raised by experiments with the rat.

From these facts the following inferences are drawn:

1. The learning process and the retention of habits are not dependent upon any finely localized structural changes within the cerebral cortex. The results are incompatible with theories of learning by changes in synaptic structure, or with any theories which assume that particular neural integrations are dependent upon definite anatomical paths specialized for them. Integration cannot be expressed in terms of connections between specific neurons.

2. The contribution of the different parts of a specialized area or of the whole cortex, in the case of non-localized functions, is qualitatively the same. There is not a summation of diverse functions, but a non-specialized dynamic function of the tissue as a whole.

3. Analysis of the maze habit indicates that its formation involves processes which are characteristic of intelligent behavior. Hence the results for the rat are generalized for cerebral function in intelligence. Data on dementia in man are suggestive of conditions similar to those found after cerebral injury in the rat.

4. The mechanisms of integration are to be sought in the dynamic relations among the parts of the nervous system rather than in details of structural differentiation. Suggestions toward a theory of the nature of these forces are presented.

52   HENRY HEAD (1861–1940) ON VIGILANCE, 1926

Henry Head, *Aphasia and Kindred Disorders of Speech,* I (New York: Macmillan, 1926), 479–487.

In 1905 Henry Head had with W. H. R. Rivers championed an evolutionary theory of cutaneous sensibility: protopathic sensibility comes first in evolution and epicritic sensibility later, whereas, when cutaneous nerves are injured, a Jacksonian dissolution occurs and a loss in epicritic sensibility precedes and is more extensive than the loss in the protopathic. Later, attacking the problems of aphasia and kindred degenerations of normal function, Head found in Hughlings Jackson the kind of evolutionary view that appealed to him and seemed best to fit the facts. In 1891 Sigmund Freud in his early monograph on the neurology of aphasia had already discovered Jackson and sought to bring him to the attention of neurologists, but neurologists were not then or immediately thereafter ready to attend to what Freud had to say. Actually Jackson's theory of dissolution seems to have been the ground for Freud's theory of regression. Head, after rediscovering Jackson, in 1915 devoted a long article to Jackson's views and published a complete bibliography of his scattered and difficult articles. The later Jacksonian theory of mental deterioration may thus be said to lead to Head's doctrine of vigilance, which specifies the way in which higher neural levels activate lower ones. These views of Head's were also influenced by Charles S. Sherrington (1857–1952), whose classical *Integrative Action of the Nervous System* (1906) shows how the simple reflex comes under the control of higher nervous levels.

The neurologist grows so familiar with the various forms of loss of function, produced by an organic lesion of the central nervous system, that he is liable to consider them all in anatomical or regional terms. Motor paralysis, defective sensibility, or disorders of speech seem to him to be equally explicable, provided he can discover the exact limits of the destruction of tissue to which they are due. He is liable to forget that, between these structural defects and the affection of sensibility or of speech, lie intermediate vital processes of profound importance. These have a determinant action on the form assumed by the manifestations. The effect of the structural changes does not depend only on their extent and severity; the mode and rapidity of onset and the general condition of the nervous system at the time are of even greater importance in determining the nature of the loss of function.

Now there is little difficulty in correlating a motor paralysis with destruction of cells and fibres in the spinal cord or even in the brain; we know that, so long as they live, they govern the function of muscular movement. Even loss of sensation seems easily explicable; for the neurologist is liable to consider mainly the destination of afferent impulses and to ignore the fact that sensation is a process of mental activity. So many of the duties exercised by afferent impulses are concerned with control and guidance of motion or other acts, which are performed outside consciousness, that the essentially psychical nature of sensation is neglected or forgotten. To many "afferent" and "sensory" are equivalent terms.

But when we consider disorders of speech, the mental aspect of the loss of function cannot be ignored. There are consequently three factors which equally demand recognition. First, the extent and nature of the anatomical destruction; secondly, the physiological disturbance it evokes, and lastly, the nature of the psychical manifestations we speak of as aphasia.

It is the second term of this chain that is habitually slurred over. The loss of tissue can be determined by anatomical examination and the clinical manifestations are patent to anyone who will take the trouble to observe them; but no intervening link is discoverable between them. Few to-day go the length of those stiff-necked anatomists who assert that single words are located in single cells or cell complexes; or that each psychical group of functions possesses an independent anatomical substratum, formed of certain cortical centres in combination with appropriate association paths. Yet, few have attempted to consider the steps by which a structural change evokes a physiological disorder which in turn is manifested as some defect of a specific psychical process.

We are accustomed to associate a violent injury to the head with loss of consciousness and other symptoms of concussion; empirically we are equally familiar with that curious unwitting state, which precedes the patient's complete recovery, during which he is liable to act apparently reasonably, but in a purely automatic manner. But we make no attempt to explain how a local lesion of the brain can abolish certain specific mental activities, leaving other independent functions of the mind intact.

In reality, all these phenomena are aspects of the same problem, though familiarity with the facts has staled the wonder they ought to excite within us. For in the phenomena of aphasia, we are face to face with the relation of body and mind in a form capable of experimental examination, and we are compelled to ask ourselves how a diminution in physiological activity can be associated with specific psychical defects.

·   ·   ·   ·   ·

[Reflexes] reach a marvellous complexity and perfection in the decerebrate animal. Here the nature of the movement of ear or tongue depends on the form and quality of the stimulus rather than on its strength. The response is truly discriminative, but can be abolished by any condition inimical to a high state of neural vitality, such as sepsis, haemorrhage into the wound, or the lighter degrees of chloroform narcosis. Apart, however, from such hostile influences, the decerebrate preparation may occasionally be found with its limbs in a state of flexion, as if asleep; any contact, or in some instances even a noise, will rouse it, and all four extremities at once reassume the characteristic posture of tonic extension. If we were dealing with an intact animal, we should say that these differences were associated with the presence or absence of consciousness.

But such an explanation of the reactions of the spinal or decerebrate preparation is unsatisfactory, and I would suggest that we employ the word "vigilance" for this state of high-grade physiological efficiency. Shortly after division, the spinal cord of man is in a condition of low vigilance. It is still excitable, for scratching the sole of the foot may produce a movement of the toes. Under favourable conditions the vigilance of the cord increases, until at last it is in a state of intense readiness to respond to excitation. In the cat the reaction may now assume a form of a series of complex movements, involving both lower extremities and necessitating coordinated excitation and inhibition. The tone of the flexor muscles has increased profoundly and, even in the extensors, has become sufficient for reappearance of a knee-jerk.

As revealed in the decerebrate preparation of Sherrington and his fellow-workers, vigilance is expressed in heightened extensor postural tone and acutely differentiated responses. This high state of physiological efficiency differs from a pure condition of raised excitability; for although the threshold value of the stimulus is not of necessity lowered, it is associated not only with an increased reaction but with highly adapted responses. These may vary profoundly according to circumstances, which are not inherent in the nature of the stimulus.

2. AUTOMATIC ACTIONS

If we did not know that the whole of the brain had been removed, we should say that the actions of the decerebrate animal were directed by consciousness. It initiates no movement spontaneously, but purposive adaptation is evident in every response. The character of its purring varies with the nature of the auditory stimulus; irritation of the ear excites a variety of highly differentiated movements, each of which is accurately designed to the end in view. Thus, every one of these actions may be purely automatic. Their formal recognition, if present, would be accessory to a mode of behaviour which can be carried into execution without it.

Both the decerebrate preparation and the normal cat react in an equally discriminating manner, when a flea or a drop of water is placed in the ear. The reactive significance of each stimulus is registered, and suitably adapted responses are prepared on a purely physiological level. But, whenever a reaction shows these characteristics, it demands a high state of vigilance in those parts of the nervous system necessary for its performance. The more highly differentiated the act, the greater degree of vigilance does it require and the more easily can it be abolished by toxic influences, such as chloroform, or by other conditions unfavourable to physiological activity. Experimentally we can watch the various responses disappear step by step under the influence of the narcosis.

Many actions acquired during the life of the individual tend to be carried out on the physiological level. A child has to be taught not to wet the bed at night, and this control of a spinal reflex becomes so completely automatic, that it is maintained even in the deepest sleep. But anything which lowers the vigilance of the central nervous system, such as dyspnoea or a debilitating illness, is liable to disturb this control and the bed is again wetted at night.

The same law is illustrated in many aptitudes, such as flying, shooting and out-door games, which are acquired and maintained by conscious effort. They never reach a high degree of perfection until the necessary movements are carried out unwittingly. The counter-stroke of an expert lawn-tennis player is determined at the moment when the ball leaves the opponent's racquet, whilst the tyro waits until it has bounded on his side of the net; a good shot watches the bird and pays no attention to his gun. Moreover, anything which at the moment concentrates the mind on the various stages of a mechanical action, diminishes the likelihood that it will be carried out successfully. The performer must concentrate on the goal or intention of his desires, and trust to habitual skill for its mechanical execution. But all these aptitudes are profoundly influenced by conditions of general health or by anything which lowers physiological vitality. A common cold, a gastro-intestinal attack or even mental worry may materially diminish any form of mechanical skill.

Every automatic act demands retentiveness and can be disturbed by vital states, which have nothing to do with consciousness. What wonder that the complex powers demanded by speech, reading and writing, can be affected by a lesion, which diminishes neural vitality. Vigilance is lowered and the specific mental aptitudes die out as an electric lamp is extinguished, when the voltage falls below the necessary level. The centres involved in those automatic processes, which form an essential part of the conscious act, may continue to live on at a lower vital level, as when under the influence of chloroform; they do not cease entirely to function, but the vigilance necessary for the performance of their high-grade activities has been abolished by the fall of neural potency.

VIII PSYCHOPHYSIOLOGICAL ISOMORPHISM

Psychophysiological isomorphism, also often called psychophysical isomorphism, is the theory that patterns of perception and of cerebral excitation show a one-to-one topological correspondence in which the spatial and temporal orders of items and events in the conscious and cerebral fields are the same, although spatial and temporal intervals between items and events, while they may correspond in their orders, do not agree in their magnitudes. This view was accepted as an inescapable postulate long before there was empirical evidence to support it. It is illustrated here by excerpts from Hering (1878), G. E. Müller (1896), Wertheimer (1912), and Köhler (1920).

53 EWALD HERING (1834–1918): ANTICIPATION OF PSYCHOPHYSIOLOGICAL ISOMORPHISM, 1878

Ewald Hering, *Zur Lehre vom Lichtsinne* (Vienna, 1878), pp. 74–80. Translated for this book by Don Cantor.

The Gestalt psychologists have argued that there must be a one-to-one topological correspondence between the relations (spatial, temporal, qualitative) of the perceptual pattern and the corresponding relations of the underlying excitatory pattern in the cerebral cortex. This view has a considerable history, having been posited by R. H. Lotze (1852), especially with respect to qualitative relations, by H. Grassmann (1853) with respect to series of colors, by G. T. Fechner (1860) in general, by Ernst Mach (1865) in general, by Ewald Hering (1878) for color theory, by G. E. Müller (1896) in his psychophysical axioms, by Max Wertheimer (1912) for perceived movement, and by Wolfgang Köhler (1920) in general. Here follow the relevant passages from Hering. Excerpts from G. E. Müller, Wertheimer, and Köhler come next.

It will not be disputed that light causes chemical changes in the nervous apparatus of the visual organ. That which is called fatigue or, more generally, variation in this apparatus' capacity to be stimulated is universally considered to be a matter of chemical change in the excitable substance. Even Fechner, who tried further to develop the resonance theory of the stimulation of the retina by light as it was originated by the physicists Herschel, Melloni, and Seebeck, felt himself called upon to take into consideration the chemical effect of light on the neural substance.

The chemical reactions aroused by light in the organ of sight were at first thought to be located in the retina. If, however, it is certain parts of the brain that take part in creating the sensations and images of

253

sight, then it must be that the chemical reactions in the retina in turn call forth chemical reactions in the nerves of sight, and that the latter then [stimulate chemical changes] in the brain substance. Because . . . we do not know whether the entire nervous substance of the organ of sight or only a part of it is to be regarded as the real psychophysical visual substance, nor, in the latter case, which part, we must temporarily be satisfied with the general assumption that vibrations in the ether produce some sort of chemical change in the organ of sight which, whether the chain of the chemical reaction be long or short and made of similar or dissimilar parts, leads finally to sensation.

For the rest, whatever has been imagined all along to be the type and place of process that takes place in the organ of sight, one error has been made by everyone: it has always been thought that only the sensation of brightness or of white—I wish . . . to avoid dealing with color entirely—is determined and sustained by certain alterations in the visual substance. The sensation of darkness or black has been totally ignored as to its physiological or psychophysical correlates . . . The facts as now developed force us to abandon this asymmetrical approach, to examine the sensations of sight, and to give the same attention to both of its principal variables, [the perception of] dark or black as well as [that of] bright or white.

I have shown how all sensations of the black-white series of sensations appear related in two ways, to have two types of moments in common, that is to say, sensations of brightness and darkness, and of black and white. I have shown further how each member of this series can be characterized by the relation in which these two moments are contained in the given sensation. If we now ask about the physical correlate of these sensations, about the psychophysical or psychochemical processes that underlie them, then the assumption that the physical correlate of the sensations of deepest blackness is nothing more than the lowest grade in intensity of the same process of which the highest intensity determines the sensation of greatest brightness or whiteness seems not only invalid of itself, but even full of contradictions and very awkward. Such an assumption calls for one and the same kind of psychophysical process for two qualities of sensation that are basically different. Our entire psychophysics, however, is based on the assumption that physical and psychological events are, in a certain sense, parallel, and especially that different qualities of sensation correspond to different qualities or forms of psychophysical processes.[1]

[1]This is often ignored, though it should really be self-evident to anyone who assumes a consistent functional relation between psychological and physical, sensation and neural, process. It seems to me that even Fechner, though guided by the same assumption, makes

We must give up the point of view that now prevails if we wish to avoid introducing with the first step into this difficult area a hypothesis that is in unresolved contradiction to the basic assumptions of all psychophysics and in any case creates a poor precedent for other entirely arbitrary and theoretically improbable suppositions. We can do this the more easily inasmuch as another offers itself that is entirely in agreement with the above-mentioned assumptions of psychophysics and at the same time satisfies much better than the present one the demands that may be placed upon such hypotheses from the standpoint of general nerve physiology. It is as follows: *Two different characteristics of the chemical process in the visual substance correspond to the two qualities of sensation which we call white, or bright, and black, or dark; and the relation of the intensities of the two psychophysical processes corresponds to [1] the relation of the distinctness or intensity with which each of the two sensations enters into [the perception of] the transition from pure white to pure black or [2] the relation in which pure white and pure black appear mixed.*

After a little thought, one can easily agree that this assumption is clearly the simplest possible, for it presents the simplest thinkable formula for the functional relation between physical and psychological events.

The formula is also sufficient for all the demands that can be made by general nerve physiology. We must assume a substance in the neural apparatus of sight that undergoes a change under the influence of incoming light; and this change, though one may if he wishes describe it physically, is nonetheless, as nerve physiology must suppose, at the same time a chemical process. If the effect of the light comes to a stop, the changed (more or less "fatigued") substance sooner or later returns to its original condition. This return can again be nothing other than a chemical alteration in the opposite direction. If one chooses to regard the chemical alteration that takes place under the influence of the light as a partial decomposition, one may call the return to the earlier condition a restitution; if one chooses to regard the decomposition as a process of splitting, the restitution is to be seen as synthetic; and so on.

---

too little use of it. Mach calls this basic assumption of all psychophysics simply "a fertile principle of psychophysical research." It is, however, more: it is the *conditio sine qua non* of all such research if that research is to bear fruit. Mach remarks: "Every psychological process corresponds to a physical one, and vice versa. Equal psychological processes correspond to equal physical ones, unequal to unequal. If a psychic process is resolved in a purely psychological way into a number of qualities *a*, *b*, and *c*, then these correspond to an equally great number of different physical processes α, β, and γ. All details of the psychical correspond to details of the physical."

This latter process, through which the living organic substance replaces that which has been lost through stimulation or activity, is usually called *Assimilation,* and I wish to retain this expression. It is generally assumed that each living and excitable organism produces certain chemical substances when active or stimulated. Analogously, I wish to call the creation of these products [of decomposition] *Dissimilation.*

The foregoing propositions, dealing with Assimilation . . . and Dissimilation . . . of the organic substance, are taken from the experience of general physiology and, in particular, nerve physiology. They have thus been developed quite independently of our hypothesis. Granted that they are correct, it is entirely incomprehensible that only the one type of chemical process in the apparatus of sight, namely dissimilation, should have a psychophysical significance, and the other, the assimilation process, none. The general view that only the chemical process that takes place under the direct influence of light, dissimilation, is sensed is clearly one-sided and unjustified. On the other hand, it seems entirely appropriate to attach to both kinds of chemical process the same value in respect to sensation.

If my hypothesis is correct, the sense of sight presents us with a way to observe exactly the process of nourishment in the seeing substance and its two principal parts, dissimilation and assimilation. Henceforth we shall be dealing not only with the idea that a complex of sensations is communicated by the human eye, which then makes them into pictures with the help of correct or false judgments or influence, *but with the proposition that whatever comes into our consciousness as a visual sensation is the physical expression or conscious correlate of the change of materials in the visual substance.*

For this change of substance, then, we have a reagent of great sensitivity, namely our consciousness. To be sure, it tells us at first nothing about the character of the chemical combination or decomposition. Probably, however, it reveals to us the entire course of change in assimilation and dissimilation, the law of their dependence on one another and on the vibrations of the ether, the rise and fall of the susceptibility of the visual substance to stimulation, and the dependence of this change in susceptibility on assimilation and dissimilation. So the topic of vision will for the first time become a truly integrated portion of physiology, whereas it has been until now necessarily [merely] the subject of a debate that was physical and philosophical rather than truly physiological.

·   ·   ·   ·   ·

[Heretofore] we have made full use of our sense perceptions to know the outer world and to make it useful to us. Now let us use them also to try to understand the material events within our own bodies so that, *first of all, we examine with their help things we do not perceive, like the outer things, through media only, but directly—that is to say, the change of material substance in our nervous systems.*

54   GEORG ELIAS MÜLLER (1850–1934) ON THE PSYCHOPHYSICAL AXIOMS, 1896

G. E. Müller, "Zur Psychophysik der Gesichtsempfindungen," *Zeitschrift für Psychologie 10,* 1–4 (1896). Translated for this book by Don Cantor.

G. E. Müller was for forty years (1881–1920) one of the great leaders of German experimental psychology at Göttingen; influenced by his predecessor at Göttingen, R. H. Lotze, and in color theory by the then outstanding phenomenologist of color, Ewald Hering, Müller undertook to lay down the fundamental psychophysical axioms for the correspondence of perceptual relations to cerebral-excitatory relations. Here follow his first four axioms. There was also a fifth axiom that had to do with the production of intermediates by the mixture of extremes. By *psychophysical* Müller meant *psychophysiological,* the correspondence between perception and neural excitation, as the word is used in the term *psychophysical parallelism.*

THE PSYCHOPHYSICAL AXIOMS AND THEIR USE IN [THE STUDY OF] VISUAL PERCEPTION

The Four First Axioms of Psychophysics

Psychophysics does not simply assume the validity of the axioms that are taught in physics and chemistry and are related to the behavior of the materials of those sciences. Rather, it builds upon them and upon certain axioms of its own. The latter deal with the relation between psychological conditions and the material events that correspond to them. At present, one can specify five such axioms of psychophysics. The following are the first four . . .

1. Every state of consciousness is based upon a material event, a so-called psychophysical process; the existence of the state of consciousness is bound up with the occurrence of the material event. (The definition of a psychophysical process asserts that every state of consciousness corresponds to a psychophysical process . . .)

2. An equivalence, similarity, or difference in the character of the sensations corresponds to an equivalence, similarity, or difference in the character of the psychophysical process, and vice versa. Indeed, a larger or smaller similarity of sensations corresponds to a larger or smaller

similarity in the psychophysical process, and vice versa. (The same is true of psychological conditions other than sensations, but here and in what follows they may be ignored.)

3. If the changes that occur in a sensation, or the differences in a series of sensations, have the same direction, then the changes occurring in the psychophysical process, or the differences in the psychophysical process, also have the same direction.

Similarly, changes or differences having the same direction in psychophysical processes always correspond to changes or differences of the same direction in sensations. If then a sensation can be changed in direction *x,* the psychophysical process upon which it is based must also be susceptible to change in that direction, and vice versa.

4. The directions in which a sensation can be altered are of various types. If the direction in which the sensation is altered leads to zero—that is to say, if continued alteration of the sensation in this direction finally reaches a point at which the sensation entirely disappears—one says that the sensation has undergone a loss in intensity. If the alteration is in the reverse direction, one speaks of an increase in intensity.

One of the many directions leading to zero in which a sensation can be altered has a particularly important place: it is the one in which the sensation, when steadily altered, reaches zero by the shortest path—that is, by passing through the smallest possible number of intermediate sensations. If the sensation is altered in this or exactly the opposite direction, we have a pure alteration in intensity. If the sensation is altered in one of the remaining directions leading to or from zero, we have a mixed change, one that concerns the quality as well as the intensity of the sensation. Alterations in sensation are called purely qualitative when they occur in a direction which leads neither to or from zero.

The foregoing comments will serve to clarify what we mean by an alteration in the intensity or quality of a sensation . . .

Now the following proposition (fourth psychophysical axiom) is valid: each qualitative change in sensation corresponds to a qualitative change in psychophysical process and vice versa; with a raising or lowering of the intensity of a sensation, the psychophysical process increases or lessens, and vice versa. If the alteration in quality or in intensity that the sensation undergoes is a pure one, then the change in the psychophysical process is one entirely of quality or of intensity, or vice versa.

It is clear that these four axioms are related to one another and that each is supplemented by the next in a definite way. It would be easier to summarize their content in a general statement of psychophysical parallelism; it is, however, more useful to formulate and bring forward singly each of the propositions, which taken together constitute this general principle of psychophysical parallelism. For the expression *psychophysical parallelism* is much too vague and permits interpretations that we are unable to accept, or at least regard as very dubious. Furthermore, many scholars (including some of high standing) who in general recognize the parallelism between the psychological and the physical, still do, in fact, contradict one or another of our . . . psychophysical axioms [in their thinking].

55   MAX WERTHEIMER (1880–1943) ON THE ISOMORPHIC RELATION BETWEEN SEEN MOVEMENT AND CORTICAL SHORT CIRCUIT, 1912

Max Wertheimer, "Experimentelle Studien über das Sehen von Bewegung, *Zeitschrift für Psychologie 61*, 247–250 (1912). Translated for this book by Don Cantor.

Gestalt psychologists accept Wertheimer as the originator of Gestalt psychology and this paper on seen movement as the starting point of the new movement. Attempting a physiological explanation of the Phi phenomenon—the perception of continuous movement when the stimulus is discrete stroboscopic displacement—Wertheimer appealed to psychophysical parallelism and isomorphism. Here is part of what he said.

In my opinion a physiological theory has two functional interrelations with experimental research. On the one hand, it should bring together the various generalities and isolated results coherently, arranging them so that they can be deduced one from the other. On the other hand—and this seems to be an essential point—a theory should further the progress of research by suggesting, through its coherent order, concrete questions that can be answered experimentally that are, to begin with, useful in testing the theory itself and lead to further understanding of the laws of phenomena.

With these principles in mind, a scheme for a physiologic consolidation is sketched here, one that has helped me to bring results together, has raised specific questions in the course of this experiment, and has up to the present time repeatedly proved itself productive of further work. Perhaps this hypothesis reaches into difficult and as yet unknown territory, but it seems to be demanded by present circumstances. Indeed, it seems both necessary and permissible, for the hypothesis leads to concrete problems requiring experimental resolution. I limit myself to sketching its essential features; a more detailed treatment will follow, in connection with the experimental handling of other pertinent areas. Our problem concerns certain central processes, physiological "cross functions" of a special kind, which serve as physiological correlates of the phenomenon [perceived movement].

Recent research in brain physiology has made probable the supposition that when a central spot, a, is stimulated, a certain area surrounding it is physiologically affected. If two places, a and b, are stimulated, the areas around both would be so affected. The area for this process of stimulation is predetermined.

Were a stimulated and then, after a given, short time, the closely related place, b, a kind of physiological short circuit from a to b would appear. In the space between the two places there would be a specific transfer of stimulation. If, for example, the effect in the area around b were to appear when the intensity of the effect in the area around a was at the high point of its course, the stimulation would flow over. This is physiologically a specific process; its direction is determined by the fact that a and the effect around a are there first.

The closer the two places a and b are, the more favorable are the conditions for the occurrence of the Phi process.

If t, the time elapsed between the beginning of the stimulation in the two successively excited places, a and b, is too great, the effect in the area around a will have died out when that around b is beginning (stage of succession). If the elapsed time is shorter, so that the effect around a is perhaps at the high point of its curve when the effect around b begins, then the transfer of stimulation occurs. If t is very short, or if the effect around a has not yet reached sufficient intensity at the critical moment, then the effects around a and b begin too closely together in time to render possible the available short circuit (stage of inactive simultaneity).

One may think of this central fusion [under the laws] of attention in several ways, but the following formulation is always correct: a place where this attentional fusion (whether it be higher excitability, higher capacity to communicate stimuli, or a more intense stage of stimulation) is present has a greater sensitivity to stimulation. The results of [our experiments] are a simple consequence of this proposition. Directing attention to the intervening space between a and b is favorable to the occurrence of the phenomenon . . .

It is natural that the effect in the area around a place that has been stimulated should be strongest near that place itself. If the optimal conditions are lacking—if, for example, the time, t, is between what is conducive to the greatest transfer of stimulation and the stage of inactive simultaneity—then the phenomenon is most effective at the borders of the two objects, and it may be below the threshold in the midregion (dual partial movement) . . .

It is to be expected that many effects that are in themselves too weak will be stronger in summation, and also that the occurrence of a definite physiological process is made easier by its having been repeatedly present.

On the other hand, if a strong transfer in a definite direction is constantly repeated, it should be expected that afterward, when the [repeated] stimulus leading the transfer in one direction is interrupted and a backward flow occurs as a tendency toward equilibrium and in the opposite direction, the negative afterimage [of seen movement] would appear.

With successive exposures, the resulting Phi processes will, under the best conditions, continually join [the successive items] together, giving rise to a single, continuous total event . . . Here we find the bridge to seeing persistent real movement: a continuous diminution of the "distance" that corresponds directly to the physical relations of real movement. The unified transfer of the Phi process has to be taken into account, along with the reception of the stimulus itself (and the processes that this reception controls), in considering the perception of real movement. In real movement one has a much greater range of optimal times . . . a fact that may be simply explained because the shorter this "distance," the larger the range of optimal t times.

56   WOLFGANG KÖHLER (1887–   ) ON ISOMORPHISM, 1920

Wolfgang Köhler, *Die physische Gestalten in Ruhe und im stationären Zustand* (Brunswick: Vieweg, 1920), pp. 189–193. Translated for this book by Don Cantor.

Wertheimer may have been the originator of Gestalt psychology, but Köhler was its prophet—that is to say, its interpreter, systematizer, and spokesman. This brilliant, difficult book, whose scholarly quality is said to have had great influence in obtaining the Berlin chair for Köhler in 1920, contains the discussion of the principle of psychophysical isomorphism shown in this excerpt. There are other clearer expositions by Köhler of this relation in 1929, 1938, and an especially clear one in 1947 (*Gestalt Psychology*, 2nd ed., New York: Liveright, 1947, pp. 55–66), but the early one is given here in order to avoid greater anachronism than necessary among the treatments of the various topics.

176. The physical areas in which excitations constitute the physical correlates of optical-phenomenal fields form a cohesive system.

Such a sentence would mean little if the characteristics of physical systems were unknown to us. However, having observed physical examples that give these words concrete meaning, we are obliged, from the point of view of physics, to draw this conclusion: psychophysical [psychophysiological] processes in the optical system display the general properties of the Gestalten of physical space.

177. This statement means somewhat more when examined in detail.

(1) Permanent conditions form and persist for the system as a whole. The processes of each limited region are supported by those in the remainder of the system, and vice versa; they arise and exist not as independent parts but only as moments [influences] in the expanded total process.

(2) Each actual psychophysical process depends on a definite complex of conditions, including:

(a) the total configuration of the stimulus upon the retina in the particular case,

(b) the relatively constant histological and material properties of the optical-somatic system,

(c) the relatively variable conditioning factors attributed primarily to the remainder of the nervous system and secondarily to the vascular system.

As with physical Gestalten, the psychophysical states must in principle be everywhere entirely dependent on the local conditions; their local moments must then conform to the total "topography."

(3) Assuming constant conditions and an invariable state, it follows from (1) that all of the extended process constitutes a unity that is objective and could not be formed capriciously by an observer, for there is in the entire area no local moment fully independent of, or without influence on, the state of any other region. The spatial coherence of the psychophysical process corresponding to a given visual field has, therefore, a suprageometric constitution, that is it is dynamically real.

(4) Here, as in physics, a physically real unity of Gestalten does not mean disorder or undifferentiated blending; it is, rather, fully compatible with strict articulation. The type of articulation depends on the specific type of the psychophysical process and the conditions in the system when it occurs; but in every case (that is, for every actual complex of conditions) the suprageometric dynamic articulation of the process is just as much a physically real property of the larger region as are the psychophysical color reactions for a given place in the field.

(5) The psychophysical Gestalten have in common with the inorganic physical ones the following degrees of inner unity throughout their systems. The moments in the smallest areas are dependent in principle on conditions in the entire system, but this dependence is a function of distance, so that the moments are more influenced by the conditioning forms in the region to which the given small area belongs and in the surrounding areas than by the topography of more distant parts. In extreme cases, just as with the physical Gestalten, the specific articulation of limited regions is not noticeably dependent on the details of form in other regions; it is rather the "total moments" of such regions

that influence one another, and the specific articulation of the limited regions is formed according to the conditions of the system within these regions . . . Such limited and clearly coherent regions can then be relatively independent in spatial articulation and structure without impairing the context of form in the entire system, the control by which the Gestalt moments are determined. Therefore, these regions represent, in a very real sense, still more narrowly circumscribed unities within the unified total process. If we keep in mind the general assumption that the process in the entire system is of a Gestalt character, we may . . . briefly designate such narrowly circumscribed unities as spatial Gestalten . . .

(6) No matter how the spatial articulation of psychophysical Gestalten may be otherwise constituted, it indicates in any case the specific type of distribution of the intensity of a situation or process and therefore of the density of energy. Under proper conditions, the intensity of energy in different regions can be very different. Here too, however, it is the entire complex of conditions in the system that is decisive . . .

178. These characteristics of the optical-somatic field correspond to the following ones of the phenomenal field . . .

I. Phenomenal optical fields appear as self-contained, coherent unities and always have suprageometric properties. Particular phenomenal regions never appear entirely as independent "parts." In this they correspond to the moments of the physical Gestalten.

II. Phenomenal unity involves order and structure, and the specific articulation of the phenomenal fields (the correlate of the state in physical Gestalten) represents a suprasummative property of the visual field which approaches the corresponding reality in experience, as, for example, in the filling out of the field with colors.

III. Without impairing the unity of the field as a whole, phenomenal unities of limited regions may appear in it. These unities are especially firmly contained in themselves and are relatively independent as compared with the remainder of the field.

IV. Strong, closely knit regions—"Gestalten" in the narrower sense— are especially apt to set themselves off clearly from the remaining "background" of the optical field when there are proper conditions for the given stimulus complex.

One immediately draws the conclusion that the strong and penetrating properties by which certain circumscribed regions emerge spontaneously as Gestalten (in the narrower sense)—so that from a phenomenal point of view they "are there" as something entirely different from being mere "background"—has its parallel in the strongly accented intensity of process, or density of energy, of the Gestalt's psychophysical correlate. In this way the phenomenal-suprageometric constitution of the optical field corresponds throughout to the physical-suprageometric properties of the self-contained, expanded condition. We have recognized an even more striking property of the visual field, a property that is likewise suprageometric but more functional than directly phenomenal. The conditions of the stimulus in widespread areas determine how limited regions of the field of vision are seen. That is true of spatial articulation in general as well as of the momentary fixing of the living "Gestalt" area (in the narrower sense); but it is also a general property of physical spatial Gestalten that the spatial expansion and the fixing of the Gestalt energy are ordered according to the conditions that exist in larger regions of the system.

.   .   .   .   .

180. When we limit ourselves, as we have until now, to the optical field, we stand, to be sure, only at the very beginning of our examination; yet we need not conceal that which is already clear: if the theory can really be worked through, it must lead us to accept a kind of essential similarity between the Gestalt properties of psychophysical processes and those of the phenomenal field—not only in general, in that Gestalten are important in both, but also for each specific Gestalt in every single case. A widespread opinion holds that the psychophysical events and the phenomenal givens which belong to them "can be compared neither in their elements nor in the manner in which these elements are connected" (Wundt). An entirely opposite view (which probably originates with Johannes Müller) asserts, on the other hand, that consciousness is the phenomenal representation of the essential properties of psychophysical events. This conception could until now be used only in the theory of the properties of order of colors in sensation, in a rule that states that the relations of the materials in the color system correspond exactly to the relations of materials in the system of possible color processes. Ours is a different and more radical opinion: actual perception is in every case related by real structural properties to the psychophysical processes (phenomenal and physical) that belong to it; the union is not left to chance; it is subject to laws.

IX THE REFLEX

No physiological concept has influenced psychology so broadly and deeply as the concept of the reflex, the quasi-mechanical triggering of isolated movements by narrowly specified stimuli. Reflexes, with their well-defined cleavage of stimulus and response and their machinelike reliability, quite early became the model of psychological analysis among theorists who favored a mechanistic approach, for they seemed to be the atoms out of which more complex molecules of behavior were composed. Since physiologists kept turning up new reflexes, showing more and more behavior to be essentially reflexive, it was natural for some psychologists to anticipate the time when all of man's actions could be resolved into these simple, scientifically attractive units. The eight selections in the present chapter pinpoint the developments in reflexology that were significant for psychology, without including the related, but separate, history of the reflex as a part of physiology.

Descartes (1662) established the reflex as the unit of analysis for involuntary action. His description of the physiology of the reflex became a focus of physiological research for at least two generations, research that soon replaced his speculations with an account that had better empirical support. Descartes himself did not contend that all human actions are reflexive, for he held that the rational soul, which was thought to distinguish man from the animals, has freedom to regulate and to alter the connections between sensory inputs and motor outputs. Almost a century later, La Mettrie (1748) in a radical extension of Descartes's position, denied freedom

to the soul by arguing for the all-inclusiveness of the principle of the reflex. Although La Mettrie did not contribute to the empirical study of reflexes, he gave the later supporters of a mechanistic philosophy of human action an early authority that they could invoke. Another anticipator of later developments in psychology was La Mettrie's English contemporary David Hartley, who showed (1749) that the classical doctrine of the association of ideas can be blended with the concept of the reflex to produce a psychology of action. When Pavlov effected a similar blending a hundred and fifty years later, it was to influence psychology greatly (see No. 101).

Physiological knowledge of the reflex accumulated rapidly after Descartes. The demonstration that reflexes, or their likes, exist was followed by studies of their dynamic properties, such as Whytt's (1751) on the relation between the intensity of the stimulus and the magnitude of the response. Also developing rapidly was the concept of nervous conduction, as in Prochaska's (1784) description of *vis nervosa* ("nervous force"). The anatomy of the reflex, too, was being elaborated. Bell (1811) and Magendie (1822) independently discovered the separation of sensory and motor nerves in the spinal cord (see Nos. 6 and 7), making it reasonable for Marshall Hall (1843, 1850) to argue that the spinal nervous system is essentially a collection of reflex arcs.

As the physiology of the reflex became more secure, the related psychological ideas became more widely and firmly established. Thus, when Sechenov (1863), the Russian physiologist, made the reflex the all-inclusive principle for man's actions, he could speak with greater authority and impact than had La Mettrie, the author of a similar argument a century earlier. The selection from Dewey (1896) shows, however, that psychology was not to rest easy under the constraints of a system based on the reflex alone.

57 RENÉ DESCARTES (1596–1650) ON MECHANISM IN HUMAN ACTION, 1662

René Descartes, *L'Homme* (Paris, 1664). Originally published in Latin as *De homine* (Leiden, 1662). Translated for this book by Tamar March from Victor Cousin, ed., *Oeuvres de Descartes*, IV (Paris, 1824), 349–363.

The inspiration for the idea of the manlike machine in this passage probably came to Descartes from the mechanical dolls popular in his time or from the hydraulic automata in the French royal gardens. The machine's actions comprise "reflexes" that are governed by a "nervous" system as Descartes conceived of it. Descartes wrote as if the apparent capacity of such a machine to mimic some human actions testified to the plausibility of his physiological model, which did not, however, long survive experimental test by his successors. Nevertheless, the description of human action in mechanistic terms of stimulus and response was eventually to prove highly effective. In other writings, Descartes told how the rational soul of man—which interacted with the body at the pineal gland in the brain—had control over human action above and beyond the mechanics of the nervous system.

I want to discuss first the composition of nerves and muscles, and to show how, from the very fact that the [animal] spirits that are in the brain enter into some nerves, they have at the same instant the power to move some member; then, touching briefly on the subject of respiration and on other such simple and ordinary movements, I shall discuss how exterior objects act on the organs of the senses; and after that I shall explain in detail all that is done in the cavities and the pores of the brain, how the animal spirits take their course there, and which of our functions this machine can imitate by means of these spirits: for if I were to begin with the brain, and only followed the course of the spirits, as I have done for blood, my discourse, it seems to me, would not be too clear.

Let us look, for example, at nerve *A* [Fig. 2], whose exterior skin is similar to a large tube, which contains several other small tubes *b, c, k, l,* and so on, composed of an interior skin that is thinner; and these two skins are continuous with the [other] two, *K, L,* which envelop the brain *MNO*.

Note also [Fig. 3] that in each of these little tubes there is a sort of a marrow composed of several very slender filaments, which come from

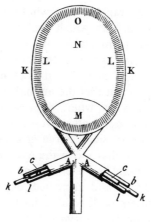

Fig. 2.

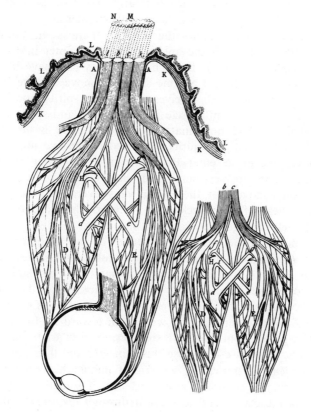

Fig. 3.

the very substance of the brain *N,* and whose extremities end, on the one side, at its interior surface, which faces its cavities, and on the other at the skin and flesh against which the tube containing them ends. But since this marrow is not useful to the movement of the members, it is enough for you to know now that it does not fill the small tubes that contain it to such an extent that the animal spirits cannot flow freely from the brain into the muscles, where these small tubes, which should be considered here as so many small nerves, lead.

Then note how the tube, or small nerve *bf* goes into muscle *D,* which I assume to be one of those that move the eye; and how, once there, it divides itself into several branches, each composed of a loose skin, which can be stretched, or widened and narrowed, depending on the amount of animal spirits that enters or leaves it; its branches or fibers are so arranged that when the animal spirits enter into them they cause the entire body of the muscle to inflate and become shorter, and consequently it pulls the eye to which it is attached; likewise the opposite, when they leave it, the muscle deflates and lengthens.

Moreover, note that, besides the tube *bf,* there is still another one, namely, *ef,* through which the animal spirits can enter into muscle *D,* and yet another, namely *dg,* through which they can leave it. And that similarly muscle *E,* which I assume serves to move the eye in the opposite way from muscle *D,* receives the animal spirits from the brain through tube *cg,* and from muscle *D* through *dg,* and sends them back to *D* through *ef.* And note that even though there is no observable passage through which the spirits contained in the two muscles *D* and *E* can leave, except to enter from one into the other, nevertheless, because their parts are very small and become incessantly even smaller by force of their movement, there are always several that escape through the skin and flesh of the muscles; but in return there are always others that come back through the two tubes *bf* and *cg.*

Finally, note that between the two tubes *bf* and *ef* there is a certain small skin *Hfi,* which separates these two tubes and serves as a door; it has two folds, *H* and *i,* so arranged that, when the animal spirits that tend to descend from *b* to *H* have more force than those that tend to go up from *e* to *i,* they lower and open this skin, thus enabling those in muscle *E* to flow very quickly with them toward *D.* But when those that tend to go up from *e* to *i* are stronger, or even when just as strong as the others, they raise and close this skin *Hfi,* and thus prevent themselves from leaving muscle *E;* whereas if neither has enough force to push it, it remains naturally ajar. And last, if sometimes these spirits contained in muscle *D* tend to leave it through *dfe,* or *dfb,* the fold *H* can be stretched and thus occlude its passage for them; and similarly, between the two tubes *cg* and *dg,* there is a small skin or valvule *g,* similar to the previous one, which remains naturally ajar and which can be closed by the spirits coming from tube *dg* and opened by those from *cg.*

Consequently it is easy to understand that, if the animal spirits in the brain do not tend, or hardly tend, to flow through tubes *bf* and *cg,* the two small skins or valvules *f* and *g* remain ajar, and thus the two muscles *D* and *E* are loose and inactive, especially since the animal spirits which they contain pass freely from one to the other, taking their course from *e* through *f* toward *d,* and conversely from *d* through *g* toward *e.* But if the spirits in the brain tend to enter with some force into the two tubes *bf* and *cg,* and if this force is equal on both sides, they immediately close the two passages *g* and *f,* and cause muscles *D* and *E* to swell as much as is possible, thus making them firmly hold and stop the eye in whatever position they find it.

Then if these spirits coming from the brain tend to flow with more force through *bf* than through *cg,* they close the small skin *g* and open *f,* and this more or less, depending on whether the spirits act more or less strongly; by this means the spirits contained in muscle [*E*] return to muscle *D* through the canal *ef;* and this faster or more slowly depending on whether the skin *f* is more or less open; so that muscle *D,* from which these spirits cannot come out, shortens, and *E* lengthens; and thus the eye is turned toward [*D*]. As on the contrary, if the spirits that are in the brain tend to flow with more force through *cg* than through *bf,* they close the small skin *f* and open *g;* so that the spirits from muscle *D* return immediately through the canal *dg* into muscle *E,* which as a result shortens and pulls the eye to its side.

For it is well known that these spirits, being like a wind or a very subtle flame, flow very quickly from one muscle into another as soon as they find some passage, although there is no other force that disposes them to it than the sole inclination they have to continue their movement according to the laws of nature. And you know besides that even though they are very mobile and subtle they nevertheless have the force to swell and tighten the muscles within which they are enclosed, just as the air in a balloon hardens it and causes the skin containing it to stretch.

Therefore it is easy to apply what I have just said about nerve *A,* and about the two muscles *D* and *E,* to all other muscles and nerves; and thus to understand how the machine of which I am speaking can be moved in all the same ways our bodies move, by the sole force of the animal spirits that flow from the brain into the nerves: because for each movement, and for its opposite, you may imagine two small nerves, or tubes, such as *bf* and *cg,* and two others such as *dg* and *ef,* and two small doors or valvules such as *Hfi* and *g.*

And as to the ways in which these tubes are inserted into the muscles, although these vary in a thousand ways, it is nevertheless not difficult to judge what they are by knowing what anatomy might teach concerning the exterior form and use of each muscle.

· · · · ·

And it is not difficult to judge from this that the animal spirits can sometimes cause movements in the members where nerves end, even though there are several where the anatomists do not note any visible nerves, as in the pupil of the eye, the heart, liver, gall bladder, spleen, and other similar [organs].

· · · · ·

Now in order to understand how the machine can be incited by exterior objects striking the organs of its senses to move its members in a thousand different ways, imagine that the small filaments, which I have already mentioned earlier as coming from the innermost part of the brain and as composing the marrow of the nerves, are so arranged in all parts serving as the organ to some sense that they can very easily be moved by the objects of the senses; and that, when they are moved somewhat strongly, they pull at the same time the parts of the brain, from whence they come, and thereby open the entrances of certain pores on the inner surface of the brain through which the animal spirits, which are in its cavities, begin immediately to take their course; the spirits go through them into the nerves and muscles, which serve to make movements in this machine very similar to those we are naturally incited to when our senses are touched in the same manner.

As, for example, if the fire A [Fig. 4] is close to the foot B, the small parts of this fire, which move, as you know, very rapidly, have the power to move along with them the part of the skin of this foot which they touch; and by this means, pulling the small filament CC, which you see attached to it, they open at the same time the entrance of the pore de, where this small filament ends, just as, by pulling one of the ends of a cord, you cause a bell attached to the other end to ring at the same time.

Now, the entrance of the pore or small passage being thus opened, the animal spirits of the cavity, F, enter and are carried by it, partly to the muscles that serve to withdraw the foot from the fire, partly to those that serve to turn the eyes and the head to look at it, and partly to those that serve to move the hands forward and to bend the body in order to bring relief.

But they can also be carried, by this same passage de, into several other muscles; and, before I stop to explain to you more exactly how

Fig. 4.

the animal spirits follow their course through the pores of the brain and how these pores are arranged, I want to discourse here specifically on . . . the senses, such as they are found in this machine, and explain how they are related to ours.

Know then, first, that there are a large number of small filaments similar to *CC*; they all begin to separate from one another at the inner surface of the brain, from whence they originate, and going from there to disperse throughout the rest of the body they serve as the organs for the sense of touch. For though ordinarily it is not they that are immediately touched by exterior objects, but the skins surrounding them, there nevertheless exists no more reason to think that it is these skins that are the organs of the sense than to think, when manipulating some object with gloved hands, that it is the gloves that serve to feel it.

.   .   .   .   .

Now I maintain that when God will unite a rational soul with this machine, as I shall assert hereafter, he will give it its principal seat in the brain, and will make it of such a nature that, according to the different ways the entrances of the pores in the interior surface of this brain will be opened by the intermission of the nerves, it will have varying sensations.

As, for example, if the small filaments that compose the marrow of these nerves are pulled with such force that they break and are separated from the part to which they were attached, so that the structure of the entire machine is somewhat less complete as a result, the movement they cause in the brain will give the soul . . . the occasion to experience the sensation of "pain."

And if they are pulled by a force almost as strong as the preceding one, but without breaking or being separated in any way from the parts to which they are attached, they will cause a movement in the brain which, giving evidence of the good constitution of the other members, will give the soul the opportunity to feel a certain corporeal pleasure that is called "tickling," and which, as you can see, being very close to pain in its cause, is its opposite in effect.

If several of these small filaments are pulled together equally, they will cause the soul to feel the surface of the body that touches the member where they end as "smooth," and they will cause it to feel it as "uneven" and "rough" if they are pulled equally.

If they are shaken just a little, independently of one another, as they continually are by the heat that the heart gives to the other members, the soul will have no sensation of it, or of any other ordinary actions; but if this movement is augmented or diminished in them by some unusual cause, its augmentation will cause the soul to have the sensation of "warmth," and its diminution that of "coldness"; and finally, according to the other diverse ways in which they will be moved, they will cause it to feel all the other qualities that belong to the sensation of touch in general, such as "humidity," "dryness," "weight," and the like.

Only it must be noted that, though they are very thin and very easy to move, they are not so to such an extent that they can bring to the brain all the smallest actions that are in nature; the smallest they do bring back to it are of the coarsest of terrestrial bodies, and there might even be some of these bodies whose parts, though quite coarse, will nonetheless slip against these small filaments so gently that they will press them or cut them completely, without their action passing to the brain; just as there are certain drugs that have the power to induce sleep or even to harm those of our members against which they are applied, without causing any sensation of it in us.

58 JULIEN OFFRAY DE LA METTRIE (1709–1751) ON THE EXTENSION OF MECHANISM TO THE HUMAN SOUL, 1748

J. O. de la Mettrie, *L'Homme machine* (Leiden, 1748). Translated by G. C. Bussey and M. W. Calkins as *Man a Machine* (Chicago and London: Open Court Publishing House, 1927), pp. 48–64.

During the eighty-odd years between Descartes and La Mettrie, numerous experiments established the reflex as a useful physiological concept. La

Mettrie's role was to complete the mechanistic picture of human action. He argued that the apparent uniqueness of human action is due to its complexity, not to the intervention of an incorporeal soul. In this way, La Mettrie, rejecting the Cartesian dualism of body and soul, was arguing for a complete monism of the body. At that time, this argument was radical enough to compel La Mettrie to escape from the heavily Catholic climate of France to the more liberal Protestant atmosphere of Prussia. La Mettrie's point of view never commanded a large following, however. Most of the philosopher-psychologists remained good dualists during the next hundred years, although there were some notable exceptions in France, such as Pierre Jean Georges Cabanis (1757–1808).

Since all the faculties of the soul depend to such a degree on the proper organization of the brain and of the whole body, that apparently they are but this organization itself, the soul is clearly an enlightened machine. For finally, even if man alone had received a share of natural law, would he be any less a machine for that? A few more wheels, a few more springs than in the most perfect animals, the brain proportionally nearer the heart and for this very reason receiving more blood—any one of a number of unknown causes might always produce this delicate conscience so easily wounded, this remorse which is no more foreign to matter than to thought, and in a word all the differences that are supposed to exist here. Could the organism then suffice for everything? Once more, yes; since thought visibly develops with our organs, why should not the matter of which they are composed be susceptible of remorse also, when once it has acquired, with time, the faculty of feeling?

The soul is therefore but an empty word, of which no one has any idea, and which an enlightened man should use only to signify the part in us that thinks. Given the least principle of motion, animated bodies will have all that is necessary for moving, feeling, thinking, repenting, or in a word for conducting themselves in the physical realm, and in the moral realm which depends upon it.

Yet we take nothing for granted; those who perhaps think that all the difficulties have not yet been removed shall now read of experiments that will completely satisfy them.

1. The flesh of all animals palpitates after death. This palpitation continues longer, the more cold blooded the animal is and the less it perspires. Tortoises, lizards, serpents, etc., are evidence of this.

2. Muscles separated from the body contract when they are stimulated.

3. The intestines keep up their peristaltic or vermicular motion for a long time.

4. According to Cowper, a simple injection of hot water reanimates the heart and the muscles.

5. A frog's heart moves for an hour or more after it has been removed from the body, especially when exposed to the sun or better still when placed on a hot table or chair. If this movement seems totally lost, one has only to stimulate the heart, and that hollow muscle beats again. Harvey made this same observation on toads.

6. Bacon of Verulam in his treatise "Sylva Sylvarum" cites the case of a man convicted of treason, who was opened alive, and whose heart thrown into hot water leaped several times, each time less high, to the perpendicular height of two feet.

7. Take a tiny chicken still in the egg, cut out the heart and you will observe the same phenomena as before, under almost the same conditions. The warmth of the breath alone reanimates an animal about to perish in the air pump.

· · · · ·

9. A drunken soldier cut off with one stroke of his sabre an Indian rooster's head. The animal remained standing, then walked, and ran: happening to run against a wall, it turned around, beat its wings still running, and finally fell down. As it lay on the ground, all the muscles of this rooster kept on moving. That is what I saw myself, and almost the same phenomena can easily be observed in kittens or puppies with their heads cut off.

10. Polyps do more than move after they have been cut in pieces. In a week they regenerate to form as many animals as there are pieces. I am sorry that these facts speak against the naturalists' system of generation; or rather I am very glad of it, for let this discovery teach us never to reach a general conclusion even on the ground of all known (and most decisive) experiments.

Here we have many more facts than are needed to prove, in an incontestable way, that each tiny fibre or part of an organized body moves by a principle which belongs to it. Its activity, unlike voluntary motions, does not depend in any way on the nerves, since the movements in question occur in parts of the body which have no connection with the circulation. But if this force is manifested even in sections of fibres the heart, which is a composite of peculiarly connected fibres, must possess the same property. I did not need Bacon's story to persuade me of this. It was easy for me to come to this conclusion, both from the perfect analogy of the structure of the human heart with that of animals, and also from the very bulk of the human heart, in which this movement escapes our eyes only because it is smothered, and finally because in corpses all the organs are cold and lifeless. If executed criminals were dissected while their bodies are still warm, we should probably see in their hearts the same movements that are observed in the face-muscles of those that have been beheaded.

The motive principle of the whole body, and even of its parts cut in pieces, is such that it produces not irregular movements, as some have thought, but very regular ones, in warm blooded and perfect animals as well as in cold and imperfect ones. No resource therefore remains open to our adversaries but to deny thousands and thousands of facts which every man can easily verify.

.   .   .   .   .

Let us now go into some detail concerning these springs of the human machine. All the vital, animal, natural, and automatic motions are carried on by their action. Is it not in a purely mechanical way that the body shrinks back when it is struck with terror at the sight of an unforeseen precipice, that the eyelids are lowered at the menace of a blow, as some have remarked, and that the pupil contracts in broad daylight to save the retina, and dilates to see objects in darkness? Is it not by mechanical means that the pores of the skin close in winter so that the cold can not penetrate to the interior of the blood vessels, and that the stomach vomits when it is irritated by poison, by a certain quantity of opium and by all emetics, etc.? that the heart, the arteries and the muscles contract in sleep as well as in waking hours, that the lungs serve as bellows continually in exercise,—that the heart contracts more strongly than any other muscle? . . .

I shall not go into any more detail concerning all these little subordinate forces, well known to all. But there is another more subtle and marvelous force, which animates them all; it is the source of all our feelings, of all our pleasures, of all our passions, and of all our thoughts: for the brain has its muscles for thinking, as the legs have muscles for walking. I wish to speak of this impetuous principle that Hippocrates calls *enormon* (soul). This principle exists and has its seat in the brain at the origin of the nerves, by which it exercises its control over all the rest of the body. By this fact is explained all that can be explained, even to the surprising effects of maladies of the imagination . . .

Look at the portrait of the famous Pope who is, to say the least, the Voltaire of the English. The effort, the energy of his genius are imprinted upon his countenance. It is convulsed. His eyes protrude from their sockets, the eyebrows are raised with the muscles of the forehead. Why? Because the brain is in travail and all the body must share in such a laborious deliverance. If there were not an internal cord which pulled the external ones, whence could come all these phenomena? To admit a soul as explanation of them, is to be reduced to [explaining phenomena by] the operations of the Holy Spirit.

In fact, if what thinks in my brain is not a part of this organ and therefore of the whole body, why does my blood boil, and the fever of my mind pass into my veins, when lying quietly in bed? I am forming the plan of some work or carrying on an abstract calculation. Put this question to men of imagination, to great poets, to men who are enraptured by the felicitous expression of sentiment, and transported by an exquisite fancy or by the charms of nature, of truth, or of virtue! By their enthusiasm, by what they will tell you they have experienced, you will judge the cause by its effects; by that harmony which Borelli, a mere anatomist, understood better than all the Leibnizians, you will comprehend the material unity of man. In short, if the nerve-tension which causes pain occasions also the fever by which the distracted mind loses its will-power, and if, conversely, the mind too much excited, disturbs the body . . . ; if an agitation rouses my desire and my ardent wish for what, a moment ago, I cared nothing about, and if in their turn certain brain impressions excite the same longing and the same desires, then why should we regard as double what is manifestly one being? In vain you fall back on the power of the will, since for one order that the will gives, it bows a hundred times to the yoke. And what wonder that in health the body obeys, since a torrent of blood and of animal spirits forces its obedience, and since the will has as ministers an invisible legion of fluids swifter than lightning and ever ready to do its bidding! But as the power of the will is exercised by means of the nerves, it is likewise limited by them.—

Does the result of jaundice surprise you? Do you not know that the color of bodies depends on the color of the glasses through which we look at them, and that whatever is the color of the humors, such is the color of objects, at least, for us, vain playthings of a thousand illusions? But remove this color from the aqueous humor of the eye, let the bile flow through its natural filter, then the soul having new eyes, will no longer see yellow. Again, is it not thus, by removing cataract, or by injecting the Eustachian canal, that sight is restored to the blind, or hearing to the deaf? How many people, who were perhaps only clever charlatans, passed for miracle workers in the dark ages! Beautiful the soul, and powerful the will which can not act save by permission of the bodily conditions, and whose tastes change with age and fever! Should we, then, be astonished that philosophers have always had in mind the health of the body, to preserve the health of the soul, that Pythagoras gave rules for the diet as carefully as Plato forbade wine?

The regime suited to the body is always the one with which sane physicians think they must begin, when it is a question of forming the mind, and of instructing it in the knowledge of truth and virtue; but these are vain words in the disorder of illness, and in the tumult of the senses. Without the precepts of hygiene, Epictetus, Socrates, Plato, and the rest preach in vain: all ethics is fruitless for one who lacks his share of temperance; it is the source of all virtues, as intemperance is the source of all vices.

．　　．　　．　　．　　．

Grant only that organized matter is endowed with a principle of motion, which alone differentiates it from the inorganic . . . and that among animals, as I have sufficiently proved, everything depends upon the diversity of this organization: these admissions suffice for guessing the riddle of substances and of man. It [thus] appears that there is but one [type of organization] in the universe, and that man is the most perfect [example]. He is to the ape, and to the most intelligent animals, as the planetary pendulum of Huyghens is to a watch of Julien Leroy. More instruments, more wheels and more springs were necessary to mark the movements of the planets than to mark or strike the hours . . . In like fashion, it was necessary that nature should use more elaborate art in making and sustaining a machine which for a whole century could mark all motions of the heart and of the mind; for though one does not tell time by the pulse, it is at least the barometer of the warmth and the vivacity by which one may estimate the nature of the soul. I am right! The human body is a watch, a large watch constructed with such skill and ingenuity, that if the wheel which marks the seconds happens to stop, the minute wheel turns and keeps on going its round, and in the same way the quarter-hour wheel, and all the others go on running when the first wheels have stopped because rusty or, for any reason, out of order . . . And is not this the reason why the loss of sight (caused by the compression of the optic nerve and by its ceasing to convey the images of objects) no more hinders hearing, than the loss of hearing (caused by obstruction of the functions of the auditory nerve) implies the loss of sight? In the same way, finally, does not one man hear . . . without being able to say that he hears, while another who hears nothing, but whose lingual nerves are uninjured in the brain, mechanically tells of all the dreams which pass through his mind? These phenomena do not surprise enlightened physicians at all. They know what to think about man's nature, and . . . of two physicians, the better one and the one who deserves more confidence is always, in my opinion, the one who is more versed in the physique or mechanism of the human body, and who, leaving aside the soul and all the anxieties which this chimera gives to fools and to ignorant men, is seriously occupied only in pure naturalism.

Therefore let the pretended M. Charp deride philosophers who have regarded animals as machines. How different is my view! I believe that Descartes would be a man in every way worthy of respect, if, born in a century that he had not been obliged to enlighten, he had known the value of experiment and observation, and the danger of cutting loose from them. But it is none the less just for me to make an authentic reparation to this great man for all the insignificant philosophers—poor jesters, and poor imitators of Locke—who instead of laughing impudently at Descartes, might better realize that without him the field of philosophy, like the field of science without Newton, might perhaps be still unculti-vated.

This celebrated philosopher, it is true, was much deceived, and no one denies that. But at any rate he understood animal nature, he was the first to prove completely that animals are pure machines. And after a discovery of this importance demanding so much sagacity, how can we without ingratitude fail to pardon all his errors!

In my eyes, they are all atoned for by that great confession. For after all, although he extols the distinctness of the two substances, this is plainly but a trick of skill, a ruse of style, to make theologians swallow a poison, hidden in the shade of an analogy which strikes everybody else and which they alone fail to notice. For it is this, this strong analogy, which forces all scholars and wise judges to confess that these proud and vain beings, more distinguished by their pride than by the name of men however much they may wish to exalt themselves, are at bottom only animals and machines which, though upright, go on all fours. They all have this marvelous instinct, which is developed by education into mind, and which always has its seat in the brain, (or for want of that when it is lacking or hardened, in the medulla oblongata) and never in the cerebellum; for I have often seen the cerebellum injured, and other observers have found it hardened, when the soul has not ceased to fulfil its functions.

To be a machine, to feel, to think, to know how to distinguish good from bad, as well as blue from yellow, in a word, to be born with an intelligence and a sure moral instinct, and to be but an animal, are therefore characters which are no more contradictory, than to be an ape or a parrot and to be able to give oneself pleasure . . . I believe that thought is so little incompatible with organized matter, that it seems to be one of its properties on a par with electricity, the faculty of motion, impenetrability, extension, etc.

59 DAVID HARTLEY (1705–1757) ON VOLUNTARY AND INVOLUNTARY ACTION, 1749

David Hartley, *Observations on Man, His Frame, His Duty, and His Expectations* (London and Bath, 1749), bk. I, chap. 1, sect. 3.

Hartley's encyclopedic work on man includes a description of neuro-muscular physiology that resembles the reflex. Some movements, said Hartley, are automatic, or involuntary, reactions to stimuli. He went on to argue that the principle of association by contiguity could be used to explain, on the one hand, how automatic move-ments come under voluntary control and, on the other hand, how voluntary movements become automatic. Al-though Hartley's primary contribution was to the development of association-ism (No. 70), this account of move-ment was one of the early efforts to bridge the gap between involuntary and voluntary action.

PROP. 18

The Vibrations, of which an Account has been given in this Chapter, may be supposed to afford a sufficient Supply of motory Vibrations, for the Purpose of contracting the Muscles.

In order to make this appear, it will be proper to distinguish the motory Vibrations, or those which descend along the Nerves of the Muscles into their Fibres, into the . . . following Classes:

First then, we are to conceive, that those sensory Vibrations which are excited in the external Organs, and ascend towards the Brain, when they arrive, in their Ascent, at the origins of motory Nerves, as they arise from the same common Trunk, Plexus, or Ganglion, with the sensory ones affected, detach a Part of themselves at each of these Origins down the motory Nerves; which Part, by agitating the small Particles of the muscular Fibres, in the manner explained [before], excites them to Contraction.

Secondly, The Remainder of the sensory Vibrations, which arrives at the Brain, not being detached down the motory Nerves in its Ascent thither, must be diffused over the whole medullary Substance. It will therefore descend from the Brain into the whole System of motory Nerves, and excite some feeble Vibrations, at least, in them. The same may be observed of ideal Vibrations generated in the Brain by associa-tion; these must pervade the whole medullary Substance, and conse-quently affect all the motory Nerves in some Degree.

Thirdly, The Heat of the Blood, and Pulsation of the Arteries, which pass thro' the medullary Substance, must always excite, or keep up, some Vibrations in it; and these must always descend into the whole System of Muscles. And I apprehend, that from these Two last Sources, taken together, we may account for that moderate Degree of Contraction, or

279

Tendency thereto, which is observable in all the Muscles, at least in all those of healthy Adults, during Vigilance.

. . . . .

PROP. 21

The voluntary and semivoluntary Motions are deducible from Association, in the manner laid down in the last Proposition.

In order to verify this Proposition, it is necessary to inquire, What Connections each automatic Motion has gained by Association with other Motions, with ideas, or with foreign sensations . . . so as to depend upon them, *i.e.* so as to be excited no longer, in the automatic Manner [just] described, . . . but merely by the previous Introduction of the associated Motion, Idea, or Sensation. If it follows that Idea, or State of Mind (*i.e.* Set of compound Vibratiuncles), which we term the Will, directly, and without our perceiving the Intervention of any other Idea, or of any sensation or motion, it may be called voluntary, in the highest Sense of this Word. If the Intervention of other Ideas, or of Sensations and Motions (all which we are to suppose to follow the Will directly), be necessary, it is imperfectly voluntary; yet still it will be called voluntary, in the Language of Mankind, if it follow certainly and readily upon the Intervention of a single Sensation, Idea, or Motion, excited by the Power of the Will: But if more than one of these be required, or if the Motion do not follow with Certainty and Facility, it is to be esteemed less and less voluntary, semivoluntary, or scarce voluntary at all, agreeably to the Circumstances. Now, if it be found, upon a careful and impartial inquiry, that the Motions which occur every Day in common Life, and which follow the Idea called the Will, immediately or mediately, perfectly or imperfectly, do this, in proportion to the Number and Degree of Strength in the Associations, this will be sufficient Authority for ascribing all which we call voluntary in Actions to Association, agreeably to the Purport of this Proposition. And this, I think, may be verified from Facts, as far as it is reasonable to expect, in a Subject of Inquiry so novel and intricate.

In the same manner as any Action may be rendered voluntary, the Cessation from any, or a forcible Restraint upon any, may be also, viz. by proper Associations with the feeble Vibrations in which Inactivity consists, or with the strong Action of the Antagonist Muscles.

After the Actions, which are most perfectly voluntary, have been rendered so by one Set of Associations, they may, by another, be made to depend upon the most diminutive Sensations, Ideas, and Motions, such as the Mind scarce regards, or is conscious of; and which therefore it can scarce recollect the Moment after the Action is over. Hence it follows, that Association not only converts automatic Actions into voluntary, but voluntary ones into automatic. For these Actions, of which the Mind is scarce conscious, and which follow mechanically, as it were, some precedent diminutive Sensation, Idea, or Motion, and without any effort of the Mind, are rather to be ascribed to the Body than the Mind, *i.e.* are to be referred to the Head of automatic Motions. I shall call them automatic Motions of the secondary Kind, to distinguish them both from those which are originally automatic, and from the voluntary ones; and shall now give a few Instances of this double Transmutation of Motions, viz. of automatic into voluntary, and of voluntary into automatic.

The Fingers of young Children bend upon almost every Impression which is made upon the Palm of the hand, thus performing the Action of Grasping, in the original automatic Manner. After a sufficient Repetition of the motory Vibrations which concur in this Action, their Vibratiuncles are generated, and associated strongly with other Vibrations or Vibratiuncles, the most common of which, I suppose, are those excited by the Sight of a favourite Play-thing which the Child uses to grasp, and hold in his Hand. He ought, therefore, according to the Doctrine of Association, to perform and repeat the Action of Grasping, upon having such a Play-thing presented to his Sight. But it is a known Fact, that Children do this. By pursuing the same Method of Reasoning, we may see how, after a sufficient Repetition of the proper Associations, the Sound of the Words *grasp, take hold,* &c. the Sight of the Nurse's Hand in a State of Contraction, the Idea of a Hand, and particularly of the Child's own Hand, in that State, and innumerable other associated Circumstances, *i.e.* Sensations, Ideas, and Motions, will put the Child upon Grasping, till, at last, that Idea, or State of Mind which we may call the Will to grasp, is generated, and sufficiently associated with the Action to produce it instantaneously. It is therefore perfectly voluntary in this Case; and, by the innumerable Repetitions of it in this perfectly voluntary State, it comes, at last, to obtain a sufficient Connection with so many diminutive Sensations, Ideas, and Motions, as to follow them in the same Manner as originally automatic Actions do the corresponding Sensations, and consequently to be automatic secondarily. And, in the same Manner, may all the Actions performed with the Hands be explained, all those that are very familiar in Life passing from the original automatic State through the several Degrees of Voluntariness till they become perfectly voluntary, and then repassing through the

same Degrees in an inverted Order, till they become secondarily automatic on many Occasions, though still perfectly voluntary on some,
viz. whensoever an express Act of the Will is exerted.

.　　　.　　　.　　　.　　　.

And thus, from the present Proposition, and the [preceding one]
taken together, we are enabled to account for all the Motions of the
human Body, upon Principles which, tho' they may be fictitious, are, at
least, clear and intelligible. The Doctrine of Vibrations explains all the
original automatic Motions, that of Association the voluntary and secondarily automatic ones. And, if the Doctrine of Association be founded
in, and deducible from, that of Vibrations, in the manner delivered
above, then all the Sensations, Ideas, and Motions, of all Animals, will
be conducted according to the Vibrations of the small medullary Particles. Let the Reader examine this Hypothesis by the Facts, and judge
for himself. There are innumerable things, which, when properly discussed, will be sufficient Tests of it. It will be necessary, in examing the
motions, carefully to distinguish the automatic State from the voluntary
one, and to remember, that the first is not to be found pure, except in
the Motions of the new-born Infant, or such as are the excited by some
violent Irritation or Pain.

Cor. 1. The brain, not the spinal Marrow, or Nerves, is the Seat of
the Soul, as far as it presides over the voluntary Motions . . .

Cor. 2. The Hypothesis here proposed is diametrically opposite to
that of *Stahl,* and his Followers. They suppose all animal Motions to be
voluntary in their original State, whereas this Hypothesis supposes them
all to be automatic at first, *i.e.* involuntary, and to become voluntary
afterwards by Degrees. However, the *Stahlians* agree with me concerning
the near Relation of these Two sorts of Motion to each other, as also
concerning the Transition (or rather Return, according to my Hypothesis)
of voluntary Motions into involuntary ones, or into those which I call
secondarily automatic. As to final Causes, which are the chief Subject
of Inquiry amongst the *Stahlians,* they are, without doubt, every where
consulted, in the Structure and Functions of the Parts; they are also of
great Use for discovering the efficient ones. But then they ought not to
be put in the place of the efficient ones; nor should the Search after the
efficient be banished from the Study of Physic, since the Power of
the Physician, such as it is, extends to these alone. Not to mention,
that the Knowledge of the efficient Causes is equally useful for discovering
the final, as may appear from many Parts of these Observations.

Cor. 3. It may afford the Reader some Entertainment, to compare
my Hypothesis with what *Des Cartes* and *Leibnitz* have advanced, con-

cerning animal Motion, and the connection between the Soul and Body. My general Plan bears a near Relation to theirs. And it seems not improbable to me, that *Des Cartes* might have had Success in the Execution of his, as proposed in the Beginning of his Treatise on Man, had he been furnished with a proper Assemblage of Facts from Anatomy, Physiology, Pathology, and Philosophy, in general . . .

Cor. 4. I will here add Sir *Isaac Newton*'s Words, concerning Sensation and voluntary Motion, as they occur at the End of his *Principia,* both because they first led me into this Hypothesis, and because they flow from it as a Corollary. He affirms then, "both that all Sensation is performed, and also the Limbs of Animals moved in a voluntary Manner, by the Power and Actions of a certain very subtle Spirit, *i.e.* by the Vibrations of this Spirit, propagated through the Solid Capillaments of the Nerves from the external Organs of the Senses to the Brain, and from the Brain into the Muscles."

Cor. 5. It follows from the Account here given of the voluntary and semivoluntary Motions, that we must get every Day voluntary and semivoluntary Powers, in respect of our Ideas and Affections. Now this Consequence of the Doctrine of Association is also agreeable to the Fact. Thus we have a voluntary Power of attending to an Idea for a short time, of recalling one, of recollecting a Name, a Fact, &c. a semivoluntary one of quickening or restraining Affections already in Motion, and a most perfectly voluntary one of exciting moral Motives, by Reading, Reflection, &c.

60   ROBERT WHYTT (1714–1766) ON EMPIRICAL REFLEXOLOGY, 1751

Robert Whytt, *An Essay on the Vital and Other Involuntary Motions of Animals* (Edinburgh, 1751), sect. 1.

By the time Whytt summarized the state of empirical knowledge of the reflex, the Cartesian hypotheses about physiology were virtually gone and in their place were a number of incipiently modern concepts suggested by a mass of experimental data. Whytt spoke of stimuli, antagonistic muscles, sphincters, and an equivalent of muscle tone. One can now perceive the essentials of the chain of events that ultimately established the reflex as a fundamental concept: a stimulus acts on nervous tissue, leading to a muscular movement whose magnitude is in some way proportional to the strength of the stimulus.

1. A certain power or influence proceeding originally from the brain and spinal-marrow, lodged afterwards in the nerves, and by their means

conveyed into the muscles is either the immediate cause of their contraction, or at least necessary to it.

The truth of this is put beyond all reasonable doubt, by the convulsive motions and palsies affecting the muscles, when the *medulla cerebri,
medulla oblongata* and *spinalis,* are pricked, or any other ways irritated
or compressed; as well as from the observation that animals lose the
power of moving their muscles, as soon as the nerve or nerves belonging
to them are strongly compressed, cut through, or otherwise destroyed.
Of this many instances might be given: But we shall content ourselves
with mentioning one, which is too strong and unexceptionable to admit
of any evasion. When the *recurrent* nerve on one side of the *larynx* is
cut, the voice becomes remarkably weaker; when both are cut, it is
entirely and irrecoverably lost, *i.e.* the animal loses all power of moving
the muscles which serve to increase or diminish the aperture of the
glottis; for I presume it will be needless now-a-days to go about to shew,
that the tying of those nerves can only affect the voice, by rendering
these muscles paralytic.

If the brain or some part of it, were not in a manner the fountain of
sensation and motion, and more peculiarly the seat of the mind than
the other bowels or members of the body; why should a slight inflammation of its membranes cause madness, or a small compression of it
produce a palsy or apoplexy, while a like inflammation of the stomach
or liver, or a compression or obstruction of these bowels, have no such
effects? If the nerves were not immediately concerned in muscular motion,
why, upon tying to destroying them, does the member to which they
are distributed, instantly lose all power of motion and sensation? . . .

The immediate cause of muscular contraction, which, from what has
been said, appears evidently to be lodged in the brain and nerves, I
chuse to distinguish by the terms of the *power* or *influence of the nerves;*
and if, in compliance with custom, I shall at any time give it the name
of *animal* or *vital spirits,* I desire it may be understood to be without
any view of ascertaining its particular nature or manner of acting; it
being sufficient for my purpose, that the existence of such a power is
granted in general, though its peculiar nature and properties be unknown.

2. While the nervous power is immediately necessary to muscular
motion, the arterial blood seems to act only in a secondary or more
remote manner.

Muscles are immediately rendered paralytic upon tying or destroying
the nerves distributed to them. But when the arteries bestowed upon
any muscles are tied, the action of these muscles is only gradually weakened, and not totally abolished till after a considerable time . . .

However, from these experiments, it seems pretty clearly to follow, that the arterial blood no otherwise conduces to muscular motion, than as it supplies the vessels and fibres of the muscles with fluids proper for their nourishment, gives them a suitable degree of warmth, and thus preserves them in such a state, as may render them most fit to be acted upon by the nervous power. While therefore the life and nourishment of the muscles are owing to the motion of the arterial blood through their vessels, their powers of motion and sensation proceed from the nerves alone.

3. The muscles of living animals are constantly endeavouring to shorten or contract themselves. Hence such as have antagonists are always in a state of tension; and the solitary muscles, such as the sphincters, and those whose antagonists are weakened or destroyed, are always contracted, except when this natural contraction is overcome by some superior power.

4. The natural contraction of the muscles is owing partly to all their vessels being distended with fluids, which separate and stretch their smallest fibres.

As a proof of this; the muscles of animals that are in full health, and abound with proper fluids, retract themselves much more remarkably towards each extremity when cut across, than the muscles of such animals as are in a languishing state, and exhausted of their fluids; besides that, soon after death, muscles become flaccid, and, when cut transversely, retract themselves but little.

But, the natural contraction of the muscles is in a great measure to be ascribed to the influence of the nerves, which is perpetually operating upon them, though in a very gentle manner: and that to this is chiefly owing the constant contraction of the sphincters, and the tension of such muscles as are balanced by antagonists, the palsy affecting the sphincters as soon as their nerves are compressed or destroyed, and the constant contraction of such muscles whose antagonists are deprived of the nervous power, evidently demonstrate.

5. The natural contraction of the muscles arising from the constant and equable action of the nervous power on their fibres, and of the distending fluids on their vessels, is very gentle, and without any such remarkable hardness or swelling of their bellies, as happens in muscles which are contracted by an effort of the will. And although the sphincters and those muscles whose antagonists are paralytic or hindered from acting, do always remain in a state of contraction; yet at any time, by an effort of the will, they can be much more strongly contracted . . .

From what has been just now advanced, it follows, that it is not necessary, in order to the mind's acting upon the muscles, that they should be stretched or extended beyond that length to which they would naturally reduce themselves, if not prevented by the action of their antagonists.

6. As often as the influence of the nerves is determined into the muscles so as to operate more powerfully on them, they are excited into stronger contractions which are not natural, and therefore may be called violent. This extraordinary determination of the nervous influence, may be owing either to the power of the will, or to a *stimulus.*

7. Voluntary contraction is owing to the stronger action of the nervous influence upon any muscle, excited by the power of the will.

8. A *stimulus,* or any thing irritating, applied to the bare muscles of living animals, immediately excites them to contract themselves.

This appears from numberless experiments and observations; and is equally true with respect to the muscles of voluntary and involuntary motion.—The muscles of a live frog, when laid bare and pricked with a needle, are strongly convulsed.—A solution of white vitriol no sooner touches the internal surface of the stomach, than this bowel is brought into convulsive contractions.—Smoke of tobacco or acrid clysters injected by the *anus,* bring convulsive motions on the great guts.—Pricking the intestines or heart of a living animal, or applying any acrid fluid to them, remarkably increases their contraction.—Many other instances might be given of the effects of *stimuli* on the muscles of animals; but these may suffice, as we shall have occasion to treat of this matter more fully afterwards.

Whatever stretches the fibres of any muscle, or stretches them beyond their usual length, excites them into contraction almost in the same manner, as if they had been irritated by any sharp instrument, or acrid liquor. Thus the motion of the heart in pigeons newly dead, is as remarkably renewed or increased by drawing the sides of the divided *thorax* asunder, and consequently stretching the great vessels to which the heart is attached, as by pricking its fibres with a pin. In luxations, muscles, by being over-stretched, are often convulsed; and the *vesica urinaria* and *intestinum rectum,* are not only excited into convulsive contractions by the acrimony of the urine and *faeces,* but also by their bulk and weight stretching the fibres of these hollow muscles.

9. In proportion as the *stimulus* is more or less gentle, so (*cæteris paribus*) is the contraction of the muscle to which it is applied.

The truth of this proposition, like the former, is not only proved by experience, but may be deduced from reason alone; for if the irritation is to be considered as the cause, and the subsequent contraction of the muscle as the effect; then, in proportion as the cause is increased or diminished, so must be its effect.—The motions occasioned by stretching the fibres of any muscle will be greater or less, as the muscle is more or less stretched; unless it be so far extended, as quite to lose its tone, and become paralytic.—It deserves however to be observed, that the effects of different *stimuli* depend very much upon the peculiar constitution of the nerves and fibres of the muscles to which they are applied: Thus what will prove a strong *stimulus* to the nerves of one part, will more weakly affect those of another, and *vice versa* . . .

10. An irritated muscle does not remain in a contracted state, although the stimulating cause continues to act upon it; but is alternately contracted and relaxed.

Thus the *stimulus* of an emetic received into the stomach, does not occasion a continued contraction of its muscular coat; and an irritation of the lower extremity of the gullet, is followed by alternate convulsions of the diaphragm. The heart of a frog or eel taken out of the body, continues its alternate motions while a needle is fixed in it . . . After the auricle of a pigeon's heart had ceased to move, I made it renew its alternate contractions, by filling the *thorax* with warm water; and after the vibrations of a frog's heart had begun to languish, they recovered their former vigour and quickness, by exposing it to the heat of a fire.

· · · · ·

11. Irritated muscles are not only agitated with alternate motions while the stimulating cause continues to act upon them, but also for some time after it is removed; although they become gradually weaker, and are repeated more slowly. If the irritation be great, these alternate motions last longer, and follow one another more quickly; if weaker, they are repeated after longer intervals, and sooner cease; if extremely gentle, and the muscle not very sensible, perhaps only a single contraction or two will ensue.

· · · · ·

The heart of an animal newly killed, is excited into motion by blowing upon it, or touching it with the point of a pin; and this motion often lasts a great while, altho' the *stimulus* is not renewed. After a pigeon's heart had ceased to move, its vibrations were not only renewed by drawing asunder the sides of the divided *thorax,* but they continued for a considerable time.

12. The motions of muscles from a *stimulus* are quite involuntary.

Every one must be sensible of the truth of this assertion, who has ever felt any of those small convulsions or pulsatory contractions, which frequently happen in different parts of the body, and which seem to be owing to some irritation of the fibres or membranes of the muscle contracted, either from acrid particles in the fluids irritating their sensible

nerves, or from too great a distension of their tender vessels by the stag-
nation of the circulating fluids. The muscles called *acceleratores urinæ*,
though at other times entirely under the power of the will, yet, while
the *semen* continues to be poured into the beginning of the *urethra*,
they are agitated with strong convulsive contractions, which we can
neither increase nor prevent.—When the tendinous fibres of the *obliquus
inferior* of the eye, or of any other of its muscles, are gently stimulated
with the point of a file, the alternate contractions which ensue, are altogether involuntary, and can neither be accelerated, retarded, augmented,
nor diminished by the power of the will. The same thing is true of the
motions of the stomach and diaphragm, excited by emetics. From which
it follows that,

13. The power of *stimuli* in exciting the muscles of living animals into
contraction, is greater than any effort of the will.

The truth of this is still further confirmed, by the following observation.
A man aged 25, who, from a palsy of twelve years continuance, had lost
all power of motion in his left arm, after trying other remedies in vain,
at last had recourse to electricity; by every shock of which the muscles
of this arm were made to contract; and the member itself, which was very
much withered, after having been electrified for some weeks, became sensibly plumper.—If then, the voluntary muscles can, even in a palsied
state, be excited into contraction by the action of a *stimulus* on their
fibres, it follows, that when this is applied to them in a found and more
sensible state, any effort of the will to prevent their contraction, must
be vain and impotent.—Hence the muscles of voluntary as well as of involuntary motion cease to be under the power of the will, while their
nerves or sensible fibres are irritated by a *stimulus.*

14. There are therefore three kinds of contraction observable in the
muscles, all of them different from each other, *viz.* natural, voluntary,
and involuntary, from *stimulus.* The first is very gentle, equable and continued, and is owing to the causes mentioned. The second proceeds immediately from the power of the will, is always stronger, and may be
continued for a longer or shorter time, or performed with more or less
force, as one pleases. The third is strong, but suddenly followed by a
relaxation, seems to be a necessary consequence of the action of the
stimulus, upon the muscle, and cannot be affected, either as to its force
or continuance, by the power of the will.

15. The natural contraction above explained, is what we observe in
the sphincters and in muscles destitute of antagonists.

16. While the sphincters of the *anus* and bladder, and those muscles
whose antagonists are destroyed, remain always in a state of contraction,
and while such muscles as have antagonists, are kept in *equilibrio,* or

without any motion, except when the will interposes; the heart, which has no proper antagonist, is alternately contracted and dilated, without our being able, by any effort of the will, directly to hinder or promote its motions.

17. The contraction of the heart is therefore not only involuntary, but of a different kind from that of the sphincters and muscles deprived of antagonists; and seems, as to its *phænomena,* to agree with the contraction of muscles from a *stimulus.*

18. The mind may, by disuse, not only lose its power of moving even the voluntary muscles, except in a particular way, but also of exciting them into contraction at all. Of the former we have an example in the uniform motions of the eyes; and of the latter in the muscles of the external ear, and of such members as have remained long without motion.

61   GEORGE PROCHASKA (1749–1820) ON THE NERVOUS SYSTEM, 1784

George Prochaska, *De functionibus systematis nervosi* (Prague, 1784), chap. 1, sect. 8; chap. 2, sect. 3; chap. 4, sect. 1; chap. 5, sect. 4. Translated by Thomas Laycock (together with *The Principles of Physiology,* by John Augustus Unzer) as *A Dissertation on the Functions of the Nervous System* (London, 1851).

Prochaska's discussion of the nervous system contributed to the modern concept of the reflex, going beyond Whytt in the detail and precision with which he described how the nervous system is acted upon by stimuli. This process is said to occur by virtue of the *vis nervosa* ("nervous force"), which Prochaska was very careful not to endow with hypothetical properties. His characterization of the *vis nervosa* was readily expanded as knowledge of the nervous impulse increased. The sensory input and the motor output were held to have their consensus in a region of the nervous system designated as the *sensorium commune.* The important thing about Prochaska's account of this region is the principle that the connection between input and output is controlled by natural laws and is not susceptible to the action of the will. Prochaska, however, did not imply that all action takes place in this way, for he assumed that there also occur voluntary movements that are governed by the will.

CHAPTER I

SECTION VIII

At length we abandon the Cartesian method of philosophising in this part of animal physics also, and adopt the Newtonian, being persuaded that the way to truth through hypotheses and conjectures is tedious and altogether uncertain, but far more certain, more excellent, and shorter, through the inductive method. Newton designated the mysterious cause of physical attraction by the term *vis attractiva,* observed and arranged

289

its effects, and discovered the laws of motion; and thus it is necessary to act with reference to the functions of the nervous system: we will term the cause latent in the pulp of the nerves, producing its effects, and not · as yet ascertained, the *vis nervosa:* we will arange its observed effects, which are the functions of the nervous system, and discover its laws; and thus we shall be able to found a true and useful doctrine, which will undoubtedly afford a new light, and more elegant character to medical art. The illustrious Haller has already used the phrase *vis nervosa,* in designating the agent which the nerves employ in exciting muscular contractions; but the celebrated and ingenious J. A. Unzer has thrown the greatest light on the subject; for although he continues the use of the term animal spirits, that he may the more conveniently and intelligibly express himself, yet, as he himself observes, his whole system is complete without them.

.   .   .   .   .

Chapter II

SECTION III

1. *A stimulus is necessary to the action of the* vis nervosa.—

i. Although this *vis nervosa* is a property inherent in the medullary pulp, it is not the chief and sole cause that excites the actions of the nervous system, but is ever latent, and exists as a predisposing cause, until another exciting cause, which we term stimulus, is brought to bear. As the spark is latent in the steel or flint, and is not elicited, unless there be friction between the flint and steel, so the *vis nervosa* is latent, nor excites action of the nervous system until excited by an applied stimulus, which continuing to act, it continues to act, or if removed, it ceases to act, or if re-applied, it acts again.

ii. *This stimulus is divided into stimulus of the body and of the mind.*— This stimulus is double: either it is some fluid or solid body applied internally or externally to the nervous system, and termed corporeal, or mechanical stimulus; or else is a mental stimulus present in a portion of the nervous system, and by means of this portion controls the rest of the nervous system, and the rest of the body, as far as it is allowed. Whether this mental stimulus takes place through a system of occasional causes, or pre-established harmonies . . . or, as assumed by many, by a physical influx, matters little to our object; it is sufficient for us that the soul can excite the nervous system to the performance of certain actions, and this power we call a mental stimulus.

iii. *The relations of the actions of the nervous system to the* vis nervosa *and stimulus, generally considered.*—As effects are proportionate to their causes, so the operations of the nervous system are proportionate to the *vis nervosa* and the *vis stimuli*. The operations of the nervous system, for example, will be the more powerful and extensive in proportion as the *vis nervosa* is more active and the stimulus efficient: and contrarily in proportion as the *vis nervosa* is less active and the stimulus feebler, in that proportion will the operations of the nervous system be more languid. A less energetic stimulus is sufficient for a more active *vis nervosa,* just as a more powerful stimulus will compensate for a less active *vis nervosa;* so that in both cases, the effect on the operations of the nervous system may be equal. The *vis nervosa* is not, however, indifferent to the kind of stimulus, for it is more readily excited by one than by another, although they may appear to be equally forcible; nay, it sometimes responds more actively to apparently a very mild than to a very powerful stimulus. Thus the heart and intestinal canal, according to Haller, are thrown into more powerful contractions by inflated air than by water, or any poison; and, on the other hand, a drop of water getting into the trachea excites a violent cough, while the air is insensibly inspired and expired through it. I shall adduce many such examples hereafter as illustrations of idiosyncrasy.

iv. *Under what circumstances the* vis nervosa *is increased.*—It is evident that the stimulus may be greater or less, longer or shorter, more or less general, or quite local; and the same is true of the *vis nervosa.* This, in fact, differs in degree according to the difference of age, sex, temperament, climate, the condition of the body as to health or disease, and other circumstances, and in a portion of the nervous system as well as in the whole, which it will suffice to prove by a few examples.

a. In the first place, the *vis nervosa* is generally greater in childhood than adult age; for a slight stimulus at that age will act violently upon the nervous system, which scarcely affects the nerves in more advanced years, a truth abundantly proved by the testimony of celebrated men. Young animals are the more sensitive, and organs which in the newly-born are irritable, become insensible through age, and languid in motion and sensation. Sensation is more acute in the young man than in the aged. The pupil is more contractile in the infant, less readily acted on by light in the aged. And the same principle illustrates the cause of senile impotence. That the sensibility of the genital organs is diminished with age is proved by the fact, that the seminal emissions so readily excited in youth, cease to take place about the fiftieth year or somewhat later, even in able and strong men. Since the female sex has a more excitable nervous system, it is established as a general rule that the feeble, or rather the tender organisms, feel more acutely than the robust. Observations also show, that the amount of *vis nervosa* varies with the climate, since those who inhabit hot climates indulge more in ease and pleasure than the inhabitants of colder regions; Montesquieu thinks we may distinguish climates by the degrees of sensibility, just as we distinguish by degrees of latitude. Often in diseases, the sensibility or *vis nervosa* of the whole nervous system, is increased in a very remarkable manner; whence it happens, I think, that we cannot then bear a slight degree of cold in the atmosphere, on account of the shiverings and unpleasant sensations excited through the whole body, that in health we should not notice. Thus, also, a moderate draught of wine greatly increases the fever of a fever-patient, but which, if taken by a person in health, would produce no change whatever in the pulse. For the same reason it happens, that in hemicrania, gout, or any painful affection, we are impatient, and cannot endure any noise or light, or a variety of objects. All nerves that have become too sensitive can no longer tolerate even the most common impressions. If it were not altogether superfluous, many other instances of increased *vis nervosa* in the whole nervous system might be adduced.

b. Frequently an increased degree of *vis nervosa* is observed in a portion only of the nervous system, and not in the whole; in the animal organs alone, or only in the sensorium commune, or in one or other of the nerves. Thus, I imagine, there is an increased degree of *vis nervosa* in the delirious and maniacal, which keeps them fixed to their ideas. That condition of some decrepid old people, in which they are more pusillanimous and timid and ready to weep than children, seems to be referable to this increased decree of the *vis nervosa.* So also may be explained the case of a man of weak mind in health, who was rendered talented by a blow on the head, but when cured relapsed into his previous simple-mindedness. The serene state of mind of dying persons, which has been aptly compared with the crackling of a dying taper, seems dependent, for the moment at least, on increased *vis nervosa.* When the *vis nervosa* is increased in the general sensorium, it seems also to have this effect,—that external impressions made on the sensitive nerves, and transmitted to the sensorium, are too suddenly and violently reflected, and pass over into the motor nerves, and excite movement and convulsions in spite of the will, as happens in the frights of infants, and also of some adults, who are terrified by any slight crash or noise. Further, that the *vis nervosa* may be locally increased in one nerve or another, is proved by innumerable examples of contused, lacerated, wounded, and inflamed parts, a slight touch of which excites much suffering, although in the natural condition it would scarcely have been felt, and this while other sound parts of the body possess their natural sensibility.

.   .   .   .   .

v. *When the* vis nervosa *is diminished.*—The *vis nervosa* is diminished in proportion as we observe the vital powers which are dependent on the *vis nervosa,* to be diminished and weakened; and which becomes so weak in death, that the natural stimuli, as for example, the influence of the inspired air, and of the blood in the heart, can no longer excite it, and a mortal repose of all the vital and animal movements results. In this ordinary termination of life, the *vis nervosa* is undoubtedly at a minimum, but it is not quite lost, for a few sparks can still be excited, if a strong stimulus be applied to the nerves. Vesalius was taught this fact by sorrowful experience, for when dissecting a body shortly after death, he excited the heart to renewed action . . . In many experiments on frogs, I observed, that when the heart was still and could no longer be excited by a stimulus, that the muscles of the thigh continued to be slightly contracted, whenever the sciatic nerve was punctured or compressed. We hence conclude, that a certain portion of the *vis nervosa* remains for a time in the nerves after death, which, although insufficient to maintain life, is sufficient to develop movements in the heart and some muscles, if excited by a powerful stimulus. For they contract, although so feebly, that weak jerking rather than contractions are only produced, and these cease after awhile, however strongly the nerves or muscles may be stimulated. When after death no muscle responds to a stimulus, are we to conclude that all *vis nervosa* has left the nerves, or is it that it cannot display itself on account of the muscles being rendered unfit for action? We cannot determine these questions. The *vis nervosa* is also diminished by opium, according to the observations of Whytt. Haller and Sprögel found that opium destroyed the *vis irritabilis* of the stomach and intestinal canal, and since (as will be hereafter shown) irritability presupposes a *vis nervosa,* the *vis nervosa* is also diminished by it . . .

vi. *The* vis nervosa *is divisible, and exists in the nerves independently of the brain.*—*Vis nervosa* is as divisible as the nervous system, so that it remains in each portion of a bisected nerve, as if it were still entire and connected with the brain. Nor does the *vis nervosa* of the nerves require continual supplies from the brain, since nerves possess their own *vis nervosa,* which never had a connection with the brain. The experiments that prove this have long been perfectly well known; namely, that if a nerve be cut or tied, although by these means its connection with the brain be destroyed, it is still as able, if irritated, to cause the muscles to contract as if its connection with the brain were entire. Haller clearly states this fact in many places. He observes: "a nerve compressed or tied, and then irritated below the ligature, excites those muscles to convulsive contraction, to which it is distributed, just as if it was perfectly free."

And elsewhere: "if the nerve of a muscle be compressed, or tied, or divided, and then irritated, provided it be fresh and moist, the irritation will produce in the muscle to which the nerve is distributed the same movements as it would have produced, if the continuity of the nerve with the brain had remained unbroken. This proposition having been proved with regard to the voluntary nerves is here shown to be applicable to the organic nerves." . . .

From these facts it is obvious, that the *vis nervosa* remaining in the nerves after the severance of their connection with the brain, must be considered as the cause whereby the heart was able to continue its movements, in the experiments instituted by Haller and other distinguished men, after the brain and cerebellum were destroyed, the head cut off, and even all the nerves of the heart divided. For the stimulus of the blood, alternately flowing into the cavities of the heart, irritated its nerves still endowed with *vis nervosa,* although separated from the brain, and thus excited it to alternate contractions. But another interpretation has been given to these facts; and especially by Haller, namely, that it is manifest, that if the heart's action continues after decapitation, or destruction of the whole cerebrum and cerebellum, the cardiac movements are not in connection with the nerves, but with an irritability innate in the heart, and not dependent on the nerves. But the fallacy in this conclusion is most manifest, since it can only be fairly inferred that the heart can continue its action without the brain and spinal cord, but not without its own nerves, which, although entirely separated from the brain, are still united to the heart, and still as endowed with the *vis nervosa,* and as impatient of a stimulus, as when in connection with the brain.

.　　.　　.　　.　　.

Chapter IV

SECTION I

The external impressions which are made on the sensorial nerves are very quickly transmitted along the whole length of the nerves, as far as their origin; and having arrived there, they are reflected by a certain law, and pass on to certain and corresponding motor nerves, through which, being again very quickly transmitted to muscles, they excite certain and definite motions. This part, in which, as in a centre, the sensorial nerves, as well as the motor nerves, meet and communicate, and in which the impressions made on the sensorial nerves are reflected on the motor nerves, is designated by a term, now adopted by most physiologists, the *sensorium commune.*

Distinguished men have not agreed as to the seat of the sensorium commune. Bontekoe, Lancisi, De La Peyronie have placed the sensorium commune in the corpus callosum; Willis derived the perception of sensation and the source of movements from the corpora striata; Des Cartes attributed the function of the sensorium commune to the pineal gland; Vieussens to the centrum ovale; Boerhaave decided that aggregate of points to be the sensorium commune, in which all the sensory nerves terminate, and from which all the motor nerves arise, and accordingly placed it in the medulla fornicata, surrounding the cavity of the ventricles. In a later work, 'De Morbis Nervorum,' Boerhaave places the sensorium commune in the boundary line of the medullary and cortical substance, which opinion Tissot thought to be extremely probable, regarding it as confirmed by the observations of Wepfer. Mayer seems to place the sensorium commune in the medulla oblongata; that distinguished man, J. D. Metzger, appears to be also of the same opinion; the celebrated Camper said, that if the sensorium commune has a seat at all, it ought to be in the pineal gland, and in the nates and testes, and that, therefore, the opinion of Des Cartes was not so very absurd. It certainly does not appear that the whole of the cerebrum and cerebellum enters into the constitution of the sensorium commune, which portions of the nervous system seem rather to be the instruments that the soul directly uses for performing its own actions, termed animal; but the sensorium commune, properly so called, seems not improbably to extend through the medulla oblongata, the crura of the cerebrum and cerebellum, also part of the thalami optici, and the whole of the medulla spinalis; in a word it is co-extensive with the origin of the nerves. That the sensorium commune extends to the medulla spinalis is manifest from the motions exhibited by decapitated animals, which cannot take place without the consentience and intervention of the nerves arising from the medulla spinalis; for the decapitated frog, if pricked, not only withdraws the punctured part, but also creeps and leaps, which cannot be done without the consensus of the sensorial and motor nerves, the seat of which consensus must necessarily be in the medulla spinalis—the remaining portion of the sensorium commune.

The reflexion of sensorial into motor impressions, which takes place in the sensorium commune, is not performed according to mere physical laws, where the angle of reflexion is equal to the angle of incidence, and where the reaction is equal to the action; but that reflexion follows according to certain laws, writ, as it were, by nature on the medullary pulp of the sensorium, which laws we are able to know from their effects only, and in nowise to find out by our reason. The general law, however, by which the sensorium commune reflects sensorial into motor impressions, is the preservation of the individual; so that certain motor impressions follow certain external impressions calculated to injure our body, and give rise to movements having this object, namely, that the annoying cause be averted and removed from our body; and *vice versa,* internal or motor impressions follow external or sensorial impressions beneficial to us, giving rise to motions tending to the end that the agreeable condition shall be still maintained. Very many instances which might be adduced, undoubtedly prove this general law of the reflexions of the sensorium commune, of which it may be sufficient to mention a few. Irritation being made on the internal membrane of the nostrils excites sneezing, because the impression made on the olfactory nerves by the irritation is conducted along them to the sensorium commune, there by a definite law is reflected upon motor nerves going to muscles employed in respiration, and through these produces a strong expiration through the nostrils, whereby the air passing with force, the cause of the irritation is removed and ejected. In like manner it happens that when irritation is caused in the trachea by the descent of a particle of food, or a drop of fluid, the irritation excited is conducted to the sensorium commune, and there reflected on the nerves devoted to the movement of respiration, so that a violent cough is excited, a most suitable means for expelling the cause of irritation, which does not cease until the irritant be ejected. If a friend brings his finger near to our eye, although we may be persuaded that no injury is about to be done to us, nevertheless the impression carried along the optic nerve to the sensorium commune is there so reflected upon the nerves devoted to the motion of the eyelids, that the eyelids are involuntarily closed, and prevent the offensive contact of the finger with the eye. These and innumerable other examples which might be brought forward, manifestly show how much the reflexion of sensorial impressions into motorial, effected through the sensorium commune, has reference to maintaining the conservation of the body. Wherefore, Tissot justly enumerates the action of the sensorium commune amongst those powers, the sum and co-ordination of which constitute the nature of our living body.

Since the principal function of the sensorium commune thus consists in the reflexion of sensorial impressions into motor, it is to be noted, that this reflexion may take place, either with consciousness or without consciousness. The movements of the heart, stomach, and intestines, are certainly in nowise dependent on the consciousness of the soul, for whilst no muscular movement can be excited, unless a stimulus applied to the sensorial nerves passes by a peculiar reflexion to the motor nerves, and excites contraction of the muscle, it is at the same time certain that the reflexion of the impressions suitable for exciting those movements, if it takes place in the sensorium commune, is effected without consciousness. But it is a question whether these impressions, in order that they may be reflected, do really travel so far as the sensorium commune, or, without taking this long circuit, are reflected nearer in the ganglia, from whence these parts derive many nerves? This matter is further to be considered afterwards. But that reflexions of sensorial impressions into motor are effected in the sensorium commune itself while the mind is altogether unconscious, is shown by certain acts remaining in apoplectics deprived entirely of consciousness; for they have a strong pulse, breathe strongly, and also raise the hand, and very often unconsciously apply it to the affected part. The sensorium commune also acts independently of consciousness in producing the convulsive movements of epileptics, and also those which are sometimes observed in persons buried in profound sleep, namely, the retractions of pricked or irritated limbs, to say nothing of the motion of the heart and the respiratory acts. To this category also belong all those motions which remain some time in the body of a decapitated man, or other animal, and are excited when the trunk, and particularly the medulla spinalis, are irritated, which motions certainly take place without consciousness, and are regulated by the remaining portion of the sensorium commune existing in the medulla spinalis. All these actions flow from the organism, and by physical laws peculiar to the sensorium commune; and are, therefore, spontaneous and automatic. The actions taking place in the animal body, with accompanying consciousness, are either such as are independent of volition, or such as the mind can restrain and prohibit at pleasure; the former being governed by the sensorium commune alone, independently of the mind, are as much automatic as those of which the soul is unconscious. Of this character are sneezing from an irritant applied to the nostrils, cough from an irritant fallen into the trachea, vomiting from a titillation of the fauces, or after taking an emetic; the tremors and convulsions in St. Vitus's dance, and in a paroxysm of intermittent fever, &c. Those actions, however, which the soul directs or limits by its own power, even although the sensorium commune has its share in producing them, are nevertheless called animal, and not automatic, and concerning them we treat in the next chapter.

.   .   .   .   .

Chapter V

SECTION IV

There are only two kinds of muscular action in the human body, according to the cause which excites it; the one kind is termed voluntary or animal, because according as the mind commands and wills, it may be excited, increased, diminished, and arrested; the other involuntary, of which the mind is either unconscious, or if conscious, the motion is performed without its consent, and is excited only by a mechanical corporeal stimulus applied to the nervous system, for which reason it is also termed spontaneous and automatic. Nerves are necessary to produce both kinds of movement. The nerves do not act, however, without a stimulus, which is either produced by the mind willing, or, if unconscious and unwilling, by some body applied to the nerves. Whence, therefore, it is manifest, that those movements ought alone to be termed animal which depend upon the untrammelled control of the soul, and which it produces or restrains by its own free will; on the other hand, those which in no degree depend on the will, but are performed when the mind is unconscious or unwilling, cannot be termed animal, but are purely mechanical and automatic.

Observation teaches, that there are some muscles in the human body over which the mind has no control whatever, and the movements of which are purely automatic during the whole of life; these are the heart . . . œsophagus, stomach, and intestinal canal; with these may be mentioned the motion of the iris. There are other muscles which are ordinarily subject to the control of the will, and for this reason are termed voluntary, such as the muscles of the limbs, trunk, head, face, eyes, tongue, genitals, and the sphincter of the anus and urinary bladder. It sometimes happens, however, that all these muscles renounce the authority of the mind, and while it is either unconscious or unwilling, are violently agitated by some preternatural mechanical stimulus, as is seen in hysterical, epileptic, or infantile convulsions, or in those affected with St. Vitus's dance; and these movements, although performed by muscles designated voluntary, can only be termed automatic. In the fœtus *in utero* and in the newly-born these muscles are not moved voluntarily, but for the most part automatically, for at that age the cerebrum is not as yet capable of thought, until the organs of the faculty of thought being gradually evolved, the mind learns to think, and to use the muscles subjected to its control. The raising of the hand and the application of it to the head in apoplexy belong also to the class of automatic movements, also the turning of the body in sleep, and partly even somnambulism itself, which, however, it would seem is partly also to be ascribed to obscure sensations and volitions which the mind instantly forgets. In the third place, there are muscles which continually act independently of the will, being excited thereto by a mechanical stimulus only, but over which the mind possesses voluntary command, and can at will accelerate, or retard, or entirely stop their movements for a time; the action of these is termed mixed. Of this kind are the muscles of respiration, which almost constantly act automatically, but over which, however, the mind has such control, that it can accelerate, retard, or even stop the respiratory movements for a time. But if a mechanical stimulus be too powerful, then the muscles of respiration are excited into action in spite of the will; for example, if a crumb slips into the trachea, a violent cough ensues altogether uncontrollable by the will; thus also the mind cannot prevent sneezing when the pituitary membrane of the nostrils is stimulated by an acrid stimulus.

In establishing that no action and no movement can be termed animal, of which the mind is not conscious, and which does not depend upon its free will, I shall possibly seem to have restricted the influence of the soul over the body too much, since there are very distinguished men, especially of the Stahlian School, who have taught that not only every movement is directly regulated by the soul, but also other functions of the animal body; adopting this fundamental principle, that consciousness is not necessary to each function of the mind. But it is certain that as yet we know nothing more of the human soul than that it thinks, and that it cannot do this, so long as it is connected with the body, without the assistance of the brain.

62   MARSHALL HALL (1790–1857) ON THE SPINAL NERVOUS SYSTEM, 1843, 1850

The first excerpt is from Marshall Hall, *New Memoir on the Nervous System* (London, 1843), sect. 3, pp. 51–52. The second excerpt is from Marshall Hall, *Synopsis of the Diastaltic Nervous System* (London, 1850), sect. 1, pp. 4–11, 13, and sect. 2, pp. 17–20.

Marshall Hall's writings on the nervous system do not contain any important new scientific facts or theories. In fact, his attempt to set the spinal nervous system apart from the rest of the nervous system was largely unsuccessful. His contribution was, nevertheless, important because it established the reflex as a definite physiological and anatomical entity and made the study of the reflex a substantial subdivision of biology. In the first selection (1843), Hall describes the functions of the spinal system; in the second (1850), he goes on to describe the way in which it operates. Since, for Hall, the distinguishing characteristic of the spinal system is its mediation of reflexes, these two excerpts give a good account of reflexology at this time.

[EXCERPT 1]

SECTION III

•   •   •   •   •

190. The first physiological remark I would offer, relates to the complete distinction between the functions of the cerebral, the true spinal,

and the ganglionic portions of the several nervous systems. Whilst the cerebral system places us in relation with the external world *mentally,* the true spinal *appropriates* some of its materials in the *mass,* and the ganglionic performs the same office still more intimately in regard to its *atoms.*

191. Every act of ingestion, of retention, of expulsion, or of exclusion, is a reflex act, an excito-motor act, an act of the true spinal system, performed through its incident nerves, its central organ (the true spinal marrow), and its reflex motor nerves; an act of the special power seated in this system. I have always wondered that such an extensive generalization did not excite more admiration of Nature's works.

192. If we wish, then, to know what are the special acts of the true spinal system, we have only to ask—what are the acts by means of which masses of matter are ingurgitated into and expelled from the animal œconomy?

193. These acts are found to preside over two important classes of functions,—those of the *preservation of the individual* and of the *propagation of the species.* The designs of Nature in the functions of the true spinal system are, therefore, obvious.

194. These views will be made conspicuous by a careful examination of the following Table of

The Physiology of the True Spinal System

  I. *The Excited Actions—*

    1. *Of the Iris;—of the Eye-lids.*

    2. *Of the Orifices* $\begin{cases} 1.\ \textit{The Larynx.} \\ 2.\ \textit{The Pharynx.} \end{cases}$

    3. *Of the Ingestion*

      1. *Of the Food,*

        1. *In Suction;*

        2. *In Deglutition.*

      2. *Of the Air, or Respiration.*

      3. *Of the Semen, or Conception.*

    4. *Of Exclusion.*

    5. *Of the Expulsors, or of Egestion*

      1. *Of the Fæces;*

      2. *Of the Urine;*

      3. *Of the Perspiration;*

      4. *Of the Semen;*

      5. *Of the Fœtus, or Parturition.*

    6. *Of the Sphincters.*

      1. *The Cardia.*

      2. *The Valvula Coli?*

3. *The Sphincter Ani.*
4. *The Sphincter Vesicæ.*

II. *The Direct Action or Influence—*

I. *In the Tone,*
II. *In the Irritability,* } *of the Muscular System.*

. . . . . .

[EXCERPT 2]

SECTION I

II. New Terms Proposed

15. On analysing the facts which have been detailed, I observed that the following anatomical relations are essential:

1. A nerve leading *from* the point or part irritated, *to* and *into* the spinal marrow;

2. The spinal marrow *itself;* and

3. A nerve, or nerves, passing *out of* or *from* the spinal marrow,—*all in essential relation or connection with each other.*

16. On these anatomical facts I have ventured to institute a new nomenclature, descriptive of what I have hitherto designated *The Spinal System,* and expressive of these essential points. The term peristaltic . . . is familiar to us all. It may be justly extended to all the movements of the interior organs, as the heart, the stomach, the large and small intestines, the uterus, &c. These movements, it is well known, are independent of the spinal marrow. But it has been shown that a series of experimental phenomena, and it will be shown hereafter that a series of important functions, are effected by means of the series of nerves in essential connection with the spinal marrow, to which I have adverted. The action or act is performed *through* the spinal marrow as its essential centre. I propose to designate the phenomena by the term *dia*-staltic.

17. The spinal system may henceforth be designated—*The Diastaltic Nervous System,* a designation which will have the advantage of including this system in the invertebrate as well as the vertebrate tribes of animals. This system embraces a peculiar anatomy, physiology, pathology, and therapeutics.

18. Perhaps the only *purely* diastaltic function is *Respiration;* and this is variously modified by volition and influenced by emotion. But there are many other functions which partake of *both* the diastaltic and peristaltic character. Such are the functions of the immediate conduits of ingestion and of egestion; —the œsophagus, the rectum, the uterus. These functions are *dia-* and *peri-* staltic.

19. How much there is in these terms calculated to excite new and

accurate inquiry! How much to refute injurious and calumnious criticism! I have hitherto spoken of the *mode* and *course* of the diastaltic actions and functions. But I shall immediately proceed to show that the *Principle* of action in the diastaltic nervous system is as special and peculiar as its direction. This principle I long ago demonstrated to be the *vis nervosa* of Haller, the *'excitabilité'* of M. Flourens, acting in newly discovered, diastaltic, forms. Now, the term 'reflex' may have been *vaguely* used by Prochaska; but the *full* and *distinct* idea of a diastaltic action of the *vis nervosa* had occurred to no one.

20. We are much in need of other terms still, to aid us in this investigation. The terms incident excitor and reflex motor have been used to designate those nerves whose influence proceeds *to* and *from* the spinal marrow. But they have never appeared to me satisfactory, and I have long wished for others more expressive and explicit. The following compounds of *odos,* a way, have appeared to competent judges very appropriate to our subject: esodic . . . will express the action *into;* exodic . . . the action *out of;* anodic . . . will express the *ascending,* cathodic . . . the *descending,* course of action; pollodic . . . and panthodic . . . will express the facts, on which I shall shortly have to dwell at considerable length, of the action of the *vis nervosa* from each *one* point of the diastaltic system, in *many* or even *all* directions, to *every* other.

. . . . .

III. Subsequent Experiments

I may now proceed with my detail:

23. If, in the severed head of the frog, the toad, the eft, the snake,—the kitten, the puppy, the young rabbit, &c. we touch the eye-lid, the eye-lash, or the conjunctiva, the eye is immediately closed. The same event occurs in the horse stunned to insensibility by the blow of the pole-axe.

24. If, in the decapitated trunk of these animals, we irritate a toe or other part of the foot of the anterior or posterior extremity, this extremity is immediately withdrawn; if we irritate the tail, or the integuments near the sphincter ani, still greater movements are produced.

25. If the brain merely be removed, in a very young animal, all these phenomena are still observed.

26. The same effects are produced by irritating the dura mater within the cranium, and other interior membranes and tissues—a fact which throws a beam of light on some pathological events.

27. By any of these irritations, an act of inspiration, if respiration were previously suspended, is especially apt to be induced.

28. Each irritation of a cutaneous or mucous surface appears to induce a peculiar, *special,* and definite movement. If in the very young kitten, deprived of cerebrum and cerebellum, the foot be irritated, it is retracted; if a finger be introduced between the lips, an act of suction is excited; if a soft substance, as milk, be inserted into the pharynx, an act of deglutition is attempted; if the border of the rectum be irritated, the sphincter is contracted. The eye-lash, the meatus of the ear, and the tufts of hair between the toes, are peculiarly excitor.

29. *Similar* phenomena are observed in the anencephalous fœtus, in the early stage of asphyxia in young animals, and in the anæsthesia induced by chloroform.

30. In the case of perfect paraplegia in the human subject, when sensibility is absolutely extinct, and voluntary movements totally abolished, diastaltic actions are excited on the application of appropriate irritants, such as tickling, a puncture, a pinch, or sudden heat or cold; of all which the patient is unconscious.

.   .   .   .   .

V. The Vis Nervosa of Haller: Its Catastaltic Law of Action

36. The *Vis Nervosa,* that power in the spinal marrow and muscular nerves, by means of which, if their tissues be irritated, muscular contraction follows, was supposed by Haller, by Bichat, by Prof. J. Müller, and, I believe, by all physiologists, to act in *one* direction only—*from above downwards.* Its action was supposed to be *cata*-staltic only.

37. As long as this view prevailed, this motor power had, and could have, no application to physiology. It was presented to us as a mere experimental fact, or, at the utmost, in its relation to pathology.

38. It was . . . sterile and without utility. Now the existence of a distinct and energetic motor power in the animal frame, without utility in the animal œconomy, would be a perfect incongruity, and contrary to every thing observed in creation.

39. By a series of oft-repeated experiments, I have demonstrated other *Laws* of action of the vis nervosa, and especially *one* which may be designated *dia*-staltic.

40. By this discovery, I have been enabled to prove the identity of the motor power in these experiments of Haller and in the experiments of Redi and of Whytt, and, disentangling the maze, to show that that double series of experiments is *not* without its application in physiology; that the latter have, in fact, their prototypes in all the acts of ingestion and of egestion in the animal œconomy, and in some instances of pathological and therapeutic actions; and the former, in certain cases of pathology. It is the first real step in the philosophy of involuntary motions.

VI. The Diastaltic Law of Action of the Vis Nervosa: Its Demonstration

41. If the spinal marrow, in the decapitated tortoise or turtle, be denuded and irritated by the point of a needle, or the galvanic current, *both* anterior and posterior extremities are moved. The *same* irritation, applied to one and the *same* point of this nervous centre, produces *all* these movements.

42. If, instead of irritating a point of the spinal marrow itself, we denude and irritate a lateral spinal nerve,—the same results, the same movements of *both* pairs of extremities are observed.

43. In the first experiment, it *is* the *vis nervosa* of Haller which acts on the *posterior* extremities. This is the general view. Are the similar and synchronous movement of the *anterior* extremities, and the similar movements of *both* anterior and posterior extremities in the second experiment, of *different* origin? But if the integument be irritated, the *same* movements still take place; and this is one of the cases of reflex or diastaltic action.

44. Lastly, if we so irritate the border of the eye-lid, the eye-lids close; or if we touch the border of the larynx, or of the sphincter ani, these orifices are closed. But these are *Functions.*

.   .   .   .   .

48. Such I believe to be the *Demonstration* of the diastaltic action, and such the application to physiology, of the *vis nervosa* of Haller. Previously a sterile experimental fact, this principle of action has now taken its place as the dynamic force presiding over the large *Class* of the *functions* of ingestion and of egestion in the animal œconomy.

49. It appears to me that the anastaltic and the diastaltic actions, in these experiments, are slower and more combined than the merely catastaltic. There is a similar difference between physiological acts, which are all diastaltic, and those pathological movements which are catastaltic.

.   .   .   .   .

VIII. Diastaltic Actions Special, Resembling Design

56. Diastaltic actions are sometimes so combined, and, as it were, concatinated, as nearly to resemble acts of volition.

57. If a toe of the posterior extremity in the frog be irritated, the limb is flexed at the first, second, and third articulations, and drawn close to the body of the animal, by those *combined* movements.

58. But frequently, besides these movements of flexion, sudden movements of extension take place; sometimes of one extremity, but sometimes also of both. These *concatenated* movements may issue in a jump

or leap even. It is only by observing that, when these movements have ceased, no further movements—*no spontaneous movements*—occur, and by a certain peculiarity, readily detectible by those who have performed a lengthened series of experiments, that their real nature is detected.

59. Movements in *all* directions from any *one* point of the system, continued movements, and concatinated movements, are observed under other circumstances.

60. *Exp.*—If a kitten be reduced to a state of asphyxia, and the nostril, the ear, a toe of the anterior or posterior extremity, the tail, or the sphincter ani, be irritated, an act of inspiration is excited.

61. If, under various circumstances, we dash cold water in the face of the human patient, an act of inspiration, or of deglutition, or an act of contraction and expulsion in the rectum, the bladder, or the uterus, may be excited.

62. Each and every part of the spinal system is bound in a bond of action with each and every other part of that system.

63. The modes of action hitherto described have been induced by excitants applied to the cutaneous surface. And, indeed, as I long ago observed . . . it is the fine *origins* of the incident or esodic nerves, spread on the cutaneous or membranous tissues, which are the most excitable.

64. Some parts of the cutaneous surface are more excitable than the rest. This extraordinary excitability is especially observed in the toes and near the sphincter ani, in the frog; and within the ear, in the eye-lid, in the sole of the foot, in the young kitten and puppy. If the integument near the sphincter ani, in the frog, be irritated, a movement of the posterior extremities, *apparently* designed to remove that irritation, occurs.

65. The final cause of all this is obvious. If an object be placed in the hand of a sleeping infant, it is grasped pretty firmly, by a mere diastaltic action. Such action appears designed to strengthen and aid, and to co-operate with, volition. If it were otherwise, each act of volition would be opposed and thwarted by diastaltic actions, and the objects of volition might be frustrated.

66. The act of inspiration presents us with the most marked example of combined action, including that of the nostril, the larynx, the intercostal muscles, and the diaphragm. This act also includes a concatenated action; for the act of inspiration is essentially *linked* with the act of expiration. This is most obvious in the case of sneezing. But it is not less so, to the attentive observer, in ordinary respiration, in which there is a sort of continued equilibrium of action. To this point I shall have to revert hereafter.

•   •   •   •   •

I. Condition of the Vis Nervosa: Static and Dynamic

78. The cerebrum and cerebellum are insensible and *in*-excitor or *a*-staltic, on being punctured or lacerated, whilst their principle of action . . . is *spontaneous* in its motor influences.

79. The spinal marrow, on the contrary, is essentially excitor, requiring the application and repetition of a stimulus for the development of each and every movement.

80. The natural condition of the spinal marrow is one of inaction, or of static equilibrium. It is by appropriate and successive stimuli that its dynamic force is made effective and manifest.

81. This statement is true in every condition of the spinal marrow. Even when its excitability is extreme, under the influence of strychnine, freedom from stimulus is freedom from all motor action.

82. Still more is this the case in the state of diminished excitability from *shock,* from chloroform, &c.

83. After the application of a stimulus and the phenomena of dynamic force, the spinal marrow again resumes its condition of static equilibrium, but with reduced excitability. The action of each stimulus is followed by this effect, and each second stimulus is accordingly less effective than the former one. The excitability is, on the other hand, restored by repose. And thus the static equilibrium and the dynamic force bear a certain relation to each other.

84. A frog, affected by shock, or placed under the influence of chloroform, may be deprived of voluntary movement, respiratory movements, and reflex actions, the circulation being also almost extinct. If it be now left at rest, respiratory movements return. If it be excited, they again cease. And thus repeatedly. The same observation applies to all other movements. Quiet is the restorer, excitement the exhauster, of the motor energies.

II. The Spinal Marrow Susceptible of Augmented Excitability

85. The degree of Excitability of the spinal marrow is, in general terms, (like Irritability of the muscular fibre), inversely as the degree of Activity or of Stimulus.

86. Augmented or restored during sleep, it is diminished during each day, by every act of volition, every act of the respiration, and by each meal.

87. But the excitability of the spinal marrow admits of intense augmentation and extreme diminution by therapeutic agents. That of the nerve admits of no such augmentation.

88. *Exp.*—The tenth part of a grain of the acetate of strychnine dissolved in distilled water, and applied over the cutaneous surface of the frog, induces the most extreme excitability, or hypererethism. The slightest stimulus induces violent tetanoid spasm. Meantime, the circulation, in the intervals of such spasms, remains unimpaired.

89. *Exp.*—On the other hand, if ten drops of chloroform be dropped on a bit of sponge and attached to the upper part of a tumbler, and this be inverted on a plate of glass, so as to enclose a frog, this animal first ceases from voluntary movements, then loses its excitability, and, lastly, its circulation.

90. Undue excitability is the usual effect of teething, of irritated esodic nerves in general, and especially in the case of a wounded nerve, as in tetanus.

91. The usual immediate effect of a convulsive seizure is augmented excitability; and therefore one seizure frequently succeeds to another. The remoter effect is diminished excitability, and the patient is frequently secure from other attacks until the excitability is slowly restored.

92. Indolence allows the excitability to become morbidly great; activity diminishes its degree or intensity. Hence the importance, in such cases, of restraining the excitability by daily exercise, limited only by approaching fatigue.

III. Relation of Irritability to the Cerebrum and Spinal Marrow

93. We are naturally led, by the consideration given in the last section, to the subject of the present one. Every act of an organ is followed by diminished energy or power. This is not only true of the nervous tissue, but of the muscular fibre. Each contraction of a muscle is followed by a diminution of the irritability of the muscular fibre. If, on the contrary, all stimulus be removed, the irritability exists in its maximum degree.

94. But, for the perfect state of the muscular irritability, it is essential that the muscle should have remained in connection, through the nerves, with the spinal marrow. *The spinal marrow is,* so far, *the source of muscular irritability.*

95. If, in experiment or disease, the influence of the brain, that is, of volition, be withdrawn from a muscle, its irritability becomes greater, comparatively, than that of the similar muscles. In cerebral paralysis, or that paralysis in which the influence of the cerebrum is removed from a limb, the muscles of that limb are more irritable, *tested* by the *mildest* galvanic influence which will produce an obvious effect, than those of the other limb.

96. But if the connection between the spinal marrow and the muscle be severed, either in an experiment or by disease, the irritability of the

muscles of the paralysed limb (and the excitability of the severed portion of nerve) is less than that of the healthy limb.

97. These conclusions are founded upon a vast number of experiments, most carefully made and observed.

98. The fact affords a *Diagnosis* between cerebral and spinal paralysis, or between the cases of paralysis in which the influence of the *cerebrum,* or of the *spinal marrow,* is *severed,* respectively—a diagnosis frequently of great importance.

63   IVAN MICHAILOVICH SECHENOV (1829–1905) ON REFLEXOLOGY AND PSYCHOLOGY, 1863

I. M. Sechenov, *Refleksy golovnogo mozga* (St. Petersburg, 1863). Translated as "Reflexes of the Brain," by A. A. Subkov in I. M. Sechenov, *Selected Works* (Moscow and Leningrad: State Publishing House for Biological and Medical Literature, 1935), pp. 264–322.

Sechenov was the father of Russian reflexology, and he carried this conception to its furthest limit. For him, all behavior—voluntary and involuntary —was reducible to reflexes. He held, moreover, that the study of behavior is adequate to an understanding of all psychology. Sechenov's views preceded the arguments for objective psychology put forth by Pavlov and Bekhterev in Russia. In America, John B. Watson formulated a similar, though independent, case for objective psychology fifty years after Sechenov. Sechenov's psychology was more physiological than Watson's, for it contained the notions of inhibition, facilitation, and integration, concepts that were later (1906) developed experimentally by the English physiologist Charles S. Sherrington, whereas Watson was concerned with characterizing the conventional psychological functions in behavioral terms.

The matter stands thus. The psychical activity of man, as is well known, expresses itself in external manifestations, and, generally, everyone,—both scientists and laymen, naturalists and people interested in questions of the spirit,—judge of psychical activity from its external expression. But the laws of the external manifestations of psychical activity have been studied very little,—even by the physiologists, on whom, as we shall see later, the obligation to do so rests. It is these laws that I wish to speak of.

Let us enter the world of phenomena which arises in the activity of the brain. People generally say that this world embraces all psychical life, and nearly everyone now accepts this idea with greater or lesser reservations. The differences in the conceptions of the various schools consists only in that some view the brain as the organ of the spirit, separating the spirit from the brain; others asseverate that spirit is the product of the activity of the brain. We are not philosophers and shall not enter into a discussion of these differences. As physiologists, we are satisfied with the fact that the brain is an organ of the spirit, i.e. a mechanism which, when brought into activity by any kind of cause, produces as a final result that series of external phenomena which we characterize as psychical activity. Every one knows how immense this psychical world is. It includes all the endless diversity of movements and sounds of which man is capable. But is it necessary to grasp all this mass of facts and not to lose sight of any of them? Yes, undoubtedly, for without this, the study of the external manifestations of psychical activity would be a pure waste of time. The task, at the first glance, appears to be impossible; but in reality it is not so, and for the following reason.

All the endless diversity of the external manifestations of the activity of the brain can be finally regarded as one phenomenon,—that of muscular movement. Be it a child laughing at the sight of toys, or Garibaldi smiling when he is persecuted for his excessive love for his fatherland; a girl trembling at the first thought of love, or Newton enunciating universal laws and writing them on paper,—everywhere the final manifestation is muscular movement. In order to help the reader to reconcile himself with this thought more readily, I will remind him of the frame-work created by the mind of humanity to include all manifestations of brain activity; this frame-work is *"word and deed."* Under *deed,* the popular mind conceives, without question, every external mechanical activity of man based exclusively on the use of muscles. And under *word,* as the educated reader will realise, is understood a certain combination of sounds produced in the larynx and the cavity of the mouth, again by means of muscular movements.

Therefore, *all the external manifestations of brain activity can be attributed to muscular movement.* In this way, the question is greatly simplified. Billions of diverse phenomena, having seemingly no relationship to each other, can be reduced to the activity of several dozen muscles . . . Moreover, it becomes clear to the reader, once and for all, that absolutely all such qualities of the external manifestations of brain activity which we characterize as animation, passion, mockery, grief, joy, etc.—are nothing more than the results of a greater or lesser contraction of some group of muscles—a purely mechanical act, as everyone knows. Even the most confirmed spiritualist must concede this. Indeed, how could it be otherwise, when we know that from a soulless instrument the hand of the musician tears out sounds full of passion and life, and that the hand of the sculptor brings life into stone. Both the hand of the musician and that of the sculptor, creating life, is capable only of mechanical movement which, strictly speaking, can even be subjected to mathematical analysis and expressed by a formula. Under these conditions, how can they express passion in sound or form, if this expression is not a purely mechanical act? If the reader considers this, he will agree that the time must come when people will be able to analyze the external manifestations of the activity of the brain just as easily as the physicist now analyzes a musical chord or the phenomena presented by a falling body.

But this happy time a still far distant, and instead of empty speculation, let us turn to our essential question and see in what manner the external manifestations of the activity of the brain are developed, and to what extent they serve as an expression of psychical activity.

Assuming that the reader agrees with me that this activity always expresses itself on the surface in the form of muscular movement, my task will consist in determining the way in which muscular movements in general issue from the brain.

Let us pass directly to our subject matter. Modern science divides all muscular movements—according to their origin—into two groups: *voluntary and involuntary.* Consequently, our task is to study the origin of both groups of muscular movements. Let us begin with the second group as the simplest; and to attain more clarity we will begin, not with the cerebral hemispheres, but with the spinal cord.

.   .   .   .   .

We are now in a position to establish definitely the difference between the two types of conditions which are necessary for the manifestation of involuntary movements when the brain is intact. If the impressions come unexpectedly, the only nervous centre participating in the reflex is the nervous centre connecting the sensory and the motor nerves. When the stimulation is expected, the activity of another mechanism interferes in the phenomenon, restricting and retarding the reflex movement. Under some circumstances, this activity is stronger than the stimulation, and then the reflex (involuntary) movement does not take place. Sometimes it is the opposite: the stimulation is stronger than the obstacle, and the involuntary movement appears.

It is difficult to find a more simple and convenient explanation than the one given here; but this explanation requires a physiological foundation, because it presupposes the existence in the brain of new mechanisms, whose action, if they really exist, can probably be observed also in animals. So we shall now occupy ourselves with the question whether there is a physiological basis for accepting the existence in the human brain of mechanisms that inhibit reflex movements.

4. About twenty years ago, when physiologists still thought that the excitation of every nerve which leads to a muscle, causes this muscle to contract, Edward Weber demonstrated by direct experiments that the

excitation of the vagus nerve which innervates (among other organs) the heart, not only does not augment the activity of the latter organ, but even paralyzes it. Weber's contemporaries wondered and wondered, and it was finally decided (at least by the greater part of modern physiologists) that this abnormal action is due to the fact that the nerves end, not directly in the muscular tissue of the heart, as in the muscles of the body, but in the nervous ganglia, which lie in the tissue of the walls of the heart. Ten years after the discovery of Weber, Pflüger obtained a similar action of the splanchnic nerves on the intestine. Later, Claude Bernard suggested that the chorda tympani, the stimulation of which augments the secretion of saliva, is not only the stimulator, but also the inhibitor (generally speaking, the regulator) of salivation. Finally, Rosenthal demonstrated that the respiratory movements (essentially involuntary) are stopped or retarded by the excitation of the fibres of the superior laryngeal nerve. These facts have gradually strengthened the belief of physiologists that there exist in the animal body nervous influences by which involuntary movements are depressed. On the other hand, our daily life presents a multitude of examples where the will acts, apparently, in the same manner: we can voluntarily stop our respiratory movements in any phase, even after exhalation, when all respiratory muscles are in a state of relaxation . . .

Knowing all these facts, contemporary physiologists could not but accept the existence of mechanisms which retard reflex movements in the human body (or, more accurately, in the brain, for our will acts only through this organ).

·   ·   ·   ·   ·

There can be no doubt, therefore, that every counter-action to the excitation of sensory nerves is brought about by mechanisms that inhibit reflex movements.

The question of the origin of involuntary movements in the presence of the brain is thereby settled. In both cases (absolutely and relatively unexpected excitation of the sensory nerves), the mechanism of the origin of involuntary reflex movements must be essentially the same, and does not differ from that which exists in the spinal cord.

It is easy to convince oneself of this by comparing the structure of the apparatus producing involuntary movements in decapitated and in normal animals,—an apparatus which has lately been studied in great detail in the frog. This mechanism, for every point of the skin of the decapitated animal, consists of the cutaneous nerve a (fig. 1) beginning in the spinal brain and ending in the cell b. This cell is connected with another cell c, located in the front part of the spinal cord; the two cells

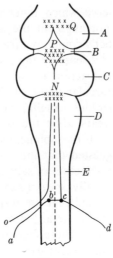

Fig. 1.

b and *c* constitute together the so-called reflex centre; the motor fibre
d arises in *c* and ends in the muscle. The reflex, as the product of the
activity of this mechanism, is nothing else than the successive excitation
of *a, b, c, d,* always beginning with the excitation of *a* in the skin. As
for the reflexes of the brain, they are produced by the activity of a
mechanism which consist of the following parts: the cutaneous fibre
o which leads to the centres of locomotion *N* (it was shown by Berezin
that the cutaneous fibres end in the brain and in the spinal cord), the
path *Nc* for the voluntary motor impulses from the brain, and finally,
the elements *c* and *d,* entering into the composition of the spinal mechanism. The function of all this apparatus is also brought about by the
excitation of *O,* that is, of the cutaneous nerve. The origin of both
reflexes is obviously identical, as long as the excitation follows the path
described; it remains identical even when the activity of the inhibitory
apparatus *P* interferes with the phenomenon, because this apparatus is
capable of inhibiting both *N* and *bc,* and is situated for both *N* and *bc*
in the brain in front of *N.* Those who consider the counter-action to
external influence as voluntary, should, naturally, suppose that the will
acts directly on *P;* however we shall later become acquainted with facts
which indicate that the retarding mechanisms can also be excited by
stimulating the sensory nerves of the skin.

5. Let us continue our study of the brain as a machine, and see what
relationship exists between the intensity of the stimulation and that of
the reflex movement—i.e. between the stimulus and its effect. We shall
again begin, as before, with the phenomena presented by the spinal

cord, because these phenomena have been studied best of all. Here we can say, on the whole, that with the gradual increase of the intensity of stimulation, the force of the movement also grows gradually and in proportion with the stimulus, spreading to a greater and greater number of muscles. For example, a weak stimulation of the skin of one of the hind legs in the decapitated frog leads to the contraction of the muscles of this leg only; with the gradual increase of the stimulus, the reflex movements spread to the front leg of the same side, and finally to the front and hind legs of the opposite side.

The same can be noticed in the cranial nerves under the condition that the brain is, so to say, inactive.

Thus, for example, a slight irritation (e.g. by means of a feather) of the skin of the face, which is innervated by the trigeminal nerve, leads in the sleeping man to the contraction of the facial muscles. If the irritation is stronger, the reflex may include the movement of the hand, and if the irritation is very strong, the man will wake up and jump from bed, i.e. the response will include movements of nearly all the muscles of the body. Consequently, here also, the augmentation of the stimulus leads to the increase and to the spreading of the reflex.

It is a different matter when the brain is active. In this case, the relationship between stimulus and effect is incomparably more complex. This problem, as far as I know, has never been scientifically analyzed. For this reason, I think it necessary to deal with it in detail. Let us analyse the cases, in animals and man, when the brain is intact, and the sensory nerves are suddenly stimulated. Hang up a frog by the lower jaw vertically in the air, wait till it has ceased to thrash about and is hanging perfectly quiet, and then touch its hind leg lightly with your finger. Very often, the frog will exhibit what is colloquially called "fright," and will begin to thrash about again, i.e. to work with all the muscles in its body. There is a popular belief that the bear responds to sudden fright (i.e. to the unexpected stimulation of the sensory nerves) by running away as fast as possible, sometimes even shedding bloody faeces. There is no doubt that a very insignificant excitation of the sensory nerves may produce in animals exceptionally strong involuntary movements. In man, this phenomenon may be expressed even more sharply; as an example, let us mention the convulsions (i.e. the reflex movements) of the whole body which are called forth by a sudden knock, or by the contact of an unfamiliar body, in hysterical women.

Apart from such extreme cases, everyone knows that sudden fright, no matter how insignificant its cause (i.e. no matter how weak the stimulation of the sensory nerves which has led to it) always calls forth strong and widespread reflex movements in man. In addition, as everyone knows, fright can be brought about by the stimulation of the spinal, as well as the cerebral, sensory nerves: it is just as easy to become frightened by the sudden contact of our body with an unfamiliar object (involving the spinal sensory nerves), as by the appearance of a strange shape before our eyes (involving the excitation of the optical nerve, which has its origin in the brain).

However this may be, it is a known fact that fright disturbs the relationship between the intensity of the stimulus and its effect, (i.e. movement) in favour of the latter. But is it possible, in this case, to assume that the development of involuntary movement during fright is machine-like? The phenomenon includes a psychical element, viz. the feeling of fright; and the reader has undoubtedly heard tales of the wonders which are sometimes accomplished under the influence of fear: people with shortness of breath run for miles without stopping, weak individuals carry tremendous weights, etc. The exceptional energy of muscular movements in such cases is usually ascribed to the moral influence of fear; but of course, no one will think that this readily explains the matter. Let us try if we cannot think of a machine where the impulses to action are insignificant but the result tremendous. If it is possible to imagine such a machine, then there is no reason to deny the machine-like character of the origin of involuntary movement during fright. Here is an example of such a machine: a wire leading from one of the poles of a powerful galvanic battery is coiled round a piece of soft iron shaped as a horse-shoe; the end of this wire is submerged in mercury. Another wire, leading from the other pole of the battery, hangs over the surface of the mercury, nearly touching it. Under the ends of the horse-shoe, there lies at some distance a piece of iron weighing several hundred pounds. As long as the circuit is disconnected the iron does not move. But it is sufficient to blow on the lead hanging over the mercury to make it touch the surface of the mercury; the circuit is thereby closed, and the horse-shoe becomes a magnet and draws to itself the heavy iron piece lying under it. The impulse applied (the blowing) is very weak; but the effect obtained (the lifting of a heavy weight) is considerable. If we set a spark to gun-powder, we see the same thing. The spark itself undoubtedly represents a certain force, which may even be approximately measured (given a definite temperature and a definite combustible substance); but this force is negligible in comparison to that produced by the explosion of the powder.

Therefore, to explain why the intensity of the movements performed during fright does not correspond to the intensity of the stimulus, it is not only possible, but even necessary to accept that the mechanism of these movements is machine-like; otherwise we should fall into an absurdity which even a spiritualist cannot commit, for we should have to admit that purely material (muscular) forces are born from moral forces.

.   .   .   .   .

This is the end of my analysis of the quantitative side of involuntary movements. The reader has seen that most external manifestations of the work of the brain can be explained by a very simple mechanical scheme. There can be no doubt that the actual phenomena are in reality much more complicated than in our scheme: in reality, most involuntary movements take place, not in a single muscular fibre or even in a single muscle, but in a whole group of muscles; whereas in our scheme we speak of the activity of single nerve fibres and a small number of nerve cells connecting these fibres. Nevertheless, this scheme explains the matter satisfactorily because it shows us the activity of the physiological elements that compose the total function of whole groups of nerves and muscles.

7. It would now be advisable to pass to the description of the qualitative side of involuntary movements; but the reader must first become acquainted with the existing scientific views concerning the manner in which the activities of the separate elements of the reflex are combined, producing an integrated reflex movement that includes more or less considerable groups of muscles. I have already pointed out that a reflex element is composed of two nerve cells and includes a primary sensory and a primary motor nerve-fibre. Consequently, the activity of such an element is limited to those muscular fibres which are connected with the given motor nerve-fibre. Anatomy shows that in the body of man or animal there is no muscle all the fibres of which are innervated by the branches of one and the same nerve-fibre. Consequently, the activity of one muscle must be due to the combined activity of several reflex elements. How does this combination take place?

Since the elements we are speaking of (the primary nerve-fibres and the nerve cells) are invisible to the naked eye, the question can be solved only by a microscopic study of the spinal cord; unfortunately, the microscope,—which has rendered such great services in the study of the animal body,—seems to be powerless in the solution of our question: up to the present it has not been able to determine the form of connection between the nerve cells. Therefore, science accepts the existence of such a connection, not as a proved fact, but as a logical necessity. *Without the help of intra-neural connections it would be impossible to explain the mechanism of even the simplest reflex.*

.   .   .   .   .

In concluding this chapter on involuntary movements, I shall briefly summarize the results of our study of these movements.

1) The basis of every involuntary movement is a more or less definite excitation of a sensory nerve.

2) A sensory excitation which produces a reflex movement may also call forth conscious sensations; such sensations, however, are not inevitable.

3) In a pure reflex devoid of the psychical element, the relationship between the force of the excitation and the intensity of the movement remains constant under given conditions.

4) If the reflex is complicated by the psychical element, this relationship may vary in both directions.

5) The reflex movement always follows rapidly after the sensory excitation.

6) The duration of the reflex movement and that of the sensory excitation correspond to each other, especially if the reflex is not complicated by psychical elements.

7) All involuntary movements are expedient from the point of view of the preservation of the integrity of the organism.

8) The types of involuntary movements described above are equally applicable to the simplest and to the most complex reflexes,—to an abrupt movement lasting one second, and to a whole series of successive reflexes.

9) The possibility of frequently repeating a reflex in the same direction depends either on the presence in the body of a definite inborn mechanism (e.g. the mechanism of sneezing, coughing etc.), or it is acquired by learning (e.g. the mechanism of walking), i.e. by means of an act in which judgement participates.

10) When the sensory reception is dulled in one, several, or all of the senses (sight, hearing, smell etc.), the mechanism of all movements originating in these senses,—whether inborn or acquired by learning, whether accompanied by psychical imagination or not,—is in every case reflex.

11) This mechanism consists of sensory and motor nerves with their cells in the cerebral centres; these cells are the beginning of the nerves, and send into the brain processes along which the latter influences the reflex movement, by strengthening or weakening it.

12) It is the activity of this mechanism that is a reflex.

13) The machine is started by the stimulation of the sensory nerves.

14) Consequently, all involuntary movements are machine like in their origin.

.   .   .   .   .

All psychical acts without exception, if they are not complicated by elements of emotion (with these we shall deal later) are developed by means of reflexes. Hence, all conscious movements (usually called voluntary), inasmuch as they arise from these acts, are reflex, in the strictest sense of this word.

The question whether voluntary movements are really based on the stimulation of sensory nerves is thereby answered affirmatively. It also becomes clear why, in the case of voluntary movements, this sensory stimulus is often unnoticeable.

The various reasons for this may be reduced to the following main ones:

1) Very frequently (if not always), some indistinct muscular, olfactory or other component is added to an association, which in all other respects is perfectly clear,—for instance, an optico-acoustic association. The essential association is then so vivid that this addition to it is hardly noticed, or even not noticed at all. Nevertheless, it exists, and it is sufficient that it should appear in our consciousness for a brief moment, to call forth the corresponding optico-acoustic combination. Here is an example: I devote my daytime to physiology; but in the evening, while going to bed, it is my habit to think of politics. It happens, of course, that among other political matters I sometimes think of the Emperor of China. This acoustic trace becomes associated with the various sensations (muscular, tactile, thermic, etc.) which I experience when lying in bed. It may happen, one day, that owing to fatigue or to the absence of work I lie down on my bed in the daytime; and lo! all of a sudden I notice that I am thinking of the Emperor of China. People usually say that there is no particular cause for such a visitation; but we see that in the given case it was called forth by the sensations of lying in bed; and now that I have written this example, I shall associate the Emperor of China with more vivid sensations, and he will become my frequent guest.

2) A series of logically connected thoughts may be associated with some idea that has absolutely no logical relation to them. And though it might seem strange to deduct that particular series of thoughts from this idea, it is nevertheless this idea that calls forth the whole series.

3) A chain of combined ideas sometimes lingers long in our consciousness. It was said above that such a chain may last a whole day, beginning with our awakening in the morning and ending when we go to sleep at night. In such cases, we sometimes find it very difficult to remember just what particular impulse has called forth a given chain of thoughts.

Nevertheless, attentive self-observation makes it possible to determine this external influence in the majority of cases.

. . . Does the mechanism of inhibition which is already known to us from our study play any role in the origin of voluntary movements? Since we have identified voluntary movements with reflexes, there is nothing irrelevant in this question.

Are there any phenomena in the conscious life of man which point to the inhibition of movements? These phenomena are so numerous and so characteristic that it is because of them that people call those movements which are performed with full consciousness, voluntary movements. Indeed, upon what is the common conception of such movements based? It is based on the fact, that under the influence of definite external and moral conditions, man can perform a certain series of movements, or can fail to do so, or finally, can perform movements of an entirely opposite character. People with a strong will may triumph over apparently irresistible involuntary movements; for example, one man will endure severe pain silently and without the slightest movement, while another one will scream and writhe. There are even men who can endure pain, performing at the same time movements which are entirely incongruous with pain, such as joking and laughing.

Consequently, even such movements which are generally called involuntary can be inhibited.

In the first chapter of this book, when discussing the origin of those involuntary movements which we perform when we are expecting some sensory stimulus, I said that these movements are best explained by assuming that there exists a special mechanism which inhibits the activity of the reflex apparatus. Experiments were also mentioned which make the presence of such mechanisms unquestionable in the brain of the frog and highly probable in the brain of man.

We must now prove this hypothesis in relation to voluntary movements.

Let us then a priori accept that in the brain of man there exist mechanisms which inhibit muscular movements. But why,—the reader will ask,—is the activity of these mechanisms so unequally expressed in different people? If the inhibition of movements is an organic feature common to all people, it would seem that this phenomenon should not present such considerable variations, as, for instance, in the case of a weak nervous woman and that of an extreme stoic; besides, it should also be present in the child. The fact is that inhibition does exist in all cases, but we must learn to inhibit movements just as much as we learn to perform these movements. No one will doubt that the new-born child possesses all the nervous centres which later regulate the acts of walking, speaking, etc.; and yet, these acts also must be learned.

So let us now see how the child learns to inhibit movements, or,—to put it more exactly,—to inhibit the last member of a reflex.

Childhood is characterised by an exceptionally wide irradiation of reflex movements in response to such external sensory stimuli, which, when applied to an adult, would call forth a relatively weak response; thus, the reflexes from the eye and ear can spread in the child to nearly all the muscles of the body. However, there' comes a time, when movements become, so to say, "grouped." One or two muscles, or a whole group of them, begin to function separately from the mass of other muscles; the expansion of movements is thereby limited, and every movement obtains a definite character. It is in this process that the inhibitory mechanisms take part. Let us take a simple case: the change from the flexion of all the fingers of the hand to the separate flexion of one finger. Supposing that the flexion of all fingers at once is the result of certain fundamental properties of the very organisation of the child (and there can be no doubt that this is really so),—it is obvious that the isolated movement of one finger can be achieved only by inhibiting the movement of all other fingers. There can be no other explanation. How does this inhibition arise? It might be possible that the bending of the fingers is prevented by the activity of the muscles that act in the opposite direction, i.e. by the contraction of the extensors. This suggestion seems very plausible at the first glance. Indeed, to prevent the movement of the other four fingers, the smallest preponderance of the contraction of their extensors over that of their flexors would be sufficient. This preponderance should, of course, be accompanied by muscular sensations, because, after all, this immobility is the result of the antagonism of two systems of muscles; but the sensation might be very weak and remain unnoticed in comparison to the vivid muscular sensations from the bent finger. In this way, the matter would be explained without the participation of special inhibitory mechanisms and would be reduced to the activity of antagonistic muscles. However, this explanation alone is not sufficient. Let us suppose that the flexion of all fingers at once is brought about by some very powerful cause. In that case, the flexion of one finger would be accompanied by a very strong tendency to bend the other four, and only a very strong activity of the antagonistic muscles could prevent their movement. The isolated flexion of one finger would therefore be accompanied by an extremely sharp muscular sensation in the other fingers. Now this never happens: a man with an ideally strong will can endure pain with perfect outward composure, i.e. without the contraction of muscles.

Consequently, without denying that movements can be prevented by the contraction of antagonistic muscles, and even accepting that this mechanism often takes part in the prevention of conscious movements,— it is necessary, nevertheless, to admit the participation, at least in some cases, of a mechanism which acts upon reflex movements in the same

319

way as the vagus nerve acts upon the heart; in other words, it is necessary to admit the existence of an activity which paralyses the muscles.

It follows from this that a psychical act has the nature of a reflex
even if it remains without external manifestation. If we accept the idea
that movements are prevented by the activity of antagonistic muscles,
then the end of the act must be a pure muscular movement. If, on the
other hand, we accept the other explanation, the end of the reflex is an
act completely equivalent to the excitation of the motor apparatus, i.e.
the motor nerve and its muscle.

As for the history of development of the capacity to prevent the end
of reflexes, the first case agrees entirely in this respect with the history
of development of grouped muscular movements in general, and the
tremendous difference in the external expression of both phenomena
(the production of movements and their prevention) is due merely to
the participation in the movement of different muscles. The capacity
to prevent movements begins with the instinctive mimicry of the child;
it is controlled by muscular sensations and their analysis; and it is acquired by means of frequent repetition. When the child has already
learned to use its muscles, i.e. when it walks and speaks (and is, consequently, able to understand what is said to it), the education of the
capacity to prevent movements continues by means of the development,
in the mind of the child, of the following type of associated conceptions:
"Do not do this or that, otherwise this or that will happen." For the
edification of the child, these admonitions are often accompanied by the
infliction of physical pain; this burdens terribly the future of the child:
under such a system of education, the morality of the motive,—which
should alone direct the activities of the child,—is concealed by the much
stronger feeling of fear, and in this way the sorrowful morale of fear
is brought in the world.

The development of the capacity to paralyse movements (let the
reader not forget that in relation to man this capacity is a mere hypothesis) is very obscure; our only guide in this matter is the sensation
which accompanies the immobility of the muscles. To understand this
better, the reader should perform the following experiment on his own
respiration: after an exhalation, let him retard the subsequent involuntary inspiratory movement. At first, he will not feel anything definite
(he is only indirectly aware that his muscles are at rest); then he feels
something which forces him to breathe, but this feeling is not located
in his muscles.

·   ·   ·   ·   ·

Now that all the properties of thought have been explained, the
reader will easily understand how we learn to separate mentally our

thoughts from the external acts to which they lead. In every man, a given sensory stimulation will in one case call forth both thought and act; another time, the act may be inhibited, and the thought seems to be the only response; a third time, the thought is again followed by activity, but the act is not the same one as the first time. This necessarily leads to the separation of thought, as something concrete, from activity which also appears in a concrete form. It is generally accepted that if one act follows another, the two acts stand in causal relationship (post hoc—ergo propter hoc); *this is why thought is generally believed to be the cause of behaviour;* and when the external sensory stimulus remains unnoticed,—which happens quite frequently,—*thought is even accepted as the initial cause of behaviour.* Add to this the extremely subjective character of thought, and you will understand how firmly man must believe in the voice of self-consciousness, when it tells him such things. In reality, however, this voice tells him the greatest of falsehoods: *the initial cause of all behaviour always lies, not in thought, but in external sensory stimulation, without which no thought is possible.*

The fact that one and the same man seems to be able to express one and the same thought by means of different acts, is another source of error. A person under the influence of some thought can deliberate on the form of his behaviour and choose one of many possible forms of conduct. This means that when a man is subjected to the action of certain external and internal conditions, there appears in him the central member of a psychical reflex, and to this central member is added (also in the form of a thought) the imagination of a possible end of the reflex. (N.B.: For the sake of brevity, I shall in future name "psychical reflex" every complete act of conscious life.) If, in the past, a middle member has had several ends because the reflex was produced in different external conditions, it is natural that these several ends should appear in imagination one after another. I shall explain later what inevitable motives dictate the choice of a definite end of the reflex.

64   JOHN DEWEY (1859–1952) AGAINST REFLEXOLOGY, 1896

John Dewey, "The reflex arc concept in psychology," *Psychological Review 3,* 357–370 (1896).

This selection contains the germ of virtually all of the modern objections against the use of the stimulus-response unit as the building block of behavior in psychological theory. Dewey pointed out that the organism is not a passive receiver of stimuli, but is active in perceiving. He insisted that behavior is continuous, not disjoined into stimuli and responses, and that the sensory and motor aspects of behavior blend continuously into each other.

To reduce all psychology to reflexes was too severe a simplification of human action, but that, said Dewey, is not to reject mechanism. Rather, it is to find the Cartesian reflex machine too poorly integrated to be right.

Let us take, for our example, the familiar child-candle instance . . . The ordinary interpretation would say the sensation of light is a stimulus to the grasping as a response, the burn resulting is a stimulus to withdrawing the hand as response and so on. There is, of course, no doubt that is a rough practical way of representing the process. But when we ask for its psychological adequacy, the case is quite different. Upon analysis, we find that we begin not with a sensory stimulus, but with a sensori-motor coördination, the optical-ocular, and that in a certain sense it is the movement which is primary, and the sensation which is secondary, the movement of body, head and eye muscles determining the quality of what is experienced. In other words, the real beginning is with the act of seeing; it is looking, and not a sensation of light. The sensory quale gives the value of the act, just as the movement furnishes its mechanism and control, but both sensation and movement lie inside, not outside the act.

Now if this act, the seeing, stimulates another act, the reaching, it is because both of these acts fall within a larger coördination; because seeing and grasping have been so often bound together to reinforce each other, to help each other out, that each may be considered practically a subordinate member of a bigger coördination. More specifically, the ability of the hand to do its work will depend, either directly or indirectly, upon its control, as well as its stimulation, by the act of vision. If the sight did not inhibit as well as excite the reaching, the latter would be purely indeterminate, it would be for anything or nothing, not for the particular object seen. The reaching, in turn, must both stimulate and control the seeing. The eye must be kept upon the candle if the arm is to do its work; let it wander and the arm takes up another task. In other words, we now have an enlarged and transformed coördination; the act is seeing no less than before, but it is now seeing-for-reaching purposes. There is still a sensori-motor circuit, one with more content or value, not a substitution of a motor response for a sensory stimulus.

Now take the affairs at its next stage, that in which the child gets burned. It is hardly necessary to point out again that this is also a sensori-motor coördination and not a mere sensation. It is worth while, however, to note especially the fact that it is simply the completion, or fulfillment, of the previous eye-arm-hand coördination and not an entirely new occurrence. Only because the heat-pain quale enters into the same circuit of experience with the optical-ocular and muscular quales, does the child learn from the experience and get the ability to avoid the experience in the future.

More technically stated, the so-called response is not merely *to* the stimulus; it is *into* it. The burn is the original seeing, the original optical-ocular experience enlarged and transformed in its value. It is no longer mere seeing; it is seeing-of-a-light-that-means-pain-when-contact-occurs. The ordinary reflex arc theory proceeds upon the more or less tacit assumption that the outcome of the response is a totally new experience; that it is, say, the substitution of a burn sensation for a light sensation through the intervention of motion. The fact is that the sole meaning of the intervening movement is to maintain, reinforce or transform (as the case may be) the original quale; that we do not have the replacing of one sort of experience by another, but the development . . . the mediation of an experience. The seeing, in a word, remains to control the reaching, and is, in turn, interpreted by the burning.

The discussion up to this point may be summarized by saying that the reflex arc idea, as commonly employed, is defective in that it assumes sensory stimulus and motor response as distinct psychical existences, while in reality they are always inside a coördination and have their significance purely from the part played in maintaining or reconstituting the coördination; and (secondly) in assuming that the quale of experience which precedes the 'motor' phase and that which succeeds it are two different states, instead of the last being always the first reconstituted, the motor phase coming in only for the sake of such mediation. The result is that the reflex arc idea leaves us with a disjointed psychology, whether viewed from the standpoint of development in the individual or in the race, or from that of the analysis of the mature consciousness. As to the former, in its failure to see that the arc of which it talks is virtually a circuit, a continual reconstitution, it breaks continuity and leaves us nothing but a series of jerks, the origin of each jerk to be sought outside the process of experience itself, in either an external pressure of 'environment,' or else in an unaccountable spontaneous variation from within the 'soul' or the 'organism.' As to the latter, failing to see the unity of activity, no matter who much it may prate of unity, it still leaves us with sensation or peripheral stimulus; idea, or central process (the equivalent of attention); and motor response, or act, as three disconnected existences, having to be somehow adjusted to each other, whether through the intervention of an extra-experimental soul, or by mechanical push and pull.

·   ·   ·   ·   ·

I hope it will not appear that I am introducing needless refinements and distinctions into what, it may be urged, is after all an undoubted fact, that movement as response follows sensation as stimulus. It is not a question of making the account of the process more complicated, though it is always wise to beware of that false simplicity which is reached by leaving out of account a large part of the problem. It is a question of finding out what stimulus or sensation, what movement and response mean; a question of seeing that they mean distinctions of flexible function only, not of fixed existence; that one and the same occurrence plays either or both parts, according to the shift of interest; and that because of this functional distinction and relationship, the supposed problem of the adjustment of one to the other, whether by superior force in the stimulus or an agency *ad hoc* in the center or the soul, is a purely self-created problem.

We may see the disjointed character of the present theory, by calling to mind that it is impossible to apply the phrase 'sensori-motor' to the occurrence as a simple phrase of description; it has validity only as a term of interpretation, only, that is, as defining various functions exercised. In terms of description, the whole process may be sensory or it may be motor, but it cannot be sensori-motor. The 'stimulus,' the excitation of the nerve ending and of the sensory nerve, the central change, are just as much, or just as little, motion as the events taking place in the motor nerve and the muscles. It is one uninterrupted, continuous redistribution of mass in motion. And there is nothing in the process, from the standpoint of description, which entitles us to call this reflex. It is redistribution pure and simple; as much so as the burning of a log, or the falling of a house or the movement of the wind. In the physical process, as physical, there is nothing which can be set off as stimulus, nothing which reacts, nothing which is response. There is just a change in the system of tensions.

The same sort of thing is true when we describe the process purely from the psychical side. It is now all sensation, all sensory quale; the motion, as psychically described, is just as much sensation as is sound or light or burn. Take the withdrawing of the hand from the candle flame as example. What we have is a certain visual-heat-pain-muscular-quale, transformed into another visual-touch-muscular-quale—the flame now being visible only at a distance, or not at all, the touch sensation being altered, etc. . . . The motion is not a certain kind of existence; it is a sort of sensory experience interpreted, just as is candle flame, or burn from candle flame. All are on a par.

But, in spite of all this, it will be urged, there is a distinction between stimulus and response, between sensation and motion. Precisely; but we ought now to be in a condition to ask of what nature is the distinction, instead of taking it for granted as a distinction somehow lying in the existence of the facts themselves. We ought to be able to see that the ordinary conception of the reflex arc theory, instead of being a case of plain science, is a survival of the metaphysical dualism, first formulated by Plato, according to which the sensation is an ambiguous dweller on the border land of soul and body, the idea (or central process) is purely psychical, and the act (or movement) purely physical. Thus the reflex arc formulation is neither physical (or physiological) nor psychological; it is a mixed materialistic-spiritualistic assumption.

If the previous descriptive analysis has made obvious the need of a reconsideration of the reflex arc idea, of the nest of difficulties and assumptions in the apparently simple statement, it is now time to undertake an explanatory analysis. The fact is that stimulus and response are not distinctions of existence, but teleological distinctions, that is, distinctions of function, or part played, with reference to reaching or maintaining an end. With respect to this teleological process, two stages should be discriminated, as their confusion is one cause of the confusion attending the whole matter. In one case, the relation represents an organization of means with reference to a comprehensive end. It represents an accomplished adaptation. Such is the case in all well developed instincts, as when we say that the contact of eggs is a stimulus to the hen to set; or the sight of corn a stimulus to pick; such also is the case with all thoroughly formed habits, as when the contact with the floor stimulates walking. In these instances there is no question of consciousness of stimulus *as* stimulus, of response *as* response. There is simply a continuously ordered sequence of acts, all adapted in themselves and in the order of their sequence, to reach a certain objective end, the reproduction of the species, the preservation of life, locomotion to a certain place. The end has got thoroughly organized into the means. In calling one stimulus, another response we mean nothing more than that such an orderly sequence of acts is taking place. The same sort of statement might be made equally well with reference to the succession of changes in a plant, so far as these are considered with reference to their adaptation to, say, producing seed. It is equally applicable to the series of events in the circulation of the blood, or the sequence of acts occurring in a self-binding reaper.

X ASSOCIATION

Although the doctrine of association states that the elements of thought are interconnected by simple rules of connection, its implications are very broad: it implies that personal experience is psychologically more important than innate endowment, that the mind is passive during thought, that the complexity of thought can be explained by uncomplicated principles with no loss of any essential properties. The doctrine is, however, far older than the recognition of its implications. It appeared first in antiquity as a minor feature of a theory of knowledge but did not achieve full development until more than two thousand years later, in the nineteenth century, when it was assimilated by the dawning science of psychology.

The earliest systematic statement of the doctrine of association was Aristotle's account of the act of remembering (ca. 350 B.C.). Ideas, he said, follow each other in memory because they are similar, contrasting, or frequently contiguous. In the two thousand years between Aristotle and Hobbes (1651), the notion of association was perpetuated by philosophers in discussions of memory, for example, St. Augustine (ca. A.D. 400). It was not, however, until the seventeenth century that the philosophical climate, provided mainly by British writers, became suitable for the growth of the notion into a rule of major significance for philosophy and psychology.

This doctrine of association has always been wedded to a theory of knowledge that holds that the only source of information about the world is sensory experience. As the theory was shaped by the school of British philosophers called "empiricists," the idea of association gradually grew in importance. The history of association thus parallels the history of the empiristic theory of knowledge. Thomas Hobbes (1651) is the first in this line, although he never developed empiricism or associationism to any great extent. His philosophical successor John Locke (1700) enlarged greatly on the empiristic view, but furthered the doctrine of association only by coining the phrase "association of ideas" and not by any significant development of its principles. Later George Berkeley, having tacitly accepted association in his first important work (1709), became explicit (1733) in applying the empiristic and associationist principles to a substantive psychological problem, the perception of space. He argued that awareness of space arises from associations between sensations of sight and touch. Although Hobbes, Locke, and Berkeley may be called assoiationists, it was not until David Hume (1739) that the doctrine of association was first definitely regarded as a central principle in the analysis of the human mind. Hume's contribution to the doctrine was to assert its importance and to propose a classification of its principles, one which differed only slightly from Aristotle's.

It was Hume's contemporary David Hartley (1749) who also recognized the importance of association and produced the first attempt at a detailed and all-inclusive theory of thought and action based on the doctrine. For at least seventy years Hartley's work was considered the authoritative work on the subject of association. Then, in 1820 Thomas Brown proposed his nine secondary laws of association, which laid the basis for an experimental science by listing such factors as amount of practice and degree of recency that might affect the strength of associations.

In the writings of James Mill (1829), the doctrine of association began to approximate its zenith as philosophical speculation. More than any of his predecessors, Mill applied associationism mechanically to the analysis of the mind, picturing all the contents of the conscious mind as being resolvable into simple sensory elements. By exposing the broad implications of the doctrine, he unwittingly produced a *reductio ad absurdum.* His son, John Stuart Mill (1843), was one of the first to recoil from this purely routine application of associative connections to mental processes. He argued that new properties may emerge in complex ideas over and above the sum of the simple ideas of which they are basically composed. Herbert Spencer (1855) was the author of yet another modification of James Mill's pure associationism. He accepted the doctrine, but

suggested the possibility that certain associations are impressed on the germ plasm and are handed down from generation to generation as hereditary endowment. Both John Stuart Mill and Spencer were sure that association is but a part, and not the whole, of the psychology of mental process. William James (1890) also saw other processes in the mind than simple associationist compounding. He argued for the existence of innate mental functions and modes of perceiving and objected to Spencer's view that these were merely inherited associations.

When psychology emerged from philosophy and physiology as a separate discipline during the latter half of the nineteenth century, the doctrine of association was an important part of its conceptual structure. Wilhelm Wundt (1896), who may properly be called the first professional psychologist, used, and thereby promoted, a version of the doctrine based on John Staurt Mill's views.

65 ARISTOTLE (384–322 B.C.) ON THE ASSOCIATIVE NATURE OF MEMORY, ca. 350 B.C.

Aristotle, *De memoria et reminiscentia,* chap. 2. Translated by J. I. Beare, in W. D. Ross, ed., *The Works of Aristotle,* III (Oxford: Clarendon Press, 1931), 451b–452b. The insertions in brackets are Beare's.

Before Aristotle, Plato proposed a theory of association, but it is rarely remembered, since it is not easily compatible with his much more famous doctrine of innate ideas. Aristotle, therefore, gets credit for the earliest extant unqualified formulation. He described a memory as an orderly sequence of thoughts (called "movements" in this selection) that recapitulate a sequence of events experienced in one's life. If such a sequence has been experienced often with little variation, then the thoughts corresponding to it are firmly connected.

If, on the other hand, a sequence has been experienced infrequently or with significant variation, then the thoughts will have only an uncertain connection. Thoughts that have become related by one's experience are examples of what was called later association by contiguity. Aristotle also mentioned that thoughts are connected by their similarity or contrast. Of his three principles, association by contiguity was to prove most stable, for the others passed in and out of vogue many times in the succeeding two thousand years. Aristotle's doctrine amounted, however, to little more than a theory of memory. It was neither a philosophical theory of knowledge nor the system of psychology it was to become in eighteenth-century Great Britain.

Acts of recollection, as they occur in experience, are due to the fact that one movement has by nature another that succeeds it in regular order.

If this order be necessary, whenever a subject experiences the former of two movements thus connected, it will [invariably] experience the latter; if, however, the order be not necessary, but customary, only in the majority of cases will the subject experience the latter of the two movements. But it is a fact that there are some movements, by a single experience of which persons take the impress of custom more deeply than they do by experiencing others many times; hence upon seeing some things but once we remember them better than others which we may have seen frequently.

Whenever, therefore, we are recollecting, we are experiencing certain of the antecedent movements until finally we experience the one after which customarily comes that which we seek. This explains why we hunt up the series [of movements], having started in thought either from a present intuition or some other, and from something either similar, or contrary, to what we seek, or else from that which is contiguous with it. Such is the empirical ground of the process of recollection; for the mnemonic movements involved in these starting-points are in some cases identical, in others, again, simultaneous, with those of the idea we seek, while in others they comprise a portion of them, so that the remnant which one experienced after that portion [and which still requires to be excited in memory] is comparatively small.

Thus, then, it is that persons seek to recollect, and thus, too it is that they recollect even without the effort of seeking to do so, viz. when the movement implied in recollection has supervened on some other which is its condition. For, as a rule, it is when antecedent movements of the classes here described have first been excited, that the particular movement implied in recollection follows. We need not examine a series of which the beginning and end lie far apart, in order to see how . . . we remember; one in which they lie near one another will serve equally well. For it is clear that the method is in each case the same, that is, one hunts up the objective series, without any previous search or previous recollection. For [there is, besides the natural order, viz. the order of the things, or events of the primary experience, also a customary order, and] by the effect of custom the mnemonic movements tend to succeed one another in a certain order. Accordingly, therefore, when one wishes to recollect, this is what he will do: he will try to obtain a beginning of movement whose sequel shall be the movement which he desires to re-awaken. This explains why attempts at recollection succeed soonest and best when they start from a beginning [of some objective series]. For, in order of succession, the mnemonic movements are to one another as the objective facts [from which they are derived]. Accordingly, things arranged in a fixed order, like the successive demonstrations in geometry, are easy to remember . . . while badly arranged subjects are remembered with difficulty.

Recollecting differs also in this respect from relearning, that one who recollects will be able, somehow, to move, solely by his own effort, to the term next after the starting-point. When one cannot do this of himself, but only by external assistance, he no longer remembers . . . It often happens that, though a person cannot recollect at the moment, yet by seeking he can do so, and discovers what he seeks. This he succeeds in doing by setting up many movements, until finally he excites one of a kind which will have for its sequel the fact he wishes to recollect. For remembering . . . is the existence, potentially, in the mind of a movement capable of stimulating it to the desired movement, and this, as has been said, in such a way that the person should be moved . . . from within himself, i.e. in consequence of movements wholly contained within himself.

But one must get hold of a starting-point. This explains why it is that persons are supposed to recollect sometimes by starting from mnemonic *loci*. The cause is that they pass swiftly in thought from one point to another, e.g. from milk to white, from white to mist, and thence to moist, from which one remembers Autumn [the 'season of mists'], if this be the season he is trying to collect.

It seems true in general that the middle point also among all things is a good mnemonic starting-point from which to reach any of them. For if one does not recollect before, he will do so when he has come to this, or, if not, nothing can help him; as, e.g. if one were to have in mind the numerical series denoted by the symbols 1, 2, 3, 4, 5, 6, 7, 8, 9. For, if he does not remember what he wants at 5, then at 5 he remembers 9; because from 5 movement in either direction is possible, to 4 or to 6. But, if it is not for one of these that he is searching, he will remember [what he *is* searching for] when he has come to 3, if he is searching for 8 or 7. But if [it is] not [for 8 or 7 that he is searching, but for one of the terms that remain], he will remember by going to 1, and so in all cases [in which one starts from a middle point]. The cause of one's some-times recollecting and sometimes not, though starting from the same point, is, that from the same starting-point a movement can be made in several directions, as, for instance, from 3 to 7 or to 4. If, then, the mind has not [when starting from 5] moved in an old path . . . it tends to move to the more customary; for [the mind having, by chance or otherwise, *missed* moving in the 'old' way] Custom now assumes the rôle of Nature. Hence the rapidity with which we recollect what we frequently think about. For as regular sequence of events is in accordance with nature, so, too, regular sequence is observed in the actualization of movements [in consciousness], and here frequency tends to produce [the regularity of] nature. And since in the realm of nature occurrences take place which are even contrary to nature, or fortuitous, the same happens *a fortiori* in the sphere swayed by custom, since in this sphere natural law is not similarly established. Hence it is that [from the same starting-point] the mind receives an impulse to move sometimes in the required direction, and at other times otherwise, [doing the latter] particularly when something else somehow deflects the mind from the right direction and attracts it to itself. This last consideration explains too how it happens that, when we want to remember a name, we remember one somewhat like it, indeed, but blunder in reference to . . . the one we intended.

Thus, then, recollection takes place.

66 THOMAS HOBBES (1588-1679) ON THE TRAIN OF THOUGHT, 1651

Thomas Hobbes, *Leviathan, or the Matter, Forme and Power of a Common-wealth Ecclesiasticall and Civill* (London, 1651), pt. 1, chap. 3.

Hobbes was primarily a political theorist. His "Leviathan" is the all-powerful state—examplified by his aspirations for the English monarchy—which men must accept if they are not to destroy one another. The need for such a state arises, he said, in the nature of man, and, in his attempt to describe this nature, Hobbes made his contribution to the doctrine of association. He argued against a belief in man's innate knowledge of the world, holding that knowledge is gained only through the senses, and that thoughts, which are but copies of past sensations, become linked by being experienced together. Thus Hobbes was restating the Aristotelian principle of association by contiguity. The course of thought, he said, may be casual or even ostensibly random, or it may be guided by some emotion or goal, but in either case it is ultimately a reflection of personal experience. Although for Hobbes this empiristic theory of knowledge served merely as a roughly sketched background for a political thesis, his successors in British philosophy were to develop it with great subtlety.

By *Consequence,* or Trayne of Thoughts, I understand that succession of one Thought to another, which is called (to distinguish it from Discourse in words) *Mentall Discourse.*

When a man thinketh on any thing whatsoever, His next Thought after, is not altogether so casuall as it seems to be. Not every Thought to every Thought succeeds indifferently. But as wee have no Imagination, whereof we have not formerly had Sense, in whole, or in parts; so we have no Transition from one Imagination to another, whereof we never had the like before in our Senses. The reason whereof is this. All Fancies are motions within us, reliques of those made in the Sense: And those motions that immediately succeeded one another in the sense, continue also together after Sense: In so much as the former comming again to take place, and to prædominant, the later followeth, by coherence of the matter moved, in such manner, as water upon a plain Table is drawn which way any one part of it is guided by the finger. But because in sense, to one and the same thing perceived, sometimes one thing, some-times another succeedeth, it comes to passe in time, that in the Imagining of any thing, there is no certainty what we shall Imagine next; Onely this is certain, it shall be something that succeeded the same before, at one time or another.

This Trayne of Thoughts, or Mentall Discourse, is of two sorts. The first is *Unguided, without Designe,* and inconstant; Wherein there is no Passionate Thought, to govern and direct those that follow, to it self, as the end and scope of some desire, or other passion: In which case the thoughts are said to wander, and seem impertinent one to another, as in a Dream. Such are Commonly the thoughts of men, that are not onely without company, but also without care of any thing; though even their Thoughts are as busie as at other times, but without harmony; as the sound which a Lute out of tune would yeeld to any man; or in tune, to one that could not play. And yet in this wild ranging of the mind, a man may oft-times perceive the way of it, and the dependance of one thought upon another. For in a Discourse of our present civill warre, what could seem more impertinent, than to ask (as one did) what was the value of a Roman Penny. Yet the Cohærence to me was manifest enough. For the Thought of the warre, introduced the Thought of the delivering up the King to his Enemies; The Thought of that, brought in the Thought of the delivering up of Christ; and that again the Thought of the 30 pence, which was the price of that treason: and thence easily followed that malicious question; and all this in a moment of time; for Thought is quick.

The second is more constant; as being *regulated* by some desire, and designe. For the impression made by such things as wee desire, or feare, is strong, and permanent, or, (if it cease for a time,) of quick return: so strong it is sometimes, as to hinder and break our sleep. From Desire, ariseth the Thought of some means we have seen produce the like of that which we ayme at; and from the thought of that, the thought of

331

means to that mean; and so continually, till we come to some beginning within our own power. And because the End, by the greatnesse of the impression, comes often to mind, in case our thoughts begin to wander, they are quickly again reduced into the way . . .

The Trayn of regulated Thoughts is of two kinds; One, when of an effect imagined, wee seek the causes, or means that produce it: and this is common to Man and Beast. The other is, when imagining any thing whatsoever, wee seek all the possible effects, that can by it be produced; that is to say, we imagine what we can do with it, when wee have it. Of which I have not at any time seen any signe, but in man onely; for this is a curiosity hardly incident to the nature of any living creature that has no other Passion but sensuall, such as are hunger, thirst, lust, and anger. In summe, the Discourse of the Mind, when it is governed by designe, is nothing but *Seeking,* or the faculty of Invention; . . . a hunting out of the causes, of some effect, present or past; or of the effects, of some present or past cause. Sometimes a man seeks what he hath lost; and from that place, and time, wherein hee misses it, his mind runs back, from place to place, and time to time, to find where, and when he had it; that is to say, to find some certain, and limited time and place, in which to begin a method of seeking. Again, from thence, his thoughts run over the same places and times, to find what action, or other occasion might make him lose it. This we call *Remembrance,* or Calling to mind: the Latines call it *Reminiscentia,* as it were a *Re-conning* of our former actions.

Sometimes a man knows a place determinate, within the compasse whereof he is to seek; and then his thoughts run over all the parts thereof, in the same manner, as one would sweep a room, to find a jewell; or as a Spaniel ranges the field, till he find a sent; or as a man should run over the Alphabet, to start a rime.

Sometime a man desires to know the event of an action; and then he thinketh of some like action past, and the events thereof one after another; supposing like events will follow like actions. As he that foresees what wil become a Criminal, re-cons what he has seen follow on the like Crime before; having this order of thoughts, The Crime, the Officer, the Prison, the Judge, and the Gallowes. Which kind of thoughts, is called *Foresight,* and *Prudence,* or *Providence;* and sometimes *Wisdome;* though such conjecture, through the difficulty of observing all circumstances, be very fallacious. But this is certain; by how much one man has more experience of things past, than another; by so much also he is more Prudent, and his expectations the seldomer faile him. The *Present* onely has a being in Nature; things *Past* have a being in the Memory onely, but things *to come* have no being at all; the *Future* being but a fiction of the mind, applying the sequels of actions Past, to the actions that are Present; which with most certainty is done by him that has most Experience; but not with certainty enough. And though it be called Prudence, when the Event answereth our Expectation; yet in its own nature, it is but Presumption. For the foresight of things to come, which is Providence, belongs onely to him by whose will they are to come. From him onely, and supernaturally, proceeds Prophecy. The best Prophet naturally is the best guesser; and the best guesser, he that is most versed and studied in the matters he guesses at: for he hath most *Signes* to guesse by.

A *Signe,* is the Event Antecedent, of the Consequent; and contrarily, the Consequent of the Antecedent, when the like Consequences have been observed, before: And the oftner they have been observed, the lesse uncertain is the Signe. And therefore he that has most experience in any kind of businesse, has most Signes, whereby to guesse at the Future time; and consequently is the most prudent: And so much more prudent than he that is new in that kind of business, as not to be equalled by any advantage of naturall and extremporary wit: though perhaps many young men think the contrary.

Neverthelesse it is not Prudence that distinguisheth man from beast. There be beasts, that at a year old observe more, and pursue that which is for their good, more prudently, than a child can do at ten.

As Prudence is a *Præsumtion* of the *Future,* contracted from the *Experience* of time *Past:* So there is a Præsumtion of things Past taken from other things (not future but) past also. For he that hath seen by what courses and degrees, a flourishing State hath first come into civil warre, and then to ruine; upon the sight of the ruines of any other State, will guesse, the like warre, and the like courses have been there also. But this conjecture, has the same incertainty almost with the conjecture of the Future; both being grounded onely upon Experience.

There is no other act of mans mind, that I can remember, naturally planted in him, so, as to need no other thing, to the exercise of it, but to be born a man, and live with the use of his five Senses. Those other Faculties, of which I shall speak by and by, and which seem proper to man onely, are acquired, and encreased by study and industry; and of most men learned by instruction, and discipline; and proceed all from the invention of Words, and Speech. For besides Sense, and Thoughts, and the Trayne of thoughts, the mind of man has no other motion; though by the help of Speech, and Method, the same Facultyes may be improved to such a height, as to distinguish men from all other living Creatures.

Whatsoever we imagine, is *Finite.* Therefore there is no Idea, or conception of any thing we call *Infinite*. No man can have in his mind an Image of infinite magnitude; nor conceive infinite swiftness, infinite time, or infinite force, or infinite power. When we say any thing is infinite, we signifie onely, that we are not able to conceive the ends, and bounds of the thing named; having no Conception of the thing, but of our own inability. And therefore the Name of *God* is used, not to make us conceive him; (for he is *Incomprebensible;* and his greatnesse, and power are unconceivable;) but that we may honour him. Also because whatsoever (as I said before,) we conceive, has been perceived first by sense, either all at once, or by parts; a man can have no thought, representing any thing, not subject to sense. No man therefore can conceive any thing, but he must conceive it in some place; and indued with some determinate magnitude; and which may be divided into parts; nor that any thing is all in this place, and all in another place at the same time; nor that two, or more things can be in one, and the same place at once: For none of these things ever have, or can be incident to Sense; but are absurd speeches, taken upon credit (without any signification at all,) from deceived Philosophers, and deceived, or deceiving Schoolemen.

67   JOHN LOCKE (1632–1704) ON DISORDERS OF THE MIND, 1700

John Locke, *An Essay concerning Humane Understanding,* 4th ed. (London, 1700), bk. II, chap. 33.

Locke is often regarded as the modern founder of the doctrine of association, for he was the first to use the expression "association of ideas." This selection, which contains virtually his entire discussion of the subject, shows, however, that he viewed association as the exception rather than as the rule. In ordinary thought, he said, ideas succeed each other by natural or rational connections, but occasionally two ideas become fortuitously associated by their contiguity when the natural connection is absent. He apparently did not recognize that association by contiguity could also account for the so-called natural connections. He used the notion of association only to explain how even reasonable people may come to hold unreasonable beliefs.

1. There is scarce any one that does not observe something that seems odd to him, and is in it self really Extravagant in the Opinions, Reasonings, and Actions of other Men. The least flaw of this kind, if at all different from his own, every one is quick-sighted enough to espie in another, and will by the Authority of Reason forwardly condemn, though he be guilty of much greater Unreasonableness in his own Tenets and Conduct, which he never perceives, and will very hardly, if at all, be convinced of.

2. This proceeds not wholly from Self-love, though that has often a great hand in it. Men of fair Minds, and not given up to the over weening of Self-flattery, are frequently guilty of it; and in many Cases one with amazement hears the Arguings, and is astonish'd at the Obstinacy of a worthy Man, who yields not to the Evidence of Reason, though laid before him as clear as Day-light.

3. This sort of Unreasonableness is usually imputed to Education and Prejudice, and for the most part truly enough, though that reaches not the bottom of the Disease, nor shews distinctly enough whence it rises, or wherein it lies. Education is often rightly assigned for the Cause, and Prejudice is a good general Name for the thing it self: But yet, I think, he ought to look a little farther who would trace this sort of Madness to the root it springs from, and so explain it, as to shew whence this flaw has its Original in very sober and rational Minds, and wherein it consists.

4. I shall be pardon'd for calling it by so harsh a name as *Madness,* when it is considered, that opposition to Reason deserves that Name, and is really Madness; and there is scarce a Man so free from it, but that if he should always on all occasions argue or do as in some cases he constantly does, would not be thought fitter for *Bedlam,* than Civil Conversation. I do not here mean when he is under the power of an unruly Passion, but in the steady calm course of his Life. That which will yet more apologize for this harsh Name, and ungrateful Imputation on the greatest part of Mankind is, that enquiring a little by the bye into the Nature of Madness . . . I found it to spring from the very same Root, and to depend on the very same Cause we are here speaking of. This considerration of the thing it self, at a time when I thought not the least on the Subject which I am now treating of, suggest'd it to me. And if this be a Weakness to which all Men are so liable; if this be a Taint which so universally infects Mankind, the greater care should be taken to lay it open under its due Name, thereby to excite the greater care in its Prevention and Cure.

5. Some of our *Ideas* have a natural Correspondence and Connexion one with another: It is the Office and Excellency of our Reason to trace these, and hold them together in that Union and Correspondence which is founded in their peculiar Beings. Besides this there is another Connexion of *Ideas* wholly owing to Chance or Custom, *Ideas* that in themselves are not at all of kin, come to be so united in some Mens Minds, that 'tis very hard to separate them, they always keep in company, and the one no sooner at any time comes into the Understanding but its Associate appears with it; and if they are more than two which are thus united, the whole gang always inseparable shew themselves together.

6. This strong Combination of *Ideas,* not ally'd by Nature, the Mind makes in it self either voluntarily, or by chance, and hence it comes in different Men to be very different, according to their different Inclinations, Educations, Interests, &c. Custom settles habits of Thinking in the Understanding, as well as of Determining in the Will, and of Motions in the Body; all which seems to be but Trains of Motion in the Animal Spirits, which once set a going continue on in the same steps they have been used to, which by often treading are worn into a smooth path, and the Motion in it becomes easy and as it were Natural. As far as we can comprehend Thinking, thus *Ideas* seem to be produced in our Minds, or if they are not, this may serve to explain their following one another in an habitual train, when once they are put into that tract, as well as it does to explain such Motions of the Body. A Musician used to any Tune will find that let it but once begin in his Head, the *Ideas* of the several Notes of it will follow one another orderly in his Understanding without any care or attention, as regularly as his Fingers move orderly over the Keys of the Organ to play out the Tune he has begun, though his unattentive Thoughts be elsewhere a wandering. Whether the natural cause of these *Ideas,* as well as of that regular Dancing of his Fingers be the Motion of his Animal Spirits: I will not determine how probable soever by this Instance it appears to be so: But this may help us a little to conceive of Intellectual Habits, and of the tying together of *Ideas.*

7. That there are such Associations of them made by Custom in the Minds of most Men, I think no Body will question who has well consider'd himself or others; and to this, perhaps, might be justly attributed most of the Sympathies and Antipathies observable in Men, which work as strongly, and produce as regular Effects as if they were Natural, and are therefore called so, though they at first had no other Original but the accidental Connexion of two *Ideas,* which either the strength of the first Impression, or future Indulgence so united, that they always afterwards kept company together in that Man's Mind, as if they were but one *Idea.* I say most of the Antipathies, I do not say all, for some of them are truly natural, depend upon our original Constitution, and are born with us; but a great part of those which are counted Natural, would have been known to be from unheeded, though, perhaps, early Impressions, or wanton Phancies at first, which would have been acknowledged the Original of them if they had been warily observed. A grown Person surfeiting with Honey, no sooner hears the Name of it, but his Phancy immediately carries Sickness and Qualms to his Stomach, and he cannot bear the very *Idea* of it; other *Ideas* of Dislike and Sickness, and Vomiting presently accompany it, and he is disturb'd, but he knows from whence to date this Weakness, and can tell how he got this Indisposition:

Had this happen'd to him, by an over dose of Honey, when a Child, all the same Effects would have followed, but the Cause would have been mistaken, and the Antipathy counted Natural.

8. I mention this not out of any great necessity there is in this present Argument, to distinguish nicely between Natural and Acquired Antipathies, but I take notice of it for another purpose (*viz.*) that those who have Children, or the charge of their Education, would think it worth their while diligently to watch, and carefully to prevent the undue Connexion of *Ideas* in the Minds of young People. This is the time most susceptible of lasting Impressions, and though those relating to the Health of the Body, are by discreet People minded and fenced against, yet I am apt to doubt, that those which relate more peculiarly to the Mind, and terminate in the Understanding, or Passions, have been much less heeded than the thing deserves, nay those relating purely to the Understanding have, as I suspect, been by most Men wholly over-look'd.

9. This wrong Connexion in our Minds of *Ideas* in themselves, loose and independent one of another has such an influence, and is of so great force to set us awry in our Actions, as well Moral as Natural, Passions, Reasoning, and Notions themselves, that, perhaps, there is not any one thing that deserves more to be looked after.

10. The *Ideas* of *Goblines* and *Sprights* have really no more to do with Darkness than Light, yet let but a foolish Maid inculcate these often on the Mind of a Child, and raise them there together, possibly he shall never be able to separate them again so long as he lives, but Darkness shall ever afterwards bring with it those frightful *Ideas,* and they shall be so joined that he can no more bear the one than the other.

11. A Man receives a sensible Injury from another, thinks on the Man and that Action over and over, and by ruminating on them strongly, or much in his Mind so cements those two *Ideas* together, that he makes them almost one; never thinks on the Man, but the Pain and Displeasure he suffered comes into his Mind with it, so that he scarce distinguishes them, but has as much an aversion for the one as the other. Thus Hatreds are often begotten from slight and almost innocent Occasions, and Quarrels propagated and continued in the World.

12. A Man has suffered Pain or Sickness in any Place, he saw his Friend die in such a Room, though these have in Nature nothing to do one with another, yet when the *Idea* of the Place occurs to his Mind, it brings (the Impression being once made) that of the Pain and Displeasure with it, he confounds them in his Mind, and can as little bear the one as the other.

13. When this Combination is settled and whilst it lasts, it is not in the power of Reason to help us, and relieve us from the Effects of it.

337

Ideas in our Minds, when they are there, will operate according to their Natures and Circumstances; and here we see the cause why Time cures certain Affections, which Reason, though in the right, and allow'd to be so, has not power over, nor is able against them to prevail with those who are apt to hearken to it in other cases. The Death of a Child, that was the daily delight of his Mothers Eyes, and joy of her Soul, rends from her Heart the whole comfort of her Life, and gives her all the torment imaginable; use the Consolations of Reason in this case, and you were as good preach Ease to one on the Rack, and hope to allay, by rational Discourses, the Pain of his Joints tearing asunder. Till time has by disuse separated the sense of that Enjoyment and its loss from the *Idea* of the Child returning to her Memory, all Representations, though never so reasonable, are in vain; and therefore some in whom the union between these *Ideas* is never dissolved, spend their Lives in Mourning, and carry an incurable Sorrow to their Graves.

14. A Friend of mine knew one perfectly cured of Madness by a very harsh and offensive Operation. The Gentleman, who was thus recovered, with great sense of Gratitude and Acknowledgment, owned the Cure all his Life after, as the greatest Obligation he could have received; but whatever Gratitude and Reason suggested to him, he could never bear the sight of the Operator: That Image brought back with it the *Idea* of that Agony which he suffer'd from his Hands, which was too mighty and intolerable for him to endure.

15. Many Children imputing the Pain they endured at School to their Books they were corrected for, so joyn those *Ideas* together, that a Book becomes their Aversion, and they are never reconciled to the study and use of them all their Lives after; and thus Reading becomes a torment to them, which otherwise possibly they might have made the great Pleasure of their Lives. There are Rooms convenient enough, that some Men cannot Study in, and fashions of Vessels, which though never so clean and commodious they cannot Drink out of, and that by reason of some accidental *Ideas* which are annex'd to them, and make them offensive; and who is there that hath not observed some Man to flag at the appearance, or in the company of some certain Person not otherwise superior to him, but because having once on some occasion got the Ascendant, the *Idea* of Authority and Distance goes along with that of the Person, and he that has been thus subjected is not able to separate them.

16. Instances of this kind are so plentiful every where, that if I add one more, it is only for the pleasant oddness of it. It is of a young Gentleman, who having learnt to Dance, and that to great Perfection, there happened to stand an old Trunk in the Room where he learnt. The *Idea*

of this remarkable piece of Housholdstuff, had so mixed it self with the turns and steps of all his Dances, that though in that Chamber he could Dance excellently well, yet it was only whilst that Trunk was there, nor could he perform well in any other place, unless that, or some such other Trunk had its due position in the Room. If this Story shall be suspected to be dressed up with some comical Circumstances, a little beyond precise Nature; I answer for my self, that I had it some Years since from a very sober and worthy Man, upon his own knowledge, as I report it; and I dare say, there are very few inquisitive Persons, who read this, who have not met with Accounts, if not Examples of this Nature, that may parallel, or at least justify this.

17. Intellectual Habits and Defects this way contracted are not less frequent and powerful, though less observed. Let the *Ideas* of Being and Matter be strongly joined either by Education or much Thought, whilst these are still combined in the Mind, what Notions, what Reasonings, will there be about separate Spirits? Let custom from the very Childhood have join'd Figure and Shape to the *Idea* of God, and what Absurdities will that Mind be liable to about the Deity?

.   .   .   .   .

18. Some such wrong and unnatural Combinations of *Ideas* will be found to establish the Irreconcilable opposition between different Sects of Philosophy and Religion; for we cannot imagine every one of their Followers to impose wilfully on himself, and knowingly refuse Truth offer'd by plain Reason. Interest though it does a great deal in the case, yet cannot be thought to work whole Societies of Men to so universal a Perverseness, as that every one of them to a Man should knowingly maintain Falshood: Some at least must be allow'd to do what all pretend to, *i.e.* to pursue Truth sincerely; and therefore there must be something that blinds their Understandings, and makes them not see the falshood of what they embrace for real Truth. That which thus captivates their Reasons, and leads Men of Sincerity blindfold from common Sence, will, when examin'd, be found to be what we are speaking of: some independent *Ideas,* of no alliance to one another, are by Education, Custom, and the constant din of their Party, so coupled in their Minds, that they always appear there together, and they can no more separate them in their Thoughts, than if they were but one *Idea,* and they operate as if they were so. This gives Sence to *Jargon,* Demonstration to Absurdities, and Consistency to Nonsense, and is the foundation of the greatest, I had almost said, of all the Errors in the World; or if it does not reach so far, it is at least the most dangerous one, since so far as it obtains it hinders Men from seeing and examining. When two things in themselves

disjoin'd appear to the sight constantly united; if the Eye sees these things rivetted which are loose, where will you begin to rectify the mistakes that follow in two *Ideas,* that they have been accustom'd so to join in their Minds, as to substitute one for the other, and, as I am apt to think, often without perceiving it themselves? This, whilst they are under the deceit of it, makes them uncapable of Conviction, and they applaud themselves as zealous Champions for Truth, when indeed they are contending for Error; and the confusion of two different *Ideas,* which a customary connexion of them in their Minds hath to them made in effect but one, fills their Heads with false Views, and their Reasonings with false Consequences.

68   GEORGE BERKELEY (1685–1753) ON ARBITRARY CONNECTIONS AMONG IDEAS, 1733

George Berkeley, *The Theory of Vision, or Visual Language, Shewing the Immediate Presence and Providence of a Deity, Vindicated and Explained* (London, 1733), sects. 39–43.

Berkeley's *New Theory of Vision* (1709) was a milestone in the development of the empiristic account of perception and objective reference (see Nos. 28 and 36) because it was a persuasive demonstration of how the perception of space results from a compounding of the elementary sensations of sight and touch. It did not, however, describe the way these sensations become compounded. In that work, Berkeley was an explicit empiricist and only a tacit associationist, but in the present selection, which is part of a rebuttal to an anonymous criticism of the *New Theory,* his associationism becomes explicit. Here he says that ideas are associated, or "suggest" each other, because they are connected in experience. Thus his associationism, like Locke's, is based on contiguity, but it goes beyond Locke's by its greater inclusiveness. For Locke, association by contiguity explained only the fortuitous or arbitrary, and not the natural, connections among ideas. For Berkeley, even natural connections may be arbitrary, and are, by implication, to be attributed to association by contiguity.

XXXIX. Ideas, which are observed to be connected with other Ideas, come to be considered as Signs, by means whereof Things, not actually perceived by Sense, are signified or suggested to the Imagination, whose Objects they are, and which alone perceives them. And as Sounds suggest other things, so Characters suggest those Sounds; and, in general, all Signs suggest the things signified, there being no Idea which may not offer to the Mind another Idea, which hath been frequently joined with it. In certain Cases, a Sign may suggest its Correlate as an Image, in others as an Effect, in others as a Cause. But where there is no such Relation of Similitude or Causality, nor any necessary Connexion whatsoever, two things by their mere Coexistence, or two Ideas, merely by being perceived together may suggest or signify one the other, their Connexion being all the while arbitrary; for it is the Connexion only, as such, that causeth this Effect.

XL. A great Number of arbitrary Signs, various and apposite, do constitute a Language. If such arbitrary Connexion be instituted by Men, it is an artificial Language; if by the Author of Nature, it is a Natural Language. Infinitely various are the Modifications of Light and Sound, whence they are each capable of supplying an endless Variety of Signs, and, accordingly, have been each employed to form Languages; the one by the arbitrary Appointment of Mankind, the other by that of God himself. A Connexion established by the Author of Nature, in the ordinary course of things, may surely be called Natural; as that made by Men will be named Artificial. And yet this doth not hinder but the one may be as arbitrary as the other. And, in Fact, there is no more Likeness to exhibit, or Necessity to infer, things tangible from the Modifications of Light, than there is in Language, to collect the Meaning from the Sound. But, such as the Connexion is of the various Tones and Articulations of Voice with their several Meanings, the same is it between the various Modes of Light and their respective Correlates; or in other Words, between the Ideas of Sight and Touch.

XLI. As to Light, and its several Modes or Colours, all thinking Men are agreed, that they are Ideas peculiar only to Sight; neither common to the Touch, nor of the same Kind with any that are perceived by that Sense. But herein lies the Mistake, that beside these, there are supposed other Ideas common to both Senses, being equally perceived by Sight and Touch, such as Extension, Size, Figure, and Motion. But that there are in reality no such common Ideas, and that the Objects of Sight, marked by those Words, are intirely different and heterogeneous from whatever is the Object of Feeling, marked by the same Names, hath been proved in the *Theory,* and seems by you admitted. Though I cannot conceive how you should in reason admit this, and at the same time contend for the received Theories, which are as much ruined, as mine is established, by this main Part and Pillar thereof.

XLII. To perceive is one thing; to judge is another. So likewise to be suggested is one thing, and to be inferred another. Things are suggested and perceived by Sense. We make Judgments and Inferences by the Understanding. What we immediately and properly perceive by Sight, is its primary Object, Light and Colours. What is suggested or perceived by Mediation thereof, are tangible Ideas, which may be considered as secondary and improper Objects of Sight. We infer Causes from Effects, Effects from Causes, and Properties one from another, where the Connection is necessary. But, how comes it to pass, that we apprehend by the Ideas of Sight certain other Ideas, which neither resemble them, nor cause them, nor are caused by them, nor have any necessary Connexion with them? The Solution of this Problem, in its full Extent, doth comprehend the whole Theory of Vision. Thus stating of the Matter placeth it on a new Foot, and in a different Light from all preceding Theories.

XLIII. To explain how the Mind or Soul of Man simply sees, is one thing, and belongs to Philosophy. To consider Particles as moving in certain Lines, Rays of Light as refracted, or reflected, or crossing, or including Angles, is quite another thing, and appertaineth to Geometry. To account for the Sense of Vision by the Mechanism of the Eye, is a third thing, which appertaineth to Anatomy and Experiments. These two latter Speculations are of use in Practice, to assist the Defects, and remedy the Distempers of Sight, agreeably to the natural Laws obtaining in this mundane System. But the former Theory is that which makes us understand the true Nature of Vision, considered as a Faculty of the Soul. Which Theory, as I have already observed, may be reduced to this simple Question, to wit, How comes it to pass, that a Set of Ideas, altogether different from tangible Ideas, should nevertheless suggest them to us, there being no necessary Connexion between them? To which the proper Answer is, That this is done in virtue of an arbitrary Connexion, instituted by the Author of Nature.

69 DAVID HUME (1711–1776) ON A PSYCHOLOGICAL ANALOGUE OF GRAVITATION, 1739

The first excerpt in the present selection is from David Hume, *Philosophical Essays concerning Human Understanding*, 2nd ed. (London, 1751), essay 2. The first edition (1748) has not been available, but presumably does not differ materially. More recent editions of this book are entitled *Enquiry concerning Human Understanding*. The second excerpt is from Hume's *Treatise of Human Nature; being An Attempt to introduce the experimental Method of Reasoning into Moral Subjects*, vol. I (London, 1739), pt. 1, sect. 4. The later work is essentially a revision and condensation of the earlier.

Hume's theory of knowledge was the first in which the doctrine of association occupied a position of prominence. He had, however, a slightly exaggerated view of his own originality, for in the *Philosophical Essays* he says, "Though it be too obvious to escape observation, that different ideas are connected together, I do not find that any philosopher has attempted to enumerate or class all the principles of association; a subject, however, that seems worthy of curiosity." The preceding selections in this chapter show that Hume failed to find predecessors only because he failed to look in the right places, but his comparison of association with gravitational attraction in the following excerpt suggests that he indeed first saw the importance of association as a psychological principle.

Hume began by setting forth his

version of empiricism. The mind, he said, contains impressions and ideas. Impressions are what we would call sensations and emotions, and ideas for Hume are merely the replicas of the impressions, differing from them only in degree of vivacity or intensity. Thus there can be nothing in the mind that did not first arise in experience. Ideas are connected by association, which takes place, Hume said, in accordance with three principles: resemblance, contiguity, and cause and effect. Hume thus rediscovered two of Aristotle's three principles, but he felt that the most important was cause and effect, the one that he added. For him, association was but a "gentle force," guiding the mind most of the time, yet not controlling it rigidly.

[EXCERPT 1]

Every one will readily allow, that there is a considerable Difference betwixt the Perceptions of the Mind, when a Man feels the Pain of excessive Heat or the Pleasure of moderate Warmth, and when he afterwards recalls to his Memory this Sensation, or anticipates it by his Imagination. These Faculties may mimick or copy the Perceptions of the Senses; but they never can reach entirely the Force and Vivacity of the original Sentiment. The utmost we say of them, even when they operate with greatest Vigour, is, that they represent their Object in so lively a Manner, that we could *almost* say we feel or see it: But, except the Mind be disorder'd by Disease or Madness, they never can arrive at such a Pitch of Vivacity as to render these Perceptions altogether undistinguishable. All the Colours of Poetry, however splendid, can never paint natural Objects in such a manner as to make the Description be taken for a real Landskip. The most lively Thought is still inferior to the dullest Sensation.

We may observe a like Distinction to run thro' all the other Perceptions of the Mind. A Man, in a Fit of Anger, is actuated in a very different Manner from one, who only thinks of that Emotion. If you tell me, that any Person is in Love, I easily understand your Meaning, and form a just Conception of his Situation; but never can mistake that Conception for the real Disorders and Agitations of the Passion. When we reflect on all our past Sentiments and Affections, our Thought is a faithful Mirror, and copies its Objects truly; but the Colours it employs are faded and dead, in comparison of those, in which our original Perceptions were cloth'd. It requires no nice Discernment nor metaphysical Head to mark the Distinction betwixt them.

Here therefore we may divide all the Perceptions of the Mind into two Classes or Species, which are distinguish'd by their different Degrees of Force and Vivacity. The less forcible and lively are commonly denominated *Thoughts* or *Ideas.* The other Species want a Name in our Language, and in most others; I suppose, because it was not requisite for any, but philosophical Purposes, to rank them under a general Term or

343

Appellation. Let us, therefore, use a little Freedom, and call them *Impressions,* employing that Word in a Sense somewhat different from the usual. By the Term, *Impressions,* then, we mean all our more lively Perceptions, when we hear, or see, or feel, or love, or hate, or desire, or will. And Impressions are contradistinguish'd from Ideas, which are the less lively Perceptions we are conscious of, when we reflect on any of these Sensations or Movements above mention'd.

Nothing, at first View, may seem more unbounded than the Thought of Man, which not only escapes all human Power and Authority, but is not even restrain'd within the Limits of Nature and Reality. To form Monsters, and join incongruous Shapes and Appearances costs it no more Trouble than to conceive the most natural and familiar Objects. And while the Body is confin'd to one Planet, along which it creeps with Pain and Difficulty; the Thought can in an Instant transport us into the most distant Regions of the Universe; or even beyond the Universe, into the unbounded Chaos, where Nature is suppos'd to lie in total Confusion. What never was seen, nor heard of may yet be conceiv'd; nor is any thing beyond the Power of Thought, except what implies an absolute Contradiction.

But tho' Thought seems to possess this unbounded Liberty, we shall find, upon a nearer Examination, that it is really confin'd within very narrow Limits, and that all this creative Power of the Mind amounts to no more than the compounding, transposing, augmenting, or diminishing the Materials afforded us by the Senses and Experience. When we think of a golden Mountain, we only join two consistent Ideas, *Gold,* and *Mountain,* with which we were formerly acquainted. A virtuous Horse we can conceive; because, from our own Feeling, we can conceive Virtue, and this we may unite to the Figure and Shape of a Horse, which is an Animal familiar to us. In short, all the Materials of thinking are deriv'd either from our outward or inward Sentiment: The Mixture and Composition of these belongs alone to the Mind and Will. Or to express myself in philosophical Language, all our Ideas or more feeble Perceptions are Copies of our Impressions or more lively ones.

To prove this, the two following Arguments will, I hope, be sufficient. First, When we analyse our Thoughts or Ideas, however compounded or sublime, we always find, that they resolve themselves into such simple Ideas as were copy'd from a precedent Feeling or Sentiment. Even those Ideas, which, at first View, seem the most wide of this Origin, are found, upon a narrower Scrutiny, to be deriv'd from it. The Idea of God, as meaning an infinitely intelligent, wise, and good Being, arises from reflecting on the Operations of our own Mind, and augmenting those Qualities of Goodness and Wisdom, without Bound or Limit. We may prosecute this Enquiry to what Length we please; where we shall always find, that every Idea we examine is copy'd from a similar Impression. Those, who would assert, that this Position is not absolutely universal and without Exception, have only one, and that an easy Method of refuting it, by producing that idea, which, in their Opinion, is not deriv'd from this Source. It will then be incumbent on us, if we would maintain our Doctrine, to produce the Impression or lively Perception, that corresponds to it.

Secondly. If it happen, from a Defect of the Organ, that a Man is not susceptible of any Species of Sensation, we always find, that he is as little susceptible of the correspondent Ideas. A blind Man can form no Notion of Colours; a deaf Man of Sounds. Restore either of them that Sense, in which he is deficient; by opening this new Inlet for his Sensations, you also open an Inlet for the Ideas, and he finds no Difficulty of conceiving these Objects. The Case is the same if the Object, proper for exciting any Sensation, has never been apply'd to the Organ. A *Laplander* or *Negro* has no Notion of the Relish of Wine. And tho' there are few or no Instances of a like Deficiency in the Mind, where a Person has never felt or is altogether incapable of a Sentiment or Passion, that belongs to his Species; yet we find the same Observation to take place in a lesser Degree. A Man of mild Manners can form no Notion of inveterate Revenge or Cruelty; nor can a selfish Heart easily conceive the Heights of Friendship and Generosity. 'Tis readily allow'd, that other Beings may possess many Senses, of which we can have no Conception; because the Ideas of them have never been introduc'd to us in the only Manner, by which an Idea can have access to the Mind, *viz.* by the actual Feeling and Sensation.

There is, however, one contradictory Phænomenon, which may prove, that 'tis not absolutely impossible for Ideas to go before their correspondent Impressions. I believe it will readily be allow'd, that the several distinct Ideas of Colours, which enter by the Eyes, or those of Sounds, which are convey'd by the Hearing, are really different from each other; tho', at the same time, resembling. Now if this be true of different Colours, it must be no less so, of the different Shades of the same Colour; and each Shade produces a distinct Idea, independent of the rest. For if this should be deny'd, 'tis possible, by the continual Gradation of Shades, to run a Colour insensibly into what is most remote from it; and if you will not allow any of the Means to be different, you cannot, without Absurdity, deny the Extremes to be the same. Suppose, therefore, a Person to have enjoy'd his Sight for thirty Years, and to have become perfectly well acquainted with Colours of all kinds, excepting one particular Shade of Blue, for Instance, which it never has been his Fortune to meet with. Let all the different Shades of that Colour, except that single one, be plac'd before him, descending gradually from the deepest to the lightest; 'tis plain, that he will perceive a Blank, where that Shade is wanting, and will be sensible, that there is a greater Distance in that Place betwixt the contiguous Colours than in any other. Now I ask, whether 'tis possible for him, from his own Imagination, to supply this Deficiency, and raise up to himself the Idea of that particular Shade, tho' it had never been convey'd to him by his Senses? I believe there are few but will be of Opinion that he can; and this may serve as a Proof, that the simple Ideas are not always, in every Instance, deriv'd from the correspondent Impressions; tho' this Instance is so particular and singular, that 'tis scarce worth our observing, and does not merit, that for it alone we should alter our general Maxim.

Here, therefore, is a Proposition, which not only seems, in itself, simple and intelligible; but, if properly employ'd, might render every Dispute equally intelligible, and banish all that Jargon, which has so long taken Possession of metaphysical Reasonings, and drawn such Disgrace upon them. All Ideas, especially abstract ones, are naturally faint and obscure: The Mind has but a slender Hold of them: They are apt to be confounded with other resembling Ideas: And when we have often employ'd any Term, tho' without a distinct Meaning, we are apt to imagine it has a determinate Idea, annex'd to it. On the contrary, all Impressions, that is, all Sensations, either outward or inward, are strong and sensible: The Limits betwixt them are more exactly determin'd: Nor is it easy to fall into any Error or Mistake with Regard to them. When therefore we entertain any Suspicion, that a philosophical Term is employ'd without any Meaning or Idea (as is but too frequent) we need but enquire, *from what Impression is that suppos'd Idea deriv'd?* And if it be impossible to assign any, this will serve to confirm our Suspicion. By bringing Ideas into so clear a Light, we may reasonably hope to remove all Dispute, that may arise, concerning their Nature and Reality.

[EXCERPT 2]

As all simple ideas may be separated by the imagination, and may be united again in what form it pleases, nothing wou'd be more unaccountable than the operations of that faculty, were it not guided by some universal principles, which render it, in some measure, uniform with itself in all times and places. Were ideas entirely loose and unconnected, chance alone wou'd join them; and 'tis impossible the same simple ideas should fall regularly into complex ones (as they commonly do) without some bond of union among them, some associating quality, by which one idea naturally introduces another. This uniting principle among ideas is not to be consider'd as an inseparable connexion; for that has been already excluded from the imagination: Nor yet are we to conclude, that without it the mind cannot join two ideas; for nothing is more free than that faculty: but we are only to regard it as a gentle force, which commonly prevails, and is the cause why, among other things, languages so nearly correspond to each other; nature in a manner pointing out to every one those simple ideas, which are most proper to be united into a complex one. The qualities, from which this association arises, and by which the mind is after this manner convey'd from one idea to another, are three, viz. *Resemblance, Contiguity* in time or place, and *Cause* and *Effect.*

I believe it will not be very necessary to prove, that these qualities produce an association among ideas, and upon the appearance of one idea naturally introduce another. 'Tis plain, that in the course of our thinking, and in the constant revolution of our ideas, our imagination runs easily from one idea to any other that *resembles* it, and that this quality alone is to the fancy a sufficient bond and association. 'Tis likewise evident, that as the senses, in changing their objects, are necessitated to change them regularly, and take them as they lie *contiguous* to each other, the imagination must by long custom acquire the same method of thinking, and run along the parts of space and time in conceiving its objects. As to the connexion, that is made by the relation of *cause and effect,* we shall have occasion afterwards to examine it to the bottom, and therefore shall not at present insist upon it. 'Tis sufficient to observe, that there is no relation, which produces a stronger connexion in the fancy, and makes one idea more readily recall another, than the relation of cause and effect betwixt their objects.

That we may understand the full extent of these relations, we must consider, that two objects are connected together in the imagination, not only when the one is immediately resembling, contiguous to, or the cause of the other, but also when there is interposed betwixt them a third object, which bears to both of them any of these relations. This may be carried on to a great length; tho' at the same time we may observe, that each remove considerably weakens the relation. Cousins in the fourth degree are connected by *causation,* if I may be allowed to use that term; but not so closely as brothers, much less as child and parent. In general we may observe, that all the relations of blood depend upon cause and effect, and are esteemed near or remote, according to the number of connecting causes interpos'd betwixt the persons.

Of the three relations above-mention'd this of causation is the most extensive. Two objects may be consider'd as plac'd in this relation, as well when one is the cause of any of the actions or motions of the other,

347

as when the former is the cause of the existence of the latter. For as that action or motion is nothing but the object itself, consider'd in a certain light, and as the object continues the same in all its different situations, 'tis easy to imagine how such an influence of objects upon one another may connect them in the imagination.

．　　．　　．　　．　　．

These are therefore the principles of union or cohesion among our simple ideas, and in the imagination supply the place of that inseparable connexion, by which they are united in our memory. Here is a kind of *Attraction,* which in the mental world will be found to have as extraordinary effects as in the natural, and to shew itself in as many and as various forms. Its effects are every where conspicuous; but as to its causes, they are mostly unknown, and must be resolv'd into *original* qualities of human nature, which I pretend not to explain. Nothing is more requisite for a true philosopher, than to restrain the intemperate desire of searching into causes, and having establish'd any doctrine upon a sufficient number of experiments, rest contented with that, when he sees a farther examination would lead him into obscure and uncertain speculations. In that case his enquiry wou'd be much better employ'd in examining the effects than the causes of his principle.

70 DAVID HARTLEY (1705-1757) ON ASSOCIATION: SUCCESSIVE AND SIMULTANEOUS, SIMPLE AND COMPLEX, 1749

David Hartley, *Observations on Man, His Frame, His Duty, and His Expectations* (London and Bath, 1749), bk. I, chap. 1, sect. 2.

Hume recognized the potential of the doctrine of association; Hartley his contemporary, attempted in an exhaustive account to demonstrate it. Hartley also undertook to provide a plausible physical basis for the formation of associations, one reflecting, perhaps, his training as a physician. Sensations, he said, are based on vibrations within the nervous system, as are ideas—the two sorts of vibrations differing only in amplitude, not in frequency, plane of oscillation, or location. Association, based solely on contiguity, takes place between sensations and ideas, ideas and ideas, and sensations or ideas and movements. Hartley spoke of contiguity in the sense of both immediate succession and simultaneity. Associations based on succession produce the temporal sequence of thought. Associations based on simultaneity produce either complex ideas by the combination of simple ones or "decomplex" ideas by the combination of complex ones. Although the earlier writers had mainly considered successive associations, simultaneous associations had at least been implied by some, notably by Berkeley in his treatment of space perception (see No. 36). Hartley, however, with his detailed, careful, and quasi-physiological theory, was to become the authority on association for eighty years, until James Mill set the pattern for the nineteenth century.

PROP. IX

Sensory vibrations, by being often repeated, beget, in the medullary Substance of the Brain, a Disposition to diminutive Vibrations, which may also be called Vibratiuncles, and Miniatures, corresponding to themselves respectively.

.   .   .   .   .   .

Since Sensations, by being often repeated, beget Ideas, it cannot but be that those Vibrations, which accompany Sensations, should beget something which may accompany Ideas in like manner; and this can be nothing but feebler Vibrations, agreeing with the sensory generating Vibrations in Kind, Place, and Line of Direction.

Or thus: . . . It appears, that some Motion must be excited in the medullary Substance, during each Sensation; . . . this Motion is determined to be a vibratory one: Since therefore some Motion must also . . . be excited in the medullary Substance during the Presence of each Idea, this Motion cannot be any other than a vibratory one: Else how should it proceed from the original Vibration attending the Sensation, in the same manner as the Idea does from the Sensation itself? It must also agree in Kind, Place, and Line of Direction, with the generating Vibration. A vibratory Motion, which recurs t times in a Second, cannot beget a diminutive one that recurs $\frac{1}{2} t$, or $2 t$ times; nor one originally impressed on the Region of the Brain corresponding to the auditory Nerves, beget diminutive Vibrations in the Region corresponding to the optic Nerves; and so of the rest. The Line of Direction must likewise be the same in the original and derivative Vibrations. It remains therefore, that each simple Idea of Sensation be attended by diminutive Vibrations of the same Kind, Place, and Line of Direction, with the original Vibrations attending the Sensation itself: Or, in the Words of the Proposition, that sensory Vibrations, by being frequently repeated, beget a Disposition to diminutive Vibrations corresponding to themselves respectively. We may add, that the vibratory Nature of the Motion which attends Ideas, may be inferred from the Continuance of some Ideas, visible ones for instance, in the Fancy for a few Moments.

.   .   .   .   .   .

PROP. X

Any Sensations A, B, C, &c. *by being associated with one another a sufficient Number of Times, get such a Power over the corresponding Ideas* a, b, c, *&c. that any one of the Sensations* A, *when impressed alone, shall be able to excite in the Mind,* b, c, *&c. the Ideas of the rest.*

Sensations may be said to be associated together, when their Impressions are either made precisely at the same Instant of Time, or in the contiguous successive Instants. We may therefore distinguish Association into Two Sorts, the synchronous, and the successive.

The Influence of Association over our Ideas, Opinions, and Affections, is so great and obvious, as scarce to have escaped the Notice of any Writer who has treated of these, though the Word *Association,* in the particular Sense here affixed to it, was first brought into Use by Mr. *Locke.* But all that has been delivered by the Antients and Moderns, concerning the Power of Habit, Custom, Example, Education, Authority, Party-prejudice, the Manner of learning the manual and liberal Arts, &c. goes upon this Doctrine as its Foundation, and may be considered as the Detail of it, in various Circumstances . . .

This Proposition, or first and simplest Case of Association, is manifest from innumerable common Observations. Thus the Names, Smells, Tastes, and tangible Qualities of natural Bodies, suggest their visible Appearances to the Fancy, *i.e.* excite their visible Ideas; and, *vice versa,* their visible Appearances impressed on the Eye raise up those Powers of reconnoitring their Names, Smells, Tastes, and tangible Qualities, which may not improperly be called their Ideas, as above noted; and in some Cases raise up Ideas, which may be compared with visible ones, in respect of Vividness. All which is plainly owing to the Association of the several sensible Qualities of Bodies with their Names, and with each other. It is remarkable, however, as being agreeable to the superior Vividness of visible and audible Ideas before taken notice of, that the Suggestion of the visible Appearance from the Name, is the most ready of any other; and, next to this, that of the Name from the visible Appearance; in which last Case, the Reality of the audible Idea, when not evident to the Fancy, may be inferred from the ready Pronunciation of the Name. For it will be shewn hereafter, that the audible Idea is most commonly a previous Requisite to Pronunciation. Other Instances of the Power of Association may be taken from compound visible and audible Impressions. Thus the Sight of Part of a large Building suggests the Idea of the rest instantaneously; and the Sound of the Words which begin a familiar Sentence, brings the remaining Part to our Memories in Order, the Association of the Parts being synchronous in the first Case, and successive in the last.

It is to be observed, that, in successive Associations, the Power of raising the Ideas is only exerted according to the Order in which the

Association is made. Thus, if the Impressions *A, B, C,* be always made in the Order of the Alphabet, *B* impressed alone will not raise *a,* but *c* only. Agreeably to which, it is easy to repeat familiar Sentences in the Order in which they always occur, but impossible to do it readily in an inverted one. The Reason of this is, that the compound Idea, *c, b, a,* corresponds to the compound Sensation *C, B, A;* and therefore requires the Impression of *C, B, A,* in the same manner as *a, b, c,* does that of *A, B, C.* This will, however, be more evident, when we come to consider the Associations of vibratory Motions, in the next Proposition.

It is also to be observed, that the Power of Association grows feebler, as the Number either of synchronous or successive Impressions is increased, and does not extend, with due Force, to more than a small one, in the first and simplest Cases. But, in complex Cases, or the Associations of Associations, of which the Memory, in its full Extent, consists, the Powers of the Mind, deducible from this Source, will be found much greater than any Person, upon his first Entrance on these Inquiries, could well imagine.

·   ·   ·   ·   ·

PROP. XII

Simple Ideas will run into complex ones, by means of Association.

·   ·   ·   ·   ·

Case 1. Let the Sensation *A* be often associated with each of the Sensations *B, C, D,* &c. *i.e.* at certain times with *B,* at certain other times with *C,* &c. it is evident . . . that *A,* impressed alone, will, at last, raise *b, c, d,* &c. all together, *i.e.* associate them with one another, provided they belong to different regions of the medullary Substance; for if any Two, or more, belong to the same Region, since they cannot exist together in their distinct Forms, *A* will raise something intermediate between them.

Case 2. If the Sensations *A, B, C, D,* &c. be associated together, according to various Combinations of Twos, or even Threes, Fours, &c. then will *A* raise *b, c, d,* &c. also *B* raise *a, c, d,* &c. . . .

It may happen, indeed, . . . that *A* may raise a particular Miniature, as *b,* preferably to any of the rest, from its being more associated with *B,* from the Novelty of the Impression of *B,* from a Tendency in the medullary Substance to favour *b,* &c. and, in like manner, that *b* may raise *c* or *d* preferably to the rest. However, all this will be over-ruled, at last, by the Recurrency of the Associations; so that any one of the Sensations will excite the Ideas of the rest, at the same Instant, *i.e.* associate them together.

Case 3. Let *A, B, C, D,* &c. represent successive Impressions, it follows
. . . that *A* will raise *b, c, d,* &c. *B* raise *c, d,* &c. And though the Ideas
do not, in this Case, rise precisely at the same Instant, yet they come
nearer together than the Sensations themselves did in their original
Impression; so that these Ideas are associated almost synchronically at
last, and successively from the first. The Ideas come nearer to one another than the Sensations, on account of their diminutive Nature, by
which all that appertains to them is contracted. And this seems to be
as agreeable to Observation as to Theory.

Case 4. All compound Impressions *A + B + C + D,* &c. after sufficient
Repetition leave compound Miniatures *a + b + c + d,* &c. which recur
every now and then from slight Causes, as well such as depend on Association, as some which are different from it. Now, in these Recurrencies
of compound Miniatures, the Parts are farther associated, and approach
perpetually nearer to each other, agreeably to what was just now observed; *i.e.* the Association becomes perpetually more close and intimate.

Case 5. When the Ideas *a, b, c, d,* &c. have been sufficiently associated
in any one or more of the foregoing Ways, if we suppose any single Idea
of these, *a* for Instance, to be raised by the Tendency of the medullary
Substance that Way, by the Association of *A* with a foreign Sensation
or Idea *X* or *x,* &c. this idea *a,* thus raised, will frequently bring in all
the rest, *b, c, d,* &c. and so associate all of them together still farther.

And, upon the Whole, it may appear to the Reader, that the simple
Ideas of Sensation must run into Clusters and Combinations, by Asso-
ciation; and that each of these will, at last, coalesce into one complex
Idea, by the Approach and Commixture of the several compounding
Parts.

It appears also from Observation, that many of our intellectual Ideas,
such as those that belong to the Heads of Beauty, Honour, moral Qualities, &c. are, in Fact, thus composed of Parts, which, by degrees, coalesce
into one complex Idea.

And as this Coalescence of simple Ideas into complex ones is thus
evinced, both by the foregoing Theory, and by Observation, so it may
be illustrated, and farther confirmed, by the similar Coalescence of
Letters into Syllables and Words, in which Association is likewise a
chief Instrument. I shall mention some of the most remarkable Particulars, relating to this Coalescence of simple Ideas into complex ones,
in the following Corollaries.

Cor. 1. If the Number of simple Ideas which compose the complex
one be very great, it may happen, that the complex Idea shall not appear
to bear any Relation to these its compounding Parts, nor to the external
Senses upon which the original Sensations, which gave Birth to the

compounding Ideas, were impressed. The Reason of this is, that each single Idea is overpowered by the Sum of all the rest, as soon as they are all intimately united together. Thus, in very compound Medicines, the several Tastes and Flavours of the separate Ingredients are lost and overpowered by the complex one of the whole Mass: So that this has a Taste and Flavour of its own, which appears to be simple and original, and like that of a natural Body. Thus also, White is vulgarly thought to be the simplest and most uncompounded of all Colours, while yet it really arises from a certain Proportion of the Seven primary Colours, with their several Shades, or Degrees. And, to resume the Illustration above-mentioned, taken from Language, it does not at all appear to Persons ignorant of the Arts of Reading and Writing, that the great Variety of complex Words of Languages can be analysed up to a few simple Sounds.

Cor. 2. One may hope, therefore, that, by pursuing and perfecting the Doctrine of Association, we may some time or other be enabled to analyse all that vast Variety of complex Ideas, which pass under the Name of Ideas of Reflection, and intellectual Ideas, into their simple compounding Parts, *i.e.* into the simple Ideas of Sensation, of which they consist. This would be greatly analogous to the Arts of Writing, and resolving the Colours of the Sun's Light, or natural Bodies, into their primary constituent ones. The complex Ideas which I here speak of, are generally excited by Words, or visible Objects; but they are also connected with other external Impressions, and depend upon them, as upon Symbols. In whatever Way we consider them, the Trains of them which are presented to the Mind seem to depend upon the then present State of the Body, the external Impressions, and the remaining Influence of prior Impressions and Associations, taken together.

Cor. 3. It would afford great Light and Clearness to the Art of Logic, thus to determine the precise Nature and Composition of the Ideas affixed to those Words which have complex Ideas, in a proper Sense, *i.e.* which excite any Combinations of simple Ideas united intimately by Association; also to explain, upon this foundation, the proper Use of those Words, which have no Ideas. For there are many Words which are mere Substitutes for other Words, and many which are only Auxiliaries.

. . . . . .

Cor. 4. As simple Ideas run into complex ones by Association, so complex Ideas run into decomplex ones by the same. But here the Varieties of the Associations, which increase with the Complexity, hinder particular ones from being so close and permanent, between the complex

Parts of decomplex Ideas, as between the simple Parts of complex ones:
To which it is analogous, in Languages, that the Letters of Words adhere
closer together than the Words of Sentences, both in Writing and Speaking.

Cor. 5. The simple Ideas of Sensation are not all equally and uniformly concerned in forming complex and decomplex Ideas; *i.e.* these
do not result from all the possible Combinations of Twos, Threes, Fours,
&c. of all the simple Ideas; but, on the contrary, some simple Ideas
occur in the complex and decomplex ones much oftener than others:
And the same holds of particular Combinations by Twos, Threes, &c.
and innumerable Combinations never occur at all in real Life, and consequently are never associated into complex or decomplex Ideas. All
which corresponds to what happens in real Languages; some Letters,
and Combinations of Letters, occur much more frequently than others,
and some Combinations never occur at all.

Cor. 6. As Persons who speak the same Language have, however,
a different Use and Extent of Words, so, tho' Mankind, in all Ages and
Nations, agree, in general, in their complex and decomplex Ideas, yet
there are many particular Differences in them; and these Differences
are greater or less, according to the Difference, or Resemblance, in Age,
Constitution, Education, Profession, Country, Age of the World, &c. *i.e.*
in their Impressions and Associations.

Cor. 7. When a Variety of Ideas are associated together, the visible
Idea, being more glaring and distinct than the rest, performs the Office
of a Symbol to all the rest, suggests them, and connects them together.
In this it somewhat resembles the first Letter of a Word, or first Word
of a Sentence, which are often made use of to bring all the rest to Mind.

Cor. 8. When Objects and Ideas, with their most common Combinations, have been often presented to the Mind, a Train of them, of a
considerable Length, may, by once occurring, leave such a Trace, as to
recur in Imagination, and in Miniature, in nearly the same Order and
Proportion as in this single Occurrence. For since each of the particular
Impressions and Ideas is familiar, there will want little more for their
Recurrency, than a few connecting Links; and even these may be, in
some measure, supplied by former similar Instances. These Considerations,
when duly unfolded, seem to me sufficient to explain the chief Phænomena
of Memory; and it will be easily seen from them, that the Memory of
Adults, and Masters in any Science, ought to be much more ready and
certain than that of Children and Novices, as it is found to be in Fact.

Cor. 9. When the Pleasure or Pain attending any Sensations, and
Ideas, is great, all the Associations belonging to them are much accelerated and strengthened. For the violent Vibrations excited in such Cases,

soon over-rule the natural Vibrations, and leave in the Brain a strong Tendency to themselves, from a few Impressions. The Associations will therefore be cemented sooner and stronger than in common Cases; which is found agreeable to the Fact.

Cor. 10. As many Words have complex Ideas annexed to them, so Sentences, which are Collections of Words, have Collections of complex Ideas, *i.e.* have decomplex Ideas. And it happens, in most Cases, that the decomplex Idea belonging to any Sentence, is not compounded merely of the complex Ideas belonging to the Words of it; but that there are also many Variations, some Oppositions, and numberless Additions. Thus Propositions, in particular, excite, as soon as heard, Assent or Dissent; which Assent and Dissent consist chiefly of additional complex Ideas, not included in the Terms of the Proposition. And it would be of the greatest Use, both in the Sciences and in common Life, thoroughly to analyse this Matter, to shew in what Manner, and by what Steps, *i.e.* by what Impressions and Associations, our Assent and Dissent, both in scientifical and moral Subjects, is formed.

71 THOMAS BROWN (1778-1820) ON THE SECONDARY LAWS OF ASSOCIATION, 1820

Thomas Brown, *Lectures on the Philosophy of the Human Mind* (Edinburgh, 1820), vol. II, lectures 35 and 37.

Brown was a member of the Scottish school of philosophy (Thomas Reid, 1710-1796; Dugald Stewart, 1753-1828) that had opposed the doctrine of association as a basic tenet of a theory of knowledge. He objected to the phrase "association of ideas," preferring Berkeley's word "suggestion." He scorned Hartley's physiologizing and criticized Hume's categories of association. His secondary laws of association were, however, a major contribution to the doctrine. Suggestion, he said, is ultimately based on contiguity, yet he spoke of there being three primary laws, Aristotle's three principles: contiguity, resemblance, and contrast. Nevertheless, knowing only these laws would not, as he pointed out, enable one to predict which of the ideas out of a set of alternatives would be suggested at any one time. For making such predictions, he offered his nine secondary laws, each of which serves to modulate the operation of the three primary laws. These secondary laws have a modern ring about them, since experiments nowadays on memory and learning often examine the very factors they describe. The third of his secondary laws, for example, stated that the strength of an association is proportional to the frequency with which it is renewed. This notion of frequency, already present in Aristotle's theory of memory, was developed by Hermann Ebbinghaus in 1885 (see No. 95) into the basis for the experimental study of association.

· · · · ·

To the threefold division, which Mr Hume has made, of the principles of association in the trains of our ideas, as consisting in *resemblance, contiguity,* and *causation,* there is an obvious objection . . . not founded on excessive *simplicity,* the love of which might more naturally be supposed to have misled him, but on its *redundancy,* according to the very principles of his own theory. *Causation,* far from being opposed to *contiguity,* so as to form a separate class, is, in truth, the most exquisite species of *proximity in time,*—and in most cases of *contiguity in place* also,—which could be adduced; because it is not a proximity depending on *casual* circumstances, and consequently liable to be *broken,* as these circumstances may exist apart,—but one which depends only on the mere existence of the two objects that are related to each other as cause and effect,—and therefore *fixed* and *never failing.* Other objects *may* sometimes be *proximate;* but a cause and effect *are* always proximate, and *must* be *proximate,* and are, indeed, classed in that relation, merely from this constant *proximity.* On his own principles, therefore, the three connexions of our ideas should indisputably be reduced to two. To speak of *resemblance, contiguity,* and *causation,* as three distinct classes, is, with Mr Hume's view of causation, and, indeed, with every view of it, as if a mathematician should divide lines into *straight, curved,* and *circular.* The inhabitants of China are said to have made a proverbial division of the human race, into men, women, and Chinese. With their view of their own importance, we understand the proud superiority of the distinction which they have made. But this sarcastic insolence would surely have been absurdity itself, if they had not intended it to express some characteristic and exclusive excellence, but had considered themselves as such ordinary *men* and *women,* as are to be found in all the other regions of the earth.

Resemblance and *contiguity in place and time,*—to which, on his own principles, Mr Hume's arrangement must be reduced—may be allowed indeed to hold a permanent rank, in whatever classification there may be formed, if any be to be formed, of the principles that regulate our trains of thought. But are there, in this case, truly distinct classes of suggestions, that are not reducible to any more common principle? or are they not all reducible to a single influence? I have already remarked the error, into which the common phrase, *Association of ideas,* has led us, by restricting, in our conception, the influence of the suggesting principle to those particular states of mind, which are exclusively denominated ideas; and it is this false restriction, which seems to me to have led to this supposition of different principles of association, to be classed in the manner proposed by Mr Hume and others, under distinct heads. All suggestion, as I conceive, may, if our analysis be sufficiently minute, be found to depend on prior coexistence, or, at least, on such immediate proximity as is itself, very probably, a modification of coexistence. For this very nice reduction, however, we must take in the influence of emotions, and other feelings, that are very different from ideas; as when an analogous object suggests an analogous object, by the influence of an emotion or sentiment, which each separately may have produced before, and which is therefore common to both. But, though a very nice analysis may lead to this reference of all our suggestions to one common influence of former proximity or coexistence of feelings, it is very convenient, in illustration of the principle, to avail ourselves of the most striking subdivisions, in which the particular instances of that proximity may be arranged; and I shall, therefore, adopt, for this purpose, the arrangement which Mr Hume has made,—if *resemblance* be allowed to comprehend every species of *analogy,* and if *contrast,* as a peculiar subdivision, be substituted for the superfluous one of causation.

.   .   .   .   .

Lecture XXXVII

.   .   .   .   .

After the remarks which I have already frequently made on this subject, I trust it is now unnecessary for me to repeat, that the term *laws,* as employed in the physics, whether of matter or of mind, is not used to denote any thing different from the phenomena themselves,—that, in short, it means nothing more than *certain circumstances of general agreement in any number of phenomena.* When Mr Hume reduced to the three orders of *resemblance, contiguity,* and *causation,* the relations on which he believed association to depend, he considered himself as stating only facts which were before familiar to every one, and *did* state only facts that were perfectly familiar. In like manner, when I reduce under a few heads those modifying circumstances, which seem to me as *secondary* laws, to guide, in every particular case, the momentary direction of the primary, my object is not to discover facts that are new, or little observed, but to arrange facts that, separately, are well known.

The *first* circumstance, which presents itself, as modifying the influence of the primary laws, in inducing one associate conception rather than another, is the length of time during which the original feelings from

357

which they flowed, continued, when they coexisted, or succeeded each other. Every one must be conscious, that innumerable objects pass before him, which are slightly observed at the time, but which form no permanent associations in the mind. The longer we dwell on objects, the more fully do we rely on our future remembrance of them.

In the *second* place, the parts of a train appear to be more closely and firmly associated, as the original feelings have been *more lively.* We remember brilliant objects, more than those which are faint and obscure. We remember for our whole lifetime, the occasions of great joy or sorrow; we forget the occasions of innumerable slight pleasures or pains, which occur to us every hour. That strong feeling of interest and curiosity, which we call attention, not only leads us to dwell longer on the consideration of certain objects, but also gives more vivacity to the objects, on which we dwell,—and in both these ways tend, as we have seen, to *fix* them, more strongly, in the mind.

In the *third* place, the parts of any train are more readily suggested, in proportion as they have been more *frequently renewed.* It is thus, we remember, after reading them three or four times over, the verses, which we could not repeat, when we had read them only once.

In the *fourth* place, the feelings are connected more strongly, in proportion as they are *more or less recent.* Immediately after reading any single line of poetry, we are able to repeat it, though we may have paid no particular attention to it;—in a very few minutes, unless when we have paid particular attention to it, we are no longer able to repeat it accurately—and in a very short time we forget it altogether. There is, indeed, one very striking exception to this law, in the case of old age: for events, which happened in youth, are then remembered, when events of the year preceding are forgotten. Yet, even in the case of extreme age,—when the time is not extended so far back,—the general law still holds; and events, which happened a few *hours* before, are remembered, when there is total forgetfulness of what happened a few days before.

In the *fifth* place, our successive feelings are associated more closely, as *each has coexisted less with other feelings.* The song, which we have never heard but from one person, can scarcely be heard again by us, without recalling that person to our memory; but there is obviously much less chance of this particular suggestion, if we have heard the same air and words frequently sung by others.

In the *sixth* place, the influence of the primary laws of suggestion is greatly modified by *original constitutional differences,* whether these are to be referred to the mind itself, or to varieties of bodily temperament. Such constitutional differences affect the primary laws in two ways,— first, by augmenting and extending the influence of all of them, as in the varieties of *the general power of remembering,* so observable in different individuals. Secondly, they modify the influence of the primary laws, by giving *greater proportional vigour* to *one set* of tendencies of suggestion than to another. It is in this modification of the suggesting principle, and the peculiar suggestions to which it gives rise, that I conceive the chief part, or, I may say, the whole of what is truly called *genius,* to consist. We have already seen, that the primary tendencies of suggestion are of various species,—some, for example, arising from mere *analogy,* others from direct contiguity or nearness in time or place of the very objects themselves,—and it is this difference of the prevailing tendency, as to these two species of suggestions, which I conceive to constitute all that is inventive in genius;—invention consisting in the suggestions of analogy, as opposed to the suggestions of grosser contiguity.

. . . . .

The *inventions* of *poetic genius* . . . are the *suggestions of analogy,*— the prevailing suggestions of *common minds,* are those of *mere contiguity;* and it is this difference of the *occasions* of suggestion, not of the *images* suggested, which forms the distinctive superiority of original genius. Any one, who has had the pleasure of reading the beautiful similé, which I have quoted to you from the Pharsalia, may, on the sight of a *decaying oak,* feel immediately the relation of analogy which this majestic trunk, still lifting as proudly to the storm, and spreading as widely its leafless arms, bears to the *decay of human grandeur,* more venerable, perhaps, in its very feebleness, than in all the magnificence of its power. The mind of *every one,* therefore, is *capable* of the *suggestion* of the *one* analogous object by the *other,* as much as the mind of Lucan. The only difference is, that, to produce this suggestion in a *common mind,* it was necessary, previously, to make the one conception successive, in point of time, to the other,—to produce, in short, a proximity of the very images that could be obtained only by a perusal of the verses, in which the images are immediately proximate:—while the suggestion, in the mind of the original *author,* though perhaps not more clear and perfect, than it was afterwards to be, in the memory of many of those who have read the similé, and felt its justness and beauty, differed, notwithstanding, in *this* most important respect, that, in *him,* it did not require such previous contiguity to produce the suggestion, but arose, by its mere *analogy,* in consequence of the greater tendency of the inventive mind to suggestions of this particular class.

. . . . .

It is the same with *inventive genius* in the *sciences* and the *severer arts,* which does not depend on the mere *knowledge* of all the *phenomena* previously observed, or of all the *applications* of them that have been made to purposes of art, but chiefly on the peculiar tendency of the mind to suggest certain *analogous* ideas, in successions, different from those ordinary successions of grosser contiguity, which occur to common minds. He may, perhaps, be called a philosopher, who knows accurately what others know, and produces, with the same means which others employ, the same effects which they produce. But *he* alone has *philosophic genius,* to whose speculations *analogous effects* suggest *analogous causes,* and who contrives practically, by the suggestions of analogy, to produce *new* effects, or to produce the *same effects* by *new* and *simpler means.*

The primary laws of association, then, it appears, as far as they operate in our *intellectual* exertions, are greatly modified by original constitutional diversities. They are not less modified by constitutional diversities of another kind. These are the diversities of what is called temper, or disposition. It is thus we speak of one person of a *gloomy,* and of another of a *cheerful* disposition; and we avoid the one, and seek the company of the other, as if with perfect confidence, that the trains of thought which rise by spontaneous suggestion to the minds of each will be *different,* and will be in accordance with that variety of character which we have supposed. To the cheerful, almost every object which they perceive, is cheerful as themselves. In the very darkness of the storm, the cloud which hides the sunshine from their eye, does not hide it from their heart: while, to the sullen, no sky is bright, and no scene is fair. There are future fogs, which to *their* eyes, pollute and darken the purest airs of spring; and spring itself is known to them less as the season which follows and repairs the desolation of winter that is past, than as the season which announces its approaching return.

The next secondary law of suggestion to which I proceed, is one akin to the last which we have considered. The primary laws are modified, not by constitutional and permanent differences only, but by differences which occur in the same individual, according to the varying emotion of the hour. As there are persons, whose general character is gloomy or cheerful, we have, in like manner, our peculiar *days* or *moments* in which we pass from one of these characters to the other, and in which our trains of thought are tinctured with the corresponding varieties. A mere change of fortune is often sufficient to alter the whole cast of sentiment. Those who are in possession of public station, and power and affluence, are accustomed to represent affairs in a favourable light: the disappointed competitors for place, to represent them in the most gloomy light; and though much of this difference may, unquestionably, be ascribed to *wilful mis-statement in both* cases, much of it is, as unquestionably, referable to that difference of colouring in which objects appear to the successful and the unsuccessful . . .

If even a *slight momentary feeling of joy or sorrow* have the power of *modifying* our suggestions, in accordance with it, emotions of a *stronger* and *lasting* kind must influence the trains of thought still more;—the meditations of every day rendering *stronger* the habitual connections of such thoughts as accord with the peculiar frame of mind. It is in this way that every passion, which has one fixed object,—such as love, jealousy, revenge, derives *nourishment* from *itself,* suggesting images that give it, in return, new force and liveliness. We see, *in every thing,* what we feel *in ourselves;* and the thoughts which external things seem to suggest, are thus, in part at least, suggested by the permanent emotion within.

· · · · ·

The temporary diversities of state, that give rise to varieties of suggestion, are not *mental* only, but *corporeal;* and this *difference of bodily state* furnishes another *secondary law,* in modification of the *primary.* I need not refer to the extreme cases of intoxication or actual delirium,— to the copious flow of follies which a little wine, or a few grains of opium, may extract from the proudest reasoner. In circumstances *less* striking, how different are the trains of thought in *health* and in *sickness,*—after a *temperate meal,* and after a *luxurious excess!* It is not to the *animal powers* only, that the burthen of digestion may become oppressive, but to the *intellectual* also; and often to the *intellectual* powers even more than to the *animal.* In that most delightful of all states, when the bodily frame has *recovered from disease,* and when, in the first walk beneath the open sunshine, amid the blossoms and balmy air of summer, there is a mixture of *corporeal* and *mental* enjoyment, in which it is not easy to discriminate what images of pleasure arise from every object, that, in other states of health, might have excited no thought or emotion whatever . . .

There is yet another principle which modifies the primary laws of suggestion with very powerful influence. This is the principle of habit. I do not speak of its influence in suggesting images, which have been already frequently suggested in a certain order,—for it would then be simpler to reduce the habit itself to the mere power of association. I speak of cases, in which the images suggested may have been of *recent acquisition,* but are suggested more readily in consequence of *general tendencies* produced by prior habits. When men of different professions observe the same circumstances, listen to the same story, or peruse the same work, their subsequent suggestions are far from being the same;

and, could the future differences of the associate feelings that are to rise, be foreseen by us at the time, we should probably be able to trace many of them to former *professional peculiarities,* which are thus always unfortunately apt to be more and more aggravated by the very suggestions to which they have themselves given rise. The most striking example, however, of the power of habit in modifying suggestion, is in the command which it gives to the orator, who has long been practised in extemporary elocution;—a command, not of words merely, but of *thoughts* and *judgments,* which, at the very moment of their sudden inspiration, appear like the long-weighed calculations of deliberative reflection. The whole divisions of his subject start before him at once; image after image, as he proceeds, arises to illustrate it; and proper words in proper places are all the while embodying his sentiments, as if without the slightest effort of his own.

In addition, then, to the primary laws of suggestion, which are founded on the mere relations of the objects or feelings to each other, it appears that there is another set of laws, the operation of which is indispensible to account for the variety in the effects of the former. To these I have given the name of *secondary laws of suggestion;*—and we have seen, accordingly, that the suggestions are various as the original feelings have been, 1*st,* Of *longer or shorter continuance;* 2*dly,* More or less lively; 3*dly,* More or less frequently present; 4*thly,* More or less recent; 5*thly,* More or less *pure,* if I may so express it, from the mixture of other feelings; 6*thly,* That they vary according to differences of original constitution; 7*thly,* According to differences of temporary emotion; 8*thly,* According to changes produced in the state of the body; and, 9*thly,* According to general tendencies produced by prior habits.

The first four laws, which relate rather to the momentary feelings themselves than to the particular frame of mind of the individual, have, it must be remembered, a double operation. When the two associate feelings have *both,* together, or in immediate succession, been of *long continuance,* very *lively, frequently renewed* in the same order, and that *recently,* the tendency to suggest each other is most powerful. But the greater tendency,—though then most remarkably exhibited,—is not confined to cases in which these laws are applicable to *both the associate feelings.* It is much increased, even when they apply only to that *one* which is second in the succession. The sight of an object which is altogether new to us,—and which, therefore, could not have formed a stronger connection with one set of objects than with another,—will more readily recal to us, by its resemblance or other relation, such objects as have been long familiar to us, than others which may have passed frequently before us, but with which we are little acquainted. The sailor sees every where some near or distant similarity to the parts of his own ship; and the phraseology, so rich in nautical metaphors, which he uses, and applies, with most rhetorical exactness, even to objects *perceived by him for the first time,* is a proof, that, for readiness of suggestion, it is not necessary that the secondary laws of suggestion should, in every particular case, have been applicable to *both* the suggesting and the suggested idea.

Even *one* of these secondary laws, alone, may be sufficient to change completely the suggestion, which would *otherwise* have arisen from the operation of the primary laws; and it is not wonderful, therefore, that when many of them, as they usually do, *concur* in one joint effect, the result in different individuals should be so various. Of the whole audience of a crowded theatre, who witness together the representation of the same piece, there are probably no two individuals, who carry away the same images, though the resemblances, contiguities, contrasts, and in general what I have called the *primary,* in opposition to the *secondary,* laws of suggestion, may have been the same to both. Some will perhaps think afterwards of the plot, and general developement of the drama; some, of the merits of the performers; some will remember little more, than that they were in a great crowd, and were very happy; a gay and dissipated young man will perhaps think only of the charms of some fascinating actress; and a young beauty will as probably carry away no remembrance so strong, as that of the eyes which were most frequently fixed upon her's.

By the consideration of these *secondary laws of suggestion,* then, the difficulty, which the consideration of the *primary laws* left unexplained, is at once removed. We see now, how *one* suggestion takes place rather than *another,* when, by the operation of the *mere primary laws,* many suggestions might arise equally; the influence of the *secondary laws* modifying this general tendency, and modifying it, of course, variously, as themselves are various.

72   JAMES MILL (1773–1836) ON MENTAL MECHANICS, 1829

James Mill, *Analysis of the Phenomena of the Human Mind* (London, 1829), chap. 3.

James Mill was a disciple of Jeremy Bentham, a utilitarian, believing as Bentham did, that morality is that which brings the greatest happiness to the greatest number. Bentham and the Mills (father and son) were convinced in respect of both political and ethical philosophy that the only measures for good and evil, in government as in personal life, are pleasure and pain.

Like Hobbes, James Mill, the father, held that political theory should be based on an understanding of human nature. Like Hartley, he believed that the primary mode of operation of the mind is association, which operates simply on the principle of contiguity. Eighty years before, Hartley had allowed the possibility that certain trains of thought, guided by the will, occur beyond the control of mere association, and Hume had spoken of association as a "gentle force"; but James Mill believed that the connections between ideas take place rigidly and inexorably in accordance with the laws of association.

Thought succeeds thought; idea follows idea, incessantly. If our senses are awake, we are continually receiving sensations, of the eye, the ear, the touch, and so forth; but not sensations alone. After sensations, ideas are perpetually excited of sensations formerly received; after those ideas, other ideas: and during the whole of our lives, a series of those two states of consciousness, called sensations, and ideas, is constantly going on. I see a horse: that is a sensation. Immediately I think of his master: that is an idea. The idea of his master makes me think of his office; he is a minister of state: that is another idea. The idea of a minister of state makes me think of public affairs; and I am led into a train of political ideas; when I am summoned to dinner. This is a new sensation, followed by the idea of dinner, and of the company with whom I am to partake it. The sight of the company and of the food are other sensations; these suggest ideas without end; other sensations perpetually intervene, suggesting other ideas: and so the process goes on.

In contemplating this train of feelings, of which our lives consist, it first of all strikes the contemplator, as of importance to ascertain, whether they occur casually and irregularly, or according to a certain order.

With respect to the SENSATIONS, it is obvious enough that they occur, according to the order established among what we call the objects of nature, whatever those objects are; to ascertain more and more of which order is the business of physical philosophy in all its branches.

Of the order established among the objects of nature, by which we mean the objects of our senses, two remarkable cases are all which here we are called upon to notice; the SYNCHRONOUS ORDER, and the SUCCESSIVE ORDER. The synchronous order, or order of simultaneous existence, is the order in space; the successive order, or order of antecedent and consequent existence, is the order in time. Thus the various objects in my room, the chairs, the tables, the books, have the synchronous order, or order in space. The falling of the spark, and the explosion of the gunpowder, have the successive order, or order in time.

According to this order, in the objects of sense, there is a synchronous, and a successive, order of our sensations. I have SYNCHRONICALLY, or at the same instant, the sight of a great variety of objects; touch of all the objects with which my body is in contact; hearing of all the sounds which are reaching my ears; smelling of all the smells which are reaching my nostrils; taste of the apple which I am eating; the sensation of resistance both from the apple which is in my mouth, and the ground on which I stand; with the sensation of motion from the act of walking. I have SUCCESSIVELY the sight of the flash from the mortar fired at a distance, the hearing of the report, the sight of the bomb, and of its motion in the air, the sight of its fall, the sight and hearing of its explosion, and lastly, the sight of all the effects of that explosion.

Among the objects which I have thus observed synchronically, or successively; that is, from which I have had synchronical or successive sensations; there are some which I have so observed frequently; others which I have so observed not frequently: in other words, of my sensations some have been frequently synchronical, others not frequently; some frequently successive, others not frequently. Thus, my sight of roast beef, and my taste of roast beef, have been frequently SYNCHRONICAL; my smell of a rose, and my sight and touch of a rose, have been frequently synchronical; my sight of a stone, and my sensations of its hardness, and weight, have been frequently synchronical. Others of my sensations have not been frequently synchronical: my sight of a lion, and the hearing of his roar; my sight of a knife, and its stabbing a man. My sight of the flash of lightning, and my hearing of the thunder, have been often SUCCESSIVE; the pain of cold, and the pleasure of heat, have been often successive; the sight of a trumpet, and the sound of a trumpet, have been often successive. On the other hand, my sight of hemlock, and my taste of hemlock, have not been often successive: and so on.

It so happens, that, of the objects from which we derive the greatest part of our sensations, most of those which are observed synchronically, are frequently observed synchronically; most of those which are observed successively, are frequently observed successively. In other words, most of our synchronical sensations, have been frequently synchronical; most of our successive sensations, have been frequently successive. Thus, most of our synchronical sensations are derived from the objects around us, the objects which we have the most frequent occasion to hear and see; the members of our family; the furniture of our houses; our food; the instruments of our occupations or amusements. In like manner, of those sensations which we have had in succession, we have had the greatest number repeatedly in succession; the sight of fire, and its warmth; the touch of snow, and its cold; the sight of food, and its taste.

Thus much with regard to the order of SENSATIONS; next with regard to the order of IDEAS.

365

As ideas are not derived from objects, we should not expect their order to be derived from the order of objects; but as they are derived from sensations, we might by analogy expect, that they would derive their order from that of the sensations; and this to a great extent is the case.

Our ideas spring up, or exist, in the order in which the sensations existed, of which they are the copies.

This is the general law of the "Association of Ideas"; by which term, let it be remembered, nothing is here meant to be expressed, but the order of occurrence.

In this law, the following things are to be carefully observed.

1. Of those sensations which occurred synchronically, the ideas also spring up synchronically. I have seen a violin, and heard the tones of the violin, synchronically. If I think of the tones of the violin, the visible appearance of the violin at the same time occurs to me. I have seen the sun, and the sky in which it is placed, synchronically. If I think of the one, I think of the other at the same time.

One of the cases of synchronical sensation, which deserves the most particular attention, is, that of the several sensations derived from one and the same object; a stone, for example, a flower, a table, a chair, a horse, a man.

From a stone I have had, synchronically, the sensation of colour, the sensation of hardness, the sensations of shape, and size, the sensation of weight. When the idea of one of these sensations occurs, the ideas of all of them occur. They exist in my mind synchronically; and their synchronical existence is called the idea of the stone; which, it is thus plain, is not a single idea, but a number of ideas in a particular state of combination.

Thus, again, I have smelt a rose, and looked at, and handled a rose, synchronically; accordingly the name rose suggests to me all those ideas synchronically; and this combination of those simple ideas is called my idea of the rose.

My idea of an animal is still more complex. The word thrush, for example, not only suggests an idea of a particular colour and shape, and size, but of song, and flight, and nestling, and eggs, and callow young, and others.

My idea of a man is the most complex of all; including not only colour, and shape, and voice, but the whole class of events in which I have observed him either the agent or the patient.

2. As the ideas of the sensations which occurred synchronically, rise synchronically, so the ideas of the sensations which occurred successively, rise successively.

Of this important case of association, or of the successive order of our ideas, many remarkable instances might be adduced. Of these none seems better adapted to the learner than the repetition of any passage, or words; the Lord's Prayer, for example, committed to memory. In learning the passage, we repeat it; that is, we pronounce the words, in successive order, from the beginning to the end. The order of the sensations is successive. When we proceed to repeat the passage, the ideas of the words also rise in succession, the preceding always suggesting the succeeding, and no other. *Our* suggests *Father, Father* suggests *which, which* suggests *art;* and so on, to the end. How remarkably this is the case, any one may convince himself, by trying to repeat backwards, even a passage with which he is as familiar as the Lord's Prayer. The case is the same with numbers. A man can go on with the numbers in the progressive order, one, two, three, &c. scarcely thinking of his act; and though it is possible for him to repeat them backward, because he is accustomed to subtraction of numbers, he cannot do so without an effort.

Of witnesses in courts of justice it has been remarked, that eye-witnesses, and ear-witnesses, always tell their story in the chronological order; in other words, the ideas occur to them in the order in which the sensations occurred; on the other hand, that witnesses, who are inventing, rarely adhere to the chronological order.

3. A far greater number of our sensations are received in the successive, than in the synchronical order. Of our ideas, also, the number is infinitely greater that rise in the successive than the synchronical order.

4. In the successive order of ideas, that which precedes, is sometimes called the suggesting, that which succeeds, the suggested idea; not that any power is supposed to reside in the antecedent over the consequent; suggesting, and suggested, mean only antecedent and consequent, with the additional idea, that such order is not casual, but, to a certain degree, permanent.

5. Of the antecedent and consequent feelings, or the suggesting, and suggested; the antecedent may be either sensations or ideas; the consequent are always ideas. An idea may be excited either by a sensation or an idea. The sight of the dog of my friend is a sensation, and it excites the idea of my friend. The idea of Professor Dugald Stewart delivering a lecture, recals the idea of the delight with which I heard him; that, the idea of the studies in which it engaged me; that, the trains of thought which succeeded; and each epoch of my mental history, the succeeding one, till the present moment; in which I am endeavouring to present to others what appears to me valuable among the innumerable ideas of which this lengthened train has been composed.

6. As there are degrees in sensations, and degrees in ideas; for one

367

sensation is more vivid than another sensation, one idea more vivid than another idea; so there are degrees in association. One association, we say, is stronger than another: First, when it is more permanent than another: Secondly, when it is performed with more certainty: Thirdly, when it is performed with more facility.

It is well known, that some associations are very transient, others very permanent. The case which we formerly mentioned, that of repeating words committed to memory, affords an apt illustration. In some cases, we can perform the repetition, when a few hours, or a few days have elapsed; but not after a longer period. In others, we can perform it after the lapse of many years. There are few children in whose minds some association has not been formed between darkness and ghosts. In some this association is soon dissolved; in some it continues for life.

In some cases the association takes place with less, in some with greater certainty. Thus, in repeating words, I am not sure that I shall not commit mistakes, if they are imperfectly got; and I may at one trial repeat them right, at another wrong: I am sure of always repeating those correctly, which I have got perfectly. Thus, in my native language, the association between the name and the thing is certain; in a language with which I am imperfectly acquainted, not certain. In expressing myself in my own language, the idea of the thing suggests the idea of the name with certainty. In speaking a language with which I am imperfectly acquainted, the idea of the thing does not with certainty suggest the idea of the name; at one time it may, at another not.

That ideas are associated in some cases with more, in some with less facility, is strikingly illustrated by the same instance, of a language with which we are well, and a language with which we are imperfectly, acquainted. In speaking our own language, we are not conscious of any effort; the associations between the words and the ideas appear spontaneous. In endeavouring to speak a language with which we are imperfectly acquainted, we are sensible of a painful effort: the associations between the words and ideas being not ready, or immediate.

7. The causes of strength in association seem all to be resolvable into two; the vividness of the associated feelings; and the frequency of the association.

In general, we convey not a very precise meaning, when we speak of the vividness of sensations and ideas. We may be understood when we say that, generally speaking, the sensation is more vivid than the idea; or the primary, than the secondary feeling; though in dreams, and in delirium, ideas are mistaken for sensations. But when we say that one sensation is more vivid than another, there is much more uncertainty. We can distinguish those sensations which are pleasurable, and those which are painful, from such as are not so; and when we call the pleasurable and painful more vivid, than those which are not so, we speak intelligibly. We can also distinguish degrees of pleasure, and of pain; and when we call the sensation of the higher degree more vivid than the sensation of the lower degree, we may again be considered as expressing a meaning tolerably precise.

In calling one IDEA more vivid than another, if we confine the appellation to the ideas of such SENSATIONS as may with precision be called more or less vivid; the sensations of pleasure and pain, in their various degrees, compared with sensations which we do not call either pleasurable or painful; our language will still have a certain degree of precision. But what is the meaning which I annex to my words, when I say, that my idea of the taste of the pine-apple which I tasted yesterday is vivid; my idea of the taste of the foreign fruit which I never tasted but once in early life, is not vivid? If I mean that I can more certainly distinguish the more recent, than the more distant sensation, there is still some precision in my language; because it seems true of all my senses, that if I compare a distant sensation with a present, I am less sure of its being or not being a repetition of the same, than if I compare a recent sensation with a present one. Thus, if I yesterday had a smell of a very peculiar kind, and compare it with a present smell, I can judge more accurately of the agreement or disagreement of the two sensations, than if I compared the present with one much more remote. The same is the case with colours, with sounds, with feelings of touch, and of resistance. It is therefore sufficiently certain, that the idea of the more recent sensation affords the means of a more accurate comparison, generally, than the idea of the more remote sensation. And thus we have three cases of vividness, of which we can speak with some precision: the case of sensations, as compared with ideas; the case of pleasurable and painful sensations, and their ideas, as compared with those which are not pleasurable or painful; and the case of the more recent, compared with the more remote.

·   ·   ·   ·   ·

. . . Next, we have to consider frequency or repetition; which is the most remarkable and important cause of the strength of our associations.

Of any two sensations, frequently perceived together, the ideas are associated. Thus, at least, in the minds of Englishmen, the idea of a soldier, and the idea of a red coat are associated; the idea of a clergyman, and the idea of a black coat; the idea of a quaker, and of a broadbrimmed hat; the idea of a woman and the idea of petticoats. A peculiar taste suggests the idea of an apple; a peculiar smell the idea of a rose.

If I have heard a particular air frequently sung by a particular person, the hearing of the air suggests the idea of the person.

The most remarkable exemplification of the effect of degrees of frequency, in producing degrees of strength in the associations, is to be found in the cases in which the association is purposely and studiously contracted; the cases in which we learn something; the use of words, for example.

Every child learns the language which is spoken by those around him. He also learns it by degrees. He learns first the names of the most familiar objects; and among familiar objects, the names of those which he most frequently has occasion to name; himself, his nurse, his food, his playthings.

A sound heard once in conjunction with another sensation; the word mamma, for example, with the sight of a woman, would produce no greater effect on the child, than the conjunction of any other sensation, which once exists and is gone for ever. But if the word mamma is frequently pronounced, in conjunction with the sight of a particular woman, the sound will by degrees become associated with the sight; and as the pronouncing of the name will call up the idea of the woman, so the sight of the woman will call up the idea of the name.

The process becomes very perceptible to us, when, at years of reflection, we proceed to learn a dead or foreign language. At the first lesson, we are told, or we see in the dictionary, the meaning of perhaps twenty words. But it is not joining the word and its meaning once, that will make the word suggest its meaning to us another time. We repeat the two in conjunction, till we think the meaning so well associated with the word, that whenever the word occurs to us, the meaning will occur along with it. We are often deceived in this anticipation; and finding that the meaning is not suggested by the word, we have to renew the process of repetition, and this, perhaps, again, and again. By force of repetition the meaning is associated, at last, with every word of the language, and so perfectly, that the one never occurs to us without the other.

Learning to play on a musical instrument is another remarkable illustration of the effect of repetition in strengthening associations, in rendering those sequences, which, at first, are slow, and difficult, afterwards, rapid, and easy. At first, the learner, after thinking of each successive note, as it stands in his book, has each time to look out with care for the key or the string which he is to touch, and the finger he is to touch it with, and is every moment committing mistakes. Repetition · is well known to be the only means of overcoming these difficulties. As the repetition goes on, the sight of the note, or even the idea of the note, becomes associated with the place of the key or the string; and that of the key or the string with the proper finger. The association for a time is imperfect, but at last becomes so strong, that it is performed with the greatest rapidity, without an effort, and almost without consciousness.

In few cases is the strength of association, derived from repetition, more worthy of attention, than in performing arithmetic. All men, whose practice is not great, find the addition of a long column of numbers, tedious, and the accuracy of the operation, by no means certain. Till a man has had considerable practice, there are few acts of the mind more toilsome. The reason is, that the names of the numbers, which correspond to the different steps, do not readily occur; that is, are not strongly associated with the names which precede them. Thus, 7 added to 5, make 12; but the antecedent, 7 added to 5, is not strongly associated with the consequent 12, in the mind of the learner, and he has to wait and search till the name occurs. Thus, again, 12 and 7 make 19; 19 and 8 make 27, and so on to any amount; but if the practice of the performer has been small, the association in each instance is imperfect, and the process irksome and slow. Practice, however; that is, frequency of repetition; makes the association between each of these antecedents and its proper consequent so perfect, that no sooner is the one conceived than the other is conceived, and an expert arithmetician can tell the amount of a long column of figures, with a rapidity, which seems almost miraculous to the man whose faculty of numeration is of the ordinary standard.

8. Where two or more ideas have been often repeated together, and the association has become very strong, they sometimes spring up in such close combination as not to be distinguishable. Some cases of sensation are analogous. For example; when a wheel, on the seven parts of which the seven prismatic colours are respectively painted, is made to revolve rapidly, it appears not of seven colours, but of one uniform colour, white. By the rapidity of the succession, the several sensations cease to be distinguishable; they run, as it were, together, and a new sensation, compounded of all the seven, but apparently a simple one, is the result. Ideas, also, which have been so often conjoined, that whenever one exists in the mind, the others immediately exist along with it, seem to run into one another, to coalesce, as it were, and out of many to form one idea; which idea, however in reality complex, appears to be no less simple, than any one of those of which it is compounded.

The word gold, for example, or the word iron, appears to express as simple an idea, as the word colour, or the word sound. Yet it is immediately seen, that the idea of each of those metals is made up of the separate ideas of several sensations; colour, hardness, extension, weight.

Those ideas, however, present themselves in such intimate union, that they are constantly spoken of as one, not many. We say, our idea of iron, our idea of gold; and it is only with an effort that reflecting men perform the decomposition.

The idea expressed by the term weight, appears so perfectly simple, that he is a good metaphysician, who can trace its composition. Yet it involves, of course, the idea of resistance, which we have shewn above to be compounded, and to involve the feeling attendant upon the contraction of muscles; and the feeling, or feelings, denominated Will; it involves the idea, not of resistance simply, but of resistance in a particular direction; the idea of direction, therefore, is included in it, and in that are involved the ideas of extension, and of place and motion, some of the most complicated phenomena of the human mind.

The ideas of hardness and extension have been so uniformly regarded as simple, that the greatest metaphysicians have set them down as the copies of simple sensations of touch. Hartley and Darwin, were, I believe, the first who thought of assigning to them a different origin.

We call a thing hard, because it resists compression, or separation of parts; that is, because to compress it, or separate it into parts, what we call muscular force is required. The idea, then, of muscular action, and of all the feelings which go to it, are involved in the idea of hardness.

The idea of extension is derived from the muscular feelings in what we call the motion of parts of our own bodies; as for example, the hands. I move my hand along a line; I have certain sensations; on account of these sensations, I call the line long, or extended. The idea of lines in the direction of length, breadth, and thickness, constitutes the general idea of extension. In the idea of extension, there are included three of the most complex of our ideas; motion; time, which is included in motion; and space, which is included in direction. We are not yet prepared to explain the simple ideas which compose the very complex ideas, of motion, space, and time; it is enough at present to have shewn, that in the idea of extension, which appears so very simple, a great number of ideas are nevertheless included; and that this is a case of that combination of ideas in the higher degrees of association, in which the simple ideas are so intimately blended, as to have the appearance, not of a complex, but of a simple idea.

It is to this great law of association, that we trace the formation of our ideas of what we call external objects; that is, the ideas of a certain number of sensations, received together so frequently that they coalesce as it were, and are spoken of under the idea of unity. Hence, what we call the idea of a tree, the idea of a stone, the idea of a horse, the idea of a man.

In using the names, tree, horse, man, the names of what I call objects, I am referring, and can be referring, only to my own sensations; in fact, therefore, only naming a certain number of sensations, regarded as in a particular state of combination; that is, concomitance. Particular sensations of sight, of touch, of the muscles, are the sensations, to the ideas of which, colour, extension, roughness, hardness, smoothness, taste, smell, so coalescing as to appear one idea, I give the name, idea of a tree.

To this case of high association, this blending together of many ideas, in so close a combination that they appear not many ideas, but one idea, we owe, as I shall afterwards more fully explain, the power of classification, and all the advantages of language. It is obviously, therefore, of the greatest moment, that this important phenomenon should be well understood.

9. Some ideas are by frequency and strength of association so closely combined, that they cannot be separated. If one exists, the other exists along with it, in spite of whatever effort we make to disjoin them.

For example; it is not in our power to think of colour, without thinking of extension; or of solidity, without figure. We have seen colour constantly in combination with extension, spread as it were, upon a surface. We have never seen it except in this connection. Colour and extension have been invariably conjoined. The idea of colour, therefore, uniformly comes into the mind, bringing that of extension along with it; and so close is the association, that it is not in our power to dissolve it. We cannot, if we will, think of colour, but in combination with extension. The one idea calls up the other, and retains it, so long as the other is retained.

This great law of our nature is illustrated in a manner equally striking, by the connection between the ideas of solidity and figure. We never have the sensations from which the idea of solidity is derived, but in conjunction with the sensations whence the idea of figure is derived. If we handle any thing solid, it is always either round, square, or of some other form. The ideas correspond with the sensations. If the idea of solidity rises, that of figure rises along with it. The idea of figure which rises, is, of course, more obscure than that of extension; because, figures being innumerable, the general idea is exceedingly complex, and hence, of necessity, obscure. But, such as it is, the idea of figure is always present when that of solidity is present; nor can we, by any effort, think of the one without thinking of the other at the same time.

Of all the cases of this important law of association, there is none more extraordinary than what some philosophers have called, the acquired perceptions of sight.

373

When I lift my eyes from the paper on which I am writing, I see the chairs, and tables, and walls of my room, each of its proper shape, and at its proper distance. I see, from my window, trees, and meadows, and horses, and oxen, and distant hills. I see each of its proper size, of its proper form, and at its proper distance; and these particulars appear as immediate informations of the eye, as the colours which I see by means of it.

Yet, philosophy has ascertained, that we derive nothing from the eye whatever, but sensations of colour; that the idea of extension, in which size, and form, and distance are included, is derived from sensations, not in the eye, but in the muscular part of our frame. How, then, is it, that we receive accurate information, by the eye, of size, and shape, and distance? By association merely.

The colours upon a body are different, according to its figure, its distance, and its size. But the sensations of colour, and what we may here, for brevity, call the sensations of extension, of figure, of distance, have been so often united, felt in conjunction, that the sensation of the colour is never experienced without raising the ideas of the extension, the figure, the distance, in such intimate union with it, that they not only cannot be separated, but are actually supposed to be seen. The sight, as it is called, of figure, or distance, appearing, as it does, a simple sensation, is in reality a complex state of consciousness; a sequence, in which the antecedent, a sensation of colour, and the consequent, a number of ideas, are so closely combined by association, that they appear not one idea, but one sensation.

· · · · ·

The following of one idea after another idea, or after a sensation, so certainly that we cannot prevent the combination, nor avoid having the *consequent* feeling as often as we have the *antecedent,* is a law of association, the operation of which we shall afterwards find to be extensive, and bearing a principal part in some of the most important phenomena of the human mind.

· · · · ·

10. It not unfrequently happens in our associated feelings, that the antecedent is of no importance farther than as it introduces the consequent. In these cases, the consequent absorbs all the attention, and the antecedent is instantly forgotten. Of this a very intelligible illustration is afforded by what happens in ordinary discourse. A friend arrives from a distant country, and brings me the first intelligence of the last illness, the last words, the last acts, and death of my son. The sound of the

voice, the articulation of every word, makes its sensation in my ear; but it is to the ideas that my attention flies. It is my son that is before me, suffering, acting, speaking, dying. The words which have introduced the ideas, and kindled the affections, have been as little heeded, as the respiration which has been accelerated, while the ideas were received.

It is important in respect to this case of association to remark, that there are large classes of our sensations, such as many of those in the alimentary duct, and many in the nervous and vascular systems, which serve, as antecedents, to introduce ideas, as consequents; but as the consequents are far more interesting than themselves, and immediately absorb the attention, the antecedents are habitually overlooked; and though they exercise, by the trains which they introduce, a great influence on our happiness or misery, they themselves are generally wholly unknown.

That there are connections between our ideas and certain states of the internal organs, is proved by many familiar instances. Thus, anxiety, in most people, disorders the digestion. It is no wonder, then, that the internal feelings which accompany indigestion, should excite the ideas which prevail in a state of anxiety. Fear, in most people, accelerates, in a remarkable manner, the vermicular motion of the intestines. There is an association, therefore, between certain states of the intestines, and terrible ideas; and this is sufficiently confirmed by the horrible dreams to which men are subject from indigestion; and the hypochondria, more or less afflicting, which almost always accompanies certain morbid states of the digestive organs. The grateful food which excites pleasurable sensations in the mouth, continues them in the stomach; and, as pleasures excite ideas of their causes, and these of similar causes, and causes excite ideas of their effects, and so on, trains of pleasurable ideas take their origin from pleasurable sensations in the stomach. Uneasy sensations in the stomach, produce analogous effects. Disagreeable sensations are associated with disagreeable circumstances; a train is introduced, in which, one painful idea following another, combinations, to the last degree afflictive, are sometimes introduced, and the sufferer is altogether overwhelmed by dismal associations.

In illustration of the fact, that sensations and ideas, which are essential to some of the most important operations of our minds, serve only as antecedents to more important consequents, and are themselves so habitually overlooked, that their existence is unknown, we may recur to the remarkable case which we have just explained, of the ideas introduced by the sensations of sight. The minute gradations of colour, which accompany varieties of extension, figure, and distance, are insignificant. The figure, the size, the distance, themselves, on the other hand, are matters of the greatest importance. The first having introduced the last, their work is done. The consequents remain the sole objects of attention, the antecedents are forgotten; in the present instance, not completely; in other instances, so completely, that they cannot be recognised.

11. Mr. Hume, and after him other philosophers, have said that our ideas are associated according to three principles; Contiguity in time and place, Causation, and Resemblance. The Contiguity in time and place, must mean, that of the sensations; and so far it is affirmed, that the order of the ideas follows that of the sensations. Contiguity of two sensations in time, means the successive order. Contiguity of two sensations in place, means the synchronous order. We have explained the mode in which ideas are associated, in the synchronous, as well as the successive order, and have traced the principle of contiguity to its proper source.

Causation, the second of Mr. Hume's principles, is the same with contiguity in time, or the order of succession. Causation is only a name for the order established between an antecedent and a consequent; that is, the established or constant antecedence of the one, and consequence of the other. Resemblance only remains, as an alleged principle of association, and it is necessary to inquire whether it is included in the laws which have been above expounded. I believe it will be found that we are accustomed to see like things together. When we see a tree, we generally see more trees than one; when we see an ox, we generally see more oxen than one; a sheep, more sheep than one; a man, more men than one. From this observation, I think, we may refer resemblance to the law of frequency, of which it seems to form only a particular case.

Mr. Hume makes contrast a principle of association, but not a separate one, as he thinks it is compounded of Resemblance and Causation. It is not necessary for us to show that this is an unsatisfactory account of contrast. It is only necessary to observe, that, as a case of association, it is not distinct from those which we have above explained.

A dwarf suggests the idea of a giant. How? We call a dwarf a dwarf, because he departs from a certain standard. We call a giant a giant, because he departs from the same standard. This is a case, therefore, of resemblance, that is, of frequency.

Pain is said to make us think of pleasure; and this is considered a case of association by contrast. There is no doubt that pain makes us think of relief from it; because they have been conjoined, and the great vividness of the sensations makes the association strong. Relief from pain is a species of pleasure; and one pleasure leads to think of another, from the resemblance. This is a compound case, therefore, of vividness and frequency. All other cases of contrast, I believe, may be expounded in a similar manner.

I have not thought it necessary to be tedious in expounding the observations which I have thus stated; for whether the reader supposes that resemblance is, or is not, an original principle of association, will not affect our future investigations.

12. Not only do simple ideas, by strong association, run together, and form complex ideas: but a complex idea, when the simple ideas which compose it have become so consolidated that it always appears as one, is capable of entering into combinations with other ideas, both simple and complex. Thus two complex ideas may be united together, by a strong association, and coalesce into one, in the same manner as two or more simple ideas coalesce into one. This union of two complex ideas into one, Dr. Hartley has called a duplex idea. Two also of these duplex, or doubly compounded ideas, may unite into one; and these again into other compounds, without end. It is hardly necessary to mention, that as two complex ideas unite to form a duplex one, not only two, but more than two may so unite; and what he calls a duplex idea may be compounded of two, three, four, or any number of complex ideas.

Some of the most familiar objects with which we are acquainted furnish instances of these unions of complex and duplex ideas.

Brick is one complex idea, mortar is another complex idea; these ideas, with ideas of position and quantity, compose my idea of a wall. My idea of a plank is a complex idea, my idea of a rafter is a complex idea, my idea of a nail is a complex idea. These, united with the same ideas of position and quantity, compose my duplex idea of a floor. In the same manner my complex idea of glass, and wood, and others, compose my duplex idea of a window; and these duplex ideas, united together, compose my idea of a house, which is made up of various duplex ideas. How many complex, or duplex ideas, are all united in the idea of furniture? How many more in the idea of merchandize? How many more in the idea called Every Thing?

73   JOHN STUART MILL (1806–1873) ON MENTAL
      CHEMISTRY, 1843

J. S. Mill, *A System of Logic, Ratiocinative and Inductive, being a connected View of the Principles of Evidence, and the Methods of Scientific Investigation* (London, 1843), vol. II, bk. VI, chap. 4.

James Mill believed that contiguity is the sole primary principle of association and that any idea, no matter how complex, is merely the sum of the simple ideas constituting it. His son, John Stuart Mill, disagreed with him on both counts. He added the principle of associative similarity, and he argued, by analogy with chemical union, that complex ideas may have properties that cannot be predicted from a knowledge of the simple components. The latter disagreement is not a mere quibble, for John Stuart Mill's view implies that each complex idea must be studied in its own right and that the doctrine of association, though it may tell how ideas originate, does not provide the whole story. Wilhelm Wundt's conception of creative synthesis in perception and ideas (see No. 76) was derived from this view of John Stuart Mill's.

3. The subject . . . of Psychology, is the uniformities of succession, the laws, whether ultimate or derivative, according to which one mental state succeeds another; is caused by, or at the least, is caused to follow, another. Of these laws, some are general, others more special. The following are examples of the most general laws.

First: Whenever any state of consciousness has once been excited in us, no matter by what cause; an inferior degree of the same state of consciousness, a state of consciousness resembling the former but inferior in intensity, is capable of being reproduced in us, without the presence of any such cause as excited it at first. Thus, if we have once seen or touched an object, we can afterwards think of the object although it be absent from our sight or from our touch. If we have been joyful or grieved at some event, we can think of, or remember, our past joy or grief, although no new event of a happy or a painful nature has taken place. When a poet has put together a mental picture of an imaginary object, a Castle of Indolence, a Una, or a Juliet, he can afterwards think of the ideal object he has created, without any fresh act of intellectual combination. This law is expressed by saying, in the language of Hume, that every mental *impression* has its *idea.*

Secondly: These Ideas, or secondary mental states, are excited by our impressions, or by other ideas, according to certain laws which are called Laws of Association. Of these laws the first is, that similar ideas tend to excite one another. The second is, that when two impressions have been frequently experienced (or even thought of) either simultaneously or in immediate succession, then whenever either of these impressions or the idea of it recurs, it tends to excite the idea of the other. The third law is, that greater intensity in either or both of the impressions, is equivalent, in rendering them excitable by one another, to a greater frequency of conjunction. These are the laws of Ideas: upon which I shall not enlarge in this place, but refer the reader to works professedly psychological, in particular to Mr. [James] Mill's *Analysis of the Phenomena of the Human Mind,* where the principal laws of association, both in themselves and in many of their applications, are copiously exemplified, and with a masterly hand.

These simple or elementary Laws of Mind have long been ascertained by the ordinary methods of experimental inquiry; nor could they have been ascertained in any other manner. But a certain number of elementary laws having thus been obtained, it is a fair subject of scientific inquiry how far those laws can be made to go in explaining the actual phenomena. It is obvious that complex laws of thought and feeling not only may, but must, be generated from these simple laws. And it is to be remarked, that the case is not always one of Composition of Causes: the effect of concurring causes is not always precisely the sum of the effects of those causes when separate, nor even always an effect of the same kind with them. Reverting to the distinction which occupies so prominent a place in the theory of induction; the laws of the phenomena of mind are sometimes analogous to mechanical, but sometimes also to chemical laws. When many impressions or ideas are operating in the mind together, there sometimes takes place a process, of a similar kind to chemical combination. When impressions have been so often experienced in conjunction, that each of them calls up readily and instantaneously the ideas of the whole group, those ideas sometimes melt and coalesce into one another, and appear not several ideas but one; in the same manner as when the seven prismatic colours are presented to the eye in rapid succession, the sensation produced is that of white. But as in this last case it is correct to say that the seven colours when they rapidly follow one another *generate* white, but not that they actually *are* white; so it appears to me that the Complex Idea, formed by the blending together of several simpler ones, should, when it really appears simple, (that is, when the separate elements are not consciously distinguishable in it,) be said to *result from,* or be *generated by,* the simple ideas, not to *consist* of them. Our idea of an orange really *consists* of the simple ideas of a certain colour, a certain form, a certain taste and smell, &c., because we can by interrogating our consciousness, perceive all these elements in the idea. But we cannot perceive, in so apparently simple a feeling as our perception of the shape of an object by the eye, all that multitude of ideas derived from other senses, without which it is well ascertained that no such visual perception would ever have had existence; nor, in our idea of Extension, can we discover those elementary ideas of resistance, derived from our muscular frame, in which Dr. Brown has rendered it highly probable that the idea originates. These therefore are cases of mental chemistry: in which it is proper to say that the simple ideas generate, rather than that they compose, the complex ones.

      •   •   •   •   •

. . . Even if all which this theory of mental phenomena contends for could be proved, we should not be the more enabled to resolve the laws

379

of the more complex feelings into those of the simpler ones. The generation of one class of mental phenomena from another, whenever it can be made out, is a highly interesting fact in psychological chemistry; but it no more supersedes the necessity of an experimental study of the generated phenomenon, than a knowledge of the properties of oxygen and sulphur enables us to deduce those of sulphuric acid without specific observation and experiment. Whatever, therefore, may be the final issue of the attempt to account for the origin of our judgments, our desires, or our volitions, from simpler mental phenomena, it is not the less imperative to ascertain the sequences of the complex phenomena themselves, by special study in conformity to the canons of Induction. Thus, in respect of Belief, the psychologist will always have to inquire, what beliefs we have intuitively, and according to what laws one belief produces another; what are the laws in virtue of which one thing is recognised by the mind, either rightly or erroneously, as evidence of another thing. In regard to Desire, he will examine what objects we desire naturally, and by what causes we are made to desire things originally indifferent or even disagreeable to us; and so forth. It may be remarked, that the general laws of association prevail among these more intricate states of mind, in the same manner as among the simpler ones. A desire, an emotion, an idea of the higher order of abstraction, even our judgments and volitions when they have become habitual, are called up by association, according to precisely the same laws as our simple ideas.

74   HERBERT SPENCER (1820–1903) ON INTELLIGENCE, 1855

Herbert Spencer, *The Principles of Psychology* (London, 1855), pt. 4, chaps. 2 and 3.

The principle of association enters Spencer's system of psychology as "the law of the succession of psychical changes." The orders of events in the environment vary from the persistent to the fortuitous, and the individual, animal or human, adapts to these external successions by setting up corresponding internal successions of conscious states, which vary in strength with the persistency of the external changes. The more intelligent the organism the more adequately does his consciousness correspond to the environmental uniformities, providing him with correct information about his world. Thus the orderliness of the mental life that is the mark of intelligence comes about by learning from experience. Spencer also believed that these learned successions of conscious states may be inherited as instincts. After the publication of Darwin's *Origin of Species* (1859), many persons, including Spencer himself, saw in this theory, under which a species' interaction with its environment gradually modified its form, an anticipation of an important part of Darwin's theory of evolution. Subsequently the concept of evolution became a central theme of Spencer's philosophy.

CHAPTER II

173. All Life, whether physical or psychical, being the combination of changes in correspondence with external coexistences and sequences; it results, that if the changes constituting psychical life, or intelligence, occur in succession, the law of their succession must be the law of their correspondence. That particular kind of Life which we distinguish as intelligence, including as it does the various developments of the correspondence in Space, in Time, in Speciality, in Complexity, &c.; it necessarily follows that the changes of which this intelligence consists, must, in their general mode of co-ordination, harmonize with the co-ordination of phenomena in the environment. The life is the correspondence; the progress of the life is the progress of the correspondence; the cessation of the life is the cessation of the correspondence: and hence, if there is one particular department of the life, which, more manifestly than any other, consists in the constant maintenance of the correspondence; the changes which make up this highest department of life, must, more manifestly than any other, display the correspondence. The fundamental condition of vitality, is, that the internal order shall be continually adjusted to the external order. If the internal order is altogether unrelated to the external order, there can be no adaptation between the actions going on in the organism and those going on in its environment: and life becomes impossible. If the relation of the internal order to the external order, is one of but partial adjustment; the adaptation of inner to outer actions is imperfect: and the life is proportionately low and brief. If, between the inner and the outer order, the adjustment is complete; the adaptation is complete: and the life is proportionately high and prolonged. Necessarily, then, the order of the states of consciousness is in correspondence with the order of phenomena in the environment. This is an *a priori* condition of intelligence.

Clear, however, as it is, that from this *a priori* condition of intelligence, must result the law of succession of psychical changes, an adequate expression of such law is by no means easy to find. Did the phenomena in the environment form, like the phenomena of consciousness, a succession; there would be no difficulty. The entire fact would be expressed by saying that the internal succession parallels the external succession. But the environment contains a great number of successions of phenomena, going on simultaneously. Further, the environment contains a great variety of phenomena that are not successive at all, but coexistent. Yet again, the environment is unlimited in extent, and the phenomena it contains are not only infinite in number, but insensibly pass into a relative non-existence, as the distance from the organism increases. And yet once more, the environment, relatively considered, is ever varying as the organism moves from place to place in it. How, then, can the succession of psychical changes be in any way formulated? How is it possible to express the law of a single series of internal phenomena, in terms of its correspondence with an infinity of external phenomena, both serial and non-serial, mixed in the most heterogeneous manner, and presented to the moving organism in an endless variety of fortuitous combinations?

Were it not that the inner relations *must* be in correspondence with the outer ones; and that therefore the order of the states of consciousness *must* be in some way expressible in terms of the external order; we might almost despair of finding any general law of psychical changes. Even as it is, we may be certain that any such general law cannot apply to extended portions of the series of changes. Dependent as these must in great measure be, upon the heterogeneous combinations of phenomena by which the organism is at any moment environed, and upon the new heterogeneous combinations perpetually disclosed by its movements, they can be no more formulated than the heterogeneous combinations of external phenomena can be formulated. Evidently, therefore, it must be in the constituent changes, and small groups of changes, rather than in the longer concatenations of changes, that we must look for a law.

And this is the indication given by certain still more general considerations. As on each particular link in a chain, depend the succeeding links; so, on each particular change in consciousness, depend all the succeeding changes: and hence the law of the succession of changes, must be really involved in the law of the individual change. If there occurs in consciousness a change from state A to state F, there will follow certain changes F to L, L to D, D to K, &c.; but if the first change had been from A to D, some other series of changes, D to J, J to C, C to N, would have resulted. So that, as the particular combination of subsequent changes is ever dependent upon the change occurring at each moment; and as each of these subsequent changes becomes, when it occurs, the change on which those succeeding it depend; it follows that the law of the individual change is the sole thing to be determined.

Not simply, therefore, as being the only phenomenon in the mental succession which there is any hope of formulating; but as being the phenomenon on which all other phenomena in the mental succession must hinge; the subject of our inquiry must be—the law of the connection between any two successive states of consciousness—the law of the elementary psychical change.

174. Using the expression *state of consciousness*, in its most extended sense, as meaning the psychical state of any order of creature, and also as meaning any species of psychical state, from the most simple to the

most complex; the law of the connection between any two successive states of consciousness, will become manifest on considering the *a priori* necessity to which it must conform. Each of the two states originally answers to some particular phenomenon external to consciousness. Every external phenomenon exists in certain relations to other phenomena. Hence, a correspondence between the internal order and the external order, implies that the relation between any two states of consciousness, corresponds with the relation between the two external phenomena producing them. How corresponds? The two states of consciousness occur in succession: and all successions are alike in so far as they are simply successions. In what, then, can the correspondence consist? It consists in this; that the *persistency* of the connection between the two states of consciousness, is proportionate to the *persistency* of the connection between the phenomena to which they answer. The relations between external phenomena are of all grades, from the absolutely necessary to the purely fortuitous. The relations between the answering states of consciousness must similarly be of all grades, from the absolutely necessary to the purely fortuitous. And as the correspondence becomes more complete, that is—as the intelligence becomes higher, the various grades of the one must be more and more accurately paralleled by those of the other. When any state a occurs, the tendency of some other state d, to follow it, must be strong or weak according to the degree of persistency with which A and D (the objects or attributes that produce a and d) occur together in the environment. If, in the environment, there is a more persistent occurrence of A with B than of A with D; then, the maintenance of the correspondence implies, that when a arises in consciousness, b shall follow rather than d. If there are in the environment a great variety of things in connection with which A occurs; then, when the state of consciousness a, arises, it must be followed by the state of consciousness answering to the thing most generally occurring along with A. These are manifest necessities. If the strengths of the connections between the internal states, are not proportionate to the persistencies of the relations between the answering external phenomena; there must be a failure of the correspondence—the inner order must disagree with the outer order. Psychical life, in common with life in general, being the continuous adjustment of inner to outer relations; and the occurrence of any relation between states of consciousness, being, in itself, nothing else than an exhibition of the fact, that the cohesion of the antecedent and consequent states was greater than the cohesion between the antecedent state and any other state; it follows inevitably, that, to effect the adjustment, the cohesion of the states must vary as the cohesion of the phenomena represented by them. The law of intelligence, therefore, is, that the strength of the tendency which the antecedent of any psy-chical change has to be followed by its consequent, is proportionate to the persistency of the union between the external things they symbolize.

To say, however, that this is the law of intelligence, is by no means to say that it is conformed to by any intelligence with which we are acquainted. It is the law of intelligence in the abstract; and is conformed to by existing intelligences in degrees more or less imperfect. To the extent that psychical changes fulfil this law, to such extent only do they constitute intelligence; and it is but very incompletely that even the highest orders of psychical changes do this.

.   .   .   .   .

Chapter III

.   .   .   .   .

Commencing with some lowly-endowed creature, respecting which it can be scarcely at all said, that the strength of the tendency which the antecedent of any psychical change has to be followed by its consequent, is proportionate to the persistency of the union between the external things they symbolize; we may note three several modes in which the progression shows itself. There is, first—increase in the *accuracy* with which the inner tendencies are proportioned to the outer persistencies. There is, second—increase in the *number* of cases, differing as to kind but like as to grade of complexity, in which there are inner tendencies answering to outer persistencies. And there is, third—increase in the *complexity* of the coherent states of consciousness, answering to coherent complexities in the environment. The organism is placed amidst an infinity of relations of all orders. It begins by imperfectly adjusting its actions to a few of the very simplest of these. To adjust its actions more exactly to these few simplest, is one form of advance. To adjust its actions to more and more of these simplest, is another form of advance. To adjust its actions to successive grades of the more complicated, is yet another form of advance. And to whatever stage it reaches, there are still the same three kinds of progression open to it—a perfecting of the correspondences already achieved; an achievement of other corre-spondences of the same order; and an achievement of correspondences of a higher order: all of them implying further fulfilment of the law of intelligence.

But now, what are the conditions to these several kinds of progression? Is the genesis of Intelligence explicable on any one general principle applying at once to all these modes of advance? And if so, what is this general principle?

180. As, in the environment, there exist relations of all orders of persistency, from the absolute to the fortuitous; it follows that in an intelligence displaying any high degree of correspondence, there must exist all grades of strength in the connections between states of consciousness. As a high intelligence is only thus possible, it is manifestly a condition of intelligence in general, that the antecedents and consequents of psychical changes shall admit of all degrees of cohesion. And the fundamental question to be determined, is:—How are these various degrees of cohesion adjusted?

Concerning their adjustment, there appear to be but two possible hypotheses, of which all other hypotheses can be but variations. It may on the one hand be asserted, that the strength of the tendency which each particular state of consciousness has to follow any other, is fixed beforehand by a Creator—that there is a pre-established harmony between the inner and outer relations. On the other hand it may be asserted, that the strength of the tendency which each particular state of consciousness has to follow any other, depends upon the frequency with which the two have been connected in experience—that the harmony between the inner and outer relations, arises from the fact, that the outer relations produce the inner relations. Let us briefly examine these two hypotheses.

The first receives an apparent support from the phenomena of reflex action and instinct; as also from those mental phenomena on which are based the doctrine of "forms of thought." But should these phenomena be otherwise explicable, the hypothesis must be regarded as altogether gratuitous. Of criticisms upon it, the first that may be passed, is, that it has not a single fact to rest upon. These facts that may be cited in its favour, are simply facts which we have not yet found a way to explain; and this alleged explanation of them as due to a pre-established harmony, is simply a disguised mode of shelving them as inexplicable. The theory is much upon a par with that which assigns, as the cause of any unusual phenomenon, "an interposition of Providence;" and the evidence for the one is just as illusive as that for the other. A further criticism is, that even those who lean towards this theory dare not apply it beyond a narrow range of cases. It is only where the connections between psychical states are absolute—as in the so-called forms of thought, and the instinctive actions—that they fall back upon pre-established harmony. But if we assume that the adjustment of inner relations to outer relations, has been in some cases fixed beforehand, we ought in consistency to assume that it has been in all cases fixed beforehand. If, answering to each absolutely persistent connection of phenomena in the environment, there has been provided some absolutely persistent connection between states of consciousness; why, where the outer connection is almost ab-

385

solutely persistent, and the inner connection proportionately persistent, must we not suppose a special provision here also? why must we not suppose special provisions for all the infinitely varied degrees of persistency? The hypothesis, if adopted at all, should be adopted in full. The consistent adoption of it, however, is declined, for sundry very obvious reasons. It would involve the assertion of a rigorous necessity in all thought and action—an assertion to which those leaning towards this hypothesis, are, more than any others, opposed. It would imply that at birth there is just as great a power of thinking, and of thinking correctly, as at any subsequent period. It would imply that men are equally wise concerning things of which they have had no experience, as concerning things of which they have had experience. It would altogether negative the fact, that those who have had a limited and exceptional experience come to erroneous conclusions. It would altogether negative that advance in enlightenment which characterizes human progression. In short, not only is it entirely without foundation in our positive knowledge of mental phenomena; but it necessitates the rejection of all such positive knowledge of mental phenomena as we have acquired.

While, for the first hypothesis, there is no evidence, for the second the evidence is overwhelming. The multitudinous facts commonly cited to illustrate the doctrine of association of ideas, support it. It is in harmony with the general truth, that from the ignorance of the infant the ascent is by slow steps to the knowledge of the adult. All theories and all methods of education take it for granted—are alike based on the belief that the more frequently states of consciousness are made to follow one another in a certain order, the stronger becomes their tendency to suggest one another in that order. The infinitely various phenomena of habit, are so many illustrations of the same law: and in the common sayings—"Practice makes perfect," and "Habit is second nature," we see how long-established and universal is the conviction that such a law exists. We see such a law exemplified in the fact, that men who, from being differently circumstanced, have had different experiences, reach different generalizations; and in the fact that an erroneous connection of ideas will become as firmly established as a correct one, if the external relation to which it answers has been as often repeated. It is in harmony with the familiar truths, that phenomena altogether unrelated in our experience, we have no tendency to think of together; that where a certain phenomenon has within our experience occurred in many relations, we think of it as most likely to recur in the relation in which it has most frequently occurred; that where we have had many agreeing experiences of a certain relation, we come to have a strong belief in that relation; that where a certain relation has been daily experienced throughout our whole lives, with scarcely an exception, it becomes extremely difficult for us to conceive it as otherwise—to break the connection between the states of consciousness representing it; and that where a relation has been perpetually repeated in our experience with absolute uniformity, we are entirely disabled from conceiving the negation of it—it becomes absolutely impossible for us to break the connection between the answering states of consciousness.

The only orders of psychical sequence which do not obviously come within this general law, are those which we class as reflex and instinctive—those which are as well performed on the first occasion as ever afterwards—those which are apparently established antecedent to experience. But there are not wanting facts which indicate that, rightly interpreted, the law covers all these cases too. Though it is manifest that reflex and instinctive sequences are not determined by the experiences of the *individual* organism manifesting them; yet there still remains the hypothesis that they are determined by the experiences of the *race* of organisms forming its ancestry, which by infinite repetition in countless successive generations have established these sequences as organic relations: and all the facts that are accessible to us, go to support this hypothesis. Hereditary transmission, displayed alike in all the plants we cultivate, in all the animals we breed, and in the human race, applies not only to physical but to psychical peculiarities. It is not simply that a modified form of constitution produced by new habits of life, is bequeathed to future generations; but it is that the modified nervous tendencies produced by such new habits of life, are also bequeathed: and if the new habits of life become permanent, the tendencies become permanent. This is illustrated in every creature respecting which we have the requisite experience, from man downwards. Though, among the families of a civilized society, the changes of occupation and habit from generation to generation, and the intermarriage of families having different occupations and habits, very greatly confuse the evidence of psychical transmission; yet, it needs but to consider national characters, in which these disturbing causes are averaged, to see distinctly, that mental peculiarities produced by habit become hereditary. We know that there are warlike, peaceful, nomadic, maritime, hunting, commercial races—races that are independent or slavish, active or slothful,—races that display great varieties of disposition; we know that many of these, if not all, have a common origin; and hence there can be no question that these varieties of disposition, which have a more or less evident relation to habits of life, have been gradually induced and established in successive generations, and have become organic. That is to say, the tendencies to certain combinations of psychical changes have become organic. In the domesticated animals, parallel facts are familiar to all. Not only the forms and constitutions, but the habits, of horses, oxen, sheep, pigs, fowls, have become different from what they were in their wild state. In the various breeds of dogs, all of them according to the test of species derived from one stock, the varieties of mental character and faculty permanently established by mode of life, are numerous; and the several tendencies are spontaneously manifested. A young pointer will point at a covey the first time he is taken afield. A retriever brought up abroad, has been remarked to fulfil his duty without instruction. And in such cases the implication is, that there is a bequeathed tendency for the psychical changes to take place in a special way. Even from the conduct of untamed creatures, we may gather some evidence having like implications. The birds of inhabited countries are far more difficult to approach than those of uninhabited ones. And the manifest inference is, that continued experience of human enmity has produced an organic effect upon them—has modified their instincts—has modified the connections among their psychical states.

75 WILLIAM JAMES (1842–1910) ON THE LIMITATIONS OF ASSOCIATIONISM, 1890

William James, *The Principles of Psychology* (New York, 1890), vol. II, chap. 28.

William James thought associationism had been adequately described by David Hartley, but he saw in this classical doctrine only limited relevance to an understanding of the human mind. Above and beyond its associations, said James, the mind possesses an inborn capacity to perceive certain relations and categories, a genetic endowment not to be explained in Spencer's terms as being inherited associations. James, who felt that these native capacities must have been evolved by natural selection, was utilizing a more modern version of evolutionary theory than had been available to Spencer, for he was writing after Darwin (1859) and after Weismann's theory of the continuity of the germ plasm and his denial of the inheritance of acquired characteristics (1875). By his belief in this sort of innate mental function, James was challenging a quality of associationism that had made it so appealing for so long a time—its all-inclusiveness. He was objecting to a system of psychology that is too simple and mechanical, just as his contemporary in American philosophy, John Dewey, was to do six years later, when he questioned the value of the reflex in psychological theory (see No. 64).

In this final chapter I shall treat of what has sometimes been called *psychogenesis,* and try to ascertain just how far the connections of things in the outward environment can account for our tendency to think of, and to react upon, certain things in certain ways and in no others, even though personally we have had of the things in question no experience, or almost no experience, at all. It is a familiar truth that some propositions are *necessary*. We *must* attach the predicate 'equal' to the subject 'opposite sides of a parallelogram' if we think those terms together at all, whereas we need not in any such way attach the predicate 'rainy,' for example, to the subject 'to-morrow.' The dubious sort of coupling of terms is universally admitted to be due to 'experience'; the certain sort is ascribed to the 'organic structure' of the mind. This structure is in turn supposed by the so-called *apriorists* to be of transcendental origin, or at any rate not to be explicable by experience; whilst by evolutionary empiricists it is supposed to be also due to experience, only not to the experience of the individual, but to that of his ancestors as far back as one may please to go. Our emotional and instinctive tendencies, our irresistible impulses to couple certain movements with the perception or thought of certain things, are also features of our connate mental structure, and like the necessary judgments, are interpreted by the apriorists and the empiricists in the same warring ways.

I shall try in the course of the chapter to make plain three things:

1) That, taking the word experience as it is universally understood, the experience of the race can no more account for our necessary or *a priori* judgments than the experience of the individual can;

2) That there is no good evidence for the belief that our instinctive reactions are fruits of our ancestors' education in the midst of the same environment, transmitted to us at birth.

3) That the features of our organic mental structure cannot be explained at all by our conscious intercourse with the outer environment, but must rather be understood as congenital variations, 'accidental' in the first instance, but then transmitted as fixed features of the race.

On the whole, then, the account which the apriorists give of the *facts* is that which I defend; although I should contend (as will hereafter appear) for a naturalistic view of their *cause*.

The first thing I have to say is that all schools (however they otherwise differ) must allow that the *elementary qualities* of cold, heat, pleasure, pain, red, blue, sound, silence, etc., are original, innate, or *a priori* properties of our subjective nature, even though they should require the touch of experience to waken them into actual consciousness, and should slumber, to all eternity, without it.

This is so on either of the two hypotheses we may make concerning the relation of the feelings to the realities at whose touch they become alive. For in the first place, if a feeling do *not* mirror the reality which wakens it and to which we say it corresponds, if it mirror no reality

389

whatever outside of the mind, it of course is a purely mental product. By its very definition it can be nothing else. But in the second place, even if it *do* mirror the reality exactly, still it *is* not that reality itself, it is a duplication of it, the result of a mental reaction. And that the mind should have the power of reacting in just that duplicate way can only be stated as a *harmony* between its nature and the nature of the truth outside of it, a harmony whereby it follows that the qualities of both parties match.

The originality of these *elements* is not, then, a question for dispute. *The warfare of philosophers is exclusively relative to their* FORMS OF COMBINATION. The empiricist maintains that these forms can only follow the order of combination in which the elements were originally awakened by the impressions of the external world; the apriorists insist, on the contrary, that *some* modes of combination, at any rate, follow from the natures of the elements themselves, and that no amount of experience can modify this result.

WHAT IS MEANT BY EXPERIENCE?

The phrase 'organic mental structure' names the matter in dispute. Has the mind such a structure or not? Are its contents *arranged* from the start, or is the arrangement they may possess simply due to the shuffling of them by experience in an absolutely plastic bed? Now the first thing to make sure of is that when we talk of 'experience,' we attach a definite meaning to the word. *Experience means experience of something foreign supposed to impress us,* whether spontaneously or in consequence of our own exertions and acts. Impressions, as we well know, affect certain orders of sequence and coexistence, and the mind's habits copy the habits of the impressions, so that our images of things assume a time- and space-arrangement which resembles the time- and space-arrangements outside. To uniform outer coexistences and sequences correspond constant conjunctions of ideas, to fortuitous coexistences and sequences casual conjunctions of ideas. We are sure that fire will burn and water wet us, less sure that thunder will come after lightning, not at all sure whether a strange dog will bark at us or let us go by. In these ways experience moulds us every hour, and makes of our minds a mirror of the time- and space-connections between the things in the world. The principle of habit within us so *fixes* the copy at last that we find it difficult even to imagine how the outward order could possibly be different from what it is, and we continually divine from the present what the future is to be. These habits of transition, from one thought to another, are features of mental structure which were lacking in us at birth; we can see their growth under experience's moulding finger, and we can see how often experience undoes her own work, and for an earlier order substitutes a new one. '*The order of experience,*' in this matter of the time- and space-conjunctions of things, is thus an indisputably *vera causa* of our forms of thought. It is our educator, our sovereign helper and friend; and its name, standing for something with so real and definite a use, ought to be kept sacred and encumbered with no vaguer meaning.

If *all* the connections among ideas in the mind could be interpreted as so many combinations of sense-data wrought into fixity in this way from without, then experience in the common and legitimate sense of the word would be the sole fashioner of the mind.

The empirical school in psychology has in the main contended that they can be so interpreted. Before our generation, it was the experience of the individual only which was meant. But when one nowadays says that the human mind owes its present shape to experience, he means the experience of ancestors as well. Mr. Spencer's statement of this is the earliest emphatic one.

· · · · · ·

TWO MODES OF ORIGIN OF BRAIN STRUCTURE

. . . The 'experience-philosophy' has from time immemorial been the opponent of theological modes of thought. The word experience has a halo of anti-supernaturalism about it; so that if anyone express dissatisfaction with any function claimed for it, he is liable to be treated as if he could only be animated by loyalty to the catechism, or in some way have the interests of obscurantism at heart. I am entirely certain that, on this ground alone, what I have erelong to say will make this a sealed chapter to many of my readers. "He denies experience!" they will exclaim, "denies science; believes the mind created by miracle; is a regular old partisan of innate ideas! That is enough! we'll listen to such antediluvian twaddle no more." Regrettable as is the loss of readers capable of such wholesale discipleship, I feel that a definite meaning for the word experience is even more important than their company. 'Experience' does not mean every natural, as opposed to every supernatural, cause. It means a particular sort of natural agency, alongside of which other more recondite natural agencies may perfectly well exist. With the scientific animus of anti-supernaturalism we ought to agree, but we ought to free ourselves from its verbal idols and bugbears.

Nature has many methods of producing the same effect. She may make a 'born' draughtsman or singer by tipping in a certain direction at an opportune moment the molecules of some human ovum; or she may bring forth a child ungifted and make him spend laborious but successful years at school. She may make our ears ring by the sound of a bell, or by a dose of quinine; make us see yellow by spreading a field of buttercups before our eyes, or by mixing a little santonine powder with our food; fill us with terror of certain surroundings by making them really dangerous, or by a blow which produces a pathological alteration of our brain. It is obvious that we need two words to designate these two modes of operating. *In the one case the natural agents produce perceptions which take cognizance of the agents themselves; in the other case, they produce perceptions which take cognizance of something else.* What is taught to the mind by the 'experience,' in the first case, is the *order of the experience itself*—the 'inner relation' (in Spencer's phrase) 'corresponds' to the 'outer relation' which produced it, by remembering and knowing the latter. But in the case of the *other* sort of natural agency, what is taught to the mind has nothing to do with the agency itself, but with some different outer relation altogether. A diagram will express the alternatives. *B* stands for our human brain in the midst

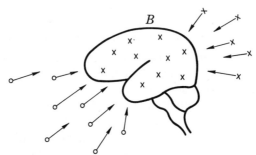

of the world. All the little *o*'s with arrows proceeding from them are natural objects (like sunsets, etc.), which impress it through the senses, and in the strict sense of the word give it *experience,* teaching it by habit and association what is the order of their ways. All the little *x*'s inside the brain and all the little *x*'s outside of it are other natural objects and processes (in the ovum, in the blood, etc.), which equally modify the brain, but mould it to no cognition of *themselves.* The *tinnitus aurium* discloses no properties of the quinine; the musical endowment teaches no embryology; the morbid dread (of solitude, perhaps) no brain-pathology; but the way in which a dirty sunset and a rainy morrow hang together in the mind copies and teaches the sequences of sunsets and rainfall in the outer world.

In zoological evolution we have two modes in which an animal race may grow to be a better match for its environment.

First, the so-called way of 'adaptation,' in which the environment may itself modify its inhabitant by exercising, hardening, and habituating him to certain sequences, and these habits may, it is often maintained, become hereditary.

Second, the way of 'accidental variation,' as Mr. Darwin termed it, in which certain young are born with peculiarities that help them and their progeny to survive. That variations of *this* sort tend to become hereditary, no one doubts.

The first mode is called by Mr. Spencer direct, the second indirect, equilibration. Both equilibrations must of course be natural and physical processes, but they belong to entirely different physical spheres. The direct influences are obvious and accessible things. The causes of variation in the young are, on the other hand, molecular and hidden. The direct influences are the animal's 'experiences,' in the widest sense of the term. Where what is influenced by them is the *mental* organism, they are *conscious* experiences, and become the *objects* as well as the causes of their effects. That is, the effect consists in a tendency of the experience itself to be remembered, or to have its elements thereafter coupled in imagination just as they were coupled in the experience. In the diagram these experiences are represented by the *o*'s exclusively. The *x*'s, on the other hand, stand for the indirect causes of mental modification—causes of which we are not immediately conscious as such, and which are not the direct *objects* of the effects they produce. Some of them are molecular accidents before birth; some of them are collateral and remote combinations, unintended combinations, one might say, of more direct effects wrought in the unstable and intricate brain-tissue. Such a result is unquestionably the susceptibility to music, which some individuals possess at the present day. It has no zoological utility; it corresponds to no object in the natural environment; it is a pure *incident* of having a hearing organ, an incident depending on such instable and inessential conditions that one brother may have it and another brother not. Just so with the susceptibility to sea-sickness, which, so far from being engendered by long experience of its 'object' (if a heaving deck can be called its object) is erelong annulled thereby. Our higher æsthetic, moral, and intellectual life seems made up of affections of this collateral and incidental sort, which have entered the mind by the back stairs, as it were, or rather have not entered the mind at all, but got surreptitiously born in the house. No one can successfully treat of psychogenesis, or the factors of mental evolution, without distinguishing between these two ways in which the mind is assailed. The way of 'experience' proper is the front door, the door of the five senses. The agents which affect the brain in this way immediately become the mind's *objects*. The other

393

agents do not. It would be simply silly to say of two men with perhaps equal effective skill in drawing, one an untaught natural genius, the other a mere obstinate plodder in the studio, that both alike owe their skill to their 'experience.' The reasons of their several skills lie in wholly disparate natural cycles of causation.

I will then, with the reader's permission, *restrict the word 'experience' to processes which influence the mind by the front-door-way of simple habits and association.* What the back-door-effects may be will probably grow clearer as we proceed; so I will pass right on to a scrutiny of the actual mental structure which we find.

THE GENESIS OF THE ELEMENTARY MENTAL CATEGORIES

We find: 1. Elementary sorts of sensation, and feelings of personal activity;

2. Emotions; desires; instincts; ideas of worth; æsthetic ideas;

3. Ideas of time and space and number;

4. Ideas of difference and resemblance, and of their degrees.

5. Ideas of causal dependence among events; of end and means; of subject and attribute.

6. Judgments affirming, denying, doubting, supposing any of the above ideas.

7. Judgments that the former judgments logically involve, exclude, or are indifferent to, each other.

Now we may postulate at the outset that all these forms of thought have a *natural* origin, if we could only get at it. That assumption must be made at the outset of every scientific investigation, or there is no temptation to proceed. But the first account of their origin which we are likely to hit upon is a snare. All these mental affections are ways of knowing objects. Most psychologists nowadays believe that the objects first, in some natural way, engendered a brain from out of their midst, and then imprinted these various cognitive affections upon it. But how? The ordinary evolutionist answer to this question is exceedingly simple-minded. The idea of most speculators seems to be that, since it suffices *now* for us to become acquainted with a complex object, that it should be simply *present* to us often enough, so it must be fair to assume universally that, with time enough given, the *mere presence* of the various objects and relations to be known must end by bringing about the latter's cognition, and that in this way all mental structure was from first to last evolved. Any ordinary Spencerite will tell you that just as the experience of blue objects wrought into our mind the color blue, and hard objects got it to feel hardness, so the presence of large and small objects in the world gave it the notion of size, moving objects made it aware of motion, and objective successions taught it time. Similarly in a world with different impressing things, the mind had to acquire a sense of difference, whilst the like parts of the world as they fell upon it kindled in it the perception of similarity. Outward sequences which sometimes held good, and sometimes failed, naturally engendered in it doubtful and uncertain forms of expectation, and ultimately gave rise to the disjunctive forms of judgment; whilst the hypothetic form, 'if *a*, then *b*,' was sure to ensue from sequences that were invariable in the outer world. On this view, if the outer order suddenly were to change its elements and modes, we should have no faculties to cognize the new order by. At most we should feel a sort of frustration and confusion. But little by little the new presence would work on us as the old one did; and in course of time another set of psychic categories would arise, fitted to take cognizance of the altered world.

This notion of the outer world inevitably building up a sort of mental duplicate of itself if we only give it time, is so easy and natural in its vagueness that one hardly knows how to start to criticise it. One thing, however, is obvious, namely that *the manner in which we now become acquainted with complex objects need not in the least resemble the manner in which the original elements of our consciousness grew up.* Now, it is true, a new sort of animal need only be present to me, to impress its image permanently on my mind; but this is because I am already in possession of categories for knowing each and all of its several attributes, and of a memory for retracing the order of their conjunction. I now have preformed categories for all possible objects. The objects need only awaken these from their slumber. But it is a very different matter to account for the categories themselves. I think we must admit that the origin of the various elementary feelings is a recondite history, even after some sort of neural tissue is there for the outer world to begin its work on. The mere existence of things to be known is even now not, as a rule, sufficient to bring about a knowledge of them. Our abstract and general discoveries usually come to us as lucky fancies; and it is only *après coup* that we find that they correspond to some reality. What immediately produced them were previous thoughts, with which, and with the brain-processes of which, that reality had naught to do.

Why may it not have been so of the original elements of consciousness, sensation, time, space, resemblance, difference, and other relations? Why may they not have come into being by the back-door method, by such physical processes as lie more in the sphere of morphological accident, of inward summation of effects, than in that of the 'sensible presence' of objects? Why may they not, in short, be pure *idiosyncrasies,* spontaneous variations, fitted by good luck (those of them which have survived) to take cognizance of objects (that is, to steer us in our active dealings with them), without being in any intelligible sense immediate derivatives from them? I think we shall find this view gain more and more plausibility as we proceed.

All these elements are subjective duplicates of outer objects. They *are* not the outer objects. The secondary qualities among them are not supposed by any educated person even to resemble the objects. Their *nature* depends more on the reacting brain than on the stimuli which touch it off. This is even more palpably true of the natures of pleasure and pain, effort, desire and aversion, and of such feelings as those of cause and substance, of denial and of doubt. Here then is a native wealth of inner forms whose origin is shrouded in mystery, and which at any rate were not simply 'impressed' from without, in any intelligible sense of the verb 'to impress.'

Their *time- and space-relations,* however, *are* impressed from without—for two outer things at least the evolutionary psychologist must believe to resemble our thoughts of them, these are the time and space in which the objects lie. *The time- and space-relations between things do stamp copies of themselves within.* Things juxtaposed in space impress us, and continue to be thought, in the relation in which they exist there. Things sequent in time, ditto. And thus, through experience in the legitimate sense of the word, there can be truly explained an immense number of our mental habitudes, many of our abstract beliefs, and all our ideas of concrete things, and of their ways of behavior. Such truths as that fire burns and water wets, that glass refracts, heat melts snow, fishes live in water and die on land, and the like, form no small part of the most refined education, and are the all-in-all of education amongst the brutes and lowest men. Here the mind is passive and tributary, a servile copy, fatally and unresistingly fashioned from without. It is the merit of the associationist school to have seen the wide scope of these effects of neighborhood in time and space; and their exaggerated applications of the principle of mere neighborhood ought not to blind us to the excellent service it has done to Psychology in their hands. As far as a large part of our thinking goes, then, it can intelligibly be formulated as a mere lot of *habits* impressed upon us from without. The degree of cohesion of our inner relations, is, in this part of our thinking, proportionate, in Mr. Spencer's phrase, to the degree of cohesion of the outer relations; the causes and the objects of our thought are one; and we are, in so far forth, what the materialistic evolutionists would have us altogether, mere offshoots and creatures of our environment, and naught besides.

But now the plot thickens, for the images impressed upon our memory by the outer stimuli are not restricted to the mere time- and space-relations, in which they originally came, but revive in various manners (dependent on the intricacy of the brain-paths and the instability of the tissue thereof), and form secondary combinations such as the *forms of judgment,* which, taken *per se,* are not congruent either with the forms in which reality exists or in those in which experiences befall us, but which may nevertheless be explained by the way in which experiences befall in a mind gifted with memory, expectation, and the possibility of feeling doubt, curiosity, belief, and denial. The conjunctions of experience befall more or less invariably, variably, or never. The idea of one term will then engender a fixed, a wavering, or a negative expectation of another, giving affirmative, the hypothetical, disjunctive, interrogative, and negative judgments, and judgments of actuality and possibility about certain things. The separation of attribute from subject in all judgments (which violates the way in which nature exists) may be similarly explained by the piecemeal order in which our perceptions come to us, a vague nucleus growing gradually more detailed as we attend to it more and more. These particular secondary mental forms have had ample justice done them by associationists from Hume downwards.

Associationists have also sought to account for discrimination, abstraction, and generalization by the rates of frequency in which attributes come to us conjoined. With much less success, I think . . .

THE GENESIS OF THE NATURAL SCIENCES

Our 'scientific' ways of thinking the outer reality are highly abstract ways. The essence of things for science is not to be what they seem, but to be atoms and molecules moving to and from each other according to strange laws. Nowhere does the account of inner relations produced by outer ones in proportion to the frequency with which the latter have been met, more egregiously break down than in the case of scientific conceptions. The order of scientific thought is quite incongruent either with the way in which reality exists or with the way in which it comes before us. Scientific thought goes by selection and emphasis exclusively. We break the solid plenitude of fact into separate essences, conceive generally what only exists particularly, and by our classifications leave nothing in its natural neighborhood, but separate the contiguous, and join what the poles divorce. The reality *exists* as a *plenum.* All its parts are contemporaneous, each is as real as any other, and each as essential for making the whole just what it is and nothing else. But we can neither experience nor think this *plenum.* What we experience, what *comes before us,* is a chaos of fragmentary impressions interrupting each other; what we *think* is an abstract system of hypothetical data and laws.

This sort of scientific algebra, little as it immediately resembles the reality given to us, turns out (strangely enough) applicable to it. That is, it yields expressions which, at given places and times, can be translated into real values, or interpreted as definite portions of the chaos that falls upon our sense. It becomes thus a practical guide to our expectations as well as a theoretic delight. But I do not see how any one with a sense for the facts can possibly call our systems immediate results of 'experience' in the ordinary sense. Every scientific conception is in the first instance a 'spontaneous variation' in some one's brain. For one that proves useful and applicable there are a thousand that perish through their worthlessness. Their genesis is strictly akin to that of the flashes of poetry and sallies of wit to which the instable brain-paths equally give rise. But whereas the poetry and wit (like the science of the ancients) are their 'own excuse for being,' and have to run the gauntlet of no farther test, the 'scientific' conceptions must prove their worth by being 'verified.' This test, however, is the cause of their *preservation,* not that of their production; and one might as well account for the origin of Artemus Ward's jokes by the 'cohesion' of subjects with predicates in proportion to the 'persistence of the outer relations' to which they 'correspond' as to treat the genesis of scientific conceptions in the same ponderously unreal way.

The most persistent outer relations which science believes in are never matters of experience at all, but have to be disengaged from under experience by a process of elimination, that is, by ignoring conditions which are always present. The *elementary* laws of mechanics, physics, and chemistry are all of this sort. The principle of uniformity in nature is of this sort; it has to be *sought* under and in spite of the most rebellious appearances; and our conviction of its truth is far more like a religious faith than like assent to a demonstration. The only cohesions which experience in the literal sense of the word produces in our mind are, as we contended some time back, the proximate laws of nature, and habitudes of concrete things, that heat melts ice, that salt preserves meat, that fish die out of water, and the like. Such 'empirical truths' as these we admitted to form an enormous part of human wisdom. The 'scientific' truths have to harmonize with these truths, or be given up as useless; but they arise in the mind in no such passive associative way as that in which the simpler truths arise. Even those experiences which are used to prove a scientific truth are for the most part artificial experiences of the laboratory gained after the truth itself has been conjectured. Instead of experiences engendering the 'inner relations,' the inner relations are what engender the experiences here.

What happens in the brain after experience has done its utmost is what happens in every material mass which has been fashioned by an outward force,—in every pudding or mortar, for example, which I may make with my hands. The fashioning from without brings the elements into collocations which set new internal forces free to exert their effects in turn. And the random irradiations and resettlements of our ideas, which *supervene upon experience,* and constitute our free mental play, are due entirely to these secondary internal processes, which vary enormously from brain to brain, even though the brains be exposed to exactly the same 'outer relations.' The higher thought-processes owe their being to causes which correspond far more to the sourings and fermentations of dough, the setting of mortar, or the subsidence of sediments in mixtures, than to the manipulations by which these physical aggregates came to be compounded. Our study of similar association and reasoning taught us that the whole superiority of man depended on the facility with which in his brain the paths worn by the most frequent outer cohesions could be ruptured. The causes of the instability, the reasons why now this point and now that become in him the seat of rupture, we saw to be entirely obscure . . . The only clear thing about the peculiarity seems to be its interstitial character, and the certainty that no mere appeal to man's 'experience' suffices to explain it.

76   WILHELM WUNDT (1832–1920) ON PSYCHOLOGICAL ANALYSIS AND CREATIVE SYNTHESIS, 1896

Wilhelm Wundt, *Grundriss der Psychologie* (Leipzig, 1896), sects. 16 and 23. Translated by C. H. Judd as *Outlines of Psychology* (London and New York, 1897).

The historical course of associationism divides after John Stuart Mill. One line continues through Spencer and James and on to Dewey, American functional psychology, and Gestalt psychology, although these last two movements continue and reinforce James's protest against associationism. The other line runs through Wundt to Külpe in his early days (1893) and to Titchener, after which it died out as a systematic issue. The line from Locke through Hume and the Mills to Wundt is clear. Wundt was the elementist *par excellence.* For him psychology studies the mental elements (sensations, images, feelings) and their combinations into *Vorstellungen* (perceptions, ideas). Our excerpt discusses the associative principles of combination. Wundt would prefer to limit the term *association* to successive association, such as occurs in learning, memory, and reproduction. Such events form, he thinks, rather loose associations. The simultaneous compounds, however, are tightly knit. They are fusions, like the union of tones in a chord, or spatial assimilations, like the visual patterns that constitute geometric illusions, or complications, like the combination of visual, tactual, and auditory elements in holding a screaming child. Such combinations operate by creative synthesis to determine new *Vorstellungen,* the psychic resultants of associative processes. Wundt's psychology is a kind of mental chemistry and was reinforced by the contemporaneous development of atomic chemistry and the formulation by Mendeleev of the periodic law of the chemical elements (1871), only three years before Wundt published the first edition of his classical mental chemistry, *Grundzüge der physiologischen Psychologie.*

SECTION 16

. . . . .

3. . . . The concept of association can gain a fixed, and in any particular case unequivocal, significance, only when association is regarded as an *elementary process* which never shows itself in the actual psychical processes except in a more or less complex form, so that the only way to find out the character of elementary association is to subject its complex products to a psychological analysis. The ordinarily so-called associations (the successive associations) are only one, and the loosest at that, of all the forms of combination. In contrast with these we have the closer combinations from which the different kinds of psychical compounds arise and to which we apply the general name *fusions,* because of the closeness of the union . . . The elementary processes from which the various compounds, the intensive, spacial, and temporal ideas, the composite feelings, the emotions, and the volitional processes arise, are, accordingly, to be considered as associative processes. For the purpose of practical discrimination, however, it will be well to limit the word "association" to those combining processes which take place between elements of *different* compounds. This narrower meaning which we give the term association in contrast with fusion, is in one respect an approach to the meaning that it had in older psychology . . . for it refers exclusively to the interconnection of psychical compounds in consciousness. It differs from the older concept, however, in two important characteristics. First it is here regarded as an *elementary process,* or, when we are dealing with complex phenomena, as a product of such elementary processes. Secondly, we recognize, just as in the case of fusions, simultaneous associations as well as successive. In fact, the former are to be looked upon as the earlier.

A. SIMULTANEOUS ASSOCIATIONS

4. Simultaneous associations made up of elements from different psychical compounds may be divided into *two* classes: into *assimilations,*

or associations between the elements of *like* compounds, and *complications,* or associations between elements of *unlike* compounds. Both may take place, in accordance with our limitation of the concept association, between those compounds only which are themselves simultaneous combinations, that is, between intensive and spacial ideas and between composite feelings.

a. Assimilations

5. *Assimilations* are a form of association that is continually met with, especially in the case of intensive and spacial ideas. It is an essential supplement to the process of formation of ideas by fusion. In the case of composite feelings this form of combination never seems to appear except where we have at the same time an assimilation of the ideational elements. It is most clearly demonstrable when certain single components of the product of an assimilation are given through external sense-impressions, while others belong to earlier ideas. In such a case the assimilation may be demonstrated by the fact that certain components of the idea which are wanting in the objective impression or are there represented by components other than those actually present in the idea itself, can be shown to arise from earlier ideas. Experience shows that of these reproduced components those are most favored which are very frequently present. Still, certain single elements of the impression are usually of more importance in determining the association than others are, so that when these dominating elements are altered, as may be the case especially with assimilations of the visual sense, the product of the assimilation undergoes a corresponding change.

6. Among intensive compounds it is especially the *auditory ideas* which are very often the results of assimilation. They also furnish the most striking examples for the principle of frequency mentioned above. Of all the auditory ideas the most familiar are the readily available *ideas of words,* for these are attended to more than other sound-impressions. As a result the hearing of words is continually accompanied by assimilations; the sound-impression is incomplete, but it is entirely filled out by earlier impressions, so that we do not notice the incompleteness. So it comes that not the correct hearing of words, but the *misunderstanding* of them, that is, the erroneous filling out of incomplete impressions through incorrect assimilations, is what generally leads us to notice the process. We may find an expression of the same fact in the ease with which any sound whatever, as, for example, the cry of an animal, the noise of water, wind, machinery, etc., can be made to sound like words almost at will.

7. In the case of *intensive feelings* we note the presence of assimilations

in the fact that impressions which are accompanied by sense-feelings and elementary aesthetic feeling, very often exercise a second direct affective influence for which we can account only when we recall certain ideas of which we are reminded by the impressions. In such cases the association is usually at first only a form of affective association, and only so long as this is true is the assimilation simultaneous. The ideational association which explains the effect is, on the contrary, a later process belonging to the forms of successive association. For this reason it is hardly possible, when we have clang-impressions or color-impressions accompanied by particular feelings, or when we have simple spacial ideas, to decide what the immediate affective influence of the impression itself is and what is that of the association. As a rule, in such cases the affective process is to be looked upon as the resultant of an immediate and an associative factor which unite to form a single, unitary total feeling in accordance with the general laws of affective fusion . . .

8. Association in the case of *spacial* ideas is of the most comprehensive character. It is not very noticeable in the sphere of *touch* when vision is present, on account of the small importance of tactual ideas in general and especially for memory. For the blind, on the other hand, it is the essential means for the rapid orientation in space which is necessary, for example, in the rapid reading of the blind-alphabet. The effects of assimilation are most strikingly evident when several tactual surfaces are concerned, because in such cases its presence is easily betrayed by the illusions which may arise in consequence of some disturbance in the usual inter-relation of the sensations. Thus, for example, when we touch a small ball with the index and middle fingers crossed, we have the idea of *two* balls. The explanation is obvious. In the ordinary position of the fingers the external impression here given actually corresponds to two balls, and the many perceptions of this kind that have been received before, exercise an assimilative action on the new impression.

9. In *visual* sense-perceptions assimilative processes play a very large part. Here they aid in the formation of ideas of the magnitude, distance, and three-dimensional character of visual objects. In this last respect they are essential supplements of immediate binocular motives for projection into depth. Thus, the correlation that exists between the ideas of the distance and magnitude of objects, as, for example, the apparent difference in the size of the sun or moon on the horizon and at the zenith, is to be explained as an effect of assimilation. The perspective of drawing and painting also depends on these influences. A picture drawn or painted on a plane surface can appear three-dimensional only on condition that the impression arouses earlier three-dimensional ideas which are assimilated with the new impression. The influence of these

assimilations is most evident in the case of unshaded drawings that can be seen either in relief or in intaglio. Observation shows that these differences in appearance are by no means accidental or dependent on the so-called "power of imagination", but that there are always elements in the immediate impression which determine completely the assimilative process. The elements that are thus operative are, above all, the sensations arising from the position and movements of the eye. Thus, for example, a linear design which can be interpreted as either a solid or a hollow prism, is seen alternately in relief and in intaglio according as we fixate in the two cases the parts of the drawing which correspond ordinarily to a solid or to a hollow object. A solid angle represented by three lines in the same plane appears in relief when the fixation-point is moved along one of the lines, starting from the apex, it appears in intaglio when the movement is in the opposite direction, from the end of the line towards the apex. In these and all like cases the assimilation is determined by the rule that in its movement over the fixation-lines of objects the eye always passes from nearer to more distant points.

In other cases the geometric optical illusions . . . which are due to the laws of ocular movements, produce secondarily certain ideas of distance, and these not infrequently eliminate the contradictions brought about in the picture by the illusions. Thus, to illustrate, an interrupted straight line appears longer than an equal uninterrupted line . . . ; as a result we tend to project the first to a greater depth than the latter. Here both lines cover just the same distances on the retina in spite of the fact that their length is perceived as different, because of the different motor energy connected with their estimation. An elimination of this contradiction is effected by means of the different ideas of distance, for when one of two lines whose retinal images are like, appears longer than the other, it must, under the ordinary conditions of vision, belong to a more distant object. Again, when one straight line is intersected at an acute angle by another, the result is an overestimation of the acute angle, which sometimes gives rise, when the line is long, to an apparent bending near the point of intersection . . . Here too the contradiction between the course of the line and the increase in the size of the angle of intersection, is often eliminated by the apparent extension of the line in the third dimension. In all these cases the perspective can be explained only as the assimilative effect of earlier ideas of corresponding character.

·   ·   ·   ·   ·

b. Complications

12. *Complications,* or the combinations between unlike psychical compounds, are no less regular components of consciousness than are as-

similations. Just as there is hardly an intensive or extensive idea or composite feeling which is not modified in some way through the processes of reciprocal assimilation with memory-elements, so almost every one of these compounds is at the same time connected with other, dissimilar compounds, with which it has some constant relations. In all cases, however, complications are different from assimilations in the fact that the unlikeness of the compounds makes the connection looser, however regular it may be, so that when one component is direct and the other reproduced, the latter can be readily distinguished at once. Still, there is another reason which makes the product of a complication appear unitary in spite of the easily recognized difference between its components. This cause is the *predominance* of *one* of the compounds, which pushes the other components into the obscurer field of consciousness.

If the complication unites a direct impression with memory-elements of disparate character, the direct impression with its assimilations is regularly the predominant component, while the reproduced elements sometimes have a noticeable influence only through their affective tone. Thus, when we speak, the auditory word-ideas are the predominant components, and in addition we have as obscure factors direct motor sensations and reproductions of the visual images of the words. In reading, on the other hand, the visual images come to the front while the rest become weaker. In general it may be said that the existence of a complication is frequently noticeable only through the peculiar coloring of the total feeling that accompanies the predominant idea. This is due to the ability of obscure ideas to have a relatively intense effect on the attention through their affective tones . . . Thus, for example, the characteristic impression of a rough surface, a dagger-point, or a gun, arises from a complication of visual and tactual impressions, and in the last case of auditory impressions as well; but as a rule such complications are noticeable only through the feelings they excite.

B. SUCCESSIVE ASSOCIATIONS

13. Successive association is by no means a process that differs essentially from the two forms of simultaneous association, assimilation and complication. It is, on the contrary, due to the same general causes as these, and differs only in the secondary characteristic that the process of combination, which in the former cases consisted, so far as immediate introspection was concerned, of a single instantaneous act, is here protracted and may therefore be readily divided into *two* acts. The first of these acts corresponds to the appearance of the *reproducing* elements, the second to the appearance of the *reproduced* elements. Here too, the

first act is often introduced by an external sense-impression, which is as a rule immediately united with an assimilation. Other reproductive elements which might enter into an assimilation or complication are held back through some inhibitory influence or other—as, for example, through other assimilations that force themselves earlier on apperception—and do not begin to exercise an influence until later. In this way we have a second act of apperception clearly distinct from the first, and differing from it in sensational content the more essentially the more numerous the new elements are which are added through the retarded assimilation and complication and the more these new elements tend to displace the earlier because of their different character.

14. In the great majority of cases the association thus formed is limited to *two* successive ideational or affective processes connected, in the manner described, through assimilations or complications. New sense-impressions or some apperceptive combinations . . . may then connect themselves with the second member of the association. Less frequently it happens that the same processes which led to the first division of an assimilation or complication into a successive process, may be repeated with the second or even with the third member, so that in this way we have a whole *associational series.* Still, this takes place generally only under exceptional conditions, especially when the normal course of apperception has been disturbed, as, for example, in the so-called "flight of ideas" of the insane. In normal cases such serial associations, that is, associations with more than two members, hardly ever appear.

.   .   .   .   .

SECTION 23

.   .   .   .   .

2. The *law of psychical resultants* finds its expression in the fact that every psychical compound shows attributes which may indeed be understood from the attributes of its elements after these elements have once been presented, but which are by no means to be looked upon as the mere sum of the attributes of these elements. A compound clang is more in its ideational and affective attributes than merely a sum of single tones. In spacial and temporal ideas the spacial and temporal arrangement is conditioned, to be sure, in a perfectly regular way by the co-operation of the elements that make up the idea, but still the arrangement itself can by no means be regarded as a property belonging to the sensational elements themselves. The nativistic theories that assume this, implicate themselves in contradictions that cannot be solved; and besides,

405

in so far as they admit subsequent changes in the original space-perceptions and time-perceptions, they are ultimately driven to the assumption of the rise, to some extent at least, of new attributes. Finally, in the apperceptive functions and in the activities of imagination and understanding, this law finds expression in a clearly recognized form. Not only do the elements united by apperceptive synthesis gain, in the aggregate idea that results from their combination, a new significance which they did not have in their isolated state, but what is of still greater importance, the aggregate idea itself is a new psychical content that was made possible, to be sure, by these elements, but was by no means contained in them. This appears most strikingly in the more complex productions of apperceptive synthesis, as, for example, in a work of art or a train of logical thought.

3. The law of psychical resultants thus expresses a principle which we may designate, in view of its results, as a *principle of creative synthesis.* This has long been recognized in the case of higher mental creations, but generally not applied to the other psychical processes. In fact, through an unjustifiable confusion with the laws of physical causality, it has even been completely reversed. A similar confusion is responsible for the notion that there is a contradiction between the principle of creative synthesis in the mental world and the general laws of the natural world, especially that of the conservation of energy. Such a contradiction is impossible from the outset because the points of view for judgment, and therefore for measurements wherever such are made, are different in the two cases, and must be different, since natural science and psychology deal, not with different contents of experience, but with one and the same content viewed from different sides . . . Physical measurements have to do with *objective masses, forces, and energies.* These are supplementary concepts which we are obliged to use in judging objective experience; and their general laws, derived as they are from experience, must not be contradicted by any single case of experience. Psychical measurements, which are concerned with the comparison of psychical components and their resultants, have to do with *subjective values and ends.* The subjective value of a whole may increase in comparison with that of its components; its purpose may be different and higher than theirs without any change in the masses, forces, and energies concerned. The muscular movements of an external volitional act, the physical processes that accompany sense-perception, association, and apperception, all follow invariably the principle of the conservation of energy. But the mental values and ends that these energies represent may be very different in quantity even while the quantity of these energies remains the same.

XI EVOLUTION AND INDIVIDUAL DIFFERENCES

Throughout the *Origin of Species* (1859), Charles Darwin used three types of evidence for his theory of evolution. First, he collected examples of *variation* from individual to individual in both structure and process, because without variation the evolutionary process cannot isolate optimal configurations. Second, he studied the geological record of past organisms for evidence of *continuity* in structure, since this is an implication of a doctrine that holds that species evolved one from another. And third, he attempted to show that the structures and processes favored by evolution perform some *function* useful to the organism. The concepts of *variation, continuity,* and *function* were each to influence all of biological science. This chapter and the next two outline their influences on psychology.

It was the importance of *variation* in the evolutionary scheme that led Francis Galton (1869) to study the distribution of ability among human beings. He showed that people differ enormously in their intellectual powers and that exceptionally great ability seems to run in families. Later (1883) he devised tests to measure sensory discrimination, hoping in this way to obtain an objective index of mental capacity that was free of the biases inherent in subjective evaluations. Thus the psychology of individual differences, started by Galton, was soon carried forward by James McKeen Cattell (1890), who constructed more elaborate and detailed tests of the sort originated by Galton, and who coined the expression "mental test" as a generic term for these techniques.

In the 1890's, interest in mental testing burgeoned, not only because it suited the evolutionary point of view, but also because it seemed to be a way for academic psychology to make a practical contribution to society by providing it with a means of distinguishing among the capacities of its members. In France, the leader of this new movement was Alfred Binet, who, with his student Victor Henri, wrote a programmatic paper (1895) setting the pattern of French investigations in this field for over a decade. Binet, unlike Galton and Cattell, believed that tests of sensory discrimination would not reveal anything about a person's general mental capacity. He felt, instead, that mental tests should engage the person in the very activities that compose intellectual behavior: reading, memorizing, calculating, judging, and so on. The modern era in mental testing began when the issue dividing Binet and Galton was settled.

Binet's views were soon (1897) to be reinforced by Hermann Ebbinghaus, who was commissioned by the municipal authorities of Breslau in Germany to construct methods for determining whether school children became mentally fatigued during the course of the school day. After considering the various sorts of mental test that were in use at the time, Ebbinghaus concluded that the most sensitive was one that he invented himself, in which the child is required to fill in missing letters in a sample of written prose. The success of this "completion test" was tangible support to Binet's conception of testing. In America, Stella Emily Sharp, working in E. B. Titch-

407

ener's laboratory at Cornell University, did an experiment (1899) designed specifically to compare the two views of mental testing, and once again it was found that Binet seemed to be on the right track. A little later (1901), Clark Wissler, in Cattell's own laboratory at Columbia University, found little support for the approach favored by his mentor. With these and other studies at hand, it was not long before Binet's arguments were generally accepted by psychologists.

Wissler's study has a dual role in the history of mental testing, for it also contains an early application of the technique of mathematical correlation. The tradition of mathematical sophistication in this branch of psy-chology may have been implicit in Galton's flair for creative mathematical reasoning, but it was not achieved until others took it in hand, among them Wissler. In 1904, Charles E. Spearman made his important contribution to this tradition with the extensive correlational study that led to his famous two-factor theory of intelligence.

By 1912, when William Stern published his book, mental testing had passed out of its infancy. It was now securely quantitative, and it had proved itself socially useful. Much of what was lacking in maturity Stern supplied, including, in particular, the concept of the intelligence quotient.

77 CHARLES ROBERT DARWIN (1809–1882) ON THE THEORY OF EVOLUTION, 1859

Charles Darwin, *On the Origin of Species by Means of Natural Selection, or the Preservation of Favoured Races in the Struggle for Life* (London, 1859), chap. 7.

Although Darwin did not originate the modern theory of organic evolution, he developed it to the point where it was accepted by most informed people. He argued that the primary mechanism of evolution is natural selection: the perpetuation of successful forms and the elimination of unsuccessful ones by the mortal competition among organisms for limited supplies of food. The idea of the "struggle for survival," as well as the vivid phrase describing it, Darwin found in Thomas Malthus's *Essay on Population* (1789), in which Malthus had attacked the optimistic faith of his contemporaries in the perfectibility of human society. Malthus had shown that, whereas human populations tend to grow in geometric progressions, increase in food supply is arithmetic, and that therefore populations must inevitably live on the brink of starvation, with only the strongest individuals, or the most cunning, surviving. Darwin, generalizing this Malthusian principle to all living organisms, produced the idea of natural selection. Natural selection was not, however, the only mechanism of evolution for Darwin. In addition, he subscribed to the Lamarckian doctrine (J. B. Lamarck, *Philosophie Zoologique,* 1809) that bodily changes wrought by the experiences of an organism during its lifetime may be transmitted by heredity to subsequent generations—a feature of his theory that is often overlooked.

The *Origin of Species* is largely, but not exclusively, concerned with the evolution of anatomical structures. The evolution of function, or behavior, Darwin discussed in a short chapter entitled "Instinct," from which the following excerpt is taken, and which contains, in spite of its brevity, the three essential ingredients of the theory of evolution—function, variation, and continuity. We know, however, from comments by Darwin's contemporaries that in an early draft of this epoch-making book the chapter on instinct was much longer and crammed with the typically Darwinian abundance of evidence, and, of course, Darwin extended the account of his theories later in *The Descent of Man* (1871) and *The Expression of the Emotions in Man and Animals* (1872).

The subject of instinct might have been worked into the previous chapters; but I have thought that it would be more convenient to treat the subject separately, especially as so wonderful an instinct as that of the hive-bee making its cells will probably have occurred to many readers, as a difficulty sufficient to overthrow my whole theory. I must premise, that I have nothing to do with the origin of the primary mental powers, any more than I have with that of life itself. We are concerned only with the diversities of instinct and of the other mental qualities of animals within the same class.

I will not attempt any definition of instinct. It would be easy to show that several distinct mental actions are commonly embraced by this term; but every one understands what is meant, when it is said that instinct impels the cuckoo to migrate and to lay her eggs in other birds' nests. An action, which we ourselves should require experience to enable us to perform, when performed by an animal, more especially by a very young one, without any experience, and when performed by many individuals in the same way, without their knowing for what purpose it is performed, is usually said to be instinctive . . .

Frederick Cuvier and several of the older metaphysicians have compared instinct with habit. This comparison gives, I think, a remarkably accurate notion of the frame of mind under which an instinctive action is performed, but not of its origin. How unconsciously many habitual actions are performed, indeed not rarely in direct opposition to our conscious will! yet they may be modified by the will or reason. Habits easily become associated with other habits, and with certain periods of time and states of the body. When once acquired, they often remain constant throughout life. Several other points of resemblance between instincts and habits could be pointed out. As in repeating a well-known song, so in instincts, one action follows another by a sort of rhythm; if a person be interrupted in a song, or in repeating anything by rote, he is generally forced to go back to recover the habitual train of thought: so P. Huber found it was with a caterpillar, which makes a very complicated hammock; for if he took a caterpillar which had completed its hammock up to, say, the sixth stage of construction, and put it into a hammock completed up only to the third stage, the caterpillar simply re-performed the fourth, fifth, and sixth stages of construction. If, however, a caterpillar were taken out of a hammock made up, for instance, to the third stage, and were put into one finished up to the sixth stage, so that much of its work was already done for it, far from feeling the benefit of this, it was much embarrassed, and, in order to complete its hammock, seemed forced to start from the third stage, where it had left off, and thus tried to complete the already finished work.

If we suppose any habitual action to become inherited—and I think it can be shown that this does sometimes happen—then the resemblance between what originally was a habit and an instinct becomes so close as not to be distinguished. If Mozart, instead of playing the pianoforte at three years old with wonderfully little practice, had played a tune with no practice at all, he might truly be said to have done so instinctively. But it would be the most serious error to suppose that the greater number of instincts have been acquired by habit in one generation, and then transmitted by inheritance to succeeding generations. It can be clearly shown that the most wonderful instincts with which we are acquainted, namely, those of the hive-bee and of many ants, could not possibly have been thus acquired.

It will be universally admitted that instincts are as important as corporeal structure for the welfare of each species, under its present conditions of life. Under changed conditions of life, it is at least possible that slight modifications of instinct might be profitable to a species; and if it can be shown that instincts do vary ever so little, then I can see no difficulty in natural selection preserving and continually accumulating variations of instinct to any extent that may be profitable. It is thus, as I believe, that all the most complex and wonderful instincts have originated. As modifications of corporeal structure arise from, and are increased by, use or habit, and are diminished or lost by disuse, so I do not doubt it has been with instincts. But I believe that the effects of habit are of quite subordinate importance to the effects of the natural selection of what may be called accidental variations of instincts;—that is of variations produced by the same unknown causes which produce slight deviations of bodily structure.

No complex instinct can possibly be produced through natural selection, except by the slow and gradual accumulation of numerous, slight, yet profitable, variations. Hence, as in the case of corporeal structures, we ought to find in nature, not the actual transitional gradations by which each complex instinct has been acquired—for these could be found only in the lineal ancestors of each species—but we ought to find in the collateral lines of descent some evidence of such gradations; or we ought at least to be able to show that gradations of some kind are possible; and this we certainly can do. I have been surprised to find, making allowance for the instincts of animals having been but little observed except in Europe and North America, and for no instinct being known amongst extinct species, how very generally gradations, leading to the most complex instincts, can be discovered. The canon of "Natura non facit saltum" applies with almost equal force to instincts as to bodily organs. Changes of instinct may sometimes be facilitated by the same species having different instincts at different periods of life, or at different seasons of the year, or when placed under different circumstances, &c.; in which case either one or the other instinct might be preserved by natural selection. And such instances of diversity of instinct in the same species can be shown to occur in nature.

.   .   .   .   .

As some degree of variation in instincts under a state of nature, and the inheritance of such variations, are indispensable for the action of natural selection, as many instances as possible ought to have been here given; but want of space prevents me. I can only assert, that instincts certainly do vary—for instance, the migratory instinct, both in extent and direction, and in its total loss. So it is with the nests of birds, which vary partly in dependence on the situations chosen, and on the nature and temperature of the country inhabited, but often from causes wholly unknown to us: Audubon has given several remarkable cases of differences in nests of the same species in the northern and southern United States. Fear of any particular enemy is certainly an instinctive quality, as may be seen in nestling birds, though it is strengthened by experience, and by the sight of fear of the same enemy in other animals. But fear of man is slowly acquired, as I have elsewhere shown, by various animals inhabiting desert islands; and we may see an instance of this, even in England, in the greater wildness of all our large birds than of our small birds; for the large birds have been most persecuted by man. We may safely attribute the greater wildness of our large birds to this cause; for in uninhabited islands large birds are not more fearful than small; and the magpie, so wary in England, is tame in Norway, as is the hooded crow in Egypt.

That the general disposition of individuals of the same species, born in a state of nature, is extremely diversified, can be shown by a multitude of facts. Several cases also, could be given, of occasional and strange habits in certain species, which might, if advantageous to the species, give rise, through natural selection, to quite new instincts. But I am well aware that these general statements, without facts given in detail, can produce but a feeble effect on the reader's mind. I can only repeat my assurance, that I do not speak without good evidence.

The possibility, or even probability, of inherited variations of instinct in a state of nature will be strengthened by briefly considering a few cases under domestication. We shall thus also be enabled to see the respective parts which habit and the selection of so-called accidental variations have played in modifying the mental qualities of our domestic animals. A number of curious and authentic instances could be given of the inheritance of all shades of disposition and tastes, and likewise of the oddest tricks, associated with certain frames of mind or periods of time. But let us look to the familiar case of the several breeds of dogs: it cannot be doubted that young pointers (I have myself seen a striking instance) will sometimes point and even back other dogs the very first time that they are taken out; retrieving is certainly in some degree inherited by retrievers; and a tendency to run round, instead of at, a flock of sheep, by shepherd-dogs. I cannot see that these actions, performed without experience by the young, and in nearly the same manner by each individual, performed with eager delight by each breed, and without the end being known,—for the young pointer can no more know that he points to aid his master, than the white butterfly knows why she lays her eggs on the leaf of the cabbage,—I cannot see that these actions differ essentially from true instincts. If we were to see one kind of wolf, when young and without any training, as soon as it scented its prey, stand motionless like a statue, and then slowly crawl forward with a peculiar gait; and another kind of wolf rushing round, instead of at, a herd of deer, and driving them to a distant point, we should assuredly call these actions instinctive. Domestic instincts, as they may be called, are certainly far less fixed or invariable than natural instincts; but they have been acted on by far less rigorous selection, and have been transmitted for an incomparably shorter period, under less fixed conditions of life.

How strongly these domestic instincts, habits, and dispositions are inherited, and how curiously they become mingled, is well shown when different breeds of dogs are crossed. Thus it is known that a cross with a bull-dog has affected for many generations the courage and obstinacy of greyhounds; and a cross with a greyhound has given to a whole family of shepherd-dogs a tendency to hunt hares. These domestic instincts, when thus tested by crossing, resemble natural instincts, which in a like manner become curiously blended together, and for a long period exhibit traces of the instincts of either parent: for example, Le Roy describes a dog, whose great-grandfather was a wolf, and this dog showed a trace of its wild parentage only in one way, by not coming in a straight line to his master when called.

Domestic instincts are sometimes spoken of as actions which have become inherited solely from long-continued and compulsory habit, but this, I think, is not true. No one would ever have thought of teaching, or probably could have taught, the tumbler-pigeon to tumble,—an action which, as I have witnessed, is performed by young birds, that have never seen a pigeon tumble. We may believe that some one pigeon showed a slight tendency to this strange habit, and that the long-continued selection of the best individuals in successive generations made tumblers what they now are; and near Glasgow there are house-tumblers, as I hear from Mr. Brent, which cannot fly eighteen inches high without going head over heels. It may be doubted whether any one would have thought of training a dog to point, had not some one dog naturally shown a tendency in this line; and this is known occasionally to happen, as I once saw in a pure terrier. When the first tendency was once displayed, methodical selection and the inherited effects of compulsory training in each successive generation would soon complete the work; and unconscious selection is still at work, as each man tries to procure, without intending to improve the breed, dogs which will stand and hunt best. On the other hand, habit alone in some cases has sufficed; no animal is more difficult to tame than the young of the wild rabbit; scarcely any animal is tamer than the young of the tame rabbit; but I do not suppose that domestic rabbits have ever been selected for tameness; and I presume that we must attribute the whole of the inherited change from extreme wildness to extreme tameness, simply to habit and long-continued close confinement.

Natural instincts are lost under domestication: a remarkable instance of this is seen in those breeds of fowls which very rarely or never become "broody," that is, never wish to sit on their eggs. Familiarity alone prevents our seeing how universally and largely the minds of our domestic animals have been modified by domestication. It is scarcely possible to doubt that the love of man has become instinctive in the dog. All wolves, foxes, jackals, and species of the cat genus, when kept tame, are most eager to attack poultry, sheep, and pigs; and this tendency has been found incurable in dogs which have been brought home as puppies from countries, such as Tierra del Fuego and Australia, where the savages do not keep these domestic animals. How rarely, on the other hand, do our civilised dogs, even when quite young, require to be taught not to attack poultry, sheep, and pigs! No doubt they occasionally do make an attack, and are then beaten; and if not cured, they are destroyed; so that habit, with some degree of selection, has probably concurred in civilising by inheritance our dogs. On the other hand, young chickens have lost, wholly by habit, that fear of the dog and cat which no doubt was originally instinctive in them, in the same way as it is so plainly instinctive in young pheasants, though reared under a hen. It is not that chickens have lost all fear, but fear only of dogs and cats, for if the hen gives the danger-chuckle, they will run (more especially young turkeys) from under her, and conceal themselves in the surrounding grass or thickets; and this is evidently done for the instinctive purpose of allowing, as we see in wild ground-birds, their mother to fly away. But this instinct retained by our chickens has become useless under domestication, for the mother-hen has almost lost by disuse the power of flight.

Hence, we may conclude, that domestic instincts have been acquired and natural instincts have been lost partly by habit, and partly by man selecting and accumulating during successive generations, peculiar mental habits and actions, which at first appeared from what we must in our ignorance call an accident. In some cases compulsory habit alone has sufficed to produce such inherited mental changes; in other cases compulsory habit has done nothing, and all has been the result of selection, pursued both methodically and unconsciously; but in most cases, probably, habit and selection have acted together.

78 FRANCIS GALTON (1822–1911) ON THE INHERITANCE OF INTELLIGENCE, 1869

Francis Galton, *Hereditary Genius: An Inquiry into Its Laws and Consequences* (London, 1869), chap. 3.

Galton inferred from the theory of evolution put forth by Charles Darwin, his elder half cousin, that human beings may vary in their genetic mental endowments and, moreover, that these variations are inheritable. The outcome of these notions appeared ten years after Darwin's *Origin of Species* in this study of the inheritance of exceptional intellect. As an example of the inequality of mental endowments, Galton described the great variation in mathematical ability shown by the students at Cambridge University, who worked competitively under comparable motivation, and who were after three years carefully ranked. The present excerpt gives the broad range of this distribution of ability and includes Galton's argument that such variation must be apportioned to human beings in accordance with the normal distribution—the same statistical law that Quetelet (1796–1874), the Belgian astronomer, had used to describe the variation in simple bodily measurements. As proof that mental capacity is inherited, Galton then presented the genealogies of families containing men eminent in various fields: legal, military, scientific, artistic, and so on, and showed that eminence tends to run in families. Today, one objects to his use of eminence as a measure of intellectual capacity and to his assumption that what he was measuring was biological inheritance rather than what is due to the social and economic environment from which eminent men are apt to emerge, but Galton's work nevertheless set up the problem of mental inheritance which is still one of psychology's preoccupations.

I have no patience with the hypothesis occasionally expressed, and often implied, especially in tales written to teach children to be good, that babies are born pretty much alike, and that the sole agencies in creating differences between boy and boy, and man and man, are steady application and moral effort. It is in the most unqualified manner that I object to pretensions of natural equality. The experiences of the nursery, the school, the University, and of professional careers, are a chain of proofs to the contrary. I acknowledge freely the great power of education and social influences in developing the active powers of the mind, just as I acknowledge the effect of use in developing the muscles of a black-smith's arm, and no further. Let the blacksmith labour as he will, he will find there are certain feats beyond his power that are well within the strength of a man of herculean make, even although the latter may have led a sedentary life. Some years ago, the Highlanders held a grand gathering in Holland Park, where they challenged all England to com-pete with them in their games of strength. The challenge was accepted, and the well-trained men of the hills were beaten in the foot-race by a youth who was stated to be a pure Cockney, the clerk of a London banker.

Everybody who has trained himself to physical exercises discovers the extent of his muscular powers to a nicety. When he begins to walk, to row, to use the dumb bells, or to run, he finds to his great delight that his thews strengthen, and his endurance of fatigue increases day after day. So long as he is a novice, he perhaps flatters himself there is hardly an assignable limit to the education of his muscles; but the daily gain is soon discovered to diminish, and at last it vanishes altogether. His maximum performance becomes a rigidly determinate quantity. He learns to an inch, how high or how far he can jump, when he has attained the highest state of training. He learns to half a pound, the force he can exert on the dynamometer, by compressing it. He can strike a blow against the machine used to measure impact, and drive its index to a certain graduation, but no further. So it is in running, in rowing, in walking, and in every other form of physical exertion. There is a definite limit to the muscular powers of every man, which he cannot by any education or exertion overpass.

This is precisely analogous to the experience that every student has had of the working of his mental powers. The eager boy, when he first goes to school and confronts intellectual difficulties, is astonished at his progress. He glories in his newly-developed mental grip and growing capacity for application, and, it may be, fondly believes it to be within his reach to become one of the heroes who have left their mark upon the history of the world. The years go by; he competes in the examinations of school and college, over and over again with his fellows, and soon finds his place among them. He knows he can beat such and such of his competitors; that there are some with whom he runs on equal terms, and others whose intellectual feats he cannot even approach. Probably his vanity still continues to tempt him, by whispering in a new strain. It tells him that classics, mathematics, and other subjects taught in universities, are mere scholastic specialities, and no test of the more valuable intellectual powers. It reminds him of numerous instances of persons who had been unsuccessful in the competitions of youth, but who had shown powers in after-life that made them the foremost men of their age. Accordingly, with newly furbished hopes, and with all the ambition of twenty-two years of age, he leaves his University and enters a larger field of competition. The same kind of experience awaits him here that he has already gone through. Opportunities occur—they occur to every man—and he finds himself incapable of grasping them. He tries, and is tried in many things. In a few years more, unless he is incurably blinded by self-conceit, he learns precisely of what performances he is capable, and what other enterprises lie beyond his compass. When he reaches mature life, he is confident only within certain limits, and knows, or ought to know, himself just as he is probably judged of by the world, with all his unmistakeable weakness and all his undeniable strength. He is no longer tormented into hopeless efforts by the fallacious promptings of overweening vanity, but he limits his undertakings to matters below the level of his reach, and finds true moral repose in an honest conviction that he is engaged in as much good work as his nature has rendered him capable of performing.

There can hardly be a surer evidence of the enormous difference between the intellectual capacity of men, than the prodigious differences in the numbers of marks obtained by those who gain mathematical honours at Cambridge. I therefore crave permission to speak at some length upon this subject, although the details are dry and of little general interest. There are between 400 and 450 students who take their degrees in each year, and of these, about 100 succeed in gaining honours in mathematics, and are ranged by the examiners in strict order of merit. About the first forty of those who take mathematical honours are distinguished by the title of wranglers, and it is a decidedly creditable thing to be even a low wrangler; it will secure a fellowship in a small college.

.   .   .   .   .

. . . The lowest man in the list of honours gains less than 300 marks; the lowest wrangler gains about 1,500 marks; and the senior wrangler, in one of the lists now before me, gained more than 7,500 marks. Consequently, the lowest wrangler has more than five times the merit of the lowest junior optime, and less than one-fifth the merit of the senior wranger. [See the table for the distribution of marks.]

SCALE OF MERIT AMONG THE MEN WHO OBTAIN MATHEMATICAL HONOURS AT CAMBRIDGE.

Number of marks obtained by candidates	Number of candidates in the two years, taken together, who obtained these marks
under 500	24
500 to 1,000	74
1,000 to 1,500	38
1,500 to 2,000	21
2,000 to 2,500	11
2,500 to 3,000	8
3,000 to 3,500	11
3,500 to 4,000	5
4,000 to 4,500	2
4,500 to 5,000	1
5,000 to 5,500	3
5,500 to 6,000	1
6,000 to 6,500	0
6,500 to 7,000	0
7,000 to 7,500	0
7,500 to 8,000	1
	200

The results of two years are thrown into a single table.
The total number of marks obtainable in each year was 17,000.

The precise number of marks obtained by the senior wrangler in the more remarkable of these two years was 7,634; by the second wrangler in the same year, 4,123; and by the lowest man in the list of honours, only 237. Consequently, the senior wrangler obtained nearly twice as many marks as the second wrangler, and more than thirty-two times as many as the lowest man.

.   .   .   .   .

I have not cared to occupy myself much with people whose gifts are below the average, but they would be an interesting study. The number of idiots and imbeciles among the twenty million inhabitants of England and Wales is approximately estimated at 50,000, or as 1 in 400. Dr. Seguin, a great French authority on these matters, states that more than thirty per cent. of idiots and imbeciles, put under suitable instruction, have been taught to conform to social and moral law, and rendered capable of order, of good feeling, and of working like *the third* of an average man. He says that more than forty per cent. have become capable of the ordinary transactions of life, under friendly control; of understanding moral and social abstractions, and of working like *two-thirds* of a man. And, lastly, that from twenty-five to thirty per cent. come nearer and nearer to the standard of manhood, till some of them will defy the scrutiny of good judges, when compared with ordinary young men and women. In the order next above idiots and imbeciles are a large number of milder cases scattered among private families and kept out of sight, the existence of whom is, however, well known to relatives and friends; they are too silly to take a part in general society, but are easily amused with some trivial, harmless occupation. Then comes a class of whom the Lord Dundreary of the famous play may be considered a representative; and so, proceeding through successive grades, we gradually ascend to mediocrity. I know two good instances of hereditary silliness short of imbecility, and have reason to believe I could easily obtain a large number of similar facts.

To conclude, the range of mental power between—I will not say the highest Caucasian and the lowest savage—but between the greatest and least of English intellects, is enormous. There is a continuity of natural ability reaching from one knows not what height, and descending to one can hardly say what depth. I propose in this chapter to range men according to their natural abilities, putting them into classes separated by equal degrees of merit, and to show the relative number of individuals included in the several classes. Perhaps some persons might be inclined to make an offhand guess that the number of men included in the several classes would be pretty equal. If he thinks so, I can assure him he is most egregiously mistaken.

The method I shall employ for discovering all this, is an application of the very curious theoretical law of "deviation from an average." First, I will explain the law, and then I will show that the production of natural intellectual gifts comes justly within its scope.

The law is an exceedingly general one. M. Quetelet, the Astronomer-Royal of Belgium, and the greatest authority on vital and social statistics, has largely used it in his inquiries. He has also constructed numerical tables, by which the necessary calculations can be easily made, whenever it is desired to have recourse to the law.

.   .   .   .   .

. . . Suppose a million . . . men to stand in turns, with their backs against a vertical board of sufficient height, and their heights to be dotted off upon it. The board would then present the appearance shown in the diagram. The line of average height is that which divides the dots into two equal parts, and stands, in the case we have assumed, at the height of sixty-six inches. The dots will be found to be ranged so

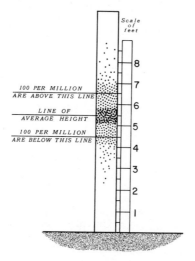

symmetrically on either side of the line of average, that the lower half of the diagram will be almost a precise reflection of the upper. Next, let a hundred dots be counted from above downwards, and let a line be drawn below them. According to the conditions, this line will stand at the height of seventy-eight inches. Using the data afforded by these two lines, it is possible, by the help of the law of deviation from an average, to reproduce, with extraordinary closeness, the entire system of dots on the board.

.   .   .   .   .

The number of grades into which we may divide ability is purely a matter of option. We may consult our convenience by sorting English-men into a few large classes, or into many small ones. I will select a system of classification that shall be easily comparable with the numbers of eminent men, as determined in the previous chapter. We have seen that 250 men per million become eminent; accordingly, I have so con-trived the classes in the following table that the two highest, F and G, together with X (which includes all cases beyond G, and which are un-classed), shall amount to about that number—namely, to 248 per million.

It will, I trust, be clearly understood that the numbers of men in the several classes in my table depend on no uncertain hypothesis. They

CLASSIFICATION OF MEN ACCORDING TO THEIR NATURAL GIFTS.

Grades of natural ability, separated by equal intervals		Numbers of men comprised in the several grades of natural ability, whether in respect to their general powers, or to special aptitudes*	
Below average	Above average	Proportionate, viz. one in	In each million of the same age
a	A	4	256,791
b	B	6	162,279
c	C	16	63,563
d	D	64	15,696
e	E	413	2,423
f	F	4,300	233
g	G	79,000	14
x (all grades below g)	X (all grades above G)	1,000,000	1
		On either side of average	500,000
		Total, both sides	1,000,000

EXAMPLE: The class F contains 1 in every 4,300 men. In other words, there are 233 of that class in each million of men. The same is true of class f . . .

*[Omitted are the estimated numbers of men of each age group and each grade of natural ability then living in England and Wales.]

are determined by the assured law of deviations from an average. It is an absolute fact that if we pick out of each million the one man who is naturally the ablest, and also the one man who is the most stupid, and divide the remaining 999,998 men into fourteen classes, the average ability in each being separated from that of its neighbours by *equal grades,* then the numbers in each of those classes will, on the average of many millions, be as is stated in the table. The table may be applied to special, just as truly as to general ability. It would be true for every examination that brought out natural gifts, whether held in painting, in music, or in statesmanship. The proportions between the different classes would be identical in all these cases, although the classes would be made up of different individuals, according as the examination differed in its purport.

79   GALTON ON MENTAL CAPACITY, 1883

Francis Galton, *Inquiries into Human Faculty and Its Development* (London, 1883), pp. 27–32.

In this book, Galton pursued his interest in human mentality and took several great strides toward modern psychology. The variation in human mental capacity that he described and documented in his earlier book, *Hereditary Genius* (1869), he now tried to measure directly. He argued that people vary in intelligence because they vary in the fineness of their sensory discriminations—a plausible thesis for someone schooled in the dominant British philosophy, which held that all knowledge is derived from the senses. With characteristic ingenuity, he devised methods to assess this fineness quickly and easily, and with characteristic energy tested his methods, with results as described in the excerpt that follows. Although his argument has not prevailed, the attempt at direct measurement was soon to stimulate others to find more adequate methods. This measurement of ability is, however, only a fraction of what posterity has found valuable in Galton's book. Here is an analysis of the varieties of mental imagery, based on investigations using the questionnaire—perhaps the first use in psychology of this now standard technique. Here is the definitive statement of the antithesis between nature and nurture as determinants of human psychology. Here is the classical argument in favor of the practice of eugenics—a term which Galton was the first to use. And, finally, here is the first description of both the technique of free word-association and its potential value in reviving forgotten experiences.

SENSITIVITY

The only information that reaches us concerning outward events appears to pass through the avenue of our senses; and the more perceptive the senses are of difference, the larger is the field upon which our judgment and intelligence can act. Sensation mounts through a series of grades of "just perceptible differences." It starts from the zero of consciousness, and it becomes more intense as the stimulus increases (though at a slower rate) up to the point when the stimulus is so strong as to begin to damage the nerve apparatus. It then yields place to pain, which is another form of sensation, and which continues until the nerve apparatus is destroyed. Two persons may be equally able just to hear the same faint sound, and they may equally begin to be pained by the same loud sound, and yet they may differ as to the number of intermediate grades of sensation. The grades will be less numerous as the organisation is of a lower order, and the keenest sensation possible to it will in consequence be less intense. An artist who is capable of discriminating more differences of tint than another man is not necessarily more capable of seeing clearly in twilight, or more or less intolerant of sunshine. A musician is not necessarily able to hear very faint sounds, nor to be more startled by loud sounds than others are. A mechanic who works hard with heavy tools and has rough and grimy thumbs, insensible to very slight pressures, may yet have a singularly discriminating power of touch in respect to the pressures that he can feel.

The discriminative faculty of idiots is curiously low; they hardly distinguish between heat and cold, and their sense of pain is so obtuse that some of the more idiotic seem hardly to know what it is. In their dull lives, such pain as can be excited in them may literally be accepted with a welcome surprise. During a visit to Earlswood Asylum I saw two boys whose toe-nails had grown into the flesh and had been excised by the surgeon. This is a horrible torture to ordinary persons, but the idiot lads were said to have shown no distress during the operation; it was not necessary to hold them, and they looked rather interested at what was being done . . .

The trials I have as yet made on the sensitivity of different persons confirms the reasonable expectation that it would on the whole be highest among the intellectually ablest. At first, owing to my confusing the quality of which I am speaking with that of nervous irritability, I fancied that women of delicate nerves who are distressed by noise, sunshine, etc., would have acute powers of discrimination. But this I found not to be the case. In morbidly sensitive persons both pain and sensation are induced by lower stimuli than in the healthy, but the number of just perceptible grades of sensation between them is not necessarily different.

I found as a rule that men have more delicate powers of discrimination than women, and the business experience of life seems to confirm this view. The tuners of pianofortes are men, and so I understand are the tasters of tea and wine, the sorters of wool, and the like. These latter occupations are well salaried, because it is of the first moment to the merchant that he should be rightly advised on the real value of what he is about to purchase or to sell. If the sensitivity of women were superior to that of men, the self-interest of merchants would lead to their being always employed; but as the reverse is the case, the opposite supposition is likely to be the true one.

Ladies rarely distinguish the merits of wine at the dinner-table, and though custom allows them to preside at the breakfast-table, men think them on the whole to be far from successful makers of tea and coffee.

Blind persons are reputed to have acquired in compensation for the loss of their eyesight an increased acuteness in their other senses; I was therefore curious to make some trials with my test apparatus, which I will describe in the next chapter. I was permitted to do so on a number of boys at a large educational blind asylum, but found that, although they were anxious to do their best, their performances were by no means superior to those of other boys. It so happened that the blind lads who showed the most delicacy of touch and won the little prizes I offered to excite emulation, barely reached the mediocrity of the various sighted lads of the same age whom I had previously tested. I have made not a few observations and inquiries, and find that the guidance of the blind depends mainly on the multitude of collateral indications to which they give much heed, and not in their superior sensitivity to any one of them. Those who see do not care for so many of these collateral indications, and habitually overlook and neglect several of them. I am convinced also that not a little of the popular belief concerning the sensitivity of the blind is due to exaggerated claims on their part that have not been verified.

. . . . .

Notwithstanding many travellers' tales, I have thus far been unsuccessful in obtaining satisfactory evidence of any general large superiority of the senses of savages over those of civilised men. My own experience, so far as it goes, of Hottentots, Damaras, and some other wild races, went to show that their sense discrimination was not superior to those of white men, even as regards keenness of eyesight. An offhand observer is apt to err by assigning to a single cause what is partly due to others as well. Thus, as regards eyesight, a savage who is accustomed to watch oxen grazing at a distance becomes so familiar with their appearance and habits that he can identify particular animals and draw conclusions as to what they are doing with an accuracy that may seem to strangers to be wholly dependent on exceptional acuteness of vision.

80   JAMES McKEEN CATTELL (1860–1944) ON MENTAL TESTS, 1890

J. McK. Cattell, "Mental tests and measurements," *Mind 15*, 373–380 (1890).

The central theme of Cattell's work was the assessment of human capacity. Thus, although he received his degree under Wundt, he followed the lines of investigation laid out by Galton and, indirectly, by Darwin. In addition, Cattell was one of the entrepreneurs of early American experimental psychology, starting the laboratories at the University of Pennsylvania (1888) and Columbia University (1891), and founding, with James Mark Baldwin, the journal *Psychological Review* (1894).

Cattell coined the expression "mental test" in the paper from which the following excerpt is taken. Like Galton, he strove to establish the normal ranges of variation for human mental capacities—in particular for the simple sensory processes—and he, too, applied statistical techniques to the data he collected. This preoccupation with variation was soon to be known as the psychology of individual differences, which, in the form of the "intelligence test," was to bring academic, experimental psychology to the attention of the general public for the first time. The modern and successful form of the intelligence test was, however, neither Cattell's nor Galton's; it was the work of Binet in France and Stern in Germany during the years between 1895 and 1915.

Psychology cannot attain the certainty and exactness of the physical sciences, unless it rests on a foundation of experiment and measurement. A step in this direction could be made by applying a series of mental tests and measurements to a large number of individuals. The results would be of considerable scientific value in discovering the constancy of mental processes, their interdependence, and their variation under different circumstances. Individuals, besides, would find their tests interesting, and, perhaps, useful in regard to training, mode of life or indication of disease. The scientific and practical value of such tests would be much increased should a uniform system be adopted, so that determinations made at different times and places could be compared and combined. With a view to obtaining agreement among those interested, I venture to suggest the following series of tests and measurements, together with methods of making them.

The first series of ten tests is made in the Psychological Laboratory of the University of Pennsylvania on all who present themselves, and the complete series on students of Experimental Psychology. The results will be published when sufficient data have been collected. Meanwhile, I should be glad to have the tests, and the methods of making them, thoroughly discussed.

The following ten tests are proposed:

    I. Dynamometer Pressure.
    II. Rate of Movement.
    III. Sensation-areas.
    IV. Pressure causing Pain.
    V. Least noticeable difference in Weight.
    VI. Reaction-time for Sound.
    VII. Time for naming Colours.
    VIII. Bi-section of a 50 cm. line.
    IX. Judgment of 10 seconds time.
    X. Number of Letters remembered on once Hearing.

It will be noticed that the series begins with determinations rather bodily than mental, and proceeds through psychophysical to more purely mental measurements.

    .   .   .   .   .

Let us now consider the tests in order.

I. *Dynamometer Pressure.* The greatest possible squeeze of the hand may be thought by many to be a purely physiological quantity. It is, however, impossible to separate bodily from mental energy. The 'sense of effort' and the effects of volition on the body are among the questions most discussed in psychology and even in metaphysics. Interesting experiments may be made on the relation between volitional control or emotional excitement and dynamometer pressure. Other determinations of bodily power could be made (in the second series I have included the 'archer's pull' and pressure of the thumb and forefinger), but the squeeze of the hand seems the most convenient. It may be readily made, cannot prove injurious, is dependent on mental conditions, and allows comparison of right- and left-handed power. The experimentee should be shown how to hold the dynamometer in order to obtain the maximum pressure. I allow two trials with each hand (the order being right, left, right, left), and record the maximum pressure of each hand.

II. *Rate of Movement.* Such a determination seems to be of considerable interest, especially in connexion with the preceding. Indeed, its physiological importance is such as to make it surprising that careful measurements have not hitherto been made. The rate of movement has the same psychological bearings as the force of movement . . . I am now making experiments to determine the rate of different movements. As a general test, I suggest the quickest possible movement of the right hand and arm from rest through 50 cm . . . An electric current is closed by the first movement of the hand, and broken when the movement through 50 cm. has been completed. I measure the time the current has been closed with the Hipp chronoscope, but it may be done by any chronographic method . . .

III. *Sensation-areas.* The distance on the skin by which two points must be separated in order that they may be felt as two is a constant, interesting both to the physiologist and psychologist. Its variation in different parts of the body (from 1 to 68 mm.) was a most important discovery. What the individual variation may be, and what inferences may be drawn from it, cannot be foreseen; but anything which may throw light on the development of the idea of space deserves careful study. Only one part of the body can be tested in a series such as the present. I suggest the back of the closed right hand, between the tendons of the first and second fingers, and in a longitudinal direction. Compasses with rounded wooden or rubber tips should be used, and I suggest that the curvature have a radius of .5 mm. This experiment requires some care and skill on the part of the experimenter. The points must be touched simultaneously, and not too hard. The experimentee must turn away his head. In order to obtain exact results, a large number of experiments would be necessary, and all the tact of the experimenter will be required to determine, without undue expenditure of time, the distance at which the touches may just be distinguished.

IV. *Pressure causing Pain.* This, like the rate of movement, is a determination not hitherto much considered, and if other more important tests can be devised they might be substituted for these. But the point at which pressure causes pain may be an important constant, and in any case it would be valuable in the diagnosis of nervous diseases and in studying abnormal states of consciousness. The determination of any fixed point or quantity in pleasure or pain is a matter of great interest in theoretical and practical ethics, and I should be glad to include some such test in the present series. To determine the pressure causing pain, I use an instrument . . . which measures the pressure applied by a tip of hard rubber 5 mm. in radius. I am now determining the pressure causing pain in different parts of the body; for the present series I recommend the centre of the forehead. The pressure should be gradually increased, and the maximum read from the indicator after the experiment is complete. As a rule, the point at which the experimentee says the pressure is painful should be recorded, but in some cases it may be necessary to record the point at which signs of pain are shown. I make two trials, and record both.

V. *Least noticeable difference in Weight.* The just noticeable sensation and the least noticeable difference in sensation are psychological constants of great interest. Indeed, the measurement of mental intensity is probably the most important question with which experimental psychology has at present to deal. The just noticeable sensation can only be determined with great pains, if at all: the point usually found being in reality the least noticeable difference for faint stimuli . . . The least noticeable difference in sensation for stimuli of a given intensity can be more readily determined, but it requires some time, and consequently not more than one sense and intensity can be tested in a preliminary series. I follow Mr. Galton in selecting 'sense of effort' or weight. I use small wooden boxes, the standard one weighing 100 gms. and the others 101, 102, up to 110 gms. The standard weight and another (beginning with 105 gms.) being given to the experimentee, he is asked which is the heavier. I allow him about 10 secs. for decision. I record the point at which he is usually right, being careful to note that he is always right with the next heavier weight.

VI. *Reaction-time for Sound.* The time elapsing before a stimulus calls forth a movement should certainly be included in a series of psychophysical tests: the question to be decided is what stimulus should be chosen. I prefer sound; on it the reaction-time seems to be the shortest and most regular, and the apparatus is most easily arranged. I measure the time with a Hipp chronoscope, but various chronographic methods have been used . . . In measuring the reaction-time, I suggest that three valid reactions be taken, and the minimum recorded. Later, the average and mean variation may be calculated.

VII.   *Time for naming Colours* . . . This time may be readily measured for a single colour by means of suitable apparatus . . . but for general use sufficient accuracy may be attained by allowing the experimentee to name ten colours and taking the average. I paste coloured papers (red, yellow, green and blue) 2 cm. square, 1 cm. apart, vertically on a strip of black pasteboard. This I suddenly uncover and start a chronoscope, which I stop when the ten colours have been named. I allow two trials (the order of colours being different in each) and record the average time per colour in the quickest trial.

VIII. *Bisection of a 50 cm. Line.* The accuracy with which space and time are judged may be readily tested, and with interesting results. I follow Mr. Galton in letting the experimentee divide an ebony rule (3 cm. wide) into two equal parts by means of a movable line, but I recommend 50 cm. in place of 1 ft., as with the latter the error is so small that it is difficult to measure, and the metric system seems preferable. The amount of error in mm. (the distance from the true middle) should be recorded, and whether it is to the right or left. One trial would seem to be sufficient.

IX. *Judgment of 10 sec. Time.* This determination is easily made. I strike on the table with the end of a pencil, and again after 10 seconds, and let the experimentee in turn strike when he judges an equal interval to have elapsed. I allow only one trial and record the time, from which the amount and direction of error can be seen.

X. *Number of Letters repeated on once Hearing.* Memory and attention may be tested by determining how many letters can be repeated on hearing once. I name distinctly and at the rate of two per second six letters, and if the experimentee can repeat these after me I go on to seven, then eight, &c.; if the six are not correctly repeated after three trials (with different letters), I give five, four, &c. The maximum number of letters which can be grasped and remembered is thus determined. Consonants only should be used in order to avoid syllables.

427

81   ALFRED BINET (1857–1911) AND VICTOR HENRI (1872–1940) ON THE PSYCHOLOGY OF INDIVIDUAL DIFFERENCES, 1895

Alfred Binet and Victor Henri, "La Psychologie individuelle," *L'Année psychologique 2,* 411–465 (1895). Translated for this book by Mollie D. Boring.

Interest in mental testing spread rapidly after Galton's initial efforts. In Germany, Axel Oehrn (1889), working under the guidance of the psychiatrist Emil Kraepelin, devised tests to measure mental capacities in relation to psychopathology, and Hugo Münsterberg (1891) described several mental tests he used on children but did not give his results with them. In America, there were, in addition to Cattell, Joseph Jastrow (1890), who tested college students at the University of Wisconsin, and the anthropologist Franz Boas (1891), who took all sorts of anthropological measurements of large numbers of school children. In 1895, the American Psychological Association appointed a committee to coordinate these and the numerous other early ventures into mental testing.

Also in 1895 appeared the influential paper by Alfred Binet and his student Victor Henri, outlining a point of view toward, and a program for achieving, a science of mental measurement. Binet was the leading experimental psychologist of his generation in France. He founded France's first

psychological journal *L'Année psychologique,* in 1895, and he was an early worker in its first psychological laboratory, established by H. Beaunis at the Sorbonne in 1889. But it is not primarily for being an entrepreneur that he is remembered; it is, rather, for his being the author of a conception of mental testing that has in essence prevailed until the present. For Binet and Henri, the mental test was to have two purposes: it was to assess the variation from individual to individual in mental faculties, and also to determine the covariation among the various mental faculties. To achieve these goals, Binet and Henri favored the testing of the higher mental functions, rather than the simple sensory processes investigated by Galton, Cattell, and other early workers. During the remainder of the nineteenth century and the first few years of the twentieth, Binet and his collaborators began to build mental tests and evaluate them with large-scale trials. Eventually, by the second decade of the century, the intelligence test, in the form conceived of by Binet, had become a standard tool in the psychologist's workshop.

The aim of individual psychology, as its very name says, is to study different psychic processes in man and, in studying them, to pay attention to the individual differences in them. General psychology studies the general properties of psychic processes, those which are common to all individuals. Individual psychology, on the other hand, studies the properties of psychic processes that vary from individual to individual—it has to determine the various properties and then study how much and in what ways they vary with the individuals. Thus for a given psychic process, like memory, general psychology will be concerned with the general laws of memory; it will establish, for instance, the fact that when one wishes to retain a certain number of impressions the time necessary to fix them in memory at first increases proportionally with the number of impressions, but, beyond a certain span, the "time of acquisition" increases much more rapidly than the number of impressions. This is a general law of memory; no one can deny it. Individual psychology will try to discover the partial properties of memory that vary from individual to individual. It will study whether the size of the span, just cited, is the same for different individuals, whether it varies, to what degree it varies, and whether the variation in span is the same for different sorts of impressions. Thus, for example, when individual A can retain up to ten numbers after hearing them once whereas another person, B, can retain only seven, it asks, will this difference persist when it is not for numbers but for letters or words or colors or, in the end, other impressions, and, if it does persist, will it be as marked?

Finally, individual psychology will also have to study whether there is any relation between variations in the span of memory and variations in other psychic, or even physical, faculties of individuals. Is there any relation, for instance, between the individual's age and the size of his span or between his power of attention and the size of his span, and so forth? There are, for any particular case, many questions like the preceding. Let us try to put in order some of the questions individual psychology must resolve, and systematize them.

We can at first discern two major problems:

1. The study of how psychic processes vary from individual to individual, what the variable properties of these processes are, and to what extent they vary.

2. The study of relations among the different psychic processes in a single individual. Are some psychic processes more important than others? To what extent can the different processes be independent of each other and to what extent do they interact?

Let us examine these two problems in greater detail. Given a psychic process, we can first determine which of its properties are common to all individuals—such is the problem of general psychology. The process can then have more or less variable properties, and it is with these that we must concern ourselves. We can approach this study of variable properties from two different points of view, according to whether we consider first the psychic process studied and then the individual, or whether, instead, we deal primarily with the individual and how a psychic process in one person may differ from the same process in another. Let us take an example that will make this difference clear: is the image evoked when one hears the word *bell* spoken the same for all individuals? Study shows that some imagine the *sound* of the bell, others the *visual image* of a bell, and still others the written word *bell*, and so on. We must conclude that different individuals hearing the word *bell* have different images for it, and we could put into groups the varieties of images presented—this is the first point of view. But we could, on the other hand, ask if two individuals, A and B, have like images when they hear the word *bell*, and here the individuals govern—that is the second point of view. Thus we cannot take into account only isolated individuals but must also consider larger or smaller groups of individuals. We may wonder, for instance, whether men and women have different images upon hearing the word *bell*. We find that women more often have visual images than do men, and, further, that women generally have more detailed images than men. It is easy to see that we can restrict the different groups as much as we wish; we wonder, for instance, if there is a difference in the images that painters have, or musicians, and so on.

We next come to the psychological studies of children, of criminals, of the differences of persons in different professions, and so forth. And that leads us also to the following question: given two or more individuals, what are the differences in their psychic faculties? If we had to answer this question completely by reviewing all the psychic faculties of the individuals, it would take us several months at least; but since there is not generally so much time, we must pay close heed to certain faculties and leave out the others. Which then, we ask next, are the faculties to examine, and which the ones to omit?

And so we come to the second problem, which consists in the study of the relation of the different psychic faculties in one individual; clearly, if we determined these relations precisely, we would need only to examine certain psychic faculties first and then to deduce the others. Thus we must not only study the relation of the different psychic faculties but also decide whether any psychic faculties govern others, which ones are most important, and upon which the other faculties depend. The whole of these primordial faculties would form what might be called the character of the individual. Thus when we have determined them, we could say that the individual is characterized. We see then how important this second problem is, and we can already predict some of its practical applications. But let us hasten to say that individual psychology is still in such an embryonic state that, although much material exists on the first problem, that is to say, on the study of individual differences in the different psychic processes, there is ever so little on the second problem. We do not know of any psychologist who has tackled it and stated it in the general form proposed here. Of course, many memoirs and books on character have been written, but these authors have sought primarily to classify character. They begin with a classification of the different characters and then, as confirmation, they give examples—and that is to follow the true method in reverse. A scientific study of character, believe us, should terminate in a classification of characters and not begin with it.

.   .   .   .   .

It would take too long to review in detail the methods to follow in the researches of individual psychology; we have already discussed them at length. Let us dwell only on the method of "mental tests." We are convinced that this method will henceforth play a certain practical role. We cannot wait for scientific study to indicate the more important processes—the ones that could characterize the person. We must search with the present knowledge and methods at hand for a series of tests to apply to an individual in order to distinguish him from others and to enable us to deduce general conclusions relative to certain of his habits and faculties. This task seems possible in some degree. We must first select the cases where the persons to be examined are of like background and practice the same professions, the persons whom we would compare with each other in order to determine their more important and characteristic differences. These cases must be treated separately from the ones in which we plan to compare individuals in different professions. We must, for example, use different tests in comparing two students in the same field from those in comparing a juggler with one of these students—that, believe us, is evident. Next, the tests must be appropriate to the background and everyday occupations of the individuals. We cannot use the same tests in comparing two bricklayers as in comparing two students or two school children.

The studies of individual psychology are one of psychology's most important practical applications since their aim is knowledge of the individual, and they must be envisioned and directed toward the goal we would attain. There are, it seems, four principal routes to be pursued: the study of races, the study of children, the study of patients, and the study of criminals. Our work would thus be divided into four parts, the first principally ethnic, the second principally pedagogical, the third principally medical, and the fourth principally judicial and criminological. The number and the nature of the tests necessarily vary in these four cases. As we cannot enter into a great many details, we shall simply state the bald outlines of the subject here. We must, on some other occasion, take up the subdivisions merely mentioned here.

Let us say once more that the goal pursued is not to determine *all* the differences in psychic faculties between two or more individuals but to determine the more marked and important differences. The tests

431

must show the characteristic traits that distinguish the two individuals, not all their traits. That is a rule other authors have not considered and followed, for, if they had been clearly aware of it, they would not have set up so many tests for determining the most elementary sensations and processes. We must, therefore, deal with the superior psychic faculties. Remember, we should not be stopped by the difficulty of determining these faculties with as much precision as the elementary ones. We do not need to be so precise, for individual differences are marked.

We propose to study the following ten processes:

Memory, the nature of mental images, imagination, attention, the faculty of comprehension, suggestibility, aesthetic feeling, moral feelings, muscular strength and will power, and motor skill and perceptual skill in spatial relations.

All these, believe us, are psychic faculties that differ greatly from individual to individual; knowing their nature for one individual gives us a general idea of this person and enables us to distinguish him from others having the same background.

How, then, shall we determine the nature of these different faculties for a given individual? The method must first be simple and not take too much time. Moreover, the means of determination must be independent of the experimenter as far as possible. It must be possible to compare the results obtained by one observer with the results obtained by others.

．　　．　　．　　．　　．

Having reached the end of the tests proposed, we certainly have many gaps still to fill in. We must modify several of the tests—that is certain. Remember that this is only a first attempt, and practice and experience will have to complete and modify the tests. All the same, having obtained answers to all our earlier questions by having questioned the subject on his chief occupations, we believe it possible to obtain an idea of the whole of an individual's faculties. The preceding tests were so chosen that they could all be done in a single session of an hour or at most an hour and a half. If we used a longer time, we would add to the experiments and vary them as much as possible. Possibly these tests are not adequate for clearly characterizing the differences between two individuals of the same background, living in the same manner and having the same occupations, but we believe they can give useful information when the persons to be compared differ more markedly.

At the beginning of this study we said that the tests must be appropriate to the background of the individuals studied. We must stress some of the items more than others. For individuals in some of the professions all these questions must be examined in detail. In this study our primary concern lies in modifying the tests for the study of individual differences in school children, since that is where the tests can now principally be applied, and, moreover, it is easiest to make experiments of this sort with school children.

82   HERMANN EBBINGHAUS (1850–1909) ON THE COMPLETION TEST, 1897

Hermann Ebbinghaus, "Ueber eine neue Methode zur Prüfung geistiger Fähigkeiten und ihre Anwendung bei Schulkindern," *Zeitschrift für Psychologie und Physiologie der Sinnesorgane 13,* 401–459 (1897). Translated by William H. Everhardy at the National Institutes of Health, Bethesda, Maryland, as "A new method of testing mental capacities and its use with school children."

In this paper, Ebbinghaus described a new type of mental test that he devised to gauge the effects of mental fatigue on the performance of children in school. A child is given a sample of written prose from which letters and groups of letters are omitted and is asked to fill in what is missing. This technique is still used in many intelligence tests. Ebbinghaus argued that this "completion test" came closer to testing intelligence than did the tests of memory, calculation, or sensory discrimination that were being used at that time, and he gave the results to support this view. This secondary contribution by the inventor of the nonsense syllable (see No. 95) and the founder of the experimental study of associations was to play a significant role in favoring Binet's belief that mental tests should deal directly with the higher mental processes, over Cattell's view that they should be restricted to the simple sensory ones.

In July 1895, the municipal authorities of the city of Breslau addressed a letter of the following import to the Hygiene Section of the Silesian Society for National Culture. According to observations on their own children, the five-hour morning session in the higher schools of the city results in such fatigue and nervous irritability that an elimination of this arrangement seems desirable. There must be considered either a distribution of the work over the entire day, hence a return to afternoon classes, or a limitation of the length of instruction, especially during the hot season. In order to have a basis for their actions, the authorities requested an authoritative statement about the five-hour session, [and a committee was appointed to consider the matter].

·   ·   ·   ·   ·

The most important question with which this committee had to concern itself at first was naturally the question of the method. How could we actually set about determining the effect of lengthy instruction on the mental condition of the school children? How can an excessive demand on mental powers be recognized by fairly certain objective criteria without calling on uncertain subjective impressions always obtained in this from a few individuals?

.   .   .   .   .

. . . If you want to investigate experimentally the effect of prolonged school instruction on children, you must not prepare a certain special type of instruction and draw conclusions about other things from its results. On the contrary, you must use the instruction as it actually is, i.e., as it is established for the children by the general hourly plan of the school, and is regularly imparted to them . . .

. . . Griesbach, an Alsation educator, . . . has observed that mental fatigue reduces the sensibility of the skin, i.e., it produces a deterioration of our ability to distinguish subjectively two objectively different compass points as two when placed on the skin. As a result of exhaustion, we become inattentive, and hence two compass points which are supposed to be recognized as two on a certain skin area must be moved farther apart than in the state of mental alertness. Therefore, Griesbach used this tactile sensitivity of the skin as a means of testing the mental efficiency, and he then studied its behavior in numerous students of higher institutions of learning in the course of a school day and, as a control, also on free days. In doing this, as we said before, he left the instruction itself entirely undisturbed; he tested the selected students only very briefly when they entered the school and at the end of the different class hours. A considerable blunting of sensitivity was actually found as a result of the instruction. As a rule, it reached its maximum at the end of the third hour, when the distance between the compass points often had to be made three times as great as in the morning before classes began. In the fourth and fifth hours before noon there was frequently shown an improvement of sensitivity, which was perhaps produced by the fact that easier subjects were taught during these hours.

Originally there was a certain preference for Griesbach's method in the committee, most of whose members were doctors. It is relatively easy to understand by persons who have received training in physiology and clinical medicine, and—since the difficulties of reliable esthesiometer testing are underestimated as a rule—easy and reliable to use. But after more thorough consideration, we had to say that here the method of application of the testing procedure is excellent but that the testing procedure itself is not very well suited to the purpose in mind. Granted that there is a close relation between mental fatigue and blunting of skin sensibility, how can it be formulated? What increase of the one corresponds to a certain deterioration of the other? Is the general mental efficiency of a person only half as great as under normal conditions when his sensitivity for distinguishing compass points has been reduced to half? What should we say here? Furthermore, when is the mental fatigue indicated by reduction of the skin sensibility to be considered as serious? Are children who show a deterioration of the results extending to uselessness also to be regarded as mentally exhausted in other respects, or have they merely become incapable of performing that fixed, relatively insignificant task to which they are not accustomed? We do not know the answer to any of these questions at present. But we would have to know them in order to be able to draw valid conclusions in the desired direction from the results of Griesbach's method. Above all, in any case, one must consequently again attack those mental activities about whose behavior under special conditions one wishes to learn something.

·   ·   ·   ·   ·

In order to find a way to do this, it is first necessary to answer the questions of what is the essence of that higher intellectual activity, the power of understanding in the narrower sense, and how can it be charactertized more precisely. It is expedient to orient yourself concerning this by examples in which the capacity for it seems especially developed and in which we speak of special mental proficiency.

What is a skilled physician? One who knows much? He must have wide knowledge, of course, but if he has nothing more than knowledge, he will perhaps be good as the writer of a textbook, but he is not a skilled physician. This also entails the ability to utilize his knowledge in specific instances; he must be able to make a correct diagnosis and to help his patients with its assistance. This means that when he is faced with various essentially ambiguous symptoms, described to him in part in a very distorted manner by the patient, he must be able to judge correctly what disease he is dealing with, and then after consideration of the nature of this disease, its present state of development and the remedies recommended for it, as well as of the personality of the patient, his social position, whether he can be spared, etc., he must be able to prescribe a treatment for him which is shown to be proper by its result. Or what is a skilled general? In a purely intellectual respect (i.e., aside from the ability to . . . reach important decisions, which plays an especially great role here), a person very similar to a skilled physician, but in an entirely different field. One who knows how to combine fragmentary, uncertain and partly absolutely false information about the enemy into a correct picture of his number, position and intentions, and

435

who, after consideration of the number and position of his own troops, their ability, the nature of the topography and roads, the possibility of a continuous supply, the individual peculiarities of his subordinates and opponents, and a thousand other things, can immediately order an attack leading to victory.

Wherever one looks, mental proficiency consists of something similar; only the material on which it acts changes. This is true for a proficient scholar who knows how to fill in and level out the gaps and contradictions of historical tradition or our material view of nature by a material concept of the whole, or for an efficient merchant who manages his affairs in accordance with his finances, the needs of the public, political and economic factors, etc. It does not manifest itself in the fact that one remembers well and hence knows much about something, i.e., that one readily reacts to the occurrence of certain isolated impressions with ideas which he had previously had occasion to associate with these impressions, but it is built on this basis, which is necessary of course, as something much more complex but at the same time more free and a new creation in a certain sense. Its essence lies in the fact that a greater multiplicity of impressions existing independently side by side, which are essentially suited for awakening very heterogeneous, partly directly opposed, associations, are answered with ideas, all of which fit them at the same time and all of which join together in a homogeneous, significant, or somewhat purposeful whole. Intellectual mental proficiency consists in the elaboration of a whole, somehow having value and significance by virtue of mutual connection, correction and supplementation of the associations suggested by many different impressions. In order to characterize its nature briefly, I shall say ... that it consists in reasoning and hence that true intellectual activity is reasoning activity. Misunderstandings to which these expressions might perhaps be exposed will no longer have to be feared in the light of the foregoing.

In the deliberations of our committee it now seemed especially desirable to several members to focus the planned studies on this point and to ascertain any effects of the instruction on the ability of the pupils to reason. Of course, there were immediately raised there the difficult questions of how this should be begun, where to obtain a criterion for evaluating the reasoning activities of the intelligence at all correctly and for making them numerically comparable with each other in this way for different conditions and in different individuals. It is clear at the outset that only very simple and easy reasoning activities can be used for this, namely those which can immediately be mastered thoroughly by all individuals of the category under consideration, although, naturally, with somewhat different expenditure of time, because the test itself should not be especially demanding but should merely determine the results of other exertions. Furthermore, the data for a further evaluation would be lacking if the problems posed remained unsolved in a large number of cases. Therefore, the different reasoning tasks must be of equal value as far as possible, naturally not for all individuals, which could hardly be possible, but for the average of large groups. Then there would be obtained a simple measure of the performance in each case merely by counting the correct reasonings found . . .

After groping about for some time in the attempt to meet these requirements, I finally hit upon the idea of recommending the following procedure to the committee. In order to measure their powers of comprehension, the pupils are given prose texts which are made incomplete in the most varied ways by small omissions. Sometimes individual syllables are omitted, both at the beginning and at the end, as well as in the middle, of a word, while sometimes only parts of syllables, and sometimes entire words, are omitted. Each syllable omitted (and also each fragment of a syllable omitted) is indicated by an underline, and the pupil is now given the task of filling in the gaps in such a text as rapidly and sensibly as possible, with due consideration of the number of syllables required. In doing this he always has to keep in mind a small number of things at the same time: the letters which are present, the adaptation to the number of syllables prescribed, especially in the sense of completing the text in both its narrower and wider sense, not only in regard to what has gone before but sometimes also in regard to what is to come . . . The working time for each individual text sample is fixed at exactly five minutes, and it is then determined in each case how many syllables were filled in correctly, how many were passed over, and how many were filled in incorrectly.

Since the committee gave a favorable reception to the suggestion of such a reasoning method, as I shall call it, I worked out a number of texts in the manner described, especially for the students of a gymnasium. In order to take into account the different mental capacity of the students, I made up two groups: a more difficult one for the upper grades . . . and an easier one for the lower grades . . . But in the interest of the greatest possible similarity, I took the texts of each group from the same author . . . It was very difficult to make the different texts for the same students as equally difficult as possible, since the presence or absence of a single letter often caused considerable differences in this regard. Naturally, this goal can always be attained only in a certain general way because of the difference of the various individuals. But so far as this is the case, I sought to come close to it not only by my own repeated testing of the texts but also by submitting them to some other persons, and then I modified them where too great dissimilarities occurred.

83 STELLA EMILY SHARP (1872–1961) ON A TEST OF MENTAL TESTING, 1899

Stella E. Sharp, "Individual psychology: a study in psychological method," *American Journal of Psychology 10*, 329–391 (1899).

By the time this paper was written, the lines of battle between the followers of Binet and of Cattell in mental testing had been clearly drawn. Binet was saying that the best way to test higher mental processes is directly, while Cattell held that a better way is to use the simple sensory faculties. Miss Sharp worked in E. B. Titchener's laboratory at Cornell—the American home of the German psychology of the generalized, or typical, adult mind —where the very idea of a psychology of individual differences of any sort was foreign, but Titchener nevertheless approved this experiment and presumably also its conclusions. Miss Sharp, while concluding that the study of individual differences cannot displace the study of the generalized mind as an approach to psychology as a whole, nevertheless found that any practical value in this study of differences would be achieved by following Binet's and not Cattell's approach. This conclusion was undoubtedly in accord with Titchener's bias, for he favored the psychology of the Continent against the psychology of England, his native land.

PART II. EXPERIMENTAL

4. Description of Tests

The following experiments were undertaken during the academic year '97–'98 as a study of Individual Psychology based, in general, upon the theories, and to a large extent upon the specific suggestions of Mm. Binet and Henri, as contained in their article *La psychologie individuelle.* The theory was provisionally accepted that the complex mental processes, rather than the elementary processes, are those the variations of which give most important information in regard to the mental characteristics whereby individuals are commonly classed. It is in the complex processes, we assumed, and in those alone, that individual differences are sufficiently great to enable us to differentiate one individual from others of the same class. Many of the particular tests recommended by the French psychologists were also adopted, but were considerably modified in the general conditions of their application by the purpose of our own investigation.

The aim of this work was (1) to ascertain the practicability of the particular tests employed, and (2) to answer the more general question as to the tenability of the theory upon which they are based, in so far as this can be judged by the experiments. In other words, we desired to assure ourselves whether investigations of this kind enabled us to advance, at least, toward a solution of the problems of Individual Psychology; whether those individual variations, and those correspondences which are necessary for classifying individuals, and for estimating the relative importance of the several processes in a single individual, could thereby be discovered.

In view of these aims, and also of the criticisms of the general conditions demanded by Mm. Binet and Henri for the application of the method of 'tests,' the procedure was necessarily different from that laid down by these psychologists. To make sure that the tests give real individual differences, and not chance variations, it is necessary to apply them to the same individuals, not once, but several times, in order that it may be observed whether the variations in the different individuals maintain a constant relation to one another at various times and, consequently, under varying subjective conditions. Instead of single tests, therefore, series of similar tests for each activity were arranged. This necessitated, of course, a very large extension of the time beyond the limit allowed by the French investigators. The advantages of a short period of varied experimentation were, however, to a large degree attained. The experimental work of each subject was divided into periods of one hour each, and separated by intervals of one week. Within a single hour-period the tests were varied as much as possible. As a rule, only one or two experiments belonging to the series of a particular test were given. In this way the tedium and fatigue due to monotonous repetition of similar operations were avoided, and a fair degree of interest in the work was maintained by the subject. The additional precaution was taken of separating by intervals longer than a week the experiments which were found to be especially trying or disagreeable to the subjects; as, *e.g.,* the development of a theme, or description of a scene or event, employed as a test for constructive imagination.

.   .   .   .   .

6. Conclusion

It is not our intention to print in this place a complete summary of the results of all experiments for the different individuals. Such a summary, has, of course, been made by us; but, in the first place, it leaves too many gaps to allow a definite differentiation of each individual from all the others, owing largely to the limited bounds within which the enquiry was purposely confined, while, secondly, we have considered it best that the reader, if he will, shall make such a summary for himself, and in this way form his estimate of the value of the tests. Our aim was principally to investigate the merit of a general method: to find the value for Individual Psychology of experimentation applied to the more complex mental activities, as well as the practicability of certain specific tests, many of which had been suggested by the advocates of such experimentation.

It will be remembered that we noted above two main problems of Individual Psychology; the first problem having reference mainly to *variations* themselves, that is, to the way in which psychical processes vary in different individuals, and according to classes of individuals; the second, to the *relations* among variations. The latter, to be sure, includes the question how individuals vary in regard to psychical processes, but it goes on further to ask how these individual variations are related to each other, when the whole range of mental processes is considered . . .

The results, we believe, have shown that, while a large proportion of the tests require intrinsic modification, or a more rigid control of conditions, others have really given such information as the Individual Psychologist seeks. Thus the tests for Imagination proved to be important as forming a basis for a general classification of the individuals, according to fairly definite types; and results from other tests gave some force of confirmation to this classification, as *e.g.*, the test on Observation by description of pictures. In general, however, a lack of correspondences in the individual differences observed in the various tests was quite as noticeable as their presence. The total change in the order of subjects in the memory of single short series of words and in the recapitulation of the words of seven short series, the fact that those subjects who showed best observation of colors were not the best visualizers, are instances of this lack of correspondence, of which many others could be cited. Whether the fact indicates a relative independence of the particular mental activities under investigation, or is due simply to superficiality of testing, can hardly be decided. While, however, we do not reject the latter possibility, we incline to the belief that the former hypothesis is in a large proportion of cases the more correct.

But little result for morphological psychology can be obtained from studies of the nature of the above investigation. So many part-processes are involved in the complex activities, and the manner of their variation is so indefinite, that it is seldom possible to tell with certainty what part of the total result is due to any particular component. It is doubtful if even the most rigorous and exhaustive analysis of test-results would yield information of importance as regards the structure of mind. At all events, there is not the slightest reason to desert current laboratory methods for the 'method of tests.'

The tests employed, considered as a whole, cannot be said to yield decisive results for Individual Psychology if applied *once* only to individuals *of the same class*. This statement the above discussion of tests seems perfectly to warrant. *Series of such tests* are necessary in order to show constant individual characteristics. The tests, to be sure, (1) if enlarged in extent to cover a wider range of activities, might be useful for roughly classifying a large number of individuals of very different training, occupation, etc., provided that the greatest care were taken that the conditions in the case of each individual should be as favorable as possible. And, on the other hand, (2) certain groups of tests, especially selected for a particular purpose, and applied, once each in series, to a limited number of individuals, might yield valuable information on points which particular circumstances rendered of practical importance. As engineers, pilots, and others who have to act upon information from colored signals, are roughly tested for color blindness, so other classes might often profitably be submitted to a psychological testing of those higher activities which are especially involved in their respective lines of duty.

All this, however, is largely beside the point; much preliminary work must be done before such special investigations can be of any great worth. This appears plainly from the present investigation where the positive results have been wholly incommensurate with the labor required for the devising of tests and evaluation of results. In the present state of the science of Individual Psychology, there can be little doubt that the method of procedure employed by M. Binet is the one most productive of fruitful results: that, namely, of selecting tests, and applying them to a number of individuals and classes of individuals with a view of discovering the chief individual differences in the mental activities to which appeal is made. To this should be added, however, an exhaustive study of the results from series of similar tests given to a small number of individuals at different times and in varying circumstances, in order to discover how constant the differences are, and how much of the variation may be due to changes in mental and physical condition, environment, etc. When this procedure has been followed for tests that cover all the principal psychical activities, then the investigation of limited groups of individuals for the purpose of characterizing them in respect to their mental differences may be undertaken with hope of easy and accurate results. The previous study will have made clear the many conditions involved, and the best way of modifying the 'test method' to suit varying circumstances.

In fine, we concur with Mm. Binet and Henri in believing that individual psychical differences should be sought for in the complex rather than in the elementary processes of mind, and that the test method is the most workable one that has yet been proposed for investigating these processes. The theory of the German psychologists, who hold that the simplest mental processes are those to which the investigator should look for a clue to all the psychical differences existing among individuals, we believe would be productive of small or, at any rate, of comparatively unimportant results. Whether the anthropometrical tests so largely used by American workers in this field of psychology will lead to any such correlation of these traits with those of a purely psychical character as has been suggested by some pursuing the inquiry, is a question which must be left for the future to decide. No adequate data are as yet at hand, and (as has been stated above) the American workers have formulated no explicit theory of Individual Psychology. The method here outlined should (and may), however, be rendered more exact by modifications in accordance with the procedure of the German investigators of Individual Psychology. A combination of the principal characteristics of the two methods is, then, it seems to us, best calculated for the attainment of satisfactory results.

84 CLARK WISSLER (1870–1947) ON THE INADEQUACY OF MENTAL TESTS, 1901

Clark Wissler, "The correlation of mental and physical tests," *The Psychological Review Monograph Supplements 3*, no. 6 (whole no. 16), 4, 27, 29, 34–36 (1901).

As a member of Cattell's laboratory at Columbia University, Wissler was in an excellent position to evaluate the results of mental testing as conceived by Cattell. In the paper from which the following excerpt is taken, Wissler showed that the intercorrelations among various tests of simple sensory and mental performance were negligible. He was taking advantage of the then new mathematical technique of correlation, first proposed by Francis Galton in 1886 and later developed by Karl Pearson in 1896. Wissler also showed that although there was little correlation between the performance on these tests and the academic standing of his subjects, who were students at Columbia and Barnard, there was good correlation among the grades earned by the students in their various courses. The effect of this study was to undermine Cattell's approach to mental testing, and also to depreciate testing itself, but the depreciation was only temporary, for Binet's approach was soon to lead to fruitful results.

The tests . . . made in the psychological laboratory are as follows: length and breadth of head, strength of hands, fatigue, eyesight, color vision, hearing, perception of pitch, perception of weight, sensation areas, sensitiveness to pain, perception of size, color preference, reaction time, rate of perception, naming colors, rate of movement, accuracy of movement, perception of time, association, imagery, memory (auditory, visual, logical and retrospective). Records of stature, weight, etc., together with data concerning parentage, personal habits and health, are a part of the gymnasium tests required of all students in Columbia College.

．　　　．　　　．　　　．　　　．

TESTS OF QUICKNESS AND ACCURACY

Reaction-time and marking out the A's furnish 252 cases in common. The reaction-time for each individual is the average of five to three valid reactions, observers recording five reactions. In the A test it has been proposed to lengthen the time recorded proportionally to the numbers of A's missed. But this is unnecessary, since $r = -0.05$ in case of A times so adjusted and -0.07 for them regardless of errors. It appears, then, that the degree of correlation between these tests is approximately zero, or a chance relation. In other words an individual with a quick reaction-time is no more likely to be quick in marking out the A's than one with a slow reaction-time.

The other correlations for tests of quickness may be enumerated as follows:

	Cases	r
Reaction and naming the colors	118	0.15
Reaction and association	153	0.08
Marking A's and naming the colors	159	0.21
Movement time and naming the colors	97	0.19
Movement time and reaction	90	0.14

The remaining possible correlations in this group of tests are movement time with A's and association, and naming colors and association, but when tabulated the distributions were such as to make a very low degree of correlation certain. Thus it appears that the time required for naming the colors correlates better than any test in this series, yet the coefficients are too low to be of much significance. [The task in the A test is to cross out every one of 100 A's in a batch of 500 letters as quickly as possible.]

．　　　．　　　．　　　．　　　．

Having now considered tests of quickness, we may take up those whose results are generally expressed in terms of accuracy . . .

The tests for accuracy of movement . . . [striking with a pencil as quickly as possible every one of 100 dots printed in a 10-cm square] and the perception of weight (force of movement) correlate neither with each other nor with the test for size. Also, no correlations were found between these tests and the accuracy of estimating time intervals or of following a given rhythm. The A test may be considered here also, since it may be graded according to the number omitted, but this fails to correlate with the above.

.   .   .   .   .

COLLEGE STANDING

In general it appears that correlations in any of the foregoing tests are not of a degree sufficient for practical purposes. We do not learn much of an individual by any one or any group of them. It appears that we are dealing here with special and quite independent abilities and that the importance attributed to such measurements of elementary processes by many investigators is not justified in this case. However, it seems probable that a basis for correlation will be found somewhere, and it may turn out that though these tests do not show much intercorrelation they may individually correlate with ability in the more complex tasks of life. Research in this direction is obviously difficult for want of adequate standards and satisfactory data, and the only attempt we have made is in respect to college standing.

.   .   .   .   .

The following correlations have been calculated for the average standing:

Reaction-time	227 cases	−0.02
Marking A's	242	−0.09
Association	160	+0.08
Naming colors	112	+0.02
Logical memory	86	+0.19
Auditory memory (position)	121	+0.16

An application of the . . . method to the other mental tests gave no hope for correlation.

Here we are face to face with another cold fact: the tests of quickness seem to hold a chance relation to class standing, and ability to do well in the memory tests has but little significance. But it may be well to examine the relative standing in the different courses.

The following correlations were calculated:

Latin and mathematics	228 cases	0.58
Mathematics and rhetoric	222	0.51
Latin and rhetoric	223	0.55
French and rhetoric	122	0.30

German and rhetoric	132	0.61
German and mathematics	115	0.52
Latin and French	130	0.60
Latin and German	129	0.61
Latin and Greek	121	0.75
Gymnasium and average grade	119	0.53

Here we find a higher degree of correlation than heretofore. The languages have a correlation of 0.60 to 0.75, or a reduction of variability approximating $\frac{1}{4}$. In other cases we have a reduction of about $\frac{1}{5}$. The exceptionally low correlation for French and rhetoric is rather puzzling, but is probably due to some accidental cause. On the other hand, the high degree of correlation between Latin and Greek is according to expectation. It is interesting to note that the degree of correlation here is about the same as for stature and weight. From what has gone before it is improbable that a high degree of correlation will be found between particular courses and the separate tests. For example, with logical memory and mathematics the coefficient is 0.11 and with Latin 0.22, no significant variation from 0.19, the correlation for the average class standing.

Whatever it is that makes for correlation in class standing seems to hold generally for all courses. The gymnasium grade, which is based chiefly upon faithfulness in attendance, correlates with the average class standing to about the same degree as one course with another. We have not carried this correlation out to its full possibility, as that would take us too far afield. Yet this serves as a suggestion as to how this method of correlation may aid in the solution of a very important phase of the test question.

85   CHARLES EDWARD SPEARMAN (1863–1945) ON GENERAL INTELLIGENCE, 1904

C. E. Spearman, "'General intelligence,' objectively determined and measured," *American Journal of Psychology 15,* 201–293 (1904).

Spearman, unlike Wissler, found large, significant correlations among measures of simple sensory and computational tasks. However, he had as his subjects a heterogeneous collection of school children, whereas Wissler had used the relatively homogeneous population of Columbia College students, and the mathematics of the correlation coefficient requires heterogeneous populations if a significant correlation is to be found. In other respects, too, Spearman brought a more thorough grasp of mathematical correlation to bear on the problems of mental testing. It was his idea to take into account measures of a test's reliability and to make allowances for errors of observation. From his careful experiments and skillful use of correlational techniques, he was led to the conclusion that all intellectual tasks partake of a single capacity—general intelligence—plus whatever other capacities, specific to the test, may also be involved. This was the famous two-factor theory of intelli-gence, a theory that remained highly influential for a generation, until it was at last abandoned. It may be regarded as a necessary early stage in the development of the modern, highly sophisticated systems of mental measurements.

4. UNIVERSAL UNITY OF THE INTELLECTIVE FUNCTION

In view of this community being discovered between such diverse functions as in-school Cleverness, out-of-school Common Sense, Sensory Discrimination, and Musical Talent, we need scarcely be astonished to continually come upon it no less paramount in other forms of intellectual activity. Always in the present experiments, approximately,

$$r_{pq} / \sqrt{(r_{pp} \cdot r_{qq})} = 1.[1]$$

I have actually tested this relation in twelve pairs of such groups taken at random, and have found the average value to be precisely 1.00 for the first two decimal places with a mean deviation of only 0.05. All examination, therefore, in the different sensory, school, or other specific intellectual faculties, may be regarded as so many independently obtained estimates of the one great common Intellective Function.

Though the range of this central Function appears so universal, and that of the specific functions so vanishingly minute, the latter must not be supposed to be altogether non-existent. We can always come upon them eventually, if we sufficiently narrow our field of view and consider branches of activity closely enough resembling one another. When, for instance, in this same preparatory school we take on the one side Latin translation with Latin grammar and on the other side French prose with French dictation, then our formula gives us a new result; for the two Latin studies correlate with the French ones by an average of 0.59, while the former correlate together by 0.66 and the latter by 0.71; so that the element common to the Latin correlates with the element common to the French by $0.59 / \sqrt{(0.66 \times 0.71)} = 0.86$ only. That is to say, the two common elements by no means coincide completely this time, but only to the extent of 0.86^2 or 74%; so that in the remaining 26%, each pair must possess a community purely specific and unshared by the other pair.

[1]Where r_{pq} = the mean of the correlations between the members of the one group p with the members of the other group q.

r_{pp} = the mean of the inter-correlations of the members of the group p among themselves, and

r_{qq} = the same as regards group q.

We therefore bring our general theorem to the following form. *When-ever branches of intellectual activity are at all dissimilar, then their cor-relations with one another appear wholly due to their being all variously saturated with some common fundamental Function (or group of Functions).* This law of the Universal Unity of the Intellective Function is both theoretically and practically so momentous, that it must acquire a much vaster corroborative basis before we can accept it even as a general principle and apart from its inevitable eventual corrections and limita-tions. Discussion of the *subjective* nature of this great central Function has been excluded from the scope of the present work. But clearly, if it be mental at all, it must inevitably become one of the foundation pillars of any psychological system claiming to accord with actual fact—and the majority of prevalent theories may have a difficulty in reckoning with it.

Of its objective relations, the principal is its unique universality, seeing that it reappears always the same in all the divers forms of in-tellectual activity tested; whereas the specific factor seems in every in-stance new and wholly different from that in all the others. As regards amount, next, there seems to be an immense diversity; already in the present examples, the central factor varies from less than $\frac{1}{5}$ to over fifteen times the size of the accompanying specific one. But all cases appear equally susceptible of positive and accurate measurement; thus we are becoming able to give a precise arithmetical limitation to the famous assertion that "at bottom, the Great Man is ever the same kind of thing."

Finally, there is the exceedingly significant fact that this central Func-tion, whatever it may be, is hardly anywhere more prominent than in the simple act of discriminating two nearly identical tones; here we find a correlation exceeding 0.90, indicating the central Function to be more than four times larger than all the other influences upon individual differentiation. Not only the psychical content but also the external relations of Sensory Discrimination offer a most valuable simplicity; for it is a single monotonous act, almost independent of age, previous gen-eral education, memory, industry, and many other factors that inextricably complicate the other functions. Moreover, the specific element can to a great extent be readily eliminated by varying and combining the kind of test. For these reasons, Discrimination has unrivalled advantages for investigating and diagnosing the central Function.

5. THE HIERARCHY OF THE INTELLIGENCES

The Theorem of Intellective Unity leads us to consider a corollary proceeding from it logically, testing it critically, and at once indicating some of its important practical uses. This corollary may be termed that of the Hierarchy of the Specific Intelligences.

For if we consider the correspondences between the four branches of school study, a very remarkable uniformity may be observed. English and French, for instance, agree with one another in having a higher correlation with Classics than with Mathematics. Quite similarly, French and Mathematics agree in both having a higher correlation with Classics than with English. And the same will be found to be the case when any other pair is compared with the remainder. The whole thus forms a *perfectly constant Hierarchy* in the following order: Classics, French, English, and Mathematics. This unbroken regularity becomes especially astonishing when we regard the minuteness of the variations involved, for the four branches have average correlations of 0.77, 0.72, 0.70, and 0.67 respectively.

When in the same experimental series we turn to the Discrimination of Pitch, we find its correlations to be of slightly less magnitude (raw) but in precisely the same relative rank, being: 0.66 with Classics, 0.65 with French, 0.54 with English, and 0.45 with Mathematics. Even in the crude correlations furnished by the whole school without excluding the non-musicians, exactly the same order is repeated, though with the general diminution caused by the impurity: Classics 0.60, French 0.56, English 0.45, and Mathematics 0.39.

Just the same principle governs even Musical Talent, a faculty that is usually set up on a pedestal entirely apart. For it is not only correlated with all the other functions, but once again in precisely the same order: with Classics 0.63, with French 0.57, with English 0.51, with Mathematics 0.51, and with Discrimination 0.40. Ability for music corresponds substantially with Discrimination of tones, but nevertheless not so much as it does with algebra, irregular verbs, etc.

The actual degree of uniformity in this Hierarchy can be most conveniently and summarily judged from the following table of correlation; the values given are those actually observed (theoretical correction would modify the relative order, but in no degree affect the amount of Hierarchy or otherwise). Each number shows the correlation between the faculty vertically above and that horizontally to the left; except in the oblique line italicized, the value always becomes smaller as the eye travels either to the right or downwards.

	Classics	French	English	Mathem.	Discrim.	Music
Classics	*0.87*	0.83	0.78	0.70	0.66	0.63
French	0.83	*0.84*	0.67	0.67	0.65	0.57
English	0.78	0.67	*0.89*	0.64	0.54	0.51
Mathem.	0.70	0.67	0.64	*0.88*	0.45	0.51
Discrim.	0.66	0.65	0.54	0.45		0.40
Music	0.63	0.57	0.51	0.51	0.40	

Altogether, we have a uniformity that is very nearly perfect and far surpasses the conceivable limits of chance coincidence. When we consider that the probable error varies between about 0.01 for the ordinary studies to about 0.03 for music, it is only surprising that the deviations are not greater. The general Hierarchy becomes even more striking when compared with the oblique line, which is no measure of the central Function and where consequently the gradation abruptly and entirely vanishes.

The above correlations are raw, and therefore do not tell us either the true rank of the respective activities or the full absolute saturation of each with General Intelligence. For the former purpose we must eliminate the observational errors, and for the latter our result must further be *squared.* Thus we get:

Activity	Correlation with gen. intell.	Ratio of the common factor to the specific factor	
Classics	0.99	99 to	1
Common sense	0.98	96	4
Pitch dis.	0.94	89	11
French	0.92	84	16
Cleverness	0.90	81	19
English	0.90	81	19
Mathematics	0.86	74	26
Pitch dis. among the uncultured	0.72	52	48
Music	0.70	49	51
Light dis.	0.57	32	68
Weight dis.	0.44	19	81

It is clear how much the amount of any observable raw correlation depends upon the two very different influences: first, there is the above intellective saturation, or extent to which the considered faculty is functionally identical with General Intelligence; and secondly, there is the accuracy with which we have estimated the faculty. As regards the ordinary school studies, this accuracy is indicated by the oblique italicized line, and therefore appears about equal in all cases (not in the least following the direction of the Hierarchy); but in other cases there is a large divergence on this head, which leads to important practical consequences. Mathematics, for example, has a saturation of 74 and Common Sense has one of about 96; but in actual use the worth of these indications becomes reversed, so that a subjective impression as to a child's "brightness" is a less reliable sign than the latter's rank in the arithmetic class; almost as good as either appears a few minutes' test with a monochord.

In the above Hierarchy one of the most noticeable features is the high position of languages; to myself, at any rate, it was no small surprise to find Classics and even French placed unequivocally above English

449

(note that this term does not refer to any study of the native tongue, but merely to the aggregate of all the lessons conducted therein, such as History, Geography, Dictation, Scripture, and Repetition).

However it may be with these or any other special facts, here would seem to lie the long wanted general rational basis for public examinations. Instead of continuing ineffectively to protest that high marks in Greek syntax are no test as to the capacity of men to command troops or to administer provinces, we shall at last actually determine the precise accuracy of the various means of measuring General Intelligence, and then we shall in an equally positive objective manner ascertain the exact relative importance of this General Intelligence as compared with the other characteristics desirable for the particular post which the candidate is to assume (such as any required Specific Intelligences, also Instruction, Force of Will, Physical Constitution, Honesty, Zeal, etc.; though some of these factors cannot easily be estimated separately, there is no insuperable obstacle to weighing their *total influence* as compared with General Intelligence). Thus, it is to be hoped, we shall eventually reach our pedagogical conclusions, not by easy subjective theories, nor by the insignificant range of personal experiences, nor yet by some catchpenny exceptional cases, but rather by an adequately representative array of established facts.

86   WILLIAM STERN (1871–1938) ON THE MENTAL QUOTIENT, 1912

William Stern, *Die psychologische Methoden der Intelligenzprüfung* (Leipzig: Barth, 1912), chap. 2. Translated by Guy Montrose Whipple as *The Psychological Methods of Testing Intelligence* (Baltimore: Warwick and York, 1914), chap. 2.

William Stern was one of the early German contributors to the study of mental testing, having published one book on the subject in 1900 and a second in 1911 before he wrote, in 1912, the one from which the following excerpt comes. Stern was working at the same time that Binet and his collaborators were creating the modern form of the intelligence test. In 1905, Binet and T. Simon had used their first scale of intelligence—a series of mental tests graded in difficulty and constructed at the behest of the minister of public instruction in Paris, who wanted a way to identify the subnormal children in the Parisian schools. Because the tests were graded, it was a simple matter to find the level of test that any particular child would fail to pass. It was not, however, a simple matter to construct adequate tests, so Binet and Simon published revisions and improvements in 1908 and 1911, and this process of refinement still goes on today. By 1911 it was clear that a powerful technique was at hand, and Stern addressed himself to problems of evaluation. One of his most lasting suggestions was that the measure of intellectual development in a child be given by the quotient of the child's mental age—roughly, the level of difficulty he could handle in these graded tests—by his age in years. Stern called this the "mental quotient." Today it is multiplied by 100, and the product is known as the "intelligence quotient," or IQ. Stern's recommenda-tion was soon taken up by many investigators, in particular by Lewis M. Terman, the American psychologist who did more than anyone else to establish Binet's conception of mental testing in the United States.

2. MENTAL AGE, MENTAL RETARDATION, ADVANCE AND ARREST; MENTAL QUOTIENT

We must . . . note how the grade of intelligence of a subject can be derived from his performances in the tests.

Considering the problem schematically, we might think that the grade of intelligence could be expressed by the stage whose tests could just be passed by the child: a subject who readily passed all the tests up through the 9-year ones, but failed with the 10-year and subsequent ones, would, accordingly, possess a nine-year grade of intelligence.

But things are never quite so simple in actuality as they are in theory. The varying tests of any given age-level—we may call them *a. b. c. d. e,*—are not all equally difficult for all children, but there are, on the contrary, quite remarkable individual variations. One child passes *a* to *d*, but fails with *e*; another passes *a*, *c* and *e*, but not *b* and *d*. This is due in part to momentary fluctuations of attention, fatigue, etc., that must, of course, always be reckoned with, but in part also to qualitative differences in intelligence. The correlation between the different phases of intellectual functions is truly never so high that a positive accomplishing of test *a* must necessarily entail a like accomplishing of the approximately 'equally difficult' tests *b*, *c* and *d*.

And so it comes about that there is no hard and fast boundary between the age-level that a child passes completely and the levels that are unquestionably beyond his powers; rather is there an intermediate territory of greater or less extent within which successes and failures are scattered in irregular fashion: we shall call this the area of irregularity. It is impossible to derive a mean or average value from the data afforded by this area without proceeding in a somewhat arbitrary manner, but the formula proposed by Binet and Simon seems to have answered very well so far.

According to it, one first ascertains up to what age-level the tests are passed without failure (save that possible failure with a single test is not counted, because such failure may have been due to a momentary lapse of attention). This age-level is taken as the basis, but every five tests passed in levels above it are counted as one more year. If, then, a child should pass all tests (save a single one) to and including the six-year level and in addition three tests each in the 7th, the 8th, and the 9th year and one test also in the 10th year, these ten additional tests would be counted as two years, and the child would obtain for the net value of his intelligence, 6 + 2 years, *i.e.,* his intelligence would be rated as that of an 8-year old child.

This net value in terms of which the total intelligence of the subject is graded has, therefore, the significance of an age-designation: it indicates that the intelligence of the child tested is equivalent to the average intelligence of the children of the age stated. We thus arrive at the concept of *mental age,* which is the cardinal feature of the method of graded tests.

Now mental age must not, of course, be thought of as an absolutely unequivocal determination of a subject's intelligence, but only as a very rough quantitative characterization of its value, without any implication as to qualitative differences, because one and the same mental age can be figured from the most varied sorts of distribution of passed and failed tests. But this very thing appears to constitute an advantage, rather than a disadvantage of the concept of mental age, for it gives expression to a fundamental psychological fact . . . that, on account of the purely formal character of intelligence and the lack of complete correlation among its constituent capacities, there never is a real phenomenological equivalence between the intelligence of two persons: what we do have is rather a teleological equivalence—when measured in terms of the single function of all intelligence, namely, adaptation to new requirements. And for this equivalence of two intelligences mental age furnishes an approximate measure, despite the fact that their equivalence does not mean their identity.

.   .   .   .   .

The area of irregularity, again, affords another value in addition to mental age, viz.: the *range of irregularity*. A child whose successes and failures are strewn irregularly over test-levels from 6 to 10 years has the same mental age, to be sure, but a very different range of irregularity, when compared with another whose mixture of successes and failures lies in the 7th to the 9th years only.

.   .   .   .   .

But let us return to mental age. The full significance of this final value is disclosed only when we consider it in relation to other circumstances. It can evidently be related to other quantitative scales, like chronological age, school grade and school standing, or we can find out how it varies with certain qualitative conditions, like social level, type of school, nationality and the like.

Doubtless most significant is the relation of mental age to the actual *chronological age* of the subject, for, as already said, a certain mental level goes normally with a certain age, so that the relation of mental to chronological age indicates the amount of discrepancy between the intelligence present and that required (in the sense of a norm to be expected), and in this way affords an expression for the degree of the child's intellectual endowment.

Up to now this discrepancy has always been computed in the simple form of the difference between the two ages, which, when negative gave the absolute *mental retardation,* when positive the absolute *mental advance* of the child in terms of years. Thus, if mental retardation $= -2$, the child's mental development is two years behind the normal level of his age.

It is perfectly clear how valuable the measurement of mental retardation is, particularly in the investigation of abnormal children. It has, however, been shown recently that the simple computation of the absolute difference between the two ages is not entirely adequate for this purpose, because this difference does not mean the same thing at different ages . . . Only when children of approximately equal age-levels are under investigation can this value suffice: for all other cases the introduction of the *mental quotient* will be recommended farther on . . . This value expresses not the difference, but the ratio of mental to chronological age and is thus partially independent of the absolute magnitude of chronological age. The formula is, then: mental quotient $=$ mental age $\div$ chronological age. With children who are just at their normal level, the value is 1, with those who are advanced, the value is greater than unity, with those mentally retarded, a proper fraction. The more pronounced the feeble-mindedness, the smaller the value of the fraction.

Another and last concept that 'mental age' supplies is that of *mental arrest.* This applies only to feeble-minded individuals and means a mental age that is not exceeded, despite increase of chronological age.

XII COMPARATIVE PSYCHOLOGY

The evolutionary doctrine of *continuity* of species meant for some psychologists that the mental processes of human beings might be duplicated, in varying degrees, in lower forms. Thus the Darwinian tradition was soon to divert psychology from its ancient and exclusive preoccupation with the mind of man. One of the first scholars to make this break with the past was G. J. Romanes (1882). Since the dominant psychology of his time was concerned with the subjective contents of the human mind, it was natural for him to try to make his comparative psychology subjective for the animal mind. His problem was considerable: how is one to discover consciousness in creatures that do not speak? Romanes decided to use the observed behavior of the animal as the basis for his inferences about their subjective states. A little later, C. Lloyd Morgan (1894), although he agreed with Romanes that behavior must be used to form inferences about the animal mind, nevertheless enjoined psychologists against imputing to the animal any mental process more complex than the simplest required to account for the observed behavior. This restriction is well known as Lloyd Morgan's canon. Jacques Loeb (1899), however, went further. He proposed that comparative psychology dispense altogether with conscious processes and regard behavior—at least animal behavior—as forced movements determined by the operation of physical principles. Such movements he called tropisms. H. S. Jennings (1906), on the other hand, attributed psychological processes to animals, yet he was not arguing for the interpretation of behavior in terms of subjective states. He was saying, rather, that the objective evidence for psychological processes is as valid for the amoeba as it is for any human being other than the observer himself, and that subjective evidence from organisms other than oneself is simply inaccessible. Thus, in a way, Loeb and Jennings both represent a movement in comparative psychology away from the subjectivism of Romanes and Morgan, a movement that dominates comparative psychology even today.

This short span of time—from Romanes in 1882 to Jennings in 1906—contains the essence of how Darwin's theory of evolution created the modern field of comparative psychology. At first, psychologists expected to discover in animals the signs of incipient mentality that might be expected if the mind of man were on a continuum with that of animals. Soon, however, they were no longer so interested in finding these mental manifestations, for they had come to realize that the concept of mind, either in men or in animals, is merely an inference from observed behavior. With this new realization, psychologists began to look for the evidence of continuity in behavioral, rather than in mental, processes. In America, the most effective spokesman for this new approach was John B. Watson (see No. 94).

87   GEORGE JOHN ROMANES (1848–1894) ON COMPARATIVE PSYCHOLOGY, 1882

G. J. Romanes, *Animal Intelligence* (London, 1882), pp. 1–10.

Romanes started with two premises and was led to two conclusions. The first premise was that psychology is the study of the mind; the second was that the difference between the mind of man and of an animal is greater or smaller depending upon the degree of evolutionary separation between them. In other words, he assumed that the Darwinian continuity of species manifests itself in mental processes as well as in physiological ones and that man's mind lies on a continuum with animals'. From these premises, Romanes concluded that the study of the animal mind is appropriate for psychologists and that animals possess in varying degrees the mental characteristics of man, including, of course, consciousness. Romanes felt that knowledge of the mind's operations can be had only subjectively, with each man studying his own mind, and that such direct knowledge cannot be obtained for animals below man.

With animals, it is clearly necessary, Romanes said, to infer the mental state from the observable behavior, but he also realized that the same indirect approach is inevitable as soon as one tries to study the mental processes of any human being other than oneself. Thus, his approach was frankly analogical: one makes the inference of mental processes in others by analogy with oneself. Romanes' procedure with animals was, however, also more anecdotal than experimental, and it was to this feature that other investigators were soon to object. Later on, Loeb, Jennings, and Watson were all to challenge Romanes' primary thesis that the business of psychology is to infer subjective states from objective data. Ultimately the behavioral data themselves became the subject matter of comparative psychology and the implications of animal consciousness were ignored.

Before we begin to consider the phenomena of mind throughout the animal kingdom it is desirable that we should understand, as far as possible, what it is that we exactly mean by mind. Now, by mind we may mean two very different things, according as we contemplate it in our own individual selves, or in other organisms. For if we contemplate our own mind, we have an immediate cognizance of a certain flow of thoughts or feelings, which are the most ultimate things, and indeed the only things, of which we are cognisant. But if we contemplate mind in other persons or organisms, we have no such immediate cognizance of thoughts or feelings. In such cases we can only *infer* the existence and the nature of thoughts and feelings from the activities of the organisms which appear to exhibit them. Thus it is that we may have a subjective analysis of mind and an objective analysis of mind—the difference between the two consisting in this, that in our subjective analysis we are restricted to the limits of a single isolated mind which we call our own, and within the territory of which we have immediate cognizance of all the processes that are going on, or at any rate of all the processes that fall within the scope of our introspection. But in our objective analysis of other or foreign minds we have no such immediate cognizance; all our knowledge of their operations is derived, as it were, through the medium of ambassadors—these ambassadors being the activities of the organism. Hence it is evident that in our study of animal intelligence we are wholly restricted to the objective method. Starting from what I know subjectively of the operations of my own individual mind, and the activities which in my own organism they prompt, I proceed by analogy to infer from the observable activities of other organisms what are the mental operations that underlie them.

Now, in this mode of procedure what is the kind of activities which may be regarded as indicative of mind? I certainly do not so regard the flowing of a river or the blowing of the wind. Why? First, because the objects are too remote in kind from my own organism to admit of my drawing any reasonable analogy between them and it; and, secondly, because the activities which they present are of invariably the same kind under the same circumstances; they afford no evidence of feeling or purpose. In other words, two conditions require to be satisfied before we even begin to imagine that observable activities are indicative of mind: first, the activities must be displayed by a living organism; and secondly, they must be of a kind to suggest the presence of two elements which we recognise as the distinctive characteristics of mind as such—consciousness and choice.

So far, then, the case seems simple enough. Wherever we see a living organism apparently exerting intentional choice, we might infer that it is conscious choice, and therefore that the organism has a mind. But further reflection shows us that this is just what we cannot do; for although it is true that there is no mind without the power of conscious choice, it is not true that all apparent choice is due to mind. In our own organisms, for instance, we find a great many adaptive movements performed without choice or even consciousness coming into play at all —such, for instance, as in the beating of our hearts. And not only so, but physiological experiments and pathological lesions prove that in our own and in other organisms the mechanism of the nervous system is sufficient, without the intervention of consciousness, to produce muscular movements of a highly co-ordinate and apparently intentional character. Thus, for instance, if a man has his back broken in such a way as to sever the nervous connection between his brain and lower extremities, on pinching or tickling his feet they are drawn suddenly away from the irritation, although the man is quite unconscious of the adaptive movement of his muscles; the lower nerve-centres of the spinal cord are competent to bring about this movement of adaptive response without requiring to be directed by the brain. This non-mental operation of the lower nerve-centres in the production of apparently intentional movements is called Reflex Action, and the cases of its occurrence, even within the limits of our own organism, are literally numberless. Therefore, in view of such non-mental nervous adjustment, leading to movements which are only in appearance intentional, it clearly becomes a matter of great difficulty to say in the case of the lower animals whether any action which appears to indicate intelligent choice is not really action of the reflex kind.

On this whole subject of mind-like and yet not truly mental action I shall have much to say in my subsequent treatise, where I shall be concerned among other things with tracing the probable genesis of mind from non-mental antecedents. But here it is sufficient merely to make this general statement of the fact, that even within the experience supplied by our own organisms adaptive movements of a highly complex and therefore apparently purposive character may be performed without any real purpose, or even consciousness of their performance. It thus becomes evident that before we can predicate the bare existence of mind in the lower animals, we need some yet more definite criterion of mind than that which is supplied by the adaptive actions of a living organism, howsoever apparently intentional such actions may be. Such a criterion I have now to lay down, and I think it is one that is as practically adequate as it is theoretically legitimate.

Objectively considered, the only distinction between adaptive movements due to reflex action and adaptive movements due to mental perception, consists in the former depending on inherited mechanisms within the nervous system being so constructed as to effect *particular* adaptive movements in response to *particular* stimulations, while the latter are independent of any such inherited adjustment of special mechanisms to the exigencies of special circumstances. Reflex actions under the influence of their appropriate stimuli may be compared to the actions of a machine under the manipulations of an operator; when certain springs of action are touched by certain stimuli, the whole machine is thrown into appropriate movement; there is no room for choice, there is no room for uncertainty; but as surely as any of these inherited mechanisms are affected by the stimulus with reference to which it has been constructed to act, so surely will it act in precisely the same way as it always has acted. But the case with conscious mental adjustment is quite different. For, without at present going into the question concerning the relation of body and mind, or waiting to ask whether cases of mental adjustment are not really quite as *mechanical* in the sense of being the necessary result or correlative of a chain of physical sequences due to a physical stimulation, it is enough to point to the variable and incalculable character of mental adjustments as distinguished from the constant and foreseeable character of reflex adjustments. All, in fact, that in an objective sense we can mean by a mental adjustment is an adjustment of a kind that has not been definitely fixed by heredity as the only adjustment possible in the given circumstances of stimulation. For were there no alternative of adjustment, the case, in an animal at least, would be indistinguishable from one of reflex action.

It is, then, adaptive action by a living organism in cases where the inherited machinery of the nervous system does not furnish data for our prevision of what the adaptive action must necessarily be—it is only here that we recognise the objective evidence of mind. The criterion of mind, therefore, which I propose, and to which I shall adhere throughout the present volume, is as follows:—Does the organism learn to make new adjustments, or to modify old ones, in accordance with the results of its own individual experience? If it does so, the fact cannot be due merely to reflex action in the sense above described, for it is impossible that heredity can have provided in advance for innovations upon, or alterations of, its machinery during the lifetime of a particular individual.

. . . I may here explain that in my use of this criterion I shall always regard it as fixing only the upper limit of non-mental action; I shall never regard it as fixing the lower limit of mental action. For it is clear that long before mind has advanced sufficiently far in the scale of development to become amenable to the test in question, it has probably begun to dawn as nascent subjectivity. In other words, because a lowly organised animal does *not* learn by its own individual experience, we may not therefore conclude that in performing its natural or ancestral adaptations to appropriate stimuli consciousness, or the mind-element, is wholly absent; we can only say that this element, if present, reveals no evidence of the fact. But, on the other hand, if a lowly organised animal *does* learn by its own individual experience, we are in possession of the best available evidence of conscious memory leading to intentional adaptation. Therefore our criterion applies to the upper limit of non-mental action, not to the lower limit of mental.

Of course to the sceptic this criterion may appear unsatisfactory, since it depends, not on direct knowledge, but on inference. Here, however, it seems enough to point out, as already observed, that it is the best criterion available; and further, that scepticism of this kind is logically bound to deny evidence of mind, not only in the case of the lower animals, but also in that of the higher, and even in that of men other than the sceptic himself. For all objections which could apply to the use of this criterion of mind in the animal kingdom would apply with equal force to the evidence of any mind other than that of the individual objector. This is obvious, because, as I have already observed, the only evidence we can have of objective mind is that which is furnished by objective activities; and as the subjective mind can never become assimilated with the objective so as to learn by direct feeling the mental processes which there accompany the objective activities, it is clearly impossible to satisfy any one who may choose to doubt the validity of inference, that in any case other than his own mental processes ever do accompany objective activities. Thus it is that philosophy can supply no demonstrative refutation of idealism, even of the most extravagant form. Common sense, however, universally feels that analogy is here a safer guide to truth than the sceptical demand for impossible evidence; so that if the objective existence of other organisms and their activities is granted—without which postulate comparative psychology, like all the other sciences, would be an unsubstantial dream—common sense will always and without question conclude that the activities of organisms other than our own, when analogous to those activities of our own which we know to be accompanied by certain mental states, are in them accompanied by analogous mental states.

The theory of animal automatism, therefore, which is usually attributed to Descartes (although it is not quite clear how far this great philosopher really entertained the theory), can never be accepted by common sense; and even as a philosophical speculation it will be seen, from what has just been said, that by no feat of logic is it possible to make the theory apply to animals to the exclusion of man. The expression of fear or affection by a dog involves quite as distinctive and complex a series of neuro-muscular actions as does the expression of similar emotions by a human being; and therefore, if the evidence of corresponding mental states is held to be inadequate in the one case, it must in consistency be held similarly inadequate in the other. And likewise, of course, with all other exhibitions of mental life.

It is quite true, however, that since the days of Descartes—or rather, we might say, since the days of Joule—the question of animal automatism has assumed a new or more defined aspect, seeing that it now runs straight into the most profound and insoluble problem that has ever been presented to human thought—viz. the relation of body to mind in view of the doctrine of the conservation of energy . . . Here I desire only to make it plain that the mind of animals must be placed in the same category, with reference to this problem, as the mind of man; and that we cannot without gross inconsistency ignore or question the evidence of mind in the former, while we accept precisely the same kind of evidence as sufficient proof of mind in the latter.

And this proof, as I have endeavoured to show, is in all cases and in its last analysis the fact of a living organism showing itself able to learn by its own individual experience. Wherever we find an animal able to do this, we have the same right to predicate mind as existing in such an animal that we have to predicate it as existing in any human being other than ourselves. For instance, a dog has always been accustomed to eat a piece of meat when his organism requires nourishment, and when his olfactory nerves respond to the particular stimulus occasioned by the proximity of the food. So far, it may be said, there is no evidence of mind; the whole series of events comprised in the stimulations and muscular movements may be due to reflex action alone. But now suppose that by a number of lessons the dog has been taught not to eat the meat when he is hungry until he receives a certain verbal signal: then we have exactly the same kind of evidence that the dog's actions are prompted by mind as we have that the actions of a man are so prompted. Now we find that the lower down we go in the animal kingdom, the more we observe reflex action, or non-mental adjustment, to predominate over volitional action, or mental adjustment. That is to say, the lower down we go in the animal kingdom, the less capacity do we find for changing adjustive movements in correspondence with changed conditions; it becomes more and more hopeless to *teach* animals—that is, to establish associations of ideas; and the reason of this, of course, is that ideas or mental units become fewer and less definite the lower we descend through the structure of mind.

·　　·　　·　　·　　·

The terms sensation, perception, emotion, and volition need not here be considered. I shall use them in their ordinary psychological significations; and although I shall subsequently have to analyse each of the organic or mental states which they respectively denote, there will be no occasion in the present volume to enter upon this subject. I may, however, point out one general consideration to which I shall throughout adhere. Taking it for granted that the external indications of mental processes which we observe in animals are trustworthy, so that we are justified in inferring particular mental states from particular bodily actions, it follows that in consistency we must everywhere apply the same criteria.

For instance, if we find a dog or a monkey exhibiting marked expressions of affection, sympathy, jealousy, rage, &c., few persons are sceptical enough to doubt that the complete analogy which these expressions afford with those which are manifested by man, sufficiently prove the existence of mental states analogous to those in man of which these expressions are the outward and visible signs. But when we find an ant or a bee apparently exhibiting by its actions these same emotions, few persons are sufficiently non-sceptical not to doubt whether the outward and visible signs are here trustworthy as evidence of analogous or corresponding inward and mental states. The whole organisation of such a creature is so different from that of a man that it becomes questionable how far analogy drawn from the activities of the insect is a safe guide to the inferring of mental states—particularly in view of the fact that in many respects, such as in the great preponderance of 'instinct' over 'reason,' the psychology of an insect is demonstrably a widely different thing from that of a man. Now it is, of course, perfectly true that the less the resemblance the less is the value of any analogy built upon the resemblance, and therefore that the inference of an ant or a bee feeling sympathy or rage is not so valid as is the similar inference in the case of a dog or a monkey. Still it *is* an inference, and, so far as it goes, a valid one—being, in fact, the only inference available. That is to say, if we observe an ant or a bee apparently exhibiting sympathy or rage, we must either conclude that some psychological state resembling that of sympathy or rage is present, or else refuse to think about the subject at all; from the observable facts there is no other inference open. Therefore, having full regard to the progressive weakening of the analogy from human to brute psychology as we recede through the animal kingdom downwards from man, still, as it is the only analogy available, I shall follow it throughout the animal series.

It may not, however, be superfluous to point out that if we have full regard to this progressive weakening of the analogy, we must feel less and less certain of the real similarity of the mental states compared; so that when we get down as low as the insects, I think the most we can confidently assert is that the known facts of human psychology furnish the best available pattern of the probable facts of insect psychology. Just as the theologians tell us—and logically enough—that if there is a Divine Mind, the best, and indeed only, conception we can form of it is that which is formed on the analogy, however imperfect, supplied by the human mind; so with 'inverted anthropomorphism' we must apply a similar consideration with a similar conclusion to the animal mind. The mental states of an insect may be widely different from those of a man, and yet most probably the nearest conception that we can form of their true nature is that which we form by assimilating them to the pattern of the only mental states with which we are actually acquainted. And this consideration, it is needless to point out, has a special validity to the evolutionist, inasmuch as upon his theory there must be a psychological, no less than a physiological, continuity extending throughout the length and breadth of the animal kingdom.

461

88   CONWY LLOYD MORGAN (1852–1936) ON LLOYD MORGAN'S CANON, 1894

C. Lloyd Morgan, *An Introduction to Comparative Psychology* (London, 1894), pp. 47–59.

In many respects, Lloyd Morgan's contribution to psychology resembles Romanes'. Both relied on the concept of continuity in evolution as a justification for comparative psychology. Both believed that psychology is the study of the mind. Both saw the same obstacle confronting the study of any mind other than one's own. Both outlined a program for comparative psychology that consisted of two phases: the observation of behavior followed by an interpretation of this behavior in terms of one's own subjective knowledge of mind. The way in which Lloyd Morgan and Romanes differed is precisely the way in which Lloyd Morgan proved to be important for the subsequent development of comparative psychology. His special contribution was a cautionary dictum to comparative psychologists, which has since been referred to as "Lloyd Morgan's canon." He said that the behavior of an animal should never be interpreted in terms of a higher mental process when a simpler one would suffice. With this injunction, he was trying to restrain the undisciplined anthropomorphism—really anthropopsychism—of some of his fellow Darwinians. His justification for this principle was the theory of evolution itself. Because mind evolved from lower to higher, he said, the postulation of a higher mental process implies all the others below it in the evolutionary scale. For this reason, the proper study of animals' minds should anticipate increasing complexity from the lower to the higher forms, rather than simply assuming man's mental processes for all animals. Most comparative psychologists today still find Lloyd Morgan's advice salutary.

We are now in a position to see clearly what is the distinctive peculiarity of the study of mind in beings other than our own individual selves. Its conclusions are reached not by a singly inductive process, as in Chemistry or Physics, in Astronomy, Geology, Biology, or other purely objective science, but by a doubly inductive process. Inductions reached through the objective study of certain physical manifestations have to be interpreted in terms of inductions reached through the introspective study of mental processes. By induction I mean the observation of facts, the framing of hypotheses to comprise the facts, and the verification of the hypotheses by constant reversion to the touchstone of fact. Our conclusions concerning the mental processes of beings other than our own individual selves are, I repeat, based on a two-fold induction. First the psychologist has to reach, through induction, the laws of mind as revealed to him in his own conscious experience. Here the facts to be studied are facts of consciousness, known at first-hand to him alone among mortals; the hypotheses may logically suggest themselves, in which case they are original so far as the observer himself is concerned, or they may be derived,—that is to say, suggested to the observer by other observers; the verification of the hypotheses is again purely subjective, original or derived theories being submitted to the touchstone of individual experience. This is the one inductive process. The other is more objective. The facts to be observed are external phenomena, physical occurrences in the objective world; the hypotheses again may be either original or derived; the verification is objective, original or derived theories being submitted to the touchstone of observable phenomena. Both inductions, subjective and objective, are necessary. Neither can be omitted without renouncing the scientific method. And then finally the objective manifestations in conduct and activity have to be interpreted in terms of subjective experience. The inductions reached by the one method have to be explained in the light of inductions reached by the other method.

I am anxious to make this matter quite clear, and I will therefore endeavour to illustrate it diagrammatically. In the first diagram (Fig. 7) the line *ab* represents the conduct, activities, and other objective phe-

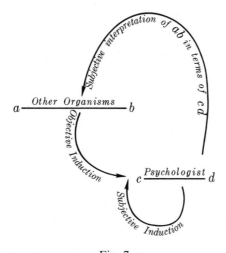

Fig. 7.

nomena exhibited by other beings or organisms than the individual psychologist, while *cd* represents the states of consciousness of which he alone has direct knowledge.

Then the diagram is intended to show how the psychologist must combine both objective induction and subjective induction, that he may reach a subjective interpretation of *ab* in terms of *cd*.

· · · · ·

Now it is idle to assert that one set of inductions is more important than the other, since both are essential. But there can be no question that the subjective inductions are in some respect more subtle and difficult and delicate than the inductions concerning objective phenomena. There can be no question that false assumptions and vague generalizations more commonly pass muster with regard to mental processes than with regard to their physical manifestations. And there can be no question that in the systematic training of the comparative psychologist the subjective aspect is not *less* important than the objective aspect.

The question now arises whether in passing from human to animal psychology any other method of interpretation is possible than that which holds good for the former. Can the zoological psychologist afford to dispense with that systematic training in introspective or subjective analysis and induction which is absolutely essential for the student of human psychology? I venture to contend that he cannot. The scheme of interpretation exhibited diagrammatically in Fig. 7 holds good I maintain as well for animal psychology as for the psychology of man.

.   .   .   .   .

Unfortunately many able men who are eminently fitted to make and record exact observations on the habits and activities of animals have not undergone the training necessary to enable them to deal with the psychological aspect of the question. The skilled naturalist or biologist is seldom also skilled in psychological analysis. Notwithstanding therefore the admirable and invaluable observations of our great naturalists, we cannot help feeling that their psychological conclusions are hardly on the same level as that reached by their conclusions in the purely biological field.

For in the study of animal psychology as a branch of scientific inquiry, it is necessary that accurate observation, and a sound knowledge of the biological relationships of animals, should go hand in hand with a thorough appreciation of the methods and results of modern psychology. The only fruitful method of procedure is the interpretation of facts observed with due care in the light of sound psychological principles.

What some of these principles are we have considered, or shall consider, in this work. There is one basal principle, however, the brief exposition of which may fitly bring to a close this chapter. It may be thus stated:—*In no case may we interpret an action as the outcome of the exercise of a higher psychical faculty, if it can be interpreted as the outcome of the exercise of one which stands lower in the psychological scale.*

To this principle several objections, none of them however of any real weight, may be raised. First there is the sentimental objection that it is ungenerous to the animal. In dealing with one's fellow-man it is ungenerous to impute to him lower motives for his actions when they may have been dictated by higher motives. Why should we adopt a different course with the poor dumb animal from that which we should adopt with our human neighbour? In the first place, it may be replied, this objection starts by assuming the very point to be proved. The scientific problem is to ascertain the limits of animal psychology. To assume that a given action may be the outcome of the exercise of either a higher or a lower faculty, and that it is more generous to adopt the former alternative, is to assume the existence of the higher faculty, which has to be proved. In the case of our neighbours we have good grounds for knowing that such and such a deed may have been dictated by either a higher or a lower motive. If we had equally good grounds for knowing that the animal was possessed of both higher and lower faculties, the scientific problem would have been solved; and the attribution of the one or the other, in any particular case, would be a purely individual matter of comparatively little general moment. In the second place, this generosity, though eminently desirable in the relations of practical social life, is not precisely the attitude which a critical scientific inquiry demands. Moreover, an ungenerous interpretation of one's neighbour's actions may lead one to express an unjust estimate of his moral character and thus to do him grave social wrong; but an ungenerous interpretation of the faculties of animals can hardly be said to be open to like practical consequences.

A second objection is, that by adopting the principle in question we may be shutting our eyes to the simplest explanation of the phenomena. Is it not simpler to explain the higher activities of animals as the direct outcome of reason or intellectual thought, than to explain them as the complex results of mere intelligence or practical sense-experience? Undoubtedly it may in many cases seem simpler. It is the apparent simplicity of the explanation that leads many people naively to adopt it. But surely the simplicity of an explanation is no necessary criterion of its truth. The explanation of the genesis of the organic world by direct creative fiat, is far simpler than the explanation of its genesis through the indirect method of evolution. The explanation of instinct and early phases of intelligence as due to inherited habit, individually acquired, is undoubtedly simpler than the explanation which Dr Weismann would substitute for it. The formation of the cañon of the Colorado by a sudden rift in the earth's crust, similar to those which opened during the Calabrian earthquakes, is simpler than its formation by the fretting of the stream during long ages under varying meteorological conditions. In these cases and in many others the simplest explanation is not the one accepted by science. Moreover, the simplicity of the explanation of the phenomena

465

of animal activity as the result of intellectual processes, can only be adopted on the assumption of a correlative complexity in the mental nature of the animal as agent. And to assume this complexity of mental nature on grounds other than those of sound induction, is to depart from the methods of scientific procedure.

But what, it may be asked, is the logical basis upon which this principle is founded? If it be true that the animal mind can only be interpreted in the light of our knowledge of human mind, why should we not use this method of interpretation freely, frankly, and fully? Is there not some contradiction in refusing to do so? For, first, it is contended that we must use the human mind as a key by which to read the brute mind, and then it is contended that this key must be applied with a difference. If we apply the key at all, should we not apply it without reservation?

This criticism might be valid if we were considering the question apart from evolution. Here evolution is postulated. The problem is this: (1) Given a number of divergently ascending grades of organisms, with divergently increasing complexity of organic structure and correlated activities: (2) granted that associated with the increasing organic complexity there is increasing mental or psychical complexity: (3) granted that in man the organic complexity, the complexity of correlated activities, and the associated mental or psychical complexity, has reached the maximum as yet attained: (4) to gauge the psychical level to which any organism has been evolved. As we have already seen, we are forced, as men, to gauge the psychical level of the animal in terms of the only mind of which we have first-hand knowledge, namely the human mind. But how are we to apply the gauge?

There would appear to be three possible methods, which are exemplified in Fig. 9. Let *a* represent the psychical stature of man, and 1, 2, 3, ascending faculties or stadia in mental development. Let *bc* represent two animals the psychical stature of each of which is to be gauged. It may be gauged first by the "method of levels," according to which the faculties or stadia are of constant value. In the diagram, *b* has not quite reached the level of the beginning of the third or highest faculty, while *c* has only just entered upon the second stadium. Secondly, it may be gauged by the "method of uniform reduction." In both *b* and *c* we have all three faculties represented in the same ratio as in *a*, but all uniformly reduced. And thirdly, it may be gauged by the "method of variation," according to which any one of the faculties 1, 2, or 3, may in *b* and *c* be either increased or reduced relatively to its development in *a*. Let us suppose, for example, that *b* represents the psychical stature of the dog. Then, according to the interpretation on the method of levels, he possesses the lowest faculty (1) in the same degree as man; in the

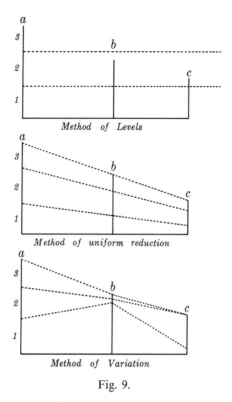

Method   of   Levels

Method   of   uniform   reduction

Method   of   Variation

Fig. 9.

faculty (2) he somewhat falls short of man; while in the highest faculty (3) he is altogether wanting. According to the interpretation on the method of uniform reduction he possesses all the faculties of man but in a reduced degree. And according to the interpretation on the method of variation he excels man in the lowest faculty, while the other two faculties are both reduced but in different degrees. The three "faculties" 1, 2, 3, are not here intended to serve any other purpose than merely to illustrate the three methods of interpretation.

On the principles of evolution we should unquestionably expect that those mental faculties which could give decisive advantage in the struggle for existence would be developed in strict accordance with the divergent conditions of life. Hence it is the third method, which I have termed the method of variation, which we should expect *a priori* to accord most nearly with observed facts. And so far as we can judge from objective observation (the only observation open to us) this would appear to be the case. Presumably there are few observers of animal habit and intelligence who would hesitate in adopting the method of variation as the most probable mode of interpretation. But note that while it is the most probable it is also the most difficult mode of interpretation. Ac-

467

cording to the method of levels the dog is just like me, without my higher faculties. According to the method of uniform reduction he is just like me, only nowise so highly developed. But according to the method of variation there are many possibilities of error in estimating the amount of such variation. Of the three methods that of variation is the least anthropomorphic, and therefore the most difficult.

In the diagram by which the method of variation is illustrated, the highest faculty 3 is in *c* reduced to zero,—in other words, is absent. It may, however, be objected that this is contrary to the principles of evolution, since the presence of any faculty in higher types involves the germ of this faculty in lower types. This criticism only holds good, however, on the assumption that the evolution of higher faculties out of lower faculties is impossible. Those evolutionists who accept this assumption as valid are logically bound to believe either (1) that all forms of animal life from the amœba upwards have all the faculties of man, only reduced in degree and range, and to interpret all animal psychology on a method of reduction (though not necessarily uniform reduction), or (2) that in the higher forms of life the introduction of the higher faculties has been effected by some means other than that of natural evolution. I am not prepared to accept the assumption as valid; and it will be part of my task in future chapters to consider how the transition from certain lower to certain higher phases of mental development may have been effected.

If this be so it is clear that any animal may be at a stage where certain higher faculties have not yet been evolved from their lower precursors; and hence we are logically bound not to assume the existence of these higher faculties until good reasons shall have been shown for such existence. In other words, we are bound to accept the principle above enunciated: that in no case is an animal activity to be interpreted as the outcome of the exercise of a higher psychical faculty, if it can be fairly interpreted as the outcome of the exercise of one which stands lower in the psychological scale.

89 JACQUES LOEB (1859–1924) ON ASSOCIATIVE MEMORY, 1899

Jacques Loeb, *Einleitung in die vergleichende Gehirnphysiologie und vergleichende Psychologie mit besonderer Berücksichtigung der wirbellosen Thiere* (Leipzig, 1899), chap. 15. The English translation, *Comparative Physiology of the Brain and Comparative Psychology* (New York: Putnam, 1900), was made by Anne Leonard Loeb and incorporated extensive changes by the author.

Loeb was an agent of the movement in comparative psychology toward objectivism (later to be called "behaviorism"). Unlike Romanes and

Lloyd Morgan, he did not favor a comparative psychology in which observed behavior is used to infer subjective experience. Instead, he argued that an account of the evolution of mental processes need contain only the objective facts of behavior. Using the tropisms of plants as a paradigm, Loeb contended that the movements of animals are in like manner automatic reactions governed by fields of physical energy. The criterion of consciousness is, he thought, "associative memory," the capacity of an organism to learn from experience. Although he and his followers discovered many instances, especially among organisms lower on the evolutionary scale, where a wholly mechanistic account of behavior is possible, it did not seem plausible to most psychologists that all behavior, human and animal, or even a significantly large part of it, could be explained in this way.

1. The most important problem in the physiology of the central nervous system is the analysis of the mechanisms which give rise to the so-called psychic phenomena. The latter appear, invariably, as a function of an elementary process, namely, the activity of the associative memory. By associative memory I mean the two following peculiarities of our central nervous system: First, that processes which occur there leave an impression or trace by which they can be reproduced even under different circumstances than those under which they originated. This peculiarity can be imitated by machines like the phonograph. Of course, we have no right to assume that the traces of processes in the central nervous system are analogous to those in the phonograph. The second peculiarity is, that two processes which occur simultaneously or in quick succession will leave traces which fuse together, so that if later one of the processes is repeated, the other will necessarily be repeated also. The odour of a rose will at the same time reproduce its visual image in our memory, or, even more than that, it will reproduce the recollection of scenes or persons who were present when the same odour made its first strong impression on us. By associative memory we mean, therefore, that mechanism by means of which a stimulus produces not only the effects which correspond to its nature and the specific structure of the stimulated organ, but which produces, in addition, such effects of other causes as at some former time may have attacked the organism almost or quite simultaneously with the given stimulus . . . The chief problem of the physiology of the brain is, then, evidently this: What is the physical character of the mechanism of associative memory? As we said . . . the answer to this question will probably be found in the field of physical chemistry.

I think it can be shown that what the metaphysician calls consciousness are phenomena determined by the mechanism of associative memory. Mach has pointed out that the consciousness of self or the ego is simply a phrase for the fact that certain constituents of memory are constantly or more frequently produced than others . . . The complex of these elements of memory is the "ego" or the "soul," or the personality of the metaphysicians. To a certain extent we are able to enumerate these constituents. They are the visual image of the body so far as it lies in the field of vision, certain sensations of touch which are repeated very frequently, the sound of our own voice, certain interests and cares, a certain feeling of comfort or discomfort according to temperament or state of health, etc. . . .

An inventory of all the memory-constituents of the ego-complex of different persons would show that the consciousness of self is not a definite unit, but, as Mach maintains, merely an artificial separation of those constituents of memory which occur most frequently in our perceptions. These will necessarily be subject to considerable variation in the same person in the different periods of life.

If we speak of loss or an interruption of consciousness, we mean a loss or an interruption of the activity of associative memory. If a faint is caused directly by lack of oxygen or indirectly by a disturbance in the circulatory system, the activity of associative memory ceases. This was proved by Speck's experiments on the effects of a low pressure of oxygen. When he breathed air with less than eight per cent. of oxygen, he soon fainted. In these experiments, he had to count the number of respirations. Before he fainted, he became confused in his counting and forgot what happened. When this disturbance in counting began to appear, he knew it was time to discontinue the experiment. When a loss of consciousness is produced by narcotics or anæsthetics, we have again to deal with an interruption in the activity of the associative memory. It is the same in the case of a deep sleep.

The metaphysician speaks of conscious sensations and conscious will. That the will is only a function of the mechanism of the associative memory can be proved. We speak of conscious volition if an idea of the resulting final complex of sensations is present before the movements causing it have taken place or have ceased. In volition three processes occur. The one is an innervation of some kind which may be caused directly or indirectly by an external stimulus. This process of innervation produces two kinds of effects. The one effect is the activity of the associative memory which produces the sensations that in former cases accompanied or followed the same innervation. The second effect is a coördinated muscular activity. It happens that in such cases the reaction-time for the memory-effect of the innervation is shorter than the time for the muscular effect. When some internal process causes us to open the window, the activity of the associative memory produces the idea of sensations which will follow or accompany the opening of the window sooner than the act of opening really occurs. As we do not realise this any more than we realise the inverted character of the retina-image, we consider the memory-effect of the innervation as the cause of the muscular effect. The common cause of both effects, the innervating process, escapes our immediate observation as our senses do not perceive it. The will of the metaphysicians is then clearly the outcome of an illusion due to the necessary incompleteness of self-observation . . . I think that we are justified in substituting the term activity of associative memory for the phrase consciousness used by the metaphysicians.

2. We have spoken of *associative* memory because the word *memory* is often applied in quite a different sense scientifically, namely, to signify any after-effect of external circumstances. For instance, the term memory has been used to account for the fact that a plant which had been cultivated in the tropics will often not endure low temperatures so well as a plant of the same species which was raised in the north. It is true in this case that preceding conditions influence the ability of the plant to react, but the process differs from the one which we have called associative memory in the lack of associative processes. No definite stimulus produces in a plant, in addition to its own effects, those of another entirely different stimulus which at some former time occurred simultaneously with the given stimulus. It is probable that the tropical plant is somewhat different chemically from the plant raised in the north. This would account for its smaller power of resistance. Further illustrations of a different use of the word memory can easily be given.

Many moths sleep during the day and wake in the evening when it becomes dark. If kept for days in a dark room, they will continue at first to do the same thing. The same is true of certain plants. One might also say in this case that the moth or the plant "remembers" the difference between day and night. It is probable, however, that internal changes take place in the organism, corresponding to the periodic change of day and night, and that these changes continue for a time in the same periodicity, when the animal is kept in the dark.

3. We will then consider the extent of associative memory in the animal kingdom instead of the extent of consciousness among animals. How can we determine whether an animal possesses the mechanism necessary for associative memory? The criteria for the existence of associative memory must form the basis of a future comparative psychology. It will require more observations than we have made at present to give absolutely unequivocal criteria. For the present, we can say that if an animal can learn, that is, if it can be trained to react in a desired way upon certain stimuli (signs), it must possess associative memory. The only fault with this criterion lies in the fact that an animal may be able

to remember (and to associate) and yet may not yield to our attempts to train it. In this case other experiments must be substituted which will prove that the animal does associate or remember.

We may conclude that associative memory is present when an animal responds upon hearing its name called, or when it can be trained upon hearing a certain sound to go to the place where it is usually fed. The optical stimulus of the place where the food is to be found and the sensations of hunger and satiety are not qualitatively the same, but they occur simultaneously in the animal. The fusion or growing together of heterogeneous but by chance simultaneous processes is a sure criterion for the existence of associative memory . . .

Associative memory probably exists in most mammals. The dog which comes when its name is called, which runs away from the whip, which welcomes its master joyfully, has associative memory. In birds, it is likewise present. The parrot learns to talk; the dove finds its way home. In lower Vertebrates, memory is also occasionally found. Tree-frogs, for example, can be trained, upon hearing a sound, to go to a certain place for food. In other frogs, *Rana esculenta,* for instance, no reaction is as yet known which proves the existence of associative memory. Some fishes evidently possess memory; in sharks, however, its existence is doubtful. With regard to the Invertebrates, the question is difficult to determine. The statements of enthusiasts who discover consciousness and resemblance to man on every side should not be too readily accepted.

90 HERBERT SPENCER JENNINGS (1868–1947) ON THE CONTINUITY OF PSYCHOLOGICAL PROCESSES, 1906

H. S. Jennings, *Behavior of the Lower Organisms* (New York: Columbia University Press, 1906), chap. 20.

Jennings, one of the prominent younger American zoologists, argued that the objective evidence for the operation of such psychological proc̆esses as discrimination, attention, or emotion is as valid for lower animal forms as it is for higher animals and man. He also claimed that this evidence suggests a complete continuity of the psychological processes throughout the phyletic animal scale. As to phyletic continuity of the subjective states that accompany these psychological processes, however, Jennings acknowledged his helplessness, for conscious processes, he said, are unknowable for any organism other than oneself, be it another man or an amoeba. Jennings is often likened to Romanes, since they both attributed psychological functions to animals somewhat freely. There are, however, two important differences that earn for Jennings a distinctive place in the history of comparative psychology. First, there is the fact that Jennings sought experimental, rather than anecdotal, evidence for his views. Second, he was content to leave the question of subjective mental states to ultimate philosophical resolution, confining himself to the study of objective psychological processes.

In describing the behavior of lower organisms we have used in the present work, so far as possible, objective terms—those having no implication of psychic or subjective qualities. We have looked at organisms as masses of matter, and have attempted to determine the laws of their movements. In ourselves we find movements and reactions resembling in some respects those of the lower organisms. We draw away from heat and cold and injurious chemicals, just as Paramecium does. Our behavior depends on physiological states, as does that of Stentor. But in ourselves there is the very interesting additional fact that these movements, reactions, and physiological states are often accompanied by subjective states,—states of consciousness. Different states of consciousness are as varied as the different possibilities of reaction; indeed, more varied. In speaking of behavior in ourselves, and as a rule in higher animals, we use terms based on these subjective states, as pleasure and pain, sensation, memory, fear, anger, reason, and the like.

The peculiarity of subjective states is that they can be perceived only by the one person directly experiencing them,—by the subject. Each of us knows directly states of consciousness only in himself. We cannot by observation and experiment detect such states in organisms outside of ourselves. But observation and experiment are the only direct means of studying behavior in the lower organisms. We can reason concerning their behavior, and through reasoning by analogy we may perhaps conclude that they also have conscious states. But reasoning by analogy, when it is afterward tested by observation and experiment, has often shown itself fallacious, so that where it cannot be tested, we must distrust its conclusiveness. Moreover, in different men it leads to different conclusions, so that it does not result in admitted certainty. Hence it seems important to keep the results of observation and experiment distinct from those of reasoning by analogy, so that we may know what is really established. On this account it is customary among most physiologists not to use, in discussing the behavior of the lower organisms, psychic terms, or those implying subjective states. This has the additional ground that the ideal of most scientific men is to explain behavior in terms of matter and energy, so that the introduction of psychic implications is considered superfluous.

While this exclusive use of objective terms has great advantages, it has one possible disadvantage. It seems to make an absolute gulf between the behavior of the lower organisms on the one hand, and that of man and higher animals on the other. From a discussion of the behavior of the lower organisms in objective terms, compared with a discussion of the behavior of man in subjective terms, we get the impression of complete discontinuity between the two.

Does such a gulf actually exist, or does it lie only in our manner of speech? We can best get evidence on this question by comparing the objective features of behavior in lower and in higher organisms. In any animal outside of man, and even in man outside of the self, the existence of perception, choice, desire, memory, emotion, intelligence, reasoning, etc., is judged from certain objective facts—certain things which the organisms do. Do we find in the lower organisms objective phenomena of a similar character, so that the same psychic names would be applied to them if found in higher organisms? Do the objective factors in the behavior of lower organisms follow laws that are similar to the laws of psychic states? Only by comparing the objective factors can we determine whether there is continuity or a gulf between the behavior of lower and higher organisms (including man), for it is only these factors that we know.

Let us then examine some of the concepts employed in discussions of the behavior of higher animals and man, determining whether there exist any corresponding phenomena in lower organisms. We shall not attempt to take into consideration the scholastic definitions of the terms used, but shall judge of them merely from the objective phenomena on which they are based.

When we say that an animal *perceives* something, or that it shows *perception* of something, we base this statement on the observation that it reacts in some way to this thing. On the same basis we could make the statement that Amœba perceives all classes of stimuli which we ourselves perceive, save sound (which is, however, essentially one form of mechanical stimulation). Perception as judged from our subjective experiences means much more: how much of this may be present in animals outside the self we cannot know.

Discrimination is a term based, so far as objective evidence goes, upon the observed fact that organisms react differently to different stimuli. In this sense Paramecium, as we have seen, discriminates acids from alkalies; Amœba discriminates a Euglena cyst from a grain of sand, and in general all lower organisms show discrimination in many phases of their behavior.

Choice is a term based objectively on the fact that the organism accepts or reacts positively to some things, while it rejects or reacts negatively or not at all to others. In this sense all lower organisms show choice, and at this we need not be surprised, for inorganic substances show a similar selectiveness. The distinctive thing about the choice of organisms is that it is regulatory; organisms on the whole choose those things which aid their normal life processes and reject those that do not. This is what justifies the use of the term "choice," as contrasted with the mere selectiveness of inorganic reactions. Choice in this regulatory sense is shown by lower organisms, as we have seen in detail in previous chapters. Choice is not perfect, from this point of view, in either lower or higher organisms. Paramecium at times accepts things that are useless or harmful to it, but perhaps on the whole less often than does man.

.   .   .   .   .

Is not what we call *attention* in higher organisms, when considered objectively, the same phenomenon that we have called the interference of one stimulus with the reaction to another? At the basis of attention lies objectively the phenomenon that the organism may react to only one stimulus even though other stimuli are present which would, if acting alone, likewise produce a response. The organism is then said to attend to the particular stimulus to which it responds. This fundamental phenomenon is clearly present in unicellular organisms. Stentor and Paramecium when reacting to contact with a solid "pay no attention" to a degree of heat or a chemical or an electric current that would produce an immediate reaction in a free individual. On the other hand, individuals reacting to heat or a chemical may not respond to contact with a mass of bacteria, to which they would under other conditions react positively. In our chapter on reaction under two or more stimuli in the infusoria, many examples of this character are given.

Indeed, attention in this objective sense seems a logical necessity for the behavior of any organism having at its command more than a single action. The characteristic responses to two present stimuli may be incompatible with each other. The organism must then react to one or the other, since it cannot react to both; it thus *attends* (objectively) to one, and not to the other. Only in case there is no reaction at all in the presence of two stimuli, or in case its reaction is precisely intermediate between those required by the two, could the basis of attention be considered lacking. An organism behaving in this way would be quickly destroyed as a result of its indecisive and ineffective behavior.

In higher animals and man we distinguish certain different conditions, —"states of feeling," "emotions," "appetites," "desires," and the like. In all cases except the self, these various states are distinguished through the fact that the organism behaves differently in the different conditions, even though the external stimuli may be the same. We find a parallel condition of affairs in the lower organisms. Here, as we have seen, the behavior under given external conditions depends largely on the physiological condition of the individual. Many illustrations of this fact are given in preceding chapters, so that we need not dwell upon it here.

In the lower organisms we can even distinguish a number of states that are parallel, so far as observation can show, with those distinguished

475

and named in higher animals and man. To begin with some of the simpler ones, the objective correlate of hunger can be distinguished at least as low in the scale as Hydra and the sea anemone. These animals, as we have seen, take food only when hungry, and if *very* hungry, will take substances as food which they otherwise reject. Doubtless hunger could be detected in still lower organisms by proper experiments. A resting condition comparable to sleep is found, as we have seen, in the flatworm . . . while there seems to be no indication of such a state in the infusoria . . . *Fatigue* can of course be distinguished in all living things, including separated muscles.

Correlative with hunger, there exists a state which corresponds so far as objective evidence goes with what we should call in higher animals a *desire* for food. Hydra when hungry opens its mouth widely when immersed in a nutritive liquid. In the flatworm, we can distinguish a certain physiological condition in which the animal moves about in an eager, searching way, as if hunting for food. Even in Amœba we find a pertinacity in the pursuit of food . . . such as we would attribute in a higher animal to a desire for it.

All the way up the scale, from Amœba and bacteria to man, we find that organisms react negatively to powerful and injurious agents. In man and higher animals such reactions are usually said to be due to *pain.* In the lower organisms the objective facts are parallel, and naturally lead to the assumption of a physiological state similar to what we have in the higher forms. As to subjective accompaniments of such a state we of course know nothing in animals other than ourselves. The essential cause of the states corresponding to pain is interference with any of the processes of which the organism is the seat, and the correlate in action of these states is a change in movement . . .

A similar basis exists for distinguishing throughout the organic series a physiological state corresponding to that accompanying pleasure in man. This is correlated with a relief from interference with the life processes, or with the uninterrupted progression of these processes.

In man and higher animals we often find a negative reaction to that which is not in itself injurious, but which is usually followed by something injurious. The sight of a wild beast is not injurious, considered by itself, but as preceding actual and injurious contact with this beast, it leads to powerful negative reactions. Such reactions are said to be due to *fear.* In fear there is then a negative reaction to a representative stimulus—one that *stands* for a really injurious stimulation. In lower organisms we find the objective indications of a parallel state of affairs. The infusoria react negatively to solutions of chemicals that are not, so far as we can determine, injurious, though they would naturally, under ordinary circumstances, be immediately followed by a solution so strong as to be injurious. Euglena reacts negatively when darkness affects only its colorless anterior end, though we have reason to believe that it is only the green part of the body which requires the light for the proper discharge of its functions. A much clearer case is seen in the sea urchin, which reacts by defensive movements when a shadow falls upon it, though shade is favorable to its normal functions. Objectively, fear has at its basis the fact that a negative reaction may be produced by a stimulus which is not in itself injurious, provided it leads to an injurious stimulation; this basis we find throughout organisms.

Sometimes higher animals and man are thrown into a "state of fear," such that they react negatively to all sorts of stimuli, that under ordinary circumstances would not cause such a reaction. A similar condition of affairs we have seen in Stentor and the flatworm. After repeated stimulation, they react negatively to all stimuli to which they react at all.

The general fact of which the reactions through fear are only a special example is the following: Organisms react appropriately to *representative* stimuli. That is, they react, not merely to stimuli that are in themselves beneficial or injurious, but to stimuli which lead to beneficial or injurious conditions. This is as true of positive as of negative reactions. It is true of Amœba when it moves toward a solid body that will give it an opportunity to creep about and obtain food. It is true of Paramecium when it settles against solids (even bits of filter paper), because usually such solids furnish a supply of bacteria. It is true of the colorless flagellate Chytridium and the white Hydra, when they move toward a source of light and thus come into the region where their prey congregate. There seems to be no general name for this positive reaction to a representative stimulus. In man we call various subjective aspects of it by different names,—foresight, anticipation, prudence, hope, etc.

The fact that lower as well as higher organisms thus react to representative stimuli is of the greatest significance. It provides the chief condition for the advance of behavior to higher planes. At the basis of reaction of this character lies the simple fact that a *change,* even though neutral in its effect, may cause reaction . . .

Related to these reactions to representative stimuli are certain other characteristics distinguished in the behavior of man and higher animals. The objective side of *memory* and what is called *habit* is shown when the behavior of an organism is modified in accordance with past stimuli received or past reactions given. If the behavior is merely changed in a way that is not regulatory, as by fatigue, we do not call this memory. In memory the reaction is modified in such a way that it is now more adequate to the conditions to be met. Habit and memory in this ob-

477

jective sense are clearly seen in the Crustacea, and in the low accœlous flatworm Convoluta . . . Something of a similar character is seen even in the protozoan Stentor. After reacting to a weak stimulus which does not lead to an injurious one it ceases to react when this stimulus is repeated, while if the weak stimulus does lead to an injurious one, the animal changes its behavior so as to react next time in a more effective way; and it repeats this more effective reaction at the next incidence of the stimulus. Habit and memory, objectively considered, are based on the law of the resolution of physiological states . . . which may be set forth in application to the present subject as follows: If a given physiological state, induced by a stimulus, is repeatedly resolved into a succeeding state, this resolution becomes easier, and may take place spontaneously, so that the reaction induced is that due primarily to the second physiological state reached. Wherever we find this law in operation, we have the ultimate basis from which habit and memory (objectively considered) are developed.

From memory in the general sense it is customary to distinguish associative memory. This is characterized objectively by the fact that the response at first given to one stimulus comes, after a time, to be transferred to another one. Examples of associative memory are seen in the experiments of Yerkes and Spaulding on crustaceans . . . It may be pointed out that the essential basis for associative memory is the same law of the resolution of physiological states which we have set forth in the last paragraph as underlying ordinary memory . . . There seems to be no difference in kind, therefore, between associative memory and other sorts; they are based on the same fundamental law. The existence of associative memory has often been considered a criterion of the existence of consciousness, but it is clear that the process underlying it is as readily conceivable in terms of matter and energy as are other physiological processes. Even in inorganic colloids, as we have seen, . . . the properties depend on the past history of the colloid, and the way in which it has reached the condition in which it is now found. If this is conceivable in terms of matter and energy, it is difficult to see why the law of the readier resolution of physiological states is not equally so.

Intelligence is commonly held to consist essentially in the modification of behavior in accordance with experience. If an organism reacts in a certain way under certain conditions, and continues this reaction no matter how disastrous the effects, we say that its behavior is unintelligent. If on the other hand it modifies its behavior in such a way as to make it more adequate, we consider the behavior as in so far intelligent. It is the "correlation of experiences and actions" that constitutes, as Hobhouse (1901) has put it, "the precise work of intelligence."

It appears clear that we find the beginnings of such adaptive changes of behavior even in the Protozoa. They are brought about through the law in accordance with which the resolution of one physiological state into another takes place more readily after repetition,—in connection with the other principle that interference with the life processes causes a change of behavior. These laws apparently form the fundamental basis of intelligent action. This fundamental basis then clearly exists even in the Protozoa; it is apparently coextensive with life. It is difficult if not impossible to draw a line separating the regulatory behavior of lower organisms from the so-called intelligent behavior of higher ones; the one grades insensibly into the other. From the lowest organisms up to man behavior is essentially regulatory in character, and what we call intelligence in higher animals is a direct outgrowth of the same laws that give behavior its regulatory character in the Protozoa.

Thus it seems possible to trace back to the lowest organisms some of the phenomena which we know, from objective evidence, to exist in the behavior of man and the higher animals, and which have received special names. It would doubtless be possible to extend this to many other phenomena. Many conditions which we can clearly distinguish in man must be followed back to a single common condition in the lower organism. But this is what we should expect. Differentiation takes place as we pass upward in the scale in these matters as in others. Because we can trace these phenomena back to conditions found in unicellular forms, it does not follow that the behavior of these organisms has as many factors and is as complex as that of higher animals. The facts are precisely parallel with what we find to be true for other functions. Amœba shows respiration, and all the essential features of respiration in man can be traced back to the condition in such an organism. Yet in man respiration is an enormously complex operation, while in Amœba it is of the simplest character possible—apparently little more than a mere interdiffusion of gases. In the case of behavior there is the same possibility of tracing all essential features back to the lower organisms, with the same great simplification as we go back.

THE QUESTION OF CONSCIOUSNESS

All that we have said thus far in the present chapter is independent of the question whether there exist in the lower organisms such subjective accompaniments of behavior as we find in ourselves, and which we call consciousness. We have asked merely whether there exist in the lower organisms objective phenomena of a character similar to what we find in the behavior of man. To this question we have been compelled to give an affirmative answer. So far as objective evidence goes, there is no difference in kind, but a complete continuity between the behavior of lower and of higher organisms.

Has this any bearing on the question of the existence of consciousness in lower animals? It is clear that objective evidence cannot give a demonstration either of the existence or of the non-existence of consciousness, for consciousness is precisely that which cannot be perceived objectively. No statement concerning consciousness in animals is open to verification or refutation by observation and experiment. There are no processes in the behavior of organisms that are not as readily conceivable without supposing them to be accompanied by consciousness as with it.

But the question is sometimes proposed: Is the behavior of lower organisms of the character which we should "naturally" expect and appreciate if they did have conscious states, of undifferentiated character, and acted under similar conscious states in a parallel way to man? Or is their behavior of such a character that it does not suggest to the observer the existence of consciousness?

If one thinks these questions through for such an organism as Paramecium, with all its limitations of sensitiveness and movement, it appears to the writer that an affirmative answer must be given to the first of the above questions, and a negative one to the second. Suppose that this animal *were* conscious to such an extent as its limitations seem to permit. Suppose that it could feel a certain degree of pain when injured; that it received certain sensations from alkali, others from acids, others from solid bodies, etc.,—would it not be natural for it to act as it does? That is, can we not, through our consciousness, *appreciate* its drawing away from things that hurt it, its trial of the environment when the conditions are bad, its attempting to move forward in various directions, till it finds one where the conditions are not bad, and the like? To the writer it seems that we can; that Paramecium in this behavior makes such an impression that one involuntarily recognizes it as a little subject acting in ways analogous to our own. Still stronger, perhaps, is this impression when observing an Amœba obtaining food . . . The writer is thoroughly convinced, after long study of the behavior of this organism, that if Amœba were a large animal, so as to come within the everyday experience of human beings, its behavior would at once call forth the attribution to it of states of pleasure and pain, of hunger, desire, and the like, on precisely the same basis as we attribute these things to the dog . . .

Of a character somewhat similar to that last mentioned is another test that has been proposed as a basis for deciding as to the consciousness of animals. This is the satisfactoriness or usefulness of the concept of consciousness in the given case. We do not usually attribute consciousness to a stone, because this would not assist us in understanding or controlling the behavior of the stone. Practically indeed it would lead us much astray in dealing with such an object. On the other hand, we usually do attribute consciousness to the dog, because this is useful; it enables us practically to appreciate, foresee, and control its actions much more readily than we could otherwise do so. If Amœba were so large as to come within our everyday ken, I believe it beyond question that we should find similar attribution to it of certain states of consciousness a practical assistance in foreseeing and controlling its behavior. Amœba is a beast of prey, and gives the impression of being controlled by the same elemental impulses as higher beasts of prey. If it were as large as a whale, it is quite conceivable that occasions might arise when the attribution to it of the elemental states of consciousness might save the unsophisticated human being from the destruction that would result from the lack of such attribution. In such a case, then, the attribution of consciousness would be satisfactory and useful. In a small way this is still true for the investigator who wishes to appreciate and predict the behavior of Amœba under his microscope.

But such impressions and suggestions of course do not demonstrate the existence of consciousness in lower organisms. Any belief on this matter can be held without conflict with the objective facts. All that experiment and observation can do is to show us whether the behavior of lower organisms is objectively similar to the behavior that in man is accompanied by consciousness. If this question is answered in the affirmative, as the facts seem to require, and if we further hold, as is commonly held, that man and the lower organisms are subdivisions of the same substance, then it may perhaps be said that objective investigation is as favorable to the view of the general distribution of consciousness throughout animals as it could well be. But the problem as to the actual existence of consciousness outside of the self is an indeterminate one; no increase of objective knowledge can ever solve it. Opinions on this subject must then be largely dominated by general philosophical considerations, drawn from other fields.

XIII FUNCTIONALISM

At the core of Darwin's theory of evolution is the concept of *function*, an assertion that, as a species is evolved, its anatomy is shaped by the requirements for survival. After Darwin, biologists saw each anatomical structure of an organism as a functioning element in an integrated and successful living system. When psychologists began to look at mental processes in this way, they were creating the new psychological movement known as functionalism.

An early example of functionalism in psychology is William James's discussion of consciousness (1890). James thought of the mind as an organ, which, like any other organ, has been evolved for the benefit of its possessor. The function of the conscious mind, said James, is to adapt a very complex organism to very complex environments. It is not enough to know the elements of sensation or the rules of the association of ideas. As a functionalist, James wished also to know how these particular elements and rules help to equip men for survival. The more traditional psychologists of the time thought that James was allowing the pure science of psychology, as practiced by Wundt and Titchener, to be contaminated by the practical temperament of America, but such complaints failed to impress many of James's American colleagues, who also found in the theory of evolution a mandate for a new psychology. John Dewey (see No. 64), for example, was also preaching this new, practical, functional psychology about this time;

and James M. Baldwin (1895), one of the pioneers in American psychology, caught in the wave of enthusiasm for the theory of evolution, viewed the human mind in the contests of both function and evolution, discussed its ontogeny, and in so doing laid the foundation for a psychology of children.

American psychologists throughout the country were adopting the functional approach, but as a self-conscious school, as opposed to a diffuse movement, functionalism found its home at the University of Chicago under the leadership of John Dewey and James R. Angell, reinforced by the philosophical sanction of G. H. Mead and A. W. Moore. In 1906, Angell gave his presidential address to the American Psychological Association, in which he brought together and systematized the principles of functionalism. Angell's student John B. Watson started at Chicago as a loyal disciple, but he soon grew dissatisfied with what he felt was a lack of objectivity in any view that made psychological fact depend on introspection, either Angell's functionalism or Titchener's introspectionism. In his paper (1913) calling for a purely objective approach to all psychology, Watson may be said to have founded the school of "behaviorism." Watson, in spite of his own denial, was, however, merely a new kind of functionalist, one who strove to bring to psychology not only the evolutionary concepts of modern biology, but also the objectivity of its methods.

91   WILLIAM JAMES (1842–1910) ON THE FUNCTION OF CONSCIOUSNESS, 1890

William James, *The Principles of Psychology* (New York, 1890), vol. I, chap. 5.

William James was the pioneer of the new psychology in America, the psychology that claimed to be scientific and experimental. James himself was no experimentalist, but he championed the scientific aspect of psychology while deploring the lack of fertility in much of the experimental work. He wrote and thought in the functional atmosphere that has characterized American psychology from the first, and he influenced the functional movement through the impression he made on John Dewey, James R. Angell, and others. His clear thinking, his gift for the happy phrase, his humane tolerance, which nevertheless did not inhibit his caustic humor when he thought he detected foolishness, won him a wide circle of admirers and disciples, but they never formed a school because James himself could not tolerate the rigidity of thought and opinion that a consistent propaganda demands. This excerpt is from his *Principles,* twelve years in the writing, where James exhibited his functionalism as he argued that it is the function of consciousness to guide the organism to the ends required for its survival. As long as the organism is simple, he said, and is capable of only a relatively small set of possible actions that can be built into it in the fashion of Descartes's reflex machine (see No. 57), it has no need of consciousness. When, however, the organism is capable of a vast set of possible actions, as is man, and may be confronted with a huge number of different environments calling for radically different adaptations, then consciousness is needed to select the course of action. James thus conceived of consciousness as an organ of the animal, one particularly suited to the needs of a complex organism in complex environments and one without which evolution could not have taken place in man.

In describing the functions of the hemispheres a short way back, we used language derived from both the bodily and the mental life, saying now that the animal made indeterminate and unforeseeable reactions, and anon that he was swayed by considerations of future good and evil; treating his hemispheres sometimes as the seat of memory and ideas in the psychic sense, and sometimes talking of them as simply a complicated addition to his reflex machinery. This sort of vacillation in the point of view is a fatal incident of all ordinary talk about these questions; but I must now settle my scores with those readers to whom I already dropped a word in passing . . . and who have probably been dissatisfied with my conduct ever since.

Suppose we restrict our view to facts of one and the same plane, and let that be the bodily plane: cannot all the outward phenomena of intelligence still be exhaustively described? Those mental images, those 'considerations,' whereof we spoke,—presumably they do not arise without neural processes arising simultaneously with them, and presumably each consideration corresponds to a process *sui generis,* and unlike all the rest. In other words, however numerous and delicately differentiated the train of ideas may be, the train of brain-events that runs alongside of it must in both respects be exactly its match, and we must postulate a neural machinery that offers a living counterpart for every shading, however fine, of the history of its owner's mind. Whatever degree of complication the latter may reach, the complication of the machinery must be quite as extreme, otherwise we should have to admit that there may be mental events to which no brain-events correspond. But such an admission as this the physiologist is reluctant to make. It would violate all his beliefs. 'No psychosis without neurosis,' is one form which the principle of continuity takes in his mind.

But this principle forces the physiologist to make still another step. If neural action is as complicated as mind; and if in the sympathetic system and lower spinal cord we see what, so far as we know, is unconscious neural action executing deeds that to all outward intent may be called intelligent; what is there to hinder us from supposing that even where we know consciousness to be there, the still more complicated neural action which we believe to be its inseparable companion is alone and of itself the real agent of whatever intelligent deeds may appear? "As actions of a certain degree of complexity are brought about by mere mechanism, why may not actions of a still greater degree of complexity be the result of a more refined mechanism?" The conception of reflex action is surely one of the best conquests of physiological theory; why not be radical with it? Why not say that just as the spinal cord is a machine with few reflexes, so the hemispheres are a machine with many, and that that is all the difference? The principle of continuity would press us to accept this view.

But what on this view could be the function of the consciousness itself? *Mechanical* function it would have none. The sense-organs would awaken the brain-cells; these would awaken each other in rational and orderly sequence, until the time for action came; and then the last brain-vibration would discharge downward into the motor tracts. But this would be a quite autonomous chain of occurrences, and whatever mind went with it would be there only as an 'epiphenomenon,' an inert spectator, a sort of 'foam, aura, or melody' as Mr. Hodgson says, whose opposition or whose furtherance would be alike powerless over the occurrences themselves. When talking, some time ago, we ought not, accordingly, *as physiologists,* to have said anything about 'considerations' as guiding the animal. We ought to have said 'paths left in the hemispherical cortex by former currents,' and nothing more.

Now so simple and attractive is this conception from the consistently physiological point of view, that it is quite wonderful to see how late it was stumbled on in philosophy, and how few people, even when it has been explained to them, fully and easily realize its import. Much of the polemic writing against it is by men who have as yet failed to take it into their imaginations. Since this has been the case, it seems worth while to devote a few more words to making it plausible, before criticising it ourselves.

To Descartes belongs the credit of having first been bold enough to conceive of a completely self-sufficing nervous mechanism which should be able to perform complicated and apparently intelligent acts. By a singularly arbitrary restriction, however, Descartes stopped short at man, and while contending that in beasts the nervous machinery was all, he held that the higher acts of man were the result of the agency of his rational soul. The opinion that beasts have no consciousness at all was of course too paradoxical to maintain itself long as anything more than a curious item in the history of philosophy. And with its abandonment the very notion that the nervous system *per se* might work the work of intelligence, which was an integral, though detachable part of the whole theory, seemed also to slip out of men's conception, until, in this century, the elaboration of the doctrine of reflex action made it possible and natural that it should again arise. But it was not till 1870, I believe, that Mr. Hodgson made the decisive step, by saying that feelings, no matter how intensely they may be present, can have no causal efficacy whatever, and comparing them to the colors laid on the surface of a mosaic, of which the events in the nervous system are represented by the stones. Obviously the stones are held in place by each other and not by the several colors which they support.

.   .   .   .   .

To comprehend completely the consequences of the dogma . . . one should unflinchingly apply it to the most complicated examples. The movements of our tongues and pens, the flashings of our eyes in conversation, are of course events of a material order, and as such their causal antecedents must be exclusively material. If we knew thoroughly the nervous system of Shakespeare, and as thoroughly all his environing conditions, we should be able to show why at a certain period of his life his hand came to trace on certain sheets of paper those crabbed little black marks which we for shortness' sake call the manuscript of Hamlet. We should understand the rationale of every erasure and alteration therein, and we should understand all this without in the slightest degree acknowledging the existence of the thoughts in Shakespeare's mind. The words and sentences would be taken, not as signs of anything beyond themselves, but as little outward facts, pure and simple. In like manner we might exhaustively write the biography of those two hundred pounds, more or less, of warmish albuminoid matter called Martin Luther, without ever implying that it felt.

But, on the other hand, nothing in all this could prevent us from giving an equally complete account of either Luther's or Shakespeare's spiritual history, an account in which every gleam of thought and emotion should find its place. The mind-history would run alongside of the body-history of each man, and each point in the one would correspond to, but not react upon, a point in the other. So the melody floats from the harp-string, but neither checks nor quickens its vibrations; so the shadow runs alongside the pedestrian, but in no way influences his steps.

Another inference, apparently more paradoxical still, needs to be made, though, as far as I am aware, Dr. Hodgson is the only writer who has explicitly drawn it. That inference is that feelings, not causing nerve-actions, cannot even cause each other. To ordinary common sense, felt pain is, as such, not only the cause of outward tears and cries, but also the cause of such inward events as sorrow, compunction, desire, or inventive thought. So the consciousness of good news is the direct producer of the feeling of joy, the awareness of premises that of the belief in conclusions. But according to the automaton-theory, each of the feelings mentioned is only the correlate of some nerve-movement whose *cause* lay wholly in a previous nerve-movement. The first nerve-movement called up the second; whatever feeling was attached to the second consequently found itself following upon the feeling that was attached to the first. If, for example, good news was the consciousness correlated with the first movement, then joy turned out to be the correlate in consciousness of the second. But all the while the items of the nerve series were the only ones in causal continuity; the items of the conscious series, however inwardly rational their sequence, were simply juxtaposed.

REASONS FOR THE THEORY

The 'conscious automaton-theory,' as this conception is generally called, is thus a radical and simple conception of the manner in which certain facts may possibly occur. But between conception and belief, proof ought to lie. And when we ask, 'What proves that all this is more than a mere conception of the possible?' it is not easy to get a sufficient reply. If we start from the frog's spinal cord and reason by continuity, saying, as that acts so intelligently, *though unconscious,* so the higher centres, *though conscious,* may have the intelligence they show quite as mechanically based; we are immediately met by the exact counterargument from continuity, an argument actually urged by such writers as Pflüger and Lewes, which starts from the acts of the hemispheres, and says: "As *these* owe *their* intelligence to the consciousness which we know to be there, so the intelligence of the spinal cord's acts must really be due to the invisible presence of a consciousness lower in degree." All arguments from continuity work in two ways, you can either level up or level down by their means; and it is clear that such arguments as these can eat each other up to all eternity.

There remains a sort of philosophic faith, bred like most faiths from an æsthetic demand. Mental and physical events are, on all hands, admitted to present the strongest contrast in the entire field of being. The chasm which yawns between them is less easily bridged over by the mind than any interval we know. Why, then, not call it an absolute chasm, and say not only that the two worlds are different, but that they are independent? This gives us the comfort of all simple and absolute formulas, and it makes each chain homogeneous to our consideration. When talking of nervous tremors and bodily actions, we may feel secure against intrusion from an irrelevant mental world. When, on the other hand, we speak of feelings, we may with equal consistency use terms always of one denomination, and never be annoyed by what Aristotle calls 'slipping into another kind.' The desire on the part of men educated in laboratories not to have their physical reasonings mixed up with such incommensurable factors as feelings is certainly very strong. I have heard a most intelligent biologist say: "It is high time for scientific men to protest against the recognition of any such thing as consciousness in a scientific investigation." In a word, feeling constitutes the 'unscientific' half of existence, and any one who enjoys calling himself a 'scientist' will be too happy to purchase an untrammelled homogeneity of terms in the studies of his predilection, at the slight cost of admitting a dualism which, in the same breath that it allows to mind an independent status of being, banishes it to a limbo of causal inertness, from whence no intrusion or interruption on its part need ever be feared.

Over and above this great postulate that matters must be kept simple, there is, it must be confessed, still another highly abstract reason for denying causal efficacity to our feelings. We can form no positive image of the *modus operandi* of a volition or other thought affecting the cerebral molecules.

Let us try to imagine an idea, say of food, producing a movement, say of carrying food to the mouth . . . What is the method of its action? Does it assist the decomposition of the molecules of the gray matter, or does it retard the process, or does it alter the direction in which the shocks are distributed? Let us imagine the molecules of the gray matter combined in such a way that they will fall into simpler combinations on the impact of an incident force. Now suppose the incident force, in the shape of a shock from some other centre, to impinge upon these molecules. By hypothesis it will decompose them, and they will fall into the simpler combination. How is the idea of food to prevent this decomposition? Manifestly it can do so only by increasing the force which binds the molecules together. Good! Try to imagine the idea of a beefsteak binding two molecules together. It is impossible. Equally impossible is it to imagine a similar idea loosening the attractive force between two molecules.[1]

This passage from an exceedingly clever writer expresses admirably the difficulty to which I allude. Combined with a strong sense of the 'chasm' between the two worlds, and with a lively faith in reflex machinery, the sense of this difficulty can hardly fail to make one turn consciousness out of the door as a superfluity so far as one's explanations go. One may bow her out politely, allow her to remain as a 'concomitant,' but one insists that matter shall hold all the power.

Having thoroughly recognized the fathomless abyss that separates mind from matter, and having so blended the very notion into his very nature that there is no chance of his ever forgetting it or failing to saturate with it all his meditations, the student of psychology has next to appreciate the association between these two orders of phenomena . . . They are associated in a manner so intimate that some of the greatest thinkers consider them different aspects of the same process . . . When the rearrangement of molecules takes place in the higher regions of the brain, a change of consciousness simultaneously occurs . . . The change of consciousness never takes place without the change in the brain; the change in the brain never . . . without the change in consciousness. But *why* the two occur together, or what the link is which connects them, we do not know, and most authorities believe that we never shall and never can know. Having firmly and tenaciously grasped these two notions, of the absolute separateness of mind and matter, and of the invariable concomitance of a mental change with a bodily change, the student will enter on the study of psychology with half his difficulties surmounted.[2]

Half his difficulties ignored, I should prefer to say. For this 'concomitance' in the midst of 'absolute separateness' is an utterly irrational notion. It is to my mind quite inconceivable that consciousness should have *nothing to do* with a business which it so faithfully attends. And the question, 'What has it to do?' is one which psychology has no right to 'surmount,' for it is her plain duty to consider it. The fact is that the whole question of interaction and influence between things is a metaphysical question, and cannot be discussed at all by those who are unwilling to go into matters thoroughly. It is truly enough hard to imagine the 'idea of a beefsteak binding two molecules together;' but since

[1]Chas. Mercier, *The Nervous System and the Mind* (1888), p. 9.
[2]Mercier, p. 11.

Hume's time it has been equally hard to imagine *anything* binding them together. The whole notion of 'binding' is a mystery, the first step towards the solution of which is to clear scholastic rubbish out of the way. Popular science talks of 'forces,' 'attractions' or 'affinities' as binding the molecules; but clear science, though she may use such words to abbreviate discourse, has no use for the conceptions, and is satisfied when she can express in simple 'laws' the bare space-relations of the molecules as functions of each other and of time. To the more curiously inquiring mind, however, this simplified expression of the bare facts is not enough; there must be a 'reason' for them, and something must 'determine' the laws. And when one seriously sits down to consider what sort of a thing one *means* when one asks for a 'reason,' one is led so far afield, so far away from popular science and its scholasticism, as to see that even such a fact as the existence or non-existence in the universe of 'the idea of a beefsteak' may not be wholly indifferent to other facts in the same universe, and in particular may have something to do with determining the distance at which two molecules in that universe shall lie apart. If this is so, then common-sense, though the intimate nature of causality and of the connection of things in the universe lies beyond her pitifully bounded horizon, has the root and gist of the truth in her hands when she obstinately holds to it that feelings and ideas are causes. However inadequate our ideas of causal efficacy may be, we are less wide of the mark when we say that our ideas and feelings have it, than the Automatists are when they say they haven't it. As in the night all cats are gray, so in the darkness of metaphysical criticism all causes are obscure. But one has no right to pull the pall over the psychic half of the subject only, as the automatists do, and to say that *that* causation is unintelligible, whilst in the same breath one dogmatizes about *material* causation as if Hume, Kant, and Lotze had never been born. One cannot thus blow hot and cold. One must be impartially *naif* or impartially critical. If the latter, the reconstruction must be thorough-going or 'metaphysical,' and will probably preserve the common-sense view that ideas are forces, in some translated form. But Psychology is a mere natural science, accepting certain terms uncritically as her data, and stopping short of metaphysical reconstruction. Like physics, she must be *naïve;* and if she finds that in her very peculiar field of study ideas *seem* to be causes, she had better continue to talk of them as such. She gains absolutely nothing by a breach with common-sense in this matter, and she loses, to say the least, all naturalness of speech. If feelings are causes, of course their effects must be furtherances and checkings of internal cerebral motions, of which in themselves we are entirely without knowledge. It is probable that for years to come we shall have to infer what

489

happens in the brain either from our feelings or from motor effects which we observe. The organ will be for us a sort of vat in which feelings and motions somehow go on stewing together, and in which innumerable things happen of which we catch but the statistical result. Why, under these circumstances, we should be asked to forswear the language of our childhood I cannot well imagine, especially as it is perfectly compatible with the language of physiology. The feelings can produce nothing absolutely new, they can only reinforce and inhibit reflex currents, and the original organization by physiological forces of these in paths must always be the ground-work of the psychological scheme.

My conclusion is that to urge the automation-theory upon us, as it is now urged, on purely *a priori* and *quasi*-metaphysical grounds, is an *unwarrantable impertinence in the present state of psychology.*

REASONS AGAINST THE THEORY

But there are much more positive reasons than this why we ought to continue to talk in psychology as if consciousness had causal efficacy. The *particulars of the distribution of consciousness,* so far as we know them, *point to its being efficacious.* Let us trace some of them.

It is very generally admitted, though the point would be hard to prove, that consciousness grows the more complex and intense the higher we rise in the animal kingdom. That of a man must exceed that of an oyster. From this point of view it seems an organ, superadded to the other organs which maintain the animal in the struggle for existence; and the presumption of course is that it helps him in some way in the struggle, just as they do. But it cannot help him without being in some way efficacious and influencing the course of his bodily history. If now it could be shown in what way consciousness *might* help him, and if, moreover, the defects of his other organs (where consciousness is most developed) are such as to make them need just the kind of help that consciousness would bring provided it *were* efficacious; why, then the plausible inference would be that it came just *because* of its efficacy—in other words, its efficacy would be inductively proved.

Now the study of the phenomena of consciousness which we shall make throughout the rest of this book will show us that consciousness is at all times primarily *a selecting agency.* Whether we take it in the lowest sphere of sense, or in the highest of intellection, we find it always doing one thing, choosing one out of several of the materials so presented to its notice, emphasizing and accentuating that and suppressing as far as possible all the rest. The item emphasized is always in close connection with some *interest* felt by consciousness to be paramount at the time.

But what are now the defects of the nervous system in those animals whose consciousness seems most highly developed? Chief among them must be *instability*. The cerebral hemispheres are the characteristically 'high' nerve-centres, and we saw how indeterminate and unforeseeable their performances were in comparison with those of the basal ganglia and the cord. But this very vagueness constitutes their advantage. They allow their possessor to adapt his conduct to the minutest alterations in the environing circumstances, any one of which may be for him a sign, suggesting distant motives more powerful than any present solicitations of sense. It seems as if certain mechanical conclusions should be drawn from this state of things. An organ swayed by slight impressions is an organ whose natural state is one of unstable equilibrium. We may imagine the various lines of discharge in the cerebrum to be almost on a par in point of permeability—what discharge a given small impression will produce may be called *accidental,* in the sense in which we say it is a matter of accident whether a rain-drop falling on a mountain ridge descend the eastern or the western slope . . . The natural law of an organ constituted after this fashion can be nothing but a law of caprice. I do not see how one could reasonably expect from it any certain pursuance of useful lines of reaction, such as the few and fatally determined performances of the lower centres constitute within their narrow sphere. The dilemma in regard to the nervous system seems, in short, to be of the following kind. We may construct one which will react infallibly and certainly, but it will then be capable of reacting to very few changes in the environment—it will fail to be adapted to all the rest. We may, on the other hand, construct a nervous system potentially adapted to respond to an infinite variety of minute features in the situation; but its fallibility will then be as great as its elaboration. We can never be sure that its equilibrium will be upset in the appropriate direction. In short, a high brain may do many things, and may do each of them at a very slight hint. But its hair-trigger organization makes of it a happy-go-lucky, hit-or-miss affair. It is as likely to do the crazy as the sane thing at any given moment. A low brain does few things, and in doing them perfectly forfeits all other use. The performances of a high brain are like dice thrown forever on a table. Unless they be loaded, what chance is there that the highest number will turn up oftener than the lowest?

All this is said of the brain as a physical machine pure and simple. *Can consciousness increase its efficiency by loading its dice?* Such is the problem.

Loading its dice would mean bringing a more or less constant pressure to bear in favor of *those* of its performances which make for the most permanent interests of the brain's owner; it would mean a constant inhibition of the tendencies to stray aside.

Well, just such pressure and such inhibition are what consciousness *seems* to be exerting all the while. And the interests in whose favor it seems to exert them are *its* interests and its alone, interests which it *creates,* and which, but for it, would have no status in the realm of being whatever. We talk, it is true, when we are darwinizing, as if the mere *body* that owns the brain had interests; we speak about the utilities of its various organs and how they help or hinder the body's survival; and we treat the survival as if it were an absolute end, existing as such in the physical world, a sort of actual *should-be,* presiding over the animal and judging his reactions, quite apart from the presence of any commenting intelligence outside. We forget that in the absence of some such super-added commenting intelligence (whether it be that of the animal itself, or only ours or Mr. Darwin's), the reactions cannot be properly talked of as 'useful' or 'hurtful' at all. Considered merely physically, all that can be said of them is that *if* they occur in a certain way survival will as a matter of fact prove to be their incidental consequence. The organs themselves, and all the rest of the physical world, will, however, all the time be quite indifferent to this consequence, and would quite as cheerfully, the circumstances changed, compass the animal's destruction. In a word, survival can enter into a purely physiological discussion only as an *hypothesis made by an onlooker* about the future. But the moment you bring a consciousness into the midst, survival ceases to be a mere hypothesis. No longer is it, "*if* survival is to occur, then so and so must brain and other organs work." It has now become an imperative decree: "Survival *shall* occur, and therefore organs *must* so work!" *Real* ends appear for the first time now upon the world's stage. The conception of consciousness as a purely cognitive form of being, which is the pet way of regarding it in many idealistic-modern as well as ancient schools, is thoroughly anti-psychological, as the remainder of this book will show. Every actually existing consciousness seems to itself at any rate to be a *fighter for ends,* of which many, but for its presence, would not be ends at all. Its powers of cognition are mainly subservient to these ends, discerning which facts further them and which do not.

Now let consciousness only be what it seems to itself, and it will help an instable brain to compass its proper ends. The movements of the brain *per se* yield the means of attaining these ends mechanically, but only out of a lot of other ends, if so they may be called, which are not the proper ones of the animal, but often quite opposed. The brain is an instrument of possibilities, but of no certainties. But the consciousness, with its own ends present to it, and knowing also well which possibilities lead thereto and which away, will, if endowed with causal efficacy, reinforce the favorable possibilities and repress the unfavorable or indifferent ones. The nerve-currents, coursing through the cells and fibres, must in this case be supposed strengthened by the fact of their awaking one consciousness and dampened by awaking another. *How* such reaction of the consciousness upon the currents may occur must remain at present unsolved: it is enough for my purpose to have shown that it may not uselessly exist, and that the matter is less simple than the brain-automatists hold.

All the facts of the natural history of consciousness lend color to this view. Consciousness, for example, is only intense when nerve-processes are hesitant. In rapid, automatic, habitual action it sinks to a minimum. Nothing could be more fitting than this, if consciousness have the teleological function we suppose; nothing more meaningless, if not. Habitual actions are certain, and being in no danger of going astray from their end, need no extraneous help. In hesitant action, there seem many alternative possibilities of final nervous discharge. The feeling awakened by the nascent excitement of each alternative nerve-tract seems by its attractive or repulsive quality to determine whether the excitement shall abort or shall become complete. Where indecision is great, as before a dangerous leap, consciousness is agonizingly intense. Feeling, from this point of view, may be likened to a cross-section of the chain of nervous discharge, ascertaining the links already laid down, and groping among the fresh ends presented to it for the one which seems best to fit the case.

The phenomena of 'vicarious function' . . . seem to form another bit of circumstantial evidence. A machine in working order acts fatally in one way. Our consciousness calls this the right way. Take out a valve, throw a wheel out of gear or bend a pivot, and it becomes a different machine, acting just as fatally in another way which we call the wrong way. But the machine itself knows nothing of wrong or right: matter has no ideals to pursue. A locomotive will carry its train through an open drawbridge as cheerfully as to any other destination.

A brain with part of it scooped out is virtually a new machine, and during the first days after the operation functions in a thoroughly abnormal manner. As a matter of fact, however, its performances become from day to day more normal, until at last a practised eye may be needed to suspect anything wrong. Some of the restoration is undoubtedly due to 'inhibitions' passing away. But if the consciousness which goes with the rest of the brain, be there not only in order to take cognizance of each functional error, but also to exert an efficient pressure to check it

493

if it be a sin of commission, and to lend a strengthening hand if it be a weakness or sin of omission,—nothing seems more natural than that the remaining parts, assisted in this way, should by virtue of the principle of habit grow back to the old teleological modes of exercise for which they were at first incapacitated. Nothing, on the contrary, seems at first sight more unnatural than that they should vicariously take up the duties of a part now lost without those *duties as such* exerting any persuasive or coercive force . . .

There is yet another set of facts which seem explicable on the supposition that consciousness has causal efficacy. *It is a well-known fact that pleasures are generally associated with beneficial, pains with detrimental, experiences.* All the fundamental vital processes illustrate this law. Starvation, suffocation, privation of food, drink and sleep, work when exhausted, burns, wounds, inflammation, the effects of poison, are as disagreeable as filling the hungry stomach, enjoying rest and sleep after fatigue, exercise after rest, and a sound skin and unbroken bones at all times, are pleasant. Mr. Spencer and others have suggested that these coincidences are due, not to any pre-established harmony, but to the mere action of natural selection which would certainly kill off in the long-run any breed of creatures to whom the fundamentally noxious experience seemed enjoyable. An animal that should take pleasure in a feeling of suffocation would, if that pleasure were efficacious enough to make him immerse his head in water, enjoy a longevity of four or five minutes. But if pleasures and pains have no efficacy, one does not see (without some such *à priori* rational harmony as would be scouted by the 'scientific' champions of the automaton-theory) why the most noxious acts, such as burning, might not give thrills of delight, and the most necessary ones, such as breathing, cause agony. The exceptions to the law are, it is true, numerous, but relate to experiences that are either not vital or not universal. Drunkenness, for instance, which though noxious, is to many persons delightful, is a very exceptional experience. But, as the excellent physiologist Fick remarks, if all rivers and springs ran alcohol instead of water, either all men would now be born to hate it or our nerves would have been selected so as to drink it with impunity . . .

. . . *A priori* analysis of both brain-action and conscious action shows us that if the latter were efficacious it would, by its selective emphasis, make amends for the indeterminateness of the former; whilst the study *a posteriori* of the *distribution* of consciousness shows it to be exactly such as we might expect in an organ added for the sake of steering a nervous system grown too complex to regulate itself. The conclusion that it is useful is, after all this, quite justifiable. But, if it is useful, it must be so through its causal efficaciousness, and the automaton-theory must succumb to the theory of common-sense. I, at any rate (pending metaphysical reconstructions not yet successfully achieved), shall have no hesitation in using the language of common-sense throughout this book.

92   JAMES MARK BALDWIN (1861–1934) ON THE PSYCHOLOGY OF CHILDREN, 1895

J. M. Baldwin, *Mental Development in the Child and the Race* (New York, 1895), pp. 1–9.

From the 1870's on, American zoologists and paleontologists were deeply concerned, not only with the evolution of animal forms in phylogenesis, but also with the ontogenesis of particular animals and with the seeming recapitulation of phylogenesis in the development of the embryo. Among the psychologists, there arose a parallel concern about the individual development of the mind of man from childhood to adulthood. The early work in child psychology was in Europe, by Bernard Perez in France (1878) and William Preyer in Germany (1882). In America, child psychology was promoted by G. Stanley Hall and James Mark Baldwin, the two most important psychological pioneers after William James and James McKeen Cattell, and they were both obsessed with a belief in the importance to psy-chology of the evolutionary doctrine. Hall's earliest contribution to psychology, during the early 1880's, was his study of the information in children's minds, a study conducted by the questionnaire method, but his major contribution was his two-volume *Adolescence* of 1904. Baldwin, his contemporary in functional psychology, was a prolific theorist. He was also an aggressive polemicist and was vigorously criticized by his contemporary experimentalists in psychology because he elaborated his theories with such assurance in the absence of a sound empirical basis. His numerous books dealing with evolutionary psychology run for over a decade, beginning in 1895 with the following argument in favor of ontogenetic, or child, psychology.

I. INFANT PSYCHOLOGY: ONTOGENESIS

No doubt we owe to the rise of the evolution idea something at least of the benefit brought about by what we may call the psychological renaissance of the last twenty-five or thirty years. The breadth of the current conception of psychology is certainly in harmony with the conceptions long ago current in other departments of scientific research; but there is a phase of this broadening of psychological inquiry strikingly brought out only when interpreted in the light of evolution doctrine. This is what we may call the genetic phase, the growth phase. The older idea of the soul was of a fixed substance, with fixed attributes. Knowledge of the soul was immediate in consciousness, and adequate; at least, as adequate as such knowledge could be made. The mind was best understood where best or most fully manifested; its higher 'faculties,' even when not in operation, were still there, but asleep.

Under such a conception, the man was father to the child. What the adult consciousness discovers in itself is true, and wherein the child lacks it falls short of the true stature of soul life. We must, therefore, if we take account of the child-mind at all, interpret it up to the revelations of the man-mind. If the adult consciousness shows the presence of principles not observable in the child consciousness, we must suppose, nevertheless, that they are really present in the child consciousness beyond the reach of our observation. The old argument was this,—and it is not too old to be found in the metaphysics of to-day,—consciousness reveals certain great ideas as simple and original: consequently they must be so. If you do not find them in the child-mind, then you must read them into it.

The genetic idea reverses all this. Instead of a fixed substance, we have the conception of a growing, developing activity. Functional psychology succeeds faculty psychology. Instead of beginning with the most elaborate exhibition of this growth and development, we shall find most instruction in the simplest activity that is at the same time the same activity. Development is a process of involution as well as of evolution, and the elements come to be hidden under the forms of complexity which they build up. Are there principles in the adult consciousness which do not appear in the child consciousness, then the adult consciousness must, if possible, be interpreted by principles present in the child consciousness; and when this is not possible, the conditions under which later principles take their rise and get their development must still be adequately explored.

Now that this genetic conception has arrived, it is astonishing that it did not arrive sooner, and it is astonishing that the 'new' psychology has hitherto made so little use of it. The difference between description and explanation is as old as science itself. What chemist long remains satisfied with a description of the substances found in nature? He is no investigator at all. His science was not born until he became an analyst. The student of philology is not content with a description, a grammar, of spoken languages: he desiderates their reduction to common vocal elements, and aims to discover the laws of their genetic development. But the mental scientist has called such description science, even when he has had examples of nature's own furnishing around him which would have confirmed or denied the results of mental analysis.

The advantages which we look to infant psychology to furnish, meet just this need of analysis; and the reason that the needed analysis is found here, is that the mind, like all other natural things, grows. This general statement may be put into concrete form under several points, which divide this branch of general psychology from others now recognized.

1. In the first place, the phenomena of the infant consciousness are simple as opposed to reflective; that is, they are the child's presentations or memories simply, not his own observations of them. In the adult consciousness the disturbing influences of inner observation is a matter of notorious moment. It is impossible for me to know exactly what I feel, for the apprehending of it through the attention alters its character. My volition also is a complex thing of alternatives, one of which is my personal pride and self-conscious egotism. But the child's emotion is as spontaneous as a spring. The effects of it in the mental life come out in action, pure and uninfluenced by calculation and duplicity and adult reserve. There is around every one of us a web of convention and prejudice of our own making. Not only do we reflect the social formalities of our environment, and thus lose the distinguishing spontaneities of childhood, but each one of us builds up his own little world of seclusion and formality with himself. We are subject not only to 'idols of the forum,' but also to 'idols of the den.'

The child, on the contrary, has not learned his own importance, his pedigree, his beauty, his social place, his religion, his paternal disgrace; and he has not observed himself through all these and countless other lenses of time, place, and circumstance. He has not yet turned himself into an idol nor the world into a temple; and we can study him apart from the complex accretions which are the later deposits of his self-consciousness.

.   .   .   .   .

2. The study of children is often the only means of testing the truth of our mental analyses. If we decide that a certain complex product is due to a union of simpler mental elements, then we may appeal to the proper period of child life to see the union taking place. The range of growth is so enormous from the infant to the adult, and the beginnings of the child's mental life are so low in the scale, in the matter of mental and moral endowment, that there is hardly a question of analysis now under debate which may not be tested by this method.

On the other hand, that such confirmation shuts out most conclusively the advocates of irreducibility in many cases, seems to admit of no question. A good example of such analysis is seen in the distinction between simple consciousness and self-consciousness. Over and over again have systems been built upon the subject-object theory of consciousness; namely, that personality, subjectivity, consciousness in any form, necessarily

497

implicated an antithesis, in consciousness, between *ego* and *non-ego*. But an example of what is thus denied may be seen upon the floor of any nursery where there is a child less than six months of age.

At this point it is that child psychology is more valuable than the study of the consciousness of animals. The latter never become men, while children do. The animals represent in some few respects a branch of the tree of growth in advance of man, while being in many other respects very far behind him. In studying animals we are always haunted by the fear that the analogy may not hold; that some element essential to the development of the human mind may not discover itself at all. Even in such a question as the localization of the motor functions of the brain, where the analogy is one of comparative anatomy and only secondarily of psychology, the monkey presents analogies with man which dogs do not. But in the study of children we may be always sure that a normal child has in him the promise of a normal man.

. . . . .

3. Again, in the study of the child-mind, we have the added advantage of a corresponding simplicity on the organic side; that is, we are able to take account of the physiological processes at a time when they are relatively simple. I say 'relatively simple,' for in reality they are enormously complex at birth, and the embryologist pushes his researches much farther back in the life history of the organism. But yet they are simple relatively to their condition after the formation of habits, motor complexes, brain connections and associations; in short, after the nervous system has been educated to its whole duty in its living environment. For example: a psychology which holds that we have a 'speech faculty,' an original mental endowment which is incapable of further reduction, may appeal to the latest physiological research and find organic confirmation, at least as far as a determination of its cerebral apparatus is concerned; but such support for the position is wanting when we return to the brain of the infant. Not only do we fail to find the series of centres into which the organic basis of speech has been divided, but even those of them which we do find have not taken up the function, either alone or together, which they perform when speech is actually realized. In other words, the primary object of each of the various centres involved is not speech, but some other and simpler function; and speech arises by development from a union of these separate functions.

We accordingly find a development of consciousness keeping pace with the development of the physical organism. The extent of possible analogies between the growth of body and that of mind may thus be estimated from below; and any outstanding facts of the inner life which cannot be correlated with facts of the physical organism get greater prominence and safer estimation.

4. In observing young children, a more direct application of the experimental method is possible. By 'experiment' here, I mean both experiment on the senses and also experiment directly on consciousness by suggestion, social influence, etc. In experimenting on adults, great difficulties arise through the fact that reactions—such as performing a voluntary movement when a signal is heard, etc.,—are broken at the centre by deliberation, habitual desire, choice, etc., and closed again by a conscious voluntary act. The subject hears a sound, identifies it, and presses a button—*if he choose* and agree to do so. What goes on in this interval between the advent of the incoming nerve process and the discharge of the outgoing nerve process? Something, at any rate, which represents a brain process of great complexity. Now, anything that fixes this sensori-motor connection or simplifies the central process, in so far gives greater certainty to the results. For this reason, experiments on reflex reactions are valuable and decisive where similar experiments on voluntary reactions are uncertain and of doubtful value. Now the fact that the child consciousness is relatively simple, and so offers a field for more fruitful experiment, is illustrated in what is said in the following pages about suggestion in infant life; it is also seen in the mechanical reactions of an infant to strong stimuli, such as bright colors, etc. Of course, this is the point where originality must be exercised in the devising and executing of experiments. After the subject is a little better developed, new experimentation will be as difficult here as in the other sciences; but at present the simplest phenomena of child life and activity are open to the investigator.

93   JAMES ROWLAND ANGELL (1869–1949) ON FUNCTIONALISM, 1906

J. R. Angell, "The province of functional psychology," *Psychological Review 14*, 61–91 (1907).

Angell came under James's influence at Harvard, where he received a master's degree in 1892, at the time when James had just turned over the Harvard psychological laboratory to the newly arrived Hugo Münsterberg but before James had asked to have his title changed back from professor of psychology to professor of philosophy. Thus Angell was reinforced in the American pattern of functional psychology by perhaps its most influential teacher. After a year at the University of Minnesota, Angell went to the University of Chicago in 1894 to take charge of the psychological laboratory in the same year that John Dewey, ten years Angell's senior, came to Chicago from Michigan. Dewey also had felt James's influence and was perhaps an even more persuasive functionalist than James. Unlike James, Dewey was an effective propagandist, and during Dewey's ten years at Chicago he and Angell may be said to have founded the Chicago school of functional psychology. Under Angell's able leadership this school quickly became an important movement in America. There was not, however, a great deal of systematic writing about its tenets, and Angell's presidential address before the American Psychological Association in 1906, from which the present excerpt is taken, is the best exposition.

Functional psychology is at the present moment little more than a point of view, a program, an ambition. It gains its vitality primarily perhaps as a protest against the exclusive excellence of another starting point for the study of the mind, and it enjoys for the time being at least the peculiar vigor which commonly attaches to Protestantism of any sort in its early stages before it has become respectable and orthodox. The time seems ripe to attempt a somewhat more precise characterization of the field of functional psychology than has as yet been offered. What we seek is not the arid and merely verbal definition which to many of us is so justly anathema, but rather an informing appreciation of the motives and ideals which animate the psychologist who pursues this path. His status in the eye of the psychological public is unnecessarily precarious. The conceptions of his purposes prevalent in non-functionalist circles range from positive and dogmatic misapprehension, through frank mystification and suspicion up to moderate comprehension. Nor is this fact an expression of anything peculiarly abstruse and recondite in his intentions. It is due in part to his own ill-defined plans, in part to his failure to explain lucidly exactly what he is about. Moreover, he is fairly numerous and it is not certain that in all important particulars he and his confrères are at one in their beliefs. The considerations which are herewith offered suffer inevitably from this personal limitation. No psychological council of Trent has as yet pronounced upon the true faith. But in spite of probable failure it seems worth while to hazard an attempt at delineating the scope of functionalist principles. I formally renounce any intention to strike out new plans; I am engaged in what is meant as a dispassionate summary of actual conditions.

Whatever else it may be, functional psychology is nothing wholly new. In certain of its phases it is plainly discernible in the psychology of Aristotle and in its more modern garb it has been increasingly in evidence since Spencer wrote his *Psychology* and Darwin his *Origin of Species*. Indeed, as we shall soon see, its crucial problems are inevitably incidental to any serious attempt at understanding mental life. All that is peculiar to its present circumstances is a higher degree of self-consciousness than it possessed before, a more articulate and persistent purpose to organize its vague intentions into tangible methods and principles.

A survey of contemporary psychological writing indicates, as was intimated in the preceding paragraph, that the task of functional psychology is interpreted in several different ways. Moreover, it seems to be possible to advocate one or more of these conceptions while cherishing abhorrence for the others. I distinguish three principal forms of the functional problem with sundry subordinate variants. It will contribute to the clarification of the general situation to dwell upon these for a moment, after which I propose to maintain that they are substantially but modifications of a single problem.

I

There is to be mentioned first the notion which derives most immediately from contrast with the ideals and purposes of structural psychology so-called. This involves the identification of functional psychology with the effort to discern and portray the typical *operations* of consciousness under actual life conditions, as over against the attempt to analyze and describe its elementary and complex *contents*. The structural psychology of sensation, *e.g.*, undertakes to determine the number and character of the various unanalyzable sensory materials, such as the varieties of color, tone, taste, etc. The functional psychology of sensation would on the other hand find its appropriate sphere of interest in the determination of the character of the various sense activities as differing in their *modus operandi* from one another and from other mental processes such as judging, conceiving, willing and the like.

.   .   .   .   .

The more extreme and ingenuous conceptions of structural psychology seem to have grown out of an unchastened indulgence in what we may call the 'states of consciousness' doctrine. I take it that this is in reality the contemporary version of Locke's 'idea.' If you adopt as your material for psychological analysis the isolated 'moment of consciousness,' it is very easy to become so absorbed in determining its constitution as to be rendered somewhat oblivious to its artificial character. The most essential quarrel which the functionalist has with structuralism in its thoroughgoing and consistent form arises from this fact and touches the feasibility and worth of the effort to get at mental process as it *is* under the conditions of actual experience rather than as it *appears* to a merely postmortem analysis. It is of course true that for introspective purposes we must in a sense always work with vicarious representatives of the particular mental processes which we set out to observe. But it makes a great difference even on such terms whether one is directing attention primarily to the discovery of the way in which such a mental process operates, and what the conditions are under which it appears, or whether one is engaged simply in tearing apart the fibers of its tissues. The latter occupation is useful and for certain purposes essential, but it often stops short of that which is as a life phenomenon the most essential, *i.e.,* the *modus operandi* of the phenomenon.

.   .   .   .   .

The fact that mental contents are evanescent and fleeting marks them off in an important way from the relatively permanent elements of anatomy. No matter how much we may talk of the preservation of psychical dispositions, nor how many metaphors we may summon to characterize the storage of ideas in some hypothetical deposit chamber of memory, the obstinate fact remains that when we are not experiencing a sensation or an idea it is, strictly speaking, non-existent. Moreover, when we manage by one or another device to secure that which we designate the same sensation or the same idea, we not only have no guarantee that our second edition is really a replica of the first, we have a good bit of presumptive evidence that from the content point of view the original never is and never can be literally duplicated.

Functions, on the other hand, persist as well in mental as in physical life. We may never have twice exactly the same idea viewed from the side of sensuous structure and composition. But there seems nothing whatever to prevent our having as often as we will contents of consciousness which *mean* the same thing. They function in one and the same practical way, however discrepant their momentary texture. The situation is rudely analogous to the biological case where very different structures may under different conditions be called on to perform identical functions; and the matter naturally harks back for its earliest analogy to the instance of protoplasm where functions seem very tentatively and imperfectly differentiated. Not only then are general functions like memory persistent, but special functions such as the memory of particular events are persistent and largely independent of the specific conscious contents called upon from time to time to subserve the functions.

.   .   .   .   .

Substantially identical with this first conception of functional psychology, but phrasing itself somewhat differently, is the view which regards the functional problem as concerned with discovering how and why conscious processes are what they are, instead of dwelling as the structuralist is supposed to do upon the problem of determining the irreducible elements of consciousness and their characteristic modes of combination. I have elsewhere defended the view that however it may be in other sciences dealing with life phenomena, in psychology at least the answer to the question 'what' implicates the answer to the questions 'how' and 'why.'

Stated briefly the ground on which this position rests is as follows: In so far as you attempt to analyze any particular state of consciousness you find that the mental elements presented to your notice are dependent upon the particular exigencies and conditions which call them forth. Not only does the affective coloring of such a psychical moment depend upon one's temporary condition, mood and aims, but the very sensations themselves are determined in their qualitative texture by the totality of circumstances subjective and objective within which they arise. You cannot get a fixed and definite color sensation for example, without keeping perfectly constant the external and internal conditions in which it appears. The particular sense quality is in short functionally determined by the necessities of the existing situation which it emerges to meet. If you inquire then deeply enough what particular sensation you have in a given case, you always find it necessary to take account of the manner in which, and the reasons why, it was experienced at all. You may of course, if you will, abstract from these considerations, but in so far as you do so, your analysis and description is manifestly partial and incomplete. Moreover, even when you do so abstract and attempt to describe certain isolable sense qualities, your descriptions are of necessity couched in terms not of the experienced quality itself, but in terms of the conditions which produced it, in terms of some other quality with which it is compared, or in terms of some more overt act to which the sense stimulation led. That is to say, the very description itself is functionalistic and must be so. The truth of this assertion can be illustrated and tested by appeal to any situation in which one is trying to reduce sensory complexes, *e.g.,* colors or sounds, to their rudimentary components.

II

A broader outlook and one more frequently characteristic of contemporary writers meets us in the next conception of the task of functional psychology. This conception is in part a reflex of the prevailing interest in the larger formulæ of biology and particularly the evolutionary hypotheses within whose majestic sweep is nowadays included the history of the whole stellar universe; in part it echoes the same philosophical call to new life which has been heard as pragmatism, as humanism, even as functionalism itself. I should not wish to commit either party by asserting that functional psychology and pragmatism are ultimately one. Indeed, as a psychologist I should hesitate to bring down on myself the avalanche of metaphysical invective which has been loosened by pragmatic writers. To be sure pragmatism has slain its thousands, but I should cherish scepticism as to whether functional psychology would the more speedily slay its tens of thousands by announcing an offensive and defensive alliance with pragmatism. In any case I only hold that the two movements spring from similar logical motivation and rely for their vitality and propagation upon forces closely germane to one another.

The functional psychologist then in his modern attire is interested not alone in the operations of mental process considered merely of and by and for itself, but also and more vigorously in mental activity as part of a larger stream of biological forces which are daily and hourly at work before our eyes and which are constitutive of the most important and most absorbing part of our world. The psychologist of this stripe is wont to take his cue from the basal conception of the evolutionary movement, *i.e.,* that for the most part organic structures and functions possess their present characteristics by virtue of the efficiency with which they fit into the extant conditions of life broadly designated the environment. With this conception in mind he proceeds to attempt some understanding of the manner in which the psychical contributes to the furtherance of the sum total of organic activities, not alone the psychical in its entirety, but especially the psychical in its particularities—mind as judging, mind as feeling, etc.

This is the point of view which instantly brings the psychologist cheek by jowl with the general biologist. It is the presupposition of every philosophy save that of outright ontological materialism that mind plays the stellar rôle in all the environmental adaptations of animals which possess it. But this persuasion has generally occupied the position of an innocuous truism or at best a jejune postulate, rather than that of a problem requiring, or permitting, serious scientific treatment. At all events, this was formerly true.

This older and more complacent attitude toward the matter is, however, being rapidly displaced by a conviction of the need for light on the exact character of the accommodatory service represented by the various great modes of conscious expression. Such an effort if successful world not only broaden the foundations for biological appreciation of the intimate nature of accommodatory process, it would also immensely enchance the psychologist's interest in the exact portrayal of conscious life. It is of course the latter consideration which lends importance to the matter from our point of view. Moreover, not a few practical consequences of value may be expected to flow from this attempt, if it achieves even a measurable degree of success. Pedagogy and mental hygiene both await the quickening and guiding counsel which can only come from a psychology of this stripe. For their purposes a strictly structural psychology is as sterile in theory as teachers and psychiatrists have found it in practice.

As a concrete example of the transfer of attention from the more general phases of consciousness as accommodatory activity to the particularistic features of the case may be mentioned the rejuvenation of interest in the quasi-biological field which we designate animal psychology. This movement is surely among the most pregnant with which we meet in our own generation. Its problems are in no sense of the merely theoretical and speculative kind, although, like all scientific endeavor, it possesses an intellectual and methodological background on which such problems loom large. But the frontier upon which it is pushing forward its explorations is a region of definite, concrete fact, tangled and confused and often most difficult of access, but nevertheless a region of fact, accessible like all other facts to persistent and intelligent interrogation.

That many of the most fruitful researches in this field have been achievements of men nominally biologists rather than psychologists in no wise affects the merits of the case. A similar situation exists in the experimental psychology of sensation where not a little of the best work has been accomplished by scientists not primarily known as psychologists.

It seems hardly too much to say that the empirical conceptions of the consciousness of the lower animals have undergone a radical alteration in the past few years by virtue of the studies in comparative psychology. The splendid investigations of the mechanism of instinct, of the facts and methods of animal orientation, of the scope and character of the several sense processes, of the capabilities of education and the range of selective accommodatory capacities in the animal kingdom, these and dozens of other similar problems have received for the first time drastic scientific examination, experimental in character wherever possible, observational elsewhere, but observational in the spirit of conservative non-anthropomorphism as earlier observations almost never were. In most cases they have to be sure but shown the way to further and more precise knowledge, yet there can be but little question that the trail which they have blazed has success at its farther end.

One may speak almost as hopefully of human genetic psychology which has been carried on so profitably in our own country. As so often in psychology, the great desideratum here, is the completion of adequate methods which will insure really stable scientific results. But already our general psychological theory has been vitalized and broadened by the results of the genetic methods thus far elaborated. These studies constantly emphasize for us the necessity of getting the longitudinal rather than the transverse view of life phenomena.

. . . .

III

The third conception which I distinguish is often in practice merged with the second, but it involves stress upon a problem logically prior perhaps to the problem raised there and so warrants separate mention. Functional psychology, it is often alleged, is in reality a form of psychophysics. To be sure, its aims and ideals are not explicitly quantitative in the manner characteristic of that science as commonly understood. But it finds its major interest in determining the relations to one another of the physical and mental portions of the organism.

It is undoubtedly true that many of those who write under functional prepossessions are wont to introduce frequent references to the physiological processes which accompany or condition mental life. Moreover, certain followers of this faith are prone to declare forthwith that psychology is simply a branch of biology and that we are in consequence entitled, if not indeed obliged, to make use where possible of biological materials.

. . . .

Whether or not one sympathizes with the views of that wing of the functionalist party to which our attention has just been directed it certainly seems a trifle unfair to cast up the mind-body difficulty in the teeth of the functionalist as such when on logical grounds he is no more guilty than any of his psychological neighbors. No courageous psychology of volition is possible which does not squarely face the mind-body problem, and in point of fact every important description of mental life contains doctrine of one kind or another upon this matter. A literally pure psychology of volition would be a sort of hanging-garden of Babylon, marvelous but inaccessible to psychologists of terrestrial habit. The functionalist is a greater sinner than others only in so far as he finds necessary and profitable a more constant insistence upon the translation of mental process into physiological process and conversely.

IV

. . . .

. . . No description of the actual circumstances attending the participation of mind in the accommodatory activities of the organism could be other than a mere empty schematism without making reference to the manner in which mental processes eventuate in motor phenomena of the physiological organism. The overt accommodatory act is, I take it, always sooner or later a muscular movement. But this fact being admitted, there is nothing for it, if one will describe accommodatory processes, but to recognize the mind-body relations and in some way give expression to their practical significance. It is only in this regard . . . that the functionalist departs a trifle in his practice and a trifle more in his theory from the rank and file of his colleagues.

94 JOHN BROADUS WATSON (1878–1958) ON BEHAVIORISM, 1913

J. B. Watson, "Psychology as the behaviorist views it," *Psychological Review 20,* 158–177 (1913).

This is the paper that founded "behaviorism" as the dominant movement in American psychology. Watson had taken his doctorate at the University of Chicago under James R. Angell, the effective leader of the functionalist school. Watson was thus imbued with the functionalist outlook, a concern with the organism's adaptation to its environment, compared to which the analysis of consciousness into its elements seemed unimportant, uninteresting, and irrelevant. Watson, moreover, was gaining his sophistication in the era when comparative psychology was still grappling unsuccessfully with the problem of how to interpret the objective data of behavior in terms of the subjective contents of the mind. It is a plausible functionalist's resolution of this problem to hold that the objective data themselves constitute the sum total of the data of psychology, as Loeb and some of the Russian physiologists (see Sechenov, No. 63, and Pavlov, No. 101) had already argued. In America, Watson took this step frankly and positively, promoting the new view so vigorously and effectively that behaviorism—the name he gave to this systematic position—quite rapidly became the representative school in what was soon to be the American tradition.

Psychology as the behaviorist views it is a purely objective experimental branch of natural science. Its theoretical goal is the prediction and control of behavior. Introspection forms no essential part of its methods, nor is the scientific value of its data dependent upon the readiness with which they lend themselves to interpretation in terms of consciousness. The behaviorist, in his efforts to get a unitary scheme of animal response, recognizes no dividing line between man and brute. The behavior of man, with all of its refinement and complexity, forms only a part of the behaviorist's total scheme of investigation.

It has been maintained by its followers generally that psychology is a study of the science of the phenomena of consciousness. It has taken as its problem, on the one hand, the analysis of complex mental states (or processes) into simple elementary constituents, and on the other the construction of complex states when the elementary constituents are given. The world of physical objects (stimuli, including here anything which may excite activity in a receptor), which forms the total phenomena of the natural scientist, is looked upon merely as means to an end. That end is the production of mental states that may be 'inspected' or 'observed.' The psychological object of observation in the case of an emotion, for example, is the mental state itself. The problem in emotion is the determination of the number and kind of elementary constituents present, their loci, intensity, order of appearance, etc. It is agreed that introspection is the method *par excellence* by means of which mental states may be manipulated for purposes of psychology. On this assumption, behavior data (including under this term everything which goes under the name of comparative psychology) have no value *per se.* They possess significance only in so far as they may throw light upon conscious states. Such data must have at least an analogical or indirect reference to belong to the realm of psychology.

Indeed, at times, one finds psychologists who are sceptical of even this analogical reference. Such scepticism is often shown by the question which is put to the student of behavior, "what is the bearing of animal work upon human psychology?" I used to have to study over this question. Indeed it always embarrassed me somewhat. I was interested in my own work and felt that it was important, and yet I could not trace any close connection between it and psychology as my questioner understood psychology. I hope that such a confession will clear the atmosphere to such an extent that we will no longer have to work under false pretences. We must frankly admit that the facts so important to us which we have been able to glean from extended work upon the senses of animals by the behavior method have contributed only in a fragmentary way to the general theory of human sense organ processes, nor have they suggested new points of experimental attack. The enormous number of experiments which we have carried out upon learning have likewise contributed little to human psychology. It seems reasonably clear that some kind of compromise must be effected: either psychology must change its viewpoint so as to take in facts of behavior, whether or not they have bearings upon the problems of 'consciousness'; or else behavior must stand alone as a wholly separate and independent science. Should human psychologists fail to look with favor upon our overtures and refuse to modify their position, the behaviorists will be driven to using human beings as subjects and to employ methods of investigation which are exactly comparable to those now employed in the animal work.

Any other hypothesis than that which admits the independent value of behavior material, regardless of any bearing such material may have upon consciousness, will inevitably force us to the absurd position of attempting to *construct* the conscious content of the animal whose behavior we have been studying. On this view, after having determined our animal's ability to learn, the simplicity or complexity of its methods of learning, the effect of past habit upon present response, the range of stimuli to which it ordinarily responds, the widened range to which it can respond under experimental conditions,—in more general terms, its various problems and its various ways of solving them,—we should still feel that the task is unfinished and that the results are worthless, until we can interpret them by analogy in the light of consciousness. Although we have solved our problem we feel uneasy and unrestful because of our definition of psychology: we feel forced to say something about the possible mental processes of our animal. We say that, having no eyes, its stream of consciousness cannot contain brightness and color sensations as we know them,—having no taste buds this stream can contain no sensations of sweet, sour, salt and bitter. But on the other hand, since it does respond to thermal, tactual and organic stimuli, its conscious content must be made up largely of these sensations; and we usually add, to protect ourselves against the reproach of being anthropomorphic, "if it has any consciousness." Surely this doctrine which calls for an analogical interpretation of all behavior data may be shown to be false: the position that the standing of an observation upon behavior is determined by its fruitfulness in yielding results which are interpretable only in the narrow realm of (really human) consciousness.

This emphasis upon analogy in psychology has led the behaviorist somewhat afield. Not being willing to throw off the yoke of consciousness he feels impelled to make a place in the scheme of behavior where the rise of consciousness can be determined. This point has been a shifting one. A few years ago certain animals were supposed to possess 'associative memory,' while certain others were supposed to lack it. One meets this search for the origin of consciousness under a good many disguises. Some of our texts state that consciousness arises at the moment when reflex and instinctive activities fail properly to conserve the organism. A perfectly adjusted organism would be lacking in consciousness. On the other hand whenever we find the presence of diffuse activity which results in habit formation, we are justified in assuming consciousness. I must confess that these arguments had weight with me when I began the study of behavior. I fear that a good many of us are still viewing behavior problems with something like this in mind. More than one student in behavior has attempted to frame criteria of the psychic—

to devise a set of objective, structural and functional criteria which, when applied in the particular instance, will enable us to decide whether such and such responses are positively conscious, merely indicative of consciousness, or whether they are purely 'physiological.' Such problems as these can no longer satisfy behavior men. It would be better to give up the province altogether and admit frankly that the study of the behavior of animals has no justification, than to admit that our search is of such a 'will o' the wisp' character. One can assume either the presence or the absence of consciousness anywhere in the phylogenetic scale without affecting the problems of behavior by one jot or one tittle; and without influencing in any way the mode of experimental attack upon them. On the other hand, I cannot for one moment assume that the paramecium responds to light; that the rat learns a problem more quickly by working at the task five times a day than once a day, or that the human child exhibits plateaux in his learning curves. These are questions which vitally concern behavior and which must be decided by direct observation under experimental conditions.

This attempt to reason by analogy from human conscious processes to the conscious processes in animals, and *vice versa:* to make consciousness, as the human being knows it, the center of reference of all behavior, forces us into a situation similar to that which existed in biology in Darwin's time. The whole Darwinian movement was judged by the bearing it had upon the origin and development of the human race. Expeditions were undertaken to collect material which would establish the position that the rise of the human race was a perfectly natural phenomenon and not an act of special creation. Variations were carefully sought along with the evidence for the heaping up effect and the weeding out effect of selection; for in these and the other Darwinian mechanisms were to be found factors sufficiently complex to account for the origin and race differentiation of man. The wealth of material collected at this time was considered valuable largely in so far as it tended to develop the concept of evolution in man. It is strange that this situation should have remained the dominant one in biology for so many years. The moment zoölogy undertook the experimental study of evolution and descent, the situation immediately changed. Man ceased to be the center of reference. I doubt if any experimental biologist today, unless actually engaged in the problem of race differentiation in man, tries to interpret his findings in terms of human evolution, or ever refers to it in his thinking. He gathers his data from the study of many species of plants and animals and tries to work out the laws of inheritance in the particular type upon which he is conducting experiments. Naturally, he follows the progress of the work upon race differentiation in man and in the descent of man, but he looks upon these as special topics, equal in importance with his own yet ones in which his interests will never be vitally engaged. It is not fair to say that all of his work is directed toward human evolution or that it must be interpreted in terms of human evolution. He does not have to dismiss certain of his facts on the inheritance of coat color in mice because, forsooth, they have little bearing upon the differentiation of the *genus homo* into separate races, or upon the descent of the *genus homo* from some more primitive stock.

In psychology we are still in that stage of development where we feel that we must select our material. We have a general place of discard for processes, which we anathematize so far as their value for psychology is concerned by saying, "this is a reflex"; "that is a purely physiological fact which has nothing to do with psychology." We are not interested (as psychologists) in getting all of the processes of adjustment which the animal as a whole employs, and in finding how these various responses are associated, and how they fall apart, thus working out a systematic scheme for the prediction and control of response in general. Unless our observed facts are indicative of consciousness, we have no use for them, and unless our apparatus and method are designed to throw such facts into relief, they are thought of in just as disparaging a way. I shall always remember the remark one distinguished psychologist made as he looked over the color apparatus designed for testing the responses of animals to monochromatic light in the attic at Johns Hopkins. It was this: "And they call this psychology!"

I do not wish unduly to criticize psychology. It has failed signally, I believe, during the fifty-odd years of its existence as an experimental discipline to make its place in the world as an undisputed natural science. Psychology, as it is generally thought of, has something esoteric in its methods. If you fail to reproduce any findings, it is not due to some fault in your apparatus or in the control of your stimulus, but it is due to the fact that your introspection is untrained. The attack is made upon the observer and not upon the experimental setting. In physics and in chemistry the attack is made upon the experimental conditions. The apparatus was not sensitive enough, impure chemicals were used, etc. In these sciences a better technique will give reproducible results. Psychology is otherwise. If you can't observe 3-9 states of clearness in attention, your introspection is poor. If, on the other hand, a feeling seems reasonably clear to you, your introspection is again faulty. You are seeing too much. Feelings are never clear.

The time seems to have come when psychology must discard all reference to consciousness; when it need no longer delude itself into thinking that it is making mental states the object of observation. We have become so enmeshed in speculative questions concerning the elements of mind, the nature of conscious content . . . that I, as an experimental student, feel that something is wrong with our premises and the types of problems which develop from them. There is no longer any guarantee that we all mean the same thing when we use the terms now current in psychology. Take the case of sensation. A sensation is defined in terms of its attributes. One psychologist will state with readiness that the attributes of a visual sensation are *quality, extension, duration,* and *intensity.* Another will add *clearness.* Still another that of *order.* I doubt if any one psychologist can draw up a set of statements describing what he means by sensation which will be agreed to by three other psychologists of different training. Turn for a moment to the question of the number of isolable sensations. Is there an extremely large number of color sensations—or only four, red, green, yellow and blue? Again, yellow, while psychologically simple, can be obtained by superimposing red and green spectral rays upon the same diffusing surface! If, on the other hand, we say that every just noticeable difference in the spectrum is a simple sensation, and that every just noticeable increase in the white value of a given color gives simple sensations, we are forced to admit that the number is so large and the conditions for obtaining them so complex that the concept of sensation is unusable, either for the purpose of analysis or that of synthesis. Titchener, who has fought the most valiant fight in this country for a psychology based upon introspection, feels that these differences of opinion as to the number of sensations and their attributes; as to whether there are relations (in the sense of elements) and on the many others which seem to be fundamental in every attempt at analysis, are perfectly natural in the present undeveloped state of psychology. While it is admitted that every growing science is full of unanswered questions, surely only those who are wedded to the system as we now have it, who have fought and suffered for it, can confidently believe that there will ever be any greater uniformity than there is now in the answers we have to such questions. I firmly believe that two hundred years from now, unless the introspective method is discarded, psychology will still be divided on the question as to whether auditory sensations have the quality of 'extension,' whether intensity is an attribute which can be applied to color, whether there is a difference in 'texture' between image and sensation and upon many hundreds of others of like character.

· · · · ·

My psychological quarrel is not with the systematic and structural psychologist alone. The last fifteen years have seen the growth of what is called functional psychology. This type of psychology decries the use of elements in the static sense of the structuralists. It throws emphasis upon the biological significance of conscious processes instead of upon the analysis of conscious states into introspectively isolable elements. I have done my best to understand the difference between functional psychology and structural psychology. Instead of clarity, confusion grows upon me. The terms sensation, perception, affection, emotion, volition are used as much by the functionalist as by the structuralist. The addition of the word 'process' ('mental act as a whole,' and like terms are frequently met) after each serves in some way to remove the corpse of 'content' and to leave 'function' in its stead. Surely if these concepts are elusive when looked at from a content standpoint, they are still more deceptive when viewed from the angle of function, and especially so when function is obtained by the introspection method. It is rather interesting that no functional psychologist has carefully distinguished between 'perception' (and this is true of the other psychological terms as well) as employed by the systematist, and 'perceptual process' as used in functional psychology. It seems illogical and hardly fair to criticize the psychology which the systematist gives us, and then to utilize his terms without carefully showing the changes in meaning which are to be attached to them . . .

One of the difficulties in the way of a consistent functional psychology is the parallelistic hypothesis. If the functionalist attempts to express his formulations in terms which make mental states really appear to function, to play some active rôle in the world of adjustment, he almost inevitably lapses into terms which are connotative of interaction. When taxed with this he replies that it is more convenient to do so and that he does it to avoid the circumlocution and clumsiness which are inherent in any thoroughgoing parallelism. As a matter of fact I believe the functionalist actually thinks in terms of interaction and resorts to parallelism only when forced to give expression to his views. I feel that *behaviorism* is the only consistent and logical functionalism. In it one avoids both the Scylla of parallelism and the Charybdis of interaction. Those time-honored relics of philosophical speculation need trouble the student of behavior as little as they trouble the student of physics. The consideration of the mind-body problem affects neither the type of problem selected nor the formulation of the solution of that problem. I can state my position here no better than by saying that I should like to bring my students up in the same ignorance of such hypotheses as one finds among the students of other branches of science.

This leads me to the point where I should like to make the argument constructive. I believe we can write a psychology . . . and . . . never use the terms consciousness, mental states, mind, content, introspectively verifiable, imagery, and the like. I believe that we can do it in a few years without running into the absurd terminology of Beer, Bethe, Von Uexküll, Nuel, and that of the so-called objective schools generally. It can be done in terms of stimulus and response, in terms of habit formation, habit integrations and the like. Furthermore, I believe that it is really worth while to make this attempt now.

The psychology which I should attempt to build up would take as a starting point, first, the observable fact that organisms, man and animal alike, do adjust themselves to their environment by means of hereditary and habit equipments. These adjustments may be very adequate or they may be so inadequate that the organism barely maintains its existence; secondly, that certain stimuli lead the organisms to make the responses. In a system of psychology completely worked out, given the response the stimuli can be predicted; given the stimuli the response can be predicted. Such a set of statements is crass and raw in the extreme, as all such generalizations must be. Yet they are hardly more raw and less realizable than the ones which appear in the psychology texts of the day. I possibly might illustrate my point better by choosing an everyday problem which anyone is likely to meet in the course of his work. Some time ago I was called upon to make a study of certain species of birds. Until I went to Tortugas I had never seen these birds alive. When I reached there I found the animals doing certain things: some of the acts seemed to work peculiarly well in such an environment, while others seemed to be unsuited to their type of life. I first studied the responses of the group as a whole and later those of individuals. In order to understand more thoroughly the relation between what was habit and what was hereditary in these responses, I took the young birds and reared them. In this way I was able to study the order of appearance of hereditary adjustments and their complexity, and later the beginnings of habit formation. My efforts in determining the stimuli which called forth such adjustments were crude indeed. Consequently my attempts to control behavior and to produce responses at will did not meet with much success. Their food and water, sex and other social relations, light and temperature conditions were all beyond control in a field study. I did find it possible to control their reactions in a measure by using the nest and egg (or young) as stimuli. It is not necessary in this paper to develop further how such a study should be carried out and how work of this kind must be supplemented by carefully controlled laboratory experiments. Had I been called upon to examine the natives of some of the Australian tribes, I should have gone about my task in the same way. I should have found the problem more difficult: the types of responses called forth by physical stimuli would have been more varied, and the number of effective stimuli larger. I should have had to determine the social setting of their lives in a far more careful way. These savages would be more influenced by the responses of each other than was the case with the birds. Furthermore, habits would have been more complex and the influences of past habits upon the present responses would have appeared more clearly. Finally, if I had been called upon to work out the psychology of the educated European, my problem would have required several lifetimes. But in the one I have at my disposal I should have followed the same general line of attack. In the main, my desire in all such work is to gain an accurate knowledge of adjustments and the stimuli calling them forth. My final reason for this is to learn general and particular methods by which I may control behavior.

XIV LEARNING

The experiments that set the pattern for most research on learning appeared within the span of less than two decades (1885–1904). What explains this sudden emergence of a new experimental discipline? It was not because of technical advance, being unlike, for example, the study of hearing, which forged rapidly ahead after the invention of electronic amplifiers and oscillators. The experimental techniques for research on learning were simple enough to have been created a century sooner. Nor was it because of a sudden interest in the learning process, for speculation about the acquisition of knowledge is antique. What seems, rather, to have happened is that the growing faith in scientific research in general and scientific psychology in particular had, at last, encouraged men to experiment on learning.

Western philosophy had developed the classical doctrine of association (see Nos. 65–76), which was, in effect, a theory of how learning takes place. The earliest experimental attempt to investigate learning (Ebbinghaus, 1885) was a transformation of the philosophical doctrine of association by contiguity into a group of experimental procedures, some of which are still used today. Calkins (1896) also created a lasting experimental technique in her research along the classical associationistic lines. In Thorndike's contribution (1898), the associationistic tradition merged with the much newer doctrine of organic evolution. Thorndike devised techniques for studying the learning process in animals and felt, on the basis of the continuity of species, that his findings would have some bearing on human learning. He

thought that the classical law of association by contiguity had to be modified, for learning, he said, takes place more readily when the learner is rewarded for a correct performance than when he merely practices. The study of animal learning was further advanced when Yerkes (1901) introduced the maze as a technique, but Small (1901) usually gets the credit for that innovation, for his maze was more interesting and his experiment more extensive.

In 1901, Thorndike and Woodworth published their first paper on the problem of how training in one skill influences performance in another, a problem about which the speculative psychology of the nineteenth century contained conflicting views. There was the doctrine of formal discipline, which held that practice in certain skills is broadly beneficial, and there was the doctrine of association, which implied that one skill influences another only to the extent that the two skills are partially identical. Educational practices in schools tended to operate on the assumption that the first alternative was correct; Woodworth and Thorndike obtained data to support the second view. Thus there was set up a standard area of research for experimental psychologists, who initiated a virtual revolution in American education.

I. P. Pavlov, distinguished for his work on digestion, announced to the world on the occasion of receiving the Nobel prize (1904) that he had undertaken a program of research on conditioned reflexes, his term for behavior that is learned and not inborn. Although Pavlov saw himself as part of a tradition in Russian physiology (see Sechenov, No. 63), his theory of learning strongly resembles the classical principle of association by contiguity, despite his explicit championing of objectivism and vigorous opposition to the mentalism of classical associationism. His dedication to a biological approach was to play a significant role in the shifting emphasis of modern psychology from mind to behavior.

The three major contributors in this early period were, therefore, Ebbinghaus, Thorndike, and Pavlov. Though they differed in important ways, they had in common the desire to reduce the complexity of thought and behavior to a simple concatenation of events. Each of them owed much to classical associationism. In Köhler (1917), however, one finds the point of view of Gestalt psychology, which formulated the antithesis to elementism in its treatment of both learning and perception.

95   HERMANN EBBINGHAUS (1850–1909) ON THE LEARNING OF NONSENSE SYLLABLES, 1885

Hermann Ebbinghaus, *Ueber das Gedächtnis* (Leipzig, 1885). Translated by Henry A. Ruger and Clara Bussenius as *Memory*, Educational Reprint No. 3 (New York: Teachers College. Columbia University, 1913), pp. 4–6, 22–24, 90–101.

With Fechner's measurement of sensation as his model of scientific psychology, Ebbinghaus transformed the classical principle of association by contiguity into an empirical hypothesis and thus initiated the experimental study of learning and memory. The logic of his approach is simple: if ideas are connected by the frequency of their contiguities, then the number of repetitions of contiguous ideas can be used as the independent variable of which memory, or learning, is a function. Yet, for all its simplicity, this work abounds in novelty. Ebbinghaus invented the procedure of memorizing a series of nonsense syllables, which allowed him to trace the formation of new associations, uncontaminated by the old associations of the subject. Not only the goals, but even the methods, of this epoch-making work are still current.

Under ordinary circumstances ... frequent repetitions are indispensable in order to make possible the reproduction of a given content. Vocabularies, discourses, and poems of any length cannot be learned by a single repetition even with the greatest concentration of attention on the part of an individual of very great ability. By a sufficient number of repetitions their final mastery is ensured, and by additional later reproductions gain in assurance and ease is secured.

Left to itself every mental content gradually loses its capacity for being revived, or at least suffers loss in this regard under the influence of time. Facts crammed at examination time soon vanish, if they were not sufficiently grounded by other study and later subjected to a sufficient review. But even a thing so early and deeply founded as one's mother tongue is noticeably impaired if not used for several years.

3. DEFICIENCIES IN OUR KNOWLEDGE CONCERNING MEMORY

The foregoing sketch of our knowledge concerning memory makes no claim to completeness. To it might be added such a series of propositions known to psychology as the following: "He who learns quickly also forgets quickly," "Relatively long series of ideas are retained better than relatively short ones," "Old people forget most quickly the things they learned last," and the like. Psychology is wont to make the picture rich with anecdote and illustration. But—and this is the main point—even if we particularise our knowledge by a most extended use of illustrative material, everything that we can say retains the indefinite, general, and comparative character of the propositions quoted above. Our information comes almost exclusively from the observation of extreme and especially striking cases. We are able to describe these quite correctly in a general way and in vague expressions of more or less. We suppose, again quite correctly, that the same influences exert themselves, although in a less degree, in the case of the inconspicuous, but a thousand fold more frequent, daily activities of memory. But if our curiosity carries us further and we crave more specific and detailed information concerning these dependencies and interdependencies, both those already mentioned and others,—if we put questions, so to speak, concerning their inner structure—our answer is silence. How does the disappearance of the ability to reproduce, forgetfulness, depend upon the length of time during which no repetitions have taken place? What proportion does the increase in the certainty of reproduction bear to the number of repetitions? How do these relations vary with the greater or less intensity of the interest in the thing to be reproduced? These and similar questions no one can answer.

This inability does not arise from a chance neglect of investigation of these relations. We cannot say that tomorrow, or whenever we wish to take time, we can investigate these problems. On the contrary this inability is inherent in the nature of the questions themselves. Although the conceptions in question—namely, degrees of forgetfulness, of certainty and interest—are quite correct, we have no means for establishing such degrees in our experience except at the extremes, and even then we cannot accurately limit those extremes. We feel therefore that we are not at all in a condition to undertake the investigation. We form certain conceptions during striking experiences, but we cannot find any realisation of them in the similar but less striking experiences of everyday life. *Vice versa* there are probably many conceptions which we have not as yet formed which would be serviceable and indispensable for a clear understanding of the facts, and their theoretical mastery.

The amount of detailed information which an individual has at his command and his theoretical elaborations of the same are mutually dependent; they grow in and through each other. It is because of the indefinite and little specialised character of our knowledge that the theories concerning the processes of memory, reproduction, and association have been up to the present time of so little value for a proper comprehension of those processes. For example, to express our ideas concerning their physical basis we use different metaphors—stored up ideas, engraved images, well-beaten paths. There is only one thing certain about these figures of speech and that is that they are not suitable.

Of course the existence of all these deficiencies has its perfectly sufficient basis in the extraordinary difficulty and complexity of the matter. It remains to be proved whether, in spite of the clearest insight into the inadequacy of our knowledge, we shall ever make any actual progress. Perhaps we shall always have to be resigned to this. But a somewhat greater accessibility than has so far been realised in this field cannot be denied to it, as I hope to prove presently. If by any chance a way to a deeper penetration into this matter should present itself, surely, considering the significance of memory for all mental phenomena, it should be our wish to enter that path at once. For at the very worst we should prefer to see resignation arise from the failure of earnest investigations rather than from persistent, helpless astonishment in the face of their difficulties.

. . . . .

11. SERIES OF NONSENSE SYLLABLES

In order to test practically, although only for a limited field, a way of penetrating more deeply into memory processes—and it is to these that the preceding considerations have been directed—I have hit upon the following method.

Out of the simple consonants of the alphabet and our eleven vowels and diphthongs all possible syllables of a certain sort were constructed, a vowel sound being placed between two consonants.

These syllables, about 2,300 in number, were mixed together and then drawn out by chance and used to construct series of different lengths, several of which each time formed the material for a test.

At the beginning a few rules were observed to prevent, in the construction of the syllables, too immediate repetition of similar sounds, but these were not strictly adhered to. Later they were abandoned and the matter left to chance. The syllables used each time were carefully laid aside till the whole number had been used, then they were mixed together and used again.

The aim of the tests carried on with these syllable series was, by means of repeated audible perusal of the separate series, to so impress them that immediately afterwards they could voluntarily just be reproduced. This aim was considered attained when, the initial syllable being given, a series could be recited at the first attempt, without hesitation, at a certain rate, and with the consciousness of being correct.

12. ADVANTAGES OF THE MATERIAL

The nonsense material, just described, offers many advantages, in part because of this very lack of meaning. First of all, it is relatively simple and relatively homogeneous. In the case of the material nearest at hand, namely poetry or prose, the content is now narrative in style, now descriptive, or now reflective; it contains now a phrase that is pathetic, now one that is humorous; its metaphors are sometimes beautiful, sometimes harsh; its rhythm is sometimes smooth and sometimes rough. There is thus brought into play a multiplicity of influences which change without regularity and are therefore disturbing. Such are associations which dart here and there, different degrees of interest, lines of verse recalled because of their striking quality or their beauty, and the like. All this is avoided with our syllables. Among many thousand combinations there occur scarcely a few dozen that have a meaning and among these there are again only a few whose meaning was realised while they were being memorised.

However, the simplicity and homogeneity of the material must not be overestimated. It is still far from ideal. The learning of the syllables calls into play the three sensory fields, sight, hearing and the muscle sense of the organs of speech. And although the part that each of these senses plays is well limited and always similar in kind, a certain complication of the results must still be anticipated because of their combined action. Again, to particularise, the homogeneity of the series of syllables falls considerably short of what might be expected of it. These series exhibit very important and almost incomprehensible variations as to the ease or difficulty with which they are learned. It even appears from this point of view as if the differences between sense and nonsense material were not nearly so great as one would be inclined *a priori* to imagine. At least I found in the case of learning by heart a few cantos from Byron's "Don Juan" no greater range of distribution of the separate numerical measures than in the case of a series of nonsense syllables in the learning of which an approximately equal time had been spent. In the former case the innumerable disturbing influences mentioned above seem to have compensated each other in producing a certain intermediate effect; whereas in the latter case the predisposition, due to the influence of the mother tongue, for certain combinations of letters and syllables must be a very heterogeneous one.

More indubitable are the advantages of our material in two other respects. In the first place it permits an inexhaustible amount of new combinations of quite homogeneous character, while different poems, different prose pieces always have something incomparable. It also makes possible a quantitative variation which is adequate and certain; whereas to break off before the end or to begin in the middle of the verse or the sentence leads to new complications because of various and unavoidable disturbances of the meaning.

. . . . .

35. ASSOCIATION ACCORDING TO TEMPORAL SEQUENCE AND ITS EXPLANATION

I shall now discuss a group of investigations made for the purpose of finding out the conditions of association. The results of these investigations are, it seems to me, theoretically of especial interest.

The non-voluntary re-emergence of mental images out of the darkness of memory into the light of consciousness takes place, as has already been mentioned, not at random and accidentally, but in certain regular forms in accordance with the so-called laws of association. General knowledge concerning these laws is as old as psychology itself, but on the other hand a more precise formulation of them has remained—characteristically enough—a matter of dispute up to the very present. Every new presentation starts out with a reinterpretation of the contents of a few lines from Aristotle, and according to the condition of our knowledge it is necessary so to do.

Of these "Laws," now—if, in accordance with usage and it is to be hoped in anticipation of the future, the use of so lofty a term is permitted in connection with formulae of so vague a character—of these laws, I say, there is one which has never been disputed or doubted. It is usually formulated as follows: Ideas which have been developed simultaneously or in immediate succession in the same mind mutually reproduce each other, and do this with greater ease in the direction of the original succession and with a certainty proportional to the frequency with which they were together.

This form of non-voluntary reproduction is one of the best verified and most abundantly established facts in the whole realm of mental events. It permeates inseparably every form of reproduction, even the so-called voluntary form. The function of the conscious will, for example, in all the numerous reproductions of the syllable-series which we have come to know, is limited to the general purpose of reproduction and to laying hold of the first member of the series. The remaining members follow automatically, so to speak, and thereby fulfill the law that things which have occurred together in a given series are reproduced in the same order.

However, the mere recognition of these evident facts has naturally not been satisfying and the attempt has been made to penetrate into the inner mechanism of which they are the result. If for a moment we try to follow up this speculation concerning the *Why*, before we have gone more than two steps we are lost in obscurities and bump up against the limits of our knowledge of the *How*.

It is customary to appeal for the explanation of this form of association to the nature of the soul. Mental events, it is said, are not passive happenings but the acts of a subject. What is more natural than that this unitary being should bind together in a definite way the contents of his acts, themselves also unified? Whatever is experienced simultaneously or in immediate succession is conceived in one act of consciousness and by that very means its elements are united and the union is naturally stronger in proportion to the number of times they are entwined by this bond of conscious unity. Whenever, now, by any chance one part only of such a related complex is revived, what else can it do than to attract to itself the remaining parts?

But this conception does not explain as much as it was intended to do. For the remaining parts of the complex are not merely drawn forth but they respond to the pull in an altogether definite direction. If the partial contents are united simply by the fact of their membership in a single conscious act and accordingly all in a similar fashion, how does it come about that a sequence of partial contents returns in precisely the same order and not in any chance combination? In order to make this intelligible, one can proceed in two ways.

In the first place it can be said that the connection of the things present simultaneously in a single conscious act is made from each member to its immediate successor but not to members further distant. This connection is in some way inhibited by the presence of intermediate members, but not by the interposition of pauses, provided that the beginning and end of the pause can be grasped in one act of consciousness. Thereby return is made to the facts, but the advantage which the whole plausible appeal to the unitary act of consciousness offered is silently abandoned. For, however much contention there may be over the number of ideas which a single conscious act may comprehend, it is quite certain that, if not always, at least in most cases, we include more than two members of a series in any one conscious act. If use is made of one feature of the explanation, the characteristic of unity, as a welcome factor, the other side, the manifoldness of the members, must be reckoned with, and the right of representation must not be denied it on assumed but unstatable grounds. Otherwise, we have only said,—and it is possible that we will have to be content with that—that it is so because there are reasons for its being so.

There is, consequently, the temptation to use this second form of statement. The ideas which are conceived in one act of consciousness are, it is true, all bound together, but not in the same way. The strength of the union is, rather, a decreasing function of the time or of the number of intervening members. It is therefore smaller in proportion as the interval which separates the individual members is greater. Let a, b, c, d be a series which has been presented in a single conscious act, then the connection of a with b is stronger than that of a with the later c; and the latter again is stronger than that with d. If a is in any way reproduced, it brings with it b and c and d, but b, which is bound to it more closely, must arise more easily and quickly than c, which is closely bound to b, etc. The series must therefore reappear in consciousness in its original form although all the members of it are connected with each other.

.   .   .   .   .

According to this conception, therefore, the associative threads, which hold together a remembered series, are spun not merely between each member and its immediate successor, but beyond intervening members to every member which stands to it in any close temporal relation. The strength of the threads varies with the distance of the members, but even the weaker of them must be considered as relatively of considerable significance.

The acceptance or rejection of this conception is clearly of great importance for our view of the inner connection of mental events, of the richness and complexity of their groupings and organisation. But it is clearly quite idle to contend about the matter if observation is limited to conscious mental life, to the registration of that which whirls around by chance on the surface of the sea of life.

For, according to the hypothesis, the threads which connect one member to its immediate successor although not the only one spun, are, however, stronger than the others. Consequently, they are, in general, as far as appearances in consciousness are concerned, the important ones, and so the only ones to be observed.

On the other hand, the methods which lie at the basis of the researches already described permit the discovery of connections of even less strength. This is done by artificially strengthening these connections until they reach a definite and uniform level of reproducibility. I have, therefore, carried on according to this method a rather large number of researches to test experimentally in the field of the syllable-series the question at issue, and to trace an eventual dependence of the strength of the association upon the sequence of the members of the series appearing in succession in consciousness.

36. METHODS OF INVESTIGATION OF ACTUAL BEHAVIOR

Researches were again carried out with six series of 16 syllables each. For greater clearness the series are designated with Roman numbers and the separate syllables with Arabic. A syllable group of the following form constituted, then, each time the material for research:

$$
\begin{array}{llll}
\text{I(1)} & \text{I(2)} & \text{I(3)} \ldots & \text{I(15)} & \text{I(16)} \\
\text{II(1)} & \text{II(2)} & \text{II(3)} \ldots & \text{II(15)} & \text{II(16)} \\
& \cdot & & & \cdot \\
& \cdot & & & \cdot \\
& \cdot & & & \cdot \\
\text{VI(1)} & & \cdot & \text{VI(15)} & \text{VI(16)}
\end{array}
$$

If I learn such a group, each series by itself, so that it can be repeated without error, and 24 hours later repeat it in the same sequence and to the same point of mastery, then the latter repetition is possible in about two thirds of the time necessary for the first. The resulting saving in work of one third clearly measures the strength of the association formed during the first learning between one member and its immediate successor.

Let us suppose now that the series are not repeated in precisely the same order in which they were learned. The syllables learned in the order I(1) I(2) I(3) · · · I(15) I(16) may for example be repeated in the order I(1) I(3) I(5) · · · I(15) I(2) I(4) I(6) · · · I(16), and the remaining series with a similar transformation. There will first be, accordingly, a set composed of all the syllables originally in the odd places and then a set of those originally in the even places, the second set immediately following the first. The new 16-syllable series, thus resulting, is then learned by heart. What will happen? Every member of the transformed series was, in the original series, separated from its present immediate neighbor by an intervening member with the exception of the middle term where there is a break. If these intervening members are actual obstructions to the associative connection, then the transformed series are as good as entirely unknown. In spite of the former learning of the series in the original sequence, no saving in work should be expected in the repetition of the transformed series. If on the other hand in the first learning threads of association are spun not merely from each member to its immediate successor but also over intervening members to more distant syllables, there would exist, already formed, certain predispositions for the new series. The syllables now in succession have already been bound together secretly with threads of a certain strength. In the learning of such a series it will be revealed that noticeably less work is required than for the learning of an altogether new series. The work, however, will be greater than in relearning a given series in unchanged order. In this case, again, the saving in work will constitute a measure of the strength of the associations existing between two members separated by a third. If from the original arrangement of the syllables new series are formed by the omission of 2, 3, or more intervening members, analogous considerations result. The derived series will either be learned without any noticeable saving of work, or a certain saving of work will result, and this will be proportionally less as the number of intervening terms increases.

On the basis of these considerations I undertook the following experiment. I constructed six series of 16 syllables each with the latter arranged by chance. Out of each group a new one was then constructed also composed of six series of 16 syllables each. These new groups were so formed that their adjacent syllables had been separated in the original series by either 1, or 2, or 3, or 7 intervening syllables.

If the separate syllables are designated by the positions which they held in the original arrangement, the following scheme results:

$$I(1) \quad I(2) \quad I(3) \ldots I(15) \quad I(16)$$
$$II(1) \quad II(2) \quad II(3) \ldots II(15) \quad II(16)$$

$$VI(1) \qquad \qquad VI(16)$$

By using the same scheme the derived groups appear as follows:

By skipping 1 syllable
$$I(1) \quad I(3) \quad I(5) \ldots I(15) \quad I(2) \quad I(4) \quad I(6) \ldots I(16)$$
$$II(1) \quad II(3) \quad II(5) \ldots II(15) \quad II(2) \quad II(4) \quad II(6) \ldots II(16)$$

$$VI(1) \; VI(3) \; \ldots \; VI(15) \; VI(2) \; VI(4) \; \ldots \; VI(16)$$

By skipping 2 syllables
$$I(1) \quad I(4) \; I(7) \; I(10) \; I(13) \; I(16) \quad I(2) \quad I(5) \; I(8) \; I(11) \; I(14) \quad I(3) \quad I(6) \; I(9) \; I(12) \; I(15)$$
$$II(1) \quad II(4) \; II(7) \; \ldots \; II(16) \; II(2) \; II(5) \; \ldots \; II(14) \; II(3) \; II(6) \; \ldots \; II(15)$$

$$VI(1) \; VI(4) \; \ldots \; VI(16) \; VI(2) \; VI(5) \; \ldots \; VI(14) \; VI(3) \; VI(6) \; \ldots \; VI(15)$$

By skipping 3 syllables

I(1)   I(5) I(9) I(13) I(2)   I(6) I(10) I(14) I(3)   I(7) I(11) I(15) I(4)   I(8) I(12)   I(16)
II(1) II(5) . . . II(2) II(6) . . .   II(3) II(7) . . .   II(4) II(8) . . . II(16)

VI(1) VI(5) . . . VI(2) VI(6) . . .   VI(3) VI(7) . .   VI(4) VI(8) . . . VI(16)

By skipping 7 syllables

I(1)   I(9) II(1) II(9) III(1) III(9) IV(1) IV(9) V(1)   V(9) VI(1) VI(9) I(2) I(10) II(2) II(10)
III(2)III(10) IV(2)IV(10) V(2) V(10) VI(2)VI(10) I(3)   I(11) II(3) II(11) III(3)III(11) IV(3)IV(11)

V(7) V(15) VI(7)VI(15) I(8)   I(16) II(8) II(16)III(8)III(16) IV(8)IV(16) V(8) V(16) VI(8)VI(16)

As a glance at this scheme will show, not all the neighboring syllables of the derived series were originally separated by the number of syllables designated. In some places in order to again obtain series of 16 syllables, greater jumps were made; but in no case was the interval less. Such places are, for example, in the series in which two syllables are skipped, the transitions from I(16) to I(2) and from I(14) to I(3). In the series in which 7 intermediates were jumped, there are seven places where there was no previous connection between successive syllables since the syllables in question came from different series and the different series, as has been often mentioned, were learned independently. The following is given in illustration: I(9) II(1), II(9) III(1), etc. The number of these breaks varies with the different kinds of derivation, but in each case is the same as the number of skipped syllables. On account of this difference, the derived series suffer from an inequality inherent in the nature of the experiment.

In the course of the experiment the skipping of more than 7 syllables was shown to be desirable, but I refrained from carrying that out. The investigations with the six 16-syllable series were carried quite far; and if series had been constructed using greater intervals, the breaks above mentioned would have had too much dominance. The derived series then contained ever fewer syllable-sequences for which an association was possible on the basis of the learning of the original arrangement; they were ever thus more incomparable.

The investigations were carried on as follows:—Each time the six series were learned in the original order and then 24 hours later in the derived and the times required were compared. On account of the limitation of the series to those described above the results are, under certain circumstances, open to a serious objection. Let it be supposed that the result is that the derived series are actually learned with a certain saving of time, then this saving is not necessarily due to the supposed cause, an association between syllables not immediately adjacent. The argument might, rather, run as follows. The syllables which are first learned in one order and after 24 hours in another are in both cases the same syllables. By means of the first learning they are impressed not merely in their definite order but also purely as individual syllables; with repetition they become to some extent familiar, at least more familiar than other syllables, which had not been learned just before. Moreover the new series have in part the same initial and final members as the old. Therefore, if they are learned in somewhat less time than the first series required, it is not to be wondered at. The basis of this does not necessarily lie in the artificial and systematic change of the arrangement, but it possibly rests merely on the identity of the syllables. If these were

Strength of Remote Associations.

1. With derivation of the series by skipping one intermediate syllable.

The original series were learned in x seconds	The corresponding derived series, in y seconds	The latter, therefore, with a saving of z seconds
$x =$	$y =$	$z =$
1187	1095	92
1220	1142	78
1139	1107	32
1428	1123	305
1279	1155	124
1245	1086	159
1390	1013	377
1254	1191	63
1335	1128	207
1266	1152	114
1259	1141	118
m 1273	1121	152

2. With derivation of the series by skipping two intermediate syllables.

$x =$	$y =$	$z =$
1400	1185	215
1213	1252	−39
1323	1245	78
1366	1103	263
1216	1066	150
1062	1003	59
1163	1161	2
1251	1204	47
1182	1086	96
1300	1076	224
1276	1339	−63
m 1250	1156	94

3. With derivation of the series by skipping three intermediate syllables.

The original series were learned in x seconds	The corresponding derived series, in y seconds	The latter, therefore, with a saving of z seconds
x =	y =	z =
1282	1347	−65
1202	1131	71
1205	1157	48
1303	1271	32
1132	1098	34
1365	1235	130
1210	1145	65
1364	1176	188
1308	1175	133
1298	1209	89
1286	1148	138
m 1269	1190	78

4. With derivation of the series by skipping seven intermediate syllables.

x =	y =	z =
1165	1086	79
1265	1295	−30
1197	1091	106
1295	1254	41
1233	1207	26
1335	1288	47
1321	1278	43
1344	1275	69
1322	1328	−6
1224	1212	12
1294	1217	77
m 1272	1230	42

5. With derivation of the series by retaining the beginning and end syllables and arranging the remainder by chance.

1305	1302	3
1181	1259	−7
1207	1237	−30
1401	1277	124
1278	1271	7
1302	1301	1
1248	1379	−131
1237	1240	−3
1355	1236	119
1214	1142	72
1147	1101	46
m 1261	1250	12

repeated on the second day in a new arrangement made entirely by chance they would probably show equally a saving in work.

In consideration of this objection and for the control of the remaining results I have introduced a further, the fifth, kind of derived series. The initial and final syllables of the original series were left in their places. The remaining 84 syllables, intermediates, were shaken up together and then, after chance drawing, were employed in the construction of new series between the original initial and final series. As a result of the learning of the original and derived series there must in this case also be revealed how much of the saving in work is to be ascribed merely to the identity of the syllable masses and to the identity of the initial and final members of the separate series.

37. RESULTS. ASSOCIATIONS OF INDIRECT SEQUENCE

For each group of original and derived series 11 double tests were instituted, 55 therefore in all. These were distributed irregularly over about 9 months. The results were as [shown in the preceding table].

To summarize the results: The new series formed by skipping 1, 2, 3 and 7 intermediate members were learned with an average saving of 152, 94, 78 and 42 seconds. In the case of the construction of a new series through a mere permutation of the syllables, there was an average saving of 12 seconds.

In order to determine the significance of these figures, it is necessary to compare them with the saving in work in my case in the relearning of an unchanged series after 24 hours. This amounted to about one third of the time necessary for the first learning in the case of 16-syllable series, therefore about 420 seconds.

This number measures the strength of the connection existing between each member and its immediate sequent, therefore the maximal effect of association under the conditions established. If this is taken as unity, then the strength of the connection of each member with the second following is a generous third and with the third following is a scant fourth.

The nature of the results obtained confirm—for myself and the cases investigated—the second conception given above . . . With repetition of the syllable series not only are the individual terms associated with their immediate sequents but connections are also established between each term and several of those which follow it beyond intervening members. To state it briefly, there seems to be an association not merely in direct but also in indirect succession. The strength of these connections decreases with the number of the intervening numbers; with a small number it was, as will be admitted, of surprising and unanticipated magnitude.

96 MARY WHITON CALKINS (1863–1930) ON THE LEARNING OF PAIRED ASSOCIATES, 1896

Mary W. Calkins, "Association: an Essay Analytic and Experimental," *The Psychological Review Monograph Supplements 1*, no. 2, 35–38, 51–52, 55–56 (1896).

In a preliminary report (1894) of this research, Miss Calkins stated that it was begun in 1892, just seven years after Ebbinghaus published his studies of association. She probably worked in ignorance of Ebbinghaus' experiments, for she makes no reference to them, but only to the classical associationistic tradition and to its elaboration by Wundt and his colleagues in Germany. Presumably, her inspiration, like Ebbinghaus', was an acceptance of the doctrine of contiguity in association and a belief in the possibility of psychological measurement, which she may have acquired from her teacher, Hugo Münsterberg. Her experiments showed frequency of repetition to be an important determiner of the strength of an association. In addition, however, she investigated strength as a function of the vividness of the material and of the position of an element in a series of the material to be learned. Although her work has neither the scope nor the originality of Ebbinghaus', she was one of the first in this new field, and she created an experimental technique that is now called the method of paired associates, which has survived to the present time. A few years later, G. E. Müller in Göttingen invented the method again, discovered Calkins, and with Alfons Pilzecker used this procedure extensively.

Experimental investigation may best supplement the purely introspective study of the nature of association by describing in relatively concrete terms the probable direction of trains of associated images. To this end there is necessary such a consideration of the so-called suggestibility of objects of consciousness as shall answer the question: what one of the numberless images which might conceivably follow upon the present percept or image will actually be associated with it?

Ordinary self-observation has long recognized that the readily associated objects are the 'interesting' ones, and has further enumerated frequency, recency, vividness or impressiveness, and primacy (the earliest position in a definite series of events) as the factors of interest, and therefore the conditions of association. A given object, then, is likely to be suggested by one with which it was frequently, recently or vividly connected, and by one with which it stood at the beginning of a series.

Logically prior to the discussion of suggestibility is the study of the suggestiveness of objects of consciousness, that is, the consideration of the question: what part of the present total content of consciousness will be associated with a following image? The suggesting object may, of course, be of varied extent. In the rare cases of 'total redintegration,' practically the entire present content is connected, as a whole, with what follows. Far more often, some one accentuated part of the total object of consciousness is the starting point of the association; and this emphasis of attention is once more upon the 'interesting' part of the entire content, that is upon some vivid, recent or repeated object, or upon one which has had the early place in a series. Finally, neither the total content of consciousness, nor a single accentuated portion of that total, but a group of these single factors or objects of consciousness may form the starting point of the association.

·   ·   ·   ·   ·

The experimental investigation whose results are here reported concerned itself with the conditions of suggestibility.

The relative significance of frequency, recency, primacy and vividness, was studied in about 2,200 experiments. This number does not include the introductory experiments undertaken in order to select satisfactory methods nor the practice experiments of each subject. There were 17 subjects, no one of whom assisted in more than 275 nor in less than 40 experiments; and the average number was 130 for each subject . . . All the subjects were entirely or comparatively ignorant of the aims and the problems of the investigation, which was not discussed until the conclusion of the work . . . Constant notes were kept of subjective experiences, but have not been reported, for none of them tended to modify the conclusions drawn from the experiments themselves except where the occurrence of natural associations made it necessary to reject entirely the results of particular experiments.

·   ·   ·   ·   ·

I. SIMPLE SERIES

a.1. Successive Arrangement. Visual Series

The method of the . . . experiments was as follows: the subjects, of whom two to eight were present at one time, sat before a white screen large enough to shield the conductor of the experiment. Through an opening, 10 cm. square, a color was shown for four seconds, followed immediately by a numeral, usually black on a white ground, for the same time. After a pause of about eight seconds, during which the subject looked steadily at the white background, another color was shown, succeeded at once by a second numeral, each exposed for four seconds. The pause of eight seconds followed, and the series of 7, 10 or 12 pairs of quickly succeeding color and numeral was continued in the same way. At the close a series was shown of the same colors in altered order, and the subject was asked, as each color appeared, to write down the suggested numerals if any such occurred. The pause between the combination-series, in which colors and numerals appeared together, and the test-series, in which the colors only were shown, was eight seconds in the case of the short series and four to six seconds in the case of the longer. Color and numeral were placed together in their position behind the opening of the screen, the numeral at first concealed by the color, which was then slipped out. There was thus a merely momentary pause between the appearance of color and of numeral. During the eight-second pauses the opening was filled by a white ground, $\frac{1}{2}$ cm. behind the screen. The subject thus saw nothing in the opening except this white ground, or the color, which filled the whole square, or the printed numeral; the movements of the experimenter were entirely concealed. The time was at first kept by following the ticks of a watch suspended close to the experimenter's ear; but in the last 1,200 tests by listening to the beats of a metronome, which rung a bell every four seconds; the metronome was enclosed in a sound-proof box, so that the subjects were not disturbed by the beats, which reached the experimenter through a rubber tube. All the series were carefully placed in order beforehand.

In the first group of experiments, some one color appeared several times in each series, once in an unimportant position with any chance numeral, but also once or more in some emphasized connection—either repeatedly with the same numeral (a 'frequent' combination), or at the very beginning or very end of a series (cases of 'primacy' and of 'recency'), or with a numeral of unusual size or color (an instance of 'vividness').

The following are representative series:

Visual series, 89. Frequency (3: 12).

I. (Combination Series.) Green, 47; brown, 73; *violet, 61* (*f*); light grey, 58; *violet, 61* (*f*); orange, 84; blue, 12; *violet, 61* (*f*); medium grey, 39; *violet, 26* (*n*); light green, 78; strawberry, 52.

II. (Test Series,) Blue, light grey, strawberry, green, *violet* (*f*), orange, brown, medium grey, light green.

Visual Series, 213. Vividness.

I. Brown, 34; peacock, 65; orange, 51; *green, 792* (*v*); blue, 19; violet, 48; *green, 27* (*n*); grey, 36; strawberry, 87; dark red, 54.

II. Blue, grey, dark red, brown, *green* (*v*), orange, strawberry, grey, peacock.

· · · · ·

II. COMPARATIVE SERIES

In showing that frequency, vividness, primacy and recentness are conditions of association these experiments have so far, of course, merely substantiated ordinary observation. The real purpose of the investigation is attained only by a comparison of these factors. Already it has appeared that the per cent. of correct 'frequency' associations is slightly the largest, and that recency is the principle of the combination in the next greatest number of cases. In order, however, to carry out the comparison under like conditions, these principles of combination were compared within the same series. To this end, long 'successive' series were arranged in which the significance of frequency was contrasted with that of vividness by showing a color three times with the same two-digit numeral (f) and once with a three-digit numeral (v); others, in which the color three times shown with a numeral (f) appeared also at the first of the series with another numeral (p). Short 'successive' series were formed in which the last color (r) had appeared once before with a three-digit numeral (v), or at the very beginning of the series (p), or twice before with a repeated numeral (f).

. . . . .

From this mass of figures a few conclusions emerge into prominence. Some of these have been already formulated, but the more important ones may be briefly stated again.

No one of these generalizations, it should be remarked, is proof against the caprice of the individual, who may have his own favorite type of association which resists opposition . . .

Frequency has been the most constant condition of suggestibility. The proportion of the frequent as compared with the normal associations is one-tenth greater than that of the vivid or of the recent. When directly compared with the vivid and the recent the proportion is still greater, though the number of associations with the contrasted numeral is larger than that of the associations with an ordinary one, because of the tendency of the repetition to accentuate the compared factor.

This significance of frequency is rather surprising. For though everybody recognizes the importance of repetition in forming associations, we are yet more accustomed to 'account for' these by referring to recent or to impressive combinations. The possibility that the prominence of frequency in our results is not fairly representative of ordinary trains of association is strengthened by the fact that it is contrasted with forms of vividness which are only two or three of many, and which do not approach the impressiveness, for instance, of richly emotional experiences. But this does not affect the importance of frequency as a corrective influence. Granted a sufficient number of repetitions, it seems possible to supplement, if not actually to supplant, associations which have been formed through impressive or through recent experiences. Moreover, the trustworthiness of the ordinary observation, which relegates frequency to a comparatively unimportant place among the factors of suggestibility, may be seriously questioned: I have found many cases, during experiments in free association in which the subject, asked to explain the association, does not always mention repetition, even when it has obviously occurred, but seems, as it were, to take it for granted. The prominence of frequency is of course of grave importance, for it means the possibility of exercising some control over the life of the imagination and of definitely combating harmful or troublesome associations.

97 EDWARD LEE THORNDIKE (1874–1949) ON ANIMAL LEARNING, 1898

E. L. Thorndike, "Animal intelligence: an experimental study of the associative processes in animals," *The Psychological Review Monograph Supplements 2*, no. 4 (whole no. 8), 1–8, 13–18, 73–74 (1898).

Like Ebbinghaus, Thorndike created a kind of experimental associationism. He departed, however, from the classical tradition in several ways. First of all, he chose to study animals, primarily cats and chickens, rather than man. That seemed a reasonable thing to do now that the continuity of species, as formulated by Darwin, had been accepted. Second, he spoke of the association between sensation and impulse—the conscious concomitant of action—rather than the association between ideas. Third, his experiments convinced him that a sensation and an impulse are most likely to become associated when the animal is satisfied by the consequences of its action, a principle that Thorndike later (1911) named the Law of Effect, because the effect of the action was thought to work retroactively to stamp in the association that led to it. Although animal trainers must have long known the value of rewards, Thorndike's Law of Effect was a genuine modification in the classical principle of association by contiguity. Finally, Thorndike suggested that an association may not require any ideational process in the animal. This suggestion of his was tentative and limited to animals, but it was opposed to the view of classical associationism, and the behaviorists who followed Thorndike extended it boldly.

Thorndike's conclusions, arose from results he obtained with a new experimental technique: an animal, placed in a box, learned to operate a latch to escape. As a measure of learning, Thorndike recorded the decreasing time taken by the animal to escape. This "puzzle box" remained standard equipment for several decades. Various adaptations of the time measure are still widely used.

This monograph is an attempt at an explanation of the nature of the process of association in the animal mind. Inasmuch as there have been no extended researches of a character similar to the present one either in subject-matter or experimental method, it is necessary to explain briefly its standpoint.

Our knowledge of the mental life of animals equals in the main our knowledge of their sense-powers, of their instincts or reactions performed without experience, and of their reactions which are built up by experience. Confining our attention to the latter we find it the opinion of the better observers and analysts that these reactions can all be explained by the ordinary associative processes without aid from abstract, conceptual, inferential thinking. These associative processes then, as present in animals' minds and as displayed in their acts, are my subject-matter. Any one familiar in even a general way with the literature of comparative psychology will recall that this part of the field has received faulty and unsuccessful treatment. The careful, minute, and solid knowledge of the sense-organs of animals finds no counterpart in the realm of associations and habits. We do not know how delicate or how complex or how permanent are the possible associations of any given group of animals. And although one would be rash who said that our present equipment of facts about instincts was sufficient or that our theories about it were surely sound, yet our notion of what occurs when a chick grabs a worm are luminous and infallible compared to our notion of what happens when a kitten runs into the house at the familiar call. The reason that they have satisfied us as well as they have is just that they are so vague. We say that the kitten associates the sound 'kitty kitty' with the experience of nice milk to drink, which does very well for a common-sense answer. It also suffices as a rebuke to those who would have the kitten ratiocinate about the matter, but it fails to tell what real mental content is present. Does the kitten feel *"sound of call, memory-image of milk in a saucer in the kitchen, thought of running into the house, a feeling, finally, of 'I will run in'?"* Does he perhaps feel only the sound of the bell and an impulse to run in, similar in quality to the impulses which make a tennis player run to and fro when playing? The word association may cover a multitude of essentially different processes, and when a writer attributes anything that an animal may do to association his statement has only the negative value of eliminating reasoning on the one hand and instinct on the other. His position is like that of a zoölogist who should today class an animal among the 'worms.' To give to the word a positive value and several definite possibilities of meaning is one aim of this investigation.

The importance to comparative psychology in general of a more scientific account of the association-process in animals is evident. Apart from the desirability of knowing all the facts we can, of whatever sort, there is the especial consideration that these associations and consequent habits have an immediate import for biological science. In the higher animals the bodily life and preservative acts are largely directed by these associations. They, and not instinct, make the animal use the best feeding grounds, sleep in the same lair, avoid new dangers and profit by new changes in nature. Their higher development in mammals is a chief factor in the supremacy of that group. This, however, is a minor consideration. The main purpose of the study of the animal mind is to learn the development of mental life down through the phylum, to trace in particular the origin of human faculty. In relation to this chief purpose of comparative psychology the associative processes assume a rôle predominant over that of sense-powers or instinct, for in a study of the associative processes lies the solution of the problem. Sense-powers and instincts have changed by addition and supersedence, but the cognitive side of consciousness has changed not only in quantity but also in quality. Somehow out of these associative processes have arisen human consciousnesses with their sciences and arts and religions. The association of ideas proper, imagination, memory, abstraction, generalization, judgment, inference, have here their source. And in the metamorphosis the instincts, impulses, emotions and sense-impressions have been transformed out of their old natures. For the origin and development of human faculty we must look to these processes of association in lower animals. Not only then does this department need treatment more, but promises to repay the worker better.

Although no work done in this field is enough like the present investigation to require an account of its results, the *method* hitherto in use invites comparison by its contrast and, as I believe, by its faults. In the first place, most of the books do not give us a psychology, but rather a *eulogy,* of animals. They have all been about animal *intelligence,* never about animal *stupidity.* Though a writer derides the notion that animals have reason, he hastens to add that they have marvellous capacity of forming associations, and is likely to refer to the fact that human beings only rarely reason anything out, that their trains of ideas are ruled mostly by association, as if, in this latter, animals were on a par with them. The history of books on animals' minds thus furnishes an illustration of the well-nigh universal tendency in human nature to find the marvelous wherever it can. We wonder that the stars are so big and so far apart, that the microbes are so small and so thick together, and for much the same reason wonder at the things animals do. They used to be wonderful because of the mysterious, God-given faculty of instinct, which could almost remove mountains. More lately they have been wondered at because of their marvellous mental powers in profiting by experience. Now imagine an astronomer tremendously eager to prove the stars as big as possible, or a bacteriologist whose great scientific desire is to demonstrate the microbes to be very, very little! Yet there has been a similar eagerness on the part of many recent writers on animal psychology to praise the abilities of animals. It cannot help leading to partiality in deductions from facts and more especially in the choice of facts for investigation. How can scientists who write like lawyers, defending animals against the charge of having no power of rationality, be at the same time impartial judges on the bench? Unfortunately the real work in this field has been done in this spirit. The level-headed thinkers who might have won valuable results have contented themselves with arguing against the theories of the eulogists. They have not made investigations of their own.

In the second place the facts have generally been derived from anecdotes. Now quite apart from such pedantry as insists that a man's word about a scientific fact is worthless unless he is a trained scientist, there are really in this field special objections to the acceptance of the testimony about animals' intelligent acts which one gets from anecdotes. Such testimony is by no means on a par with testimony about the size of a fish or the migration of birds, etc. For here one has to deal not merely with ignorant or inaccurate testimony, but also with prejudiced testimony. Human folk are as a matter of fact eager to find intelligence in animals. They like to. And when the animal observed is a pet belonging to them or their friends, or when the story is one that has been told as a story to entertain, further complications are introduced. Nor is this all. Besides commonly misstating what facts they report, they report only such facts as show the animal at his best. Dogs get lost hundreds of times and no one ever notices it or sends an account of it to a scientific magazine. But let one find his way from Brooklyn to Yonkers and the fact immediately becomes a circulating anecdote. Thousands of cats on thousands of occasions sit helplessly yowling, and no one takes thought of it or writes to his friend, the professor; but let one cat claw at the knob of a door supposedly as a signal to be let out, and straightway this cat becomes the representative of the cat-mind in all the books. The unconscious distortion of the facts is almost harmless compared to the unconscious neglect of an animal's mental life until it verges on the unusual and marvelous. It is as if some denizen of a planet where communication was by thought-transference, who was surveying humankind and reporting their psychology, should be oblivious to all our intercommunication save such as the psychical-research society has noted. If he should further misinterpret the cases of mere coincidence of thoughts as facts comparable to telepathic communication, he would not be more wrong than some of the animal psychologists. In short, the anecdotes give really the *abnormal* or *super-normal* psychology of animals.

Further, it must be confessed that these vices have been only ameliorated, not obliterated, when the observation is first hand, is made by the psychologist himself. For as men of the utmost scientific skill have failed to prove good observers in the field of spiritualistic phenomena, so biologists and psychologists before the pet terrier or hunted fox often become like Samson shorn. They, too, have looked for the intelligent and unusual and neglected the stupid and normal.

Finally, in all cases, whether of direct observation or report by good observers or bad, there have been three other defects. Only a single case is studied, and so the results are not necessarily true of the type; the observation is not repeated, nor are the conditions perfectly regulated; the previous history of the animal in question is not known. Such observations may tell us, if the observer is perfectly reliable, that a certain thing takes place, but they cannot assure us that it will take place universally among the animals of that species, or universally with the same animal. Nor can the influence of previous experience be estimated. All this refers to means of getting knowledge about what animals *do*. The next question is, "What do they *feel?*" Previous work has not furnished an answer or the material for an answer to this more important question. Nothing but carefully designed, crucial experiments can. In abandoning the old method one ought to seek above all to replace it by one which will not only tell more accurately *what they do,* and give the much-needed information *how they do it,* but also inform us *what they feel* while they act.

To remedy these defects experiment must be substituted for observation and the collection of anecdotes. Thus you immediately get rid of several of them. You can repeat the conditions at will, so as to see whether or not the animal's behavior is due to mere coincidence. A number of animals can be subjected to the same test, so as to attain typical results. The animal may be put in situations where its conduct is especially instructive. After considerable preliminary observation of animals' behavior under various conditions, I chose for my general method one which, simple as it is, possesses several other marked advantages besides those which accompany experiment of any sort. It was merely to put animals when hungry in enclosures from which they could escape by some simple act, such as pulling at a loop of cord, pressing a lever, or stepping on a platform. (A detailed description of these boxes and pens will be given later.) The animal was put in the enclosure, food was left outside in sight, and his actions observed. Besides recording his general behavior, special notice was taken of how he succeeded in doing the necessary act (in case he did succeed), and a record was kept of the time that he was in the box before performing the successful pull, or clawing, or bite. This was repeated until the animal had formed a perfect association between the sense-impression of the interior of that box and the impulse leading to the successful movement. When the association was thus perfect, the time taken to escape was, of course, practically constant and very short.

If, on the other hand, after a certain time the animal did not succeed, he was taken out, but *not fed*. If, after a sufficient number of trials, he failed to get out, the case was recorded as one of complete failure. Enough different sorts of methods of escape were tried to make it fairly sure that association in general, not association of a particular sort of impulse, was being studied. Enough animals were taken with each box or pen to make it sure that the results were not due to individual peculiarities. None of the animals used had any previous acquaintance with any of the mechanical contrivances by which the doors were opened. So far as possible the animals were kept in a uniform state of hunger, which was practically utter hunger. That is, no cat or dog was experimented on when the experiment involved any important question of fact or theory, unless I was sure that his motive was of the standard strength. With chicks this is not practicable, on account of their delicacy. But with them dislike of loneliness acts as a uniform motive to get back to the other chicks. Cats (or rather kittens), dogs and chicks were the subjects of the experiments. All were apparently in excellent health, save an occasional chick.

By this method of experimentation the animals are put in situations which call into activity their mental functions and permit them to be carefully observed. One may, by following it, observe personally more intelligent acts than are included in any anecdotal collection. And this actual vision of animals in the act of using their minds is far more fruitful than any amount of histories of what animals have done without the history of how they did it. But besides affording this opportunity for purposeful and systematic observation, our method is valuable because it frees the animal from any influence of the observer. The animal's behavior is quite independent of any factors save its own hunger, the mechanism of the box it is in, the food outside, and such general matters as fatigue, indisposition, etc. Therefore the work done by one investigator may be repeated and verified or modified by another. No personal factor is present save in the observation and interpretation. Again, our method gives some very important results which are quite uninfluenced by *any* personal factor in any way. The curves showing the progress of the formation of associations, which are obtained from the records of the times taken by the animal in successive trials, are facts which may be obtained by any observer who can tell time. They are absolute, and whatever can be deduced from them is sure. So also the question of whether an animal does or does not form a certain association requires for an answer no higher qualification in the observer than a pair of eyes. The literature of animal psychology shows so uniformly and often so sadly the influence of the personal equation that any method which can partially eliminate it deserves a trial.

Furthermore, although the associations formed are such as could not have been previously experienced or provided for by heredity, they are still not too remote from the animal's ordinary course of life. They mean simply the connection of a certain act with a certain situation and resultant pleasure, and this general type of association is found throughout the animal's life normally. The muscular movements required are all such as might often be required of the animal. And yet it will be noted that the acts required are nearly enough like the acts of the anecdotes to enable one to compare the results of experiment by this method with the work of the anecdote school.

·   ·   ·   ·   ·

The starting point for the formation of any association in these cases, then, is the set of instinctive activities which are aroused when a cat feels discomfort in the box either because of confinement or a desire for food. This discomfort, plus the sense-impression of a surrounding, confining wall, expresses itself prior to any experience, in squeezings, clawings, bitings, etc. From among these movements one is selected by success. But this is the starting point only in the case of the first box experienced. After that the cat has associated with the feeling of confinement certain impulses which have led to success more than others and are thereby strengthened. A cat that has learned to escape from *A* by clawing has when put into *C* or *G* a greater tendency to claw at things than it instinctively had at the start, and a less tendency to squeeze through holes. A very pleasant form of this decrease in instinctive impulses was noticed in the gradual cessation of howling and mewing. However, the useless instinctive impulses die out slowly, and often play an important part even after the cat has had experience with six or eight boxes. And what is important in our previous statement, namely, that the activity of an animal when first put into a new box is not directed by any appreciation of *that* box's character, but by certain general impulses to acts, is not affected by this modification. Most of this activity is determined by heredity; some of it, by previous experience.

My use of the words *instinctive* and *impulse* may cause some misunderstanding unless explained here. Let us, throughout this book, understand by instinct any reaction which an animal makes to a situation *without experience*. It thus includes unconscious as well as conscious acts. Any reaction, then, to totally new phenomena, when first experienced, will be called instinctive. Any impulse then felt will be called an instinctive impulse. Instincts include whatever the nervous system of an animal, as far as inherited, is capable of. My use of the word will, I hope, everywhere make clear what fact I mean. If the reader gets the fact meant in mind it does not in the least matter whether he would himself call such a fact instinct or not. Any one who objects to the word may substitute 'hocus-pocus' for it wherever it occurs. The definition here made will not be used to prove or disprove any theory, but simply as a signal for the reader to imagine a certain sort of fact.

The word *impulse* is used against the writer's will, but there is no better. Its meaning will probably become clear as the reader finds it in actual use, but to avoid misconception at any time I will state now that *impulse* means the consciousness accompanying a muscular innervation *apart* from *that feeling of the act which comes from seeing oneself move, from feeling one's body in a different position, etc.* It is the *direct feeling of the doing* as distinguished from the *idea of the act done* gained through eye, etc. For this reason I say 'impulse *and* act' instead of simply 'act.' Above all, it must be borne in mind that by impulse I never mean the *motive* to the act. In popular speech you may say that hunger is the impulse which makes the cat claw. That will never be the use here. The word *motive* will always denote that sort of consciousness. Any one who thinks that the act ought not to be thus subdivided into impulse and deed may feel free to use the word *act* for *impulse* or *impulse and act* throughout, if he will remember that the act in this aspect of being felt as to be done or as doing is in animals the important thing, is the thing which gets associated, while the act as done, as viewed from outside, is a secondary affair. I prefer to have a separate word, impulse, for the former, and keep the word act for the latter, which it commonly means.

Starting, then, with its store of instinctive impulses, the cat hits upon the successful movement, and gradually associates it with the sense-impression of the interior of the box until the connection is perfect, so that it performs the act as soon as confronted with the sense-impression. The formation of each association may be represented graphically by a time-curve. In these curves lengths of one millimeter along the abscissa represent successive experiences in the box, and heights of one millimeter above it each represent ten seconds of time. The curve is formed by joining the tops of perpendiculars erected along the abscissa 1 mm. apart (the first perpendicular coinciding with the y line), each perpendicular representing the time the cat was in the box before escaping. Thus, in Fig. 2 . . . the curve marked 12 in *A* shows that, in 24 experi-

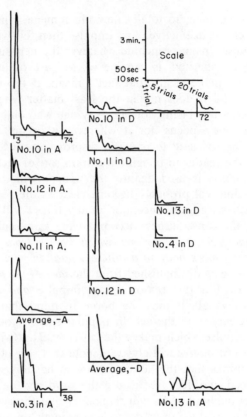

Fig. 2.

ences or trials in box *A*, cat 12 took the following times to perform the act, 160 sec., 30 sec., 90 sec., 60, 15, 28, 20, 30, 22, 11, 15, 20, 12, 10, 14, 10, 8, 8, 5, 10, 8, 6, 6, 7. A short vertical line below the abscissa denotes that an interval of approximately 24 hours elapsed before the next trial. Where the interval was longer it is designated by a figure 2 for two days, 3 for three days, etc. If the interval was shorter the number of hours is specified by 1 hr., 2 hrs., etc. In many cases the animal failed in some trial to perform the act in ten or fifteen minutes and was then taken out by me. Such failures are denoted by a break in the curve either at its start or along its course. In some cases there are short curves after the main ones. These, as shown by the figures beneath, represent the animal's mastery of the association after a very long interval of time, and may be called memory curves . . .

The time-curve is obviously a fair representation of the progress of the formation of the association, for the two essential factors in the latter are the disappearance of all activity save the particular sort which brings success with it, and perfection of that particular sort of act so that it is done precisely and at will. Of these the second is, on deeper analysis, found to be a part of the first; any clawing at a loop except the particular claw which depresses it is theoretically a useless activity. If we stick to the looser phraseology, however, no harm will be done. The combination of these two factors is inversely proportional to the time taken, provided the animal surely wants to get out at once. This was rendered almost certain by the degree of hunger. Theoretically a perfect association is formed when both factors are perfect,—when the animal, for example, does nothing but claw at the loop, and claws at it in the most useful way for the purpose. In some cases . . . neither factor ever gets perfected in a great many trials. In some cases the first factor does but the second does not, and the cat goes at the thing not always in the desirable way. In all cases there is a fraction of the time which represents getting oneself together after being dropped in the box, and realizing where one is. But for our purpose all these matters count little, and we may take the general slope of the curve as representing very fairly the progress of the association. The slope of any particular part of it may be due to accident. Thus, very often the second experience may have a higher time-point than the first, because the first few successes may all be entirely due to accidentally hitting the loop, or whatever it is, and whether the accident will happen sooner in one trial than another is then a matter of chance. Considering the general slope, it is, of course, apparent that a gradual descent—say, from initial times of 300 sec. to a constant time of 6 or 8 sec. in the course of 20 to 30 trials—represents a difficult association, while an abrupt descent, say in 5 trials, from a similar initial height, represents a very easy association.

·   ·   ·   ·   ·

Presumably the reader has already seen budding out of this dogma a new possibility, a further simplification of our theories about animal consciousness. The possibility is that animals may have *no images or memories at all, no ideas to associate.* Perhaps the entire fact of association in animals is the presence of sense-impressions with which are associated, by resultant pleasure, certain impulses, and that therefore, and therefore only, a certain situation brings forth a certain act. Returning to our analysis of the association, this theory would say that . . . the sense-impression gave rise, when accompanied by the feeling of discomfort, to the impulse . . . directly, without the intervention of any representations of the taste of the food, or the experience of being outside, or the sight of oneself doing the act. This theory might be modified so as to allow that the representations could be there, but to deny that they were necessary, were inevitably present, that the impulse was connected to the sense-impression through them. It would then claim that the effective part of the association was a direct bond between the situation and the impulse, but would not cut off the possibility of there being an aura of memories along with the process. It then becomes a minor question of interpretation which will doubtless sooner or later demand an answer. I shall not try to answer it now. The more radical question, the question of the utter exclusion of representative trains of thought, of any genuine association of *ideas* from the mental life of animals, is worth serious consideration. I confess that, although certain authentic anecdotes and certain experiments to be described soon, lead me to reject this exclusion, there are many qualities in animals' behavior which seem to back it up. If one takes his stand by a rigid application of the law of parsimony, he will find justification for this view which no experiments of mine can overthrow.

Of one thing I am sure, and that is that it is worth while to state the question and how to solve it, for although the point of view involved is far removed from that of our leading psychologists to-day, it cannot long remain so. I am sorry that I cannot pretend to give a final decision.

98   ROBERT MEARNS YERKES (1876–1956) ON THE INTELLIGENCE OF THE TURTLE, 1901

R. M. Yerkes, "The formation of habits in the turtle," *The Popular Science Monthly 58*, 519–525 (1901).

Although the experiment reported here by Yerkes was performed under Thorndike's influence, it is novel in certain respects. Yerkes moved down the phyletic scale to the turtle for his subject. He devised an experimental situation which is more a maze than a puzzle box, and, in the subsequent development of the study of animal learning, after Willard Small, the maze became for a long time the dominant device. Yerkes' dominating interest lay in the evolution of intelligence, as he demonstrated during his long and distinguished career by investigating the behavior of protozoon, crab, turtle, medusa, crawfish, frog, Daphnia, tortoise, mouse, earthworm, rat, crow, pig, ringdove, monkey, ape, and man.

Habits are determinants in human life. It is true that we are free, within limits, to form them; it is also true that, once formed, they mold our lives. In the life of the brute habit plays an even more important rôle than it does in man. The ability to survive, for example, frequently depends upon the readiness with which new feeding habits can be formed. So, too, in case of dangers habitually avoided, those individuals which form habits most quickly have the best chances of life. But it is unnecessary to emphasize the importance of habit to all living beings, for it is obvious. We have now to ask, What precisely is a habit?

A habit proves in analysis to be nothing more or less than a tendency toward a certain action or line of conduct—a tendency due to structural and functional modifications of the organism which have resulted from repetition of the action itself; for nothing can be done by the animal mechanism without resultant changes in its organization. These changes it is which influence all subsequent activities and constitute the physical basis of habit. Repetition of an act apparently leads to the formation of a track for the controlling nervous impulse—a line of least resistance, so to speak—along which the current therefore *tends* to pass. A duck when thrown into the water does not have to stop to think what to do to get out, how to move this leg and then that; it instinctively, we say, meets the situation with that combination of movements called swimming. But the duck swims almost, if not quite, as well the first time it is put into the water as it ever does. There is little profiting by experience. This simply means that the structural basis of the swimming habit is present at birth, and does not have to be formed by repetition of the action thereafter. The habit is, in other words, inherited. For man swimming is not an instinctive act; he has to learn every detail of the complex muscular process by trial; he has to establish by repetition of the activity the basis of the habit. Finally, however, the man will be able to meet the situation—water, a distant shore, and a desire to be on the shore—as the duck does—that is, habitually.

Since habits make an animal what it is in great part, the study of their formation, of the manner and rapidity of their growth, and of their permanence must be of practical as well as of scientific importance. We are rapidly realizing, as the increasing interest in animal psychology clearly indicates, that the mental life of all animal types must be understood before we can attain to a satisfactory science of psychology or give a history of the evolution of mind. To watch the progress of a habit's growth is exceedingly interesting, whether the subject be a man or one of the lower animals. Ordinarily the chief difficulties in the way of such a study are the great length of time and the constancy of observation necessary. But these obstacles may readily be avoided by making observations under artificial or experimental conditions—that is, by adapting conditions to the needs of the experiment, instead of trying to adapt one's self to natural conditions. The account which follows presents, as an example of this kind of work, observations on habit formation in the

545

common 'speckled turtle' (*Chelopus guttatus*). It has been my aim to give a brief account of the way in which a particular turtle profited by experience.

The work was undertaken to determine to what extent and with what rapidity turtles can learn; to measure as accurately as might be their intelligence. Reptiles are usually considered sluggish and unintelligent creatures, and there can be no question about the general truth of this opinion. Turtles certainly appear to be very stupid—so much so, indeed, that one would not expect much in the way of intelligent actions. Just how stupid, or better perhaps, just how intelligent they are, we shall be better able to judge after studying the habits of the animals more carefully, and collecting more evidence like the following:

The finding of the way through a labyrinth to a nest was chosen as the habit to be studied. The motives employed to get the subject to try to find its way to the nest were: first, the desire to hide in some dark, secluded place; secondly, the impulse to escape from confinement; and lastly, the desire to get to a place of comfort. Dr. Thorndike, in studying the associative processes of cats and dogs . . . used hunger as the chief motive for escape. This is unsatisfactory in the case of turtles, because they frequently do not eat well in confinement, and at best their feeding or desire for food is very irregular and hard to control as a motive in experimental work.

The method of experimentation was simple. A box three feet long, two feet wide and ten inches deep was divided into four portions by partitions, also ten inches deep, arranged as shown in Fig. 1. In each

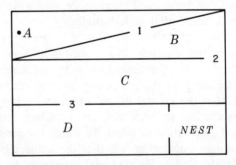

Fig. 1. Plan of labyrinth no. 1.

partition was a hole four inches long and two inches deep, just large enough to permit the turtle to pass through easily. The box is shown in ground plan by Fig. 1.

A is the space in which the animal was placed to start, the starting-point being marked by a dot (.). The corner marked nest contained a mass of damp grass and was darkened. When everything was ready for an experiment the animal was placed in *A* at the dot and allowed to wander about until it found the nest by passing through the openings marked 1, 2 and 3.

On July 20 the animal, a speckled turtle about four inches long which was found in Woods Hole, Mass., was placed in *A* for the first time. After wandering about almost constantly for thirty-five minutes, it chanced to find the nest, into which it immediately crawled, there remaining until taken out for another experiment two hours later. The observations were made from one to two hours apart, in order to avoid fatiguing the animal, and also to leave it some inducement for seeking the nest, for if it were taken out each time as soon as it got back to the comfortable corner, the game would soon lose interest. The second time the nest was reached in fifteen minutes, with much less wandering. The time for the third trip was five minutes, and for the fourth, three minutes thirty seconds. During the first three trials the courses taken were so tortuous that it seemed foolish to try to record them. There was aimless wandering from point to point within each space, and from space to space. After the third trip the routes became much more direct, and accurate records of them were obtained. Fig. 2 gives the course taken in the fourth experiment. It is fairly direct, but shows that the animal lost its way in *A* and again in *B*; having passed through 2, it took the shortest path to the nest.

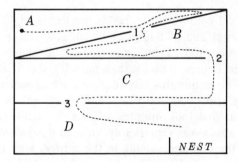

Fig. 2. Course for fourth trip.

A record of the route in connection with the time of the trip is necessary as an index of the effect of experience, because if the animal takes a direct course, with no wrong turns, but makes several halts, the time may indicate no profiting by the former acts, whereas the route will at once show that there has been improvement. Thus one record supplements the other.

These experiments were made six or eight times a day until fifty trials had been given. The tenth trip was made in three minutes five seconds, with two mistakes in turning. The time of the twentieth journey was but forty-five seconds, and that of the thirtieth, forty seconds. In the latter experiment a direct course was taken; this was also true in the case of the fiftieth trip, which was made in thirty-five seconds. Fig. 3 represents graphically the times of the first twenty experiments of this

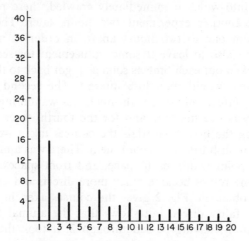

Fig. 3. Times of experiments from one to twenty.

series. The vertical column of figures at the left, 1 to 40, indicates minutes; the horizontal line of figures, 1 to 20, gives the number of trials.

That the turtle profited by experience, and that very rapidly, is evident from the figures. The average time for the first ten trips, from one to ten, was eight minutes fifty-four and a half seconds; the average time of the ten trips between thirty and forty was one minute three seconds. What at first took minutes, after a few trials required only as many seconds. There was remarkably little aimless wandering, crawling up the sides of the box and sulking in the corners after the third experiment. In fact, the animal soon began to behave as if it had the goal in mind and was intent on making directly for it. It learned with surprising quickness to make the proper turns and to take the shortest path. Three or four times I noticed it turn in the wrong direction, crawl into a corner and, as it seemed, become confused, for it then returned to the starting-point, as if to get its bearings, and started out afresh. In every case the second attempt resulted in a direct and unusually quick journey to the nest. Very frequently halts just in front of the holes were noticed. It looked as if the animal were meditating upon the course to be taken.

Had one seen a man in a similar situation he would unhesitatingly have said that the person was trying to decide which way to go. There can be little doubt, however, that the mental attitude of the turtle was extremely simple compared with a man's under similar conditions. There are those who would claim that even the turtle was thinking about its environmental conditions, but it seems far more probable that it stopped in order the better to get those sensory data by which it was enabled to follow its former course. Smell and sight furnish the most important elements in the associative processes of lower animals. This interpretation of the action is supported by the fact that it occurred most frequently after the course had been gone over a few times.

A more complex and novel labyrinth was now substituted. Its new features were a blind alley (see *F*, Fig. 4) and three inclined planes (3, 4 and 6 of Fig. 4). A plan of the labyrinth is shown in Fig. 4. At

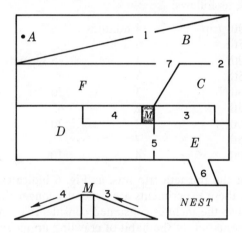

Fig. 4. Plan of labyrinth no. 2.

the left of the nest a side view of the inclines 3 and 4 is shown. Each was one foot long, and the middle point (*M*) was four inches from the floor.

Labyrinth No. 2 was used in the same way as No. 1, the turtle being placed in *A* and permitted to seek the nest, which was this time a box filled with moist sand. The inclines at first baffled the little fellow, and it was an hour and thirty-one minutes before he reached the nest. *A* and *B* seemed to offer no difficulties, but the new features—the blind alley and the inclines—were puzzles. By the fifth trial, however, these had become somewhat familiar. The route taken in this experiment has been produced in Fig. 5.

549

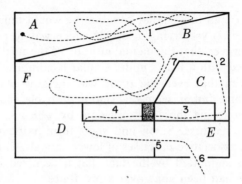

Fig. 5. Course for fifth trip.

The time of this trip was sixteen minutes. The times of some of the other trials were as follows:

10th trip	4 minutes	
15th	6	
20th	4	5 seconds
25th	3	
30th	3	20
35th	2	45
40th	4	20
45th	7	
50th	4	10

The route for the thirtieth trip was, as Fig. 6 indicates, almost direct.

The times of these experiments are generally longer than those of the first series because the inclines consumed considerable time.

During the formation of the habit of crawling up incline 3 and sliding

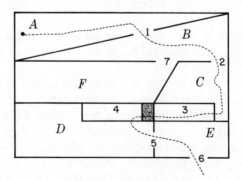

Fig. 6. Course of thirtieth trip.

down incline 4 a very interesting modification of the action occurred, namely, the shortening of the path to the sand-box by crawling over the edge of incline 4. At first the animal, after climbing up 3, would slide all the way to the bottom of 4 and would then turn toward the nest. Soon, however, it began making the turn toward the nest before reaching the bottom, thus throwing itself over the edge of 4. The turn was made earlier and earlier on 4, until finally it got to crawling over as soon as it reached the top of 3, or *M*. It always turned itself over the edge carefully, and landed, after a fall of four inches, usually on its head or back. By this process the path was shortened eight or ten inches. This action is a splendid illustration of the way in which an advantageous habit may grow by accretion, as it were, until it seems as if it must have been the result of reasoning. Some would, no doubt, hold that in this case the turtle chose the direct path because of inferences from judgments. Although this *may* be true, there is surely a sufficient explanation of the habit, as we have come to know it, in the profiting by chance experience. No one would say that the nest was at first found by inferences. It was reached because of the animal's impulse to move about, to seek escape or hiding. Had the turtle stopped to judge and draw inferences as to the way to escape, instead of persistently moving from place to place, it would probably be in the pen yet. No; the wandering impulse led by chance to the finding of satisfaction, turtle pleasure, in the nest. Because of this satisfaction, the action was impressed on the vital mechanism, so that there was a tendency (the beginning of a habit) toward repetition of it. Had the action failed to give satisfaction, the probability of its being repeated would have been merely that of chance, and not chance plus the influence of the former pleasure-giving activity. The turtle happened to crawl over the edge of the incline, and, finding that this enabled it to get to the nest quicker, it continued the act, thus forming a habit.

99   WILLARD STANTON SMALL (1870–1943) ON THE MAZE, 1901

W. S. Small, "Experimental study of the mental processes of the rat, II," *American Journal of Psychology 12,* 206–209, 218–220, 228–232 (1901).

It was Small's use of the maze, rather than Yerkes', that established it as an important instrument for experimental psychology. Small ran white rats through his maze, which was a small-scale replica of the garden maze at Hampton Court in England. He kept a record of time and errors as the rats gradually learned the pathway to the location of the food. The white rat, the complex maze baited with food, and the tabulation of time and errors soon became near-standard features of experimentation on animal learning. Although Small himself did not propose a lasting theory of the learning process, his technique and his detailed accounts of the behavior of the rats formed the basis for many investigations that came afterward.

APPARATUS AND CONDITIONS OF EXPERIMENTS

The aim in these experiments . . . was to make observations upon the free expression of the animal's mental processes, under as definitely controlled conditions as possible; and, at the same time, to minimize the inhibitive influence of restraint, confinement and unfamiliar or unnatural circumstances. Fear, which in lack of a better term, may be used to include the three influences just noted, and too great difficulties are the things most rigorously to be guarded against. On the positive side, the experiments must conform to the psycho-biological character of an animal if sane results are to be obtained.

. . . . .

The Hampton Court Maze served as model for the apparatus. The diagram given in the Encyclopedia Britannica was corrected to a rectangular form, as being easier of construction. The character of the problem was not affected. Three mazes were made. The first was as follows. The dimensions were 6 by 8 feet. The bottom was of wood, the boards being fastened together so as to make a portable whole of the apparatus. All the rest: top, sides, and partitions between galleries were of wire netting, $\frac{1}{4}$ in. mesh. The height of the sides was 4 inches; the width of galleries, the same. In the center was a large open space. The accompanying diagram gives the ground plan of the maze. The entrance is marked by the figure *O*. Figures *1* to *7* indicate seven blind alleys—seven possibilities of error. It will be observed that *4* does not lead necessarily into the *cul-de-sac 5,* but does inevitably furnish a chance for error. The letter *x* marks a dividing of the ways either of which, however, may be followed without completely losing the trail. Certain other points are indicated by letters *a, b, c,* etc., for convenience in description. *C* stands for center. A glance at the maze will be sufficient to convince one of the difficulty of the problem. This maze is designated as Maze I. The two other mazes were identical with this, except that they were made throughout of wire netting, bottom as well as top and sides . . . The floor of the maze was covered with sawdust. This was renewed whenever a new rat was introduced.

. . . . .

ANALYSES OF RESULTS

In appreciating the results of this series of experiments, . . . the [following] . . . facts come into view: . . . the initial indefiniteness of move-

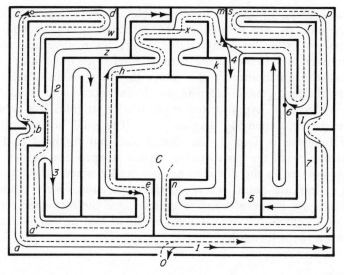

ment and the fortuitousness of success; the just observable profit from the first experiences; the gradually increasing certainty of knowledge indicated by increase of speed and definiteness, and the recognition of critical points indicated by hesitation and indecision; the lack of imitation and the improbability of following by scent; the outbreak of the instincts of play and curiosity after the edge of appetite is dulled. In addition are to be noted the further observations upon the contrast between the slow and cautious entrance into, and the rapid exit from the blind alleys, after the first few trials; the appearance of disgust on reaching the end of a blind alley; the clear indication of centrally excited sensation (images) of some kind; memory (as I have used the term); the persistence of certain errors; and the almost automatic character of the movements in the later experiments. Viewed objectively, these observations all converge towards one central consideration; the continuous and rapid improvement of the rats in threading the maze, amounting to almost perfect accuracy in the last experiments. No qualification of this view was found necessary in the light of many later experiments. Rather they all confirm it. The mental aspect is considerably more complex, the mental factors, much more difficult of analysis and evaluation; but the central fact in the process seems to be the recognition by the rats of particular parts of the maze. Deferring consideration of this side of the matter, and looking now only at the objective side the important points are . . . (1) the increase in speed, and (2) the decrease in the number of errors and in uncertainty . . . There are fluctuations both in time and number of errors, *e.g.,* Exp. 9 shows an increase over Exp. 8 in both respects. This is to be expected from the character of the animal and the confessed impossibility of completely controlling even the external conditions, not to mention the particular internal conditions in each case. Allowing for such variable factors, the relative time required and number of mistakes made furnish a fairly accurate index of the progressive acquaintance of the rats with the problem. The contrast between the first slow, blundering, accidental success and the definitely foreseen success of Exp. 8 (taken as the best) is striking. This is brought out [in the accompanying diagram] . . . The solid line indicates the course followed in Exp. 1; the dotted line that followed in Exp. 8. The arrows mark the point where the rat stopped and turned about, and are pointed in the direction he was headed when he stopped. The dots (·) indicate points where considerable pauses were made. The arrow and dot between *4* and *m* indicate that the rat, when returning from an abortive essay into *4*, went as far as that arrow, paused, then turned and went forward.

· · · · ·

THE PROCESS OF LEARNING

Whether the process of learning the way through this maze is adequately described as a gradual establishment of direct associations by profiting by chance experience depends upon the meaning attached to the phrase 'profiting by chance experience.' I wish in the remainder of this paper to attempt an analysis of the mental factors involved in the animal's solution of the problem; and to offer some suggestions upon the character of the perfected knowledge.

It will be well to résumé the facts by following the rats *A* and *B* of Series II from their introduction to the maze, throughout their experience of getting acquainted with their new environment, to the time when they have perfect mastery of the situation. In their first trial, after a lapse of 13 minutes and after many errors, returns, and delays, they find their way into *C*. Here they are rewarded for their labors by all the pleasure possible to a rat from the satisfaction of a keen-edged appetite by a good meal of bread and milk. This first success is assumed to be accidental; its realization does not depend at all upon previsioning intelligence. The animal does not foresee the end and set to work to attain that end. There is no reflection. The determining conditions in the rat's mind are more immediate in their effect. The most obvious of these are: hunger, perception of the odor of the food, curiosity, normal activity (the obverse of curiosity) and the instinctive special trait of following out tortuous passages—a definite rat-hole consciousness that acts, as it were, thygmotactically. These factors, with the inhibitive balance-spring of timidity, firmly rooted and deeply toned emotionally, constitute the relatively stable background of consciousness over which play the lights of the perceptive and discriminative processes as the animal proceeds with the task. The rat, when he enters the maze, is psychically a confused complexus of these factors. No one of them looms high above the others in the wave of consciousness. Attention is dispersed; perhaps, better, distraction prevails. Nevertheless, 'experimental reasoning' begins at once. The animal keeps constantly moving; but his activity at this stage is evidently sensori-motor (or organo-motor). Motive in the sense of ideated end is absent. The nearest approach to this is the possible idea of a definite kind of food incident to the perception of the familiar food-odor. This is not impossible. On the other hand the effect on consciousness may be only an intensification of the hunger psychosis resulting in an increase of motor activity. However this may be it probably is the animal's instinctive fondness for following out devious ways, his thygmotactic rat-hole psychosis, rather than the smell of the food that gives determinateness to his movements at first. Were this trait less imperative, the rat, when he comes near the food (*e.g.,* at *e*) would become the victim of his hunger and his perception of the position of the food—for the food at this point could be located by smell—and would spend himself stupidly in endeavoring to force an entrance. In general, however, he soon passes on, going directly away from the Canaan of his desire. Failure to get through by gnawing and digging, quickly results in a wavering and dispersion of attention. Concomitantly the perception of the odor relapses into the margin of consciousness, and the instinctive motor tendency at this juncture reasserts itself as the focal and directive influence. The first success then may be set down as the accidental issue of a trial and error series, motivated by hunger and curiosity, mediated by the sense of smell and, more largely, by this instinctive motor trait, and consummated by the pleasure of hunger satisfied. And yet the term *accidental* must be used with reservations. The rats of this series, and of all others in their first trial, seemed to profit at once by experience. By this I mean that after they had made an error once or twice, though they had not yet succeeded in reaching *C*, they would hesitate or even avoid the error when going over the ground a second time. For example, rat *A* went only a few steps into *3* the second time he reached that place, and avoided *6* and *7* completely . . . Such cases may be attributed to pure chance; the conduct of the rat, his hesitation more than his avoidance of the error, indicates, rather, recognition and selection.

It will be remembered that the rats have the entire night each time for exploration of the maze. This results in remarkable improvement in the second trial. In the succeeding experiments the improvement is continuous in the elimination of errors and in the increase in definiteness and speed. The rats soon acquire a practically perfect knowledge of the maze, so that they can make the journey quickly and accurately when they want to do so, or stroll about as they list.

How explain this improvement? What does 'profiting by chance experience' mean in this instance? how is it assimilated and how utilized?

Doubtless one factor in the process is the memory of the pleasant experience at the end. In addition to the undirected and undifferentiated motive of hunger and the motor trait of the first trial, there is, in the second, a dimly ideated end which probably becomes progressively clearer in the subsequent experience. But the essential point is certainly the recognition of the critical points along the way and the discrimination of the divergent paths at these points, leading to purposive selection of the right path. The memory of the pleasant experience at the end would be of slight avail, if the rats did not recognize the critical points and discriminate and select their paths. The animal begins by going right and wrong wholly by chance. After a few trials he comes to recognize the doubtful places, and hesitates when he comes to them, undecided which way to take. The external signs of indecision vary between standing still as if trying to think which way to go, and abortive starts each way. Sometimes to these is added standing up and sniffing in the usual manner of orientation. This movement seldom was observed after the first two or three experiments, *i.e.,* after the dilemma began to be clearly felt. At this stage, the choice of path is still about as often wrong as right. The distinguishing accidentia are acquired gradually. Progress in discrimination is marked by decrease of hesitation and in more frequent choice of the right path. The path chosen often is pursued doubtfully. If the wrong one is chosen, the error frequently is retrieved after a few steps; if it is followed to the end the return is made swiftly and the right path is taken confidently. In the final stage, errors and hesitation drop out entirely. The right path is followed from start to finish without attention to specific points *en route.*

It should be noted that the learning was slowest in connection with *o, x*, and *4*. The persistent confusion at *o* is attributable probably to its being at the entrance of the maze. At this point there is a maximum of affective excitement. The momentum of association has to be gathered as the animal goes along in the familiar path. In a remote way it may be likened to the stumbling and groping of an orator at the beginning of a familiar theme . . . The slower discrimination at *x* and *4* was due doubtless to the fact that wrong choice at these points consisted in taking the roundabout, rather than the direct path. Strictly speaking there was no error. In all the other cases, taking one path was associated ultimately with success; taking the other, with failure—disappointment. In these cases the association would seem to be between path and distance.

In such cases profiting by experience manifestly involves the processes of recognition, discrimination and choice. If the problem set were merely the selection of one effective movement out of several haphazard movements, as was the case with the puzzle-box experiments reported by Dr. Thorndike . . . then the profiting by experience could be accounted for by the fading away of the useless movements. They would drop off like dead branches from a tree, of their own weight. They would be associated with nothing—either positively or negatively. The right movement would be selected *naturally*. In the present case, however, two direct associations are formed and discriminated between, and the advantageous one selected. Recognition of the critical places is equivalent to doubt as to the right path. This doubt is the correlative in consciousness of the struggle between the two associations or 'constructs.' The positively useless or the less advantageous association does not fall away mechanically, but only in virtue of discrimination between the two constructs, and, finally, the conscious selection of the right one. In such a case as that of choice at *x* if the animal did not consciously select, there could never be any fixed association; consequently never any habitual reaction. Both ways lead to success. In a sufficient number of trials the theory of probabilities would require an equal number of selections of each path. But the short road is soon habitually selected, just as is the right path at other critical points. There is involved an elementary form of comparison and judgment; for comparison, judgment and reflection, even, are present in embryo. They all take their rise in the struggle of ideas and images, and lower down of 'constructs,' which "gives in animal, as in man, the illusion of choice and free intelligence."

100

EDWARD LEE THORNDIKE (1874–1949) AND ROBERT SESSIONS WOODWORTH (1869–1962) ON TRANSFER OF TRAINING, 1901

E. L. Thorndike and R. S. Woodworth, "The influence of improvement in one mental function upon the efficiency of other functions," *Psychological Review 8*, 247–256 (1901).

Does the mind possess faculties for dealing with categories of intellectual activity, like language or mathematics, or does it merely contain the specific associations that a person has learned? This question, which is one that divided the Scottish philosophers (Reid and Stewart) from the British associationists (such as Berkeley and Hume), was recognized by Thorndike and Woodworth as crucial for educational practices. The following paper, the first in a rapidly expanding field, reported findings that favored the associationistic view. Along with further research that supported its conclusions, it ushered in an era in American education that was characterized by an emphasis on highly practical and specific subjects. It also established transfer of training—the modern name of this venerable question—as a significant area of research for experimental psychologists.

This is the first of a number of articles reporting an inductive study of the facts suggested by the title. It will comprise a general statement of the results and of the methods of obtaining them, and a detailed account of one type of experiment.

The word function is used without any rigor to refer to the mental basis of such things as spelling, multiplication, delicacy in discrimination of size, force of movement, marking *a*'s on a printed page, observing the word *boy* in a printed page, quickness, morality, verbal memory, chess playing, reasoning, etc. Function is used for all sorts of qualities in all sorts of performances from the narrowest to the widest, *e.g.,* from attention to the word 'fire' pronounced in a certain tone, to attention to all sorts of things. By the word improvement we shall mean those changes in the workings of functions which psychologists would commonly call by that name. Its use will be clear in each case and the psychological problem will never be different even if the changes studied be not such as everyone would call improvements. For all purposes 'change' may be used instead of 'improvement' in the title. By efficiency we shall mean the status of a function which we use when comparing individuals or the same individual at different times, the status on which we would grade people in that function. By other function we mean any function differing in any respect whatever from the first. We shall at times use the word function-group to mean those cases where most psychologists would say that the same operation occurred with different data. The function *attention,* for instance, is really a vast group of functions.

Our chief method was to test the efficiency of some function or functions, then to give training in some other function or functions until a certain amount of improvement was reached, and then to test the first function or set of functions. Provided no other factors were allowed to affect the tests, the difference between the test before and the test after training measures the influence of the improvement in the trained functions on the functions tested.

·   ·   ·   ·   ·

Perhaps the most striking method of showing the influence or lack of influence of one function on another is that of testing the same function-group, using cases where there are very slightly different data. If, for instance, we test a person's ability to estimate a series of magnitudes differing each from the next very slightly, and find that he estimates one very much more accurately than its neighbors on either side, we can be sure that what he has acquired from his previous experience or from the experience of the test is not improvement in the function-group of estimating magnitudes but a lot of particular improvements in estimating particular magnitudes, improvements which may be to a large extent independent of each other.

· · · · ·

The evidence given by our experiments makes the following conclusions seem probable:

It is misleading to speak of sense discrimination, attention, memory, observation, accuracy, quickness, etc., as multitudinous separate individual functions are referred to by any one of these words. These functions may have little in common. There is no reason to suppose that any general change occurs corresponding to the words 'improvement of the attention,' or 'of the power of observation,' or 'of accuracy.'

It is even misleading to speak of these functions as exercised within narrow fields as units. For example, 'attention to words' or 'accurate discrimination of lengths' or 'observation of animals' or 'quickness of visual perception' are mythological, not real entities. The words do not mean any existing fact with anything like the necessary precision for either theoretical or practical purposes, for, to take a sample case, attention to the meaning of words does not imply equal attention to their spelling, nor attention to their spelling equal attention to their length, nor attention to certain letters in them equal attention to other letters.

The mind is, on the contrary, on its dynamic side a machine for making particular reactions to particular situations. It works in great detail, adapting itself to the special data of which it has had experience. The word *attention,* for example, can properly mean only the sum total of a lot of particular tendencies to attend to particular sorts of data, and ability to attend can properly mean only the sum total of all the particular abilities and inabilities, each of which may have an efficiency largely irrespective of the efficiencies of the rest.

Improvement in any single mental function need not improve the ability in functions commonly called by the same name. It may injure it.

Improvement in any single mental function rarely brings about equal improvement in any other function, no matter how similar, for the working of every mental function-group is conditioned by the nature of the data in each particular case.

The very slight amount of variation in the nature of the data necessary to affect the efficiency of a function-group makes it fair to infer that no change in the data, however slight, is without effect on the function. The loss in the efficiency of a function trained with certain data, as we pass to data more and more unlike the first, makes it fair to infer that there is always a point where the loss is complete, a point beyond which the influence of the training has not extended. The rapidity of this loss, that is, its amount in the case of data very similar to the data on which the function was trained, makes it fair to infer that this point is nearer than has been supposed.

The general consideration of the cases of retention or of loss of practice effect seems to make it likely that spread of practice occurs only where identical elements are concerned in the influencing and influenced function.

The particular samples of the influence of training in one function on the efficiency of other functions chosen for investigation were as follows:

1. The influence of certain special training in the estimation of magnitudes on the ability to estimate magnitudes of the same general sort, *i.e.,* lengths or areas or weights, differing in amount, in accessory qualities (such as shape, color, form) or in both. The general method was here to test the subject's accuracy of estimating certain magnitudes, *e.g.,* lengths of lines. He would, that is, guess the length of each. Then he would practice estimating lengths within certain limits until he attained a high degree of proficiency. Then he would once more estimate the lengths of the preliminary test series. Similarly with weights, areas, etc. This is apparently the sort of thing that happens in the case of a tea-taster, tobacco-buyer, wheat-taster or carpenter, who attains high proficiency in judging magnitudes or, as we ambiguously say, in delicacy of discriminating certain sense data. It is thus like common cases of sense training in actual life.

2. The influence of training in observing words containing certain combinations of letters (*e.g., s* and *e*) or some other characteristic on the general ability to observe words. The general method here was to test the subject's speed and accuracy in picking out and marking certain letters, words containing certain letters, words of a certain length, geometric figures, misspelled words, etc. He then practiced picking out and marking words of some one special sort until he attained a high degree of proficiency. He was then re-tested. The training here corresponds to a fair degree with the training one has in learning to spell, to notice forms and endings in studying foreign languages, or in fact in learning to attend to any small details.

3. The influence of special training in memorizing on the general ability to memorize. Careful tests of one individual and a group test of students confirmed Professor James' result (see Principles of Psychology, Vol. I., pp. 666–668). These tests will not be described in detail.

·   ·   ·   ·   ·

A SAMPLE EXPERIMENT

There was a series of about 125 pieces of paper cut in various shapes. (Area test series.) Of these 13 were rectangles of almost the same shape and of sizes from 20 to 90 sq. cm. (series 1), 27 others were triangles, circles, irregular figures, etc., within the same limits of size (series 2). A subject was given the whole series of areas and asked to write down the area in sq. cm. of each one. In front of him was a card on which three squares, 1, 25 and 100 sq. cm. in area, respectively, were drawn. He was allowed to look at them as much as he pleased but not to superpose the pieces of paper on them. No other means of telling the areas were present. After being thus tested the subject was given a series of paper rectangles, from 10 to 100 sq. cm. in area and of the same shape as those of series 1. These were shuffled and the subject guessed the area of one, then looked to see what it really was and recorded his error. This was continued and the pieces of paper were kept shuffled so that he could judge their area only from their intrinsic qualities. After a certain amount of improvement had been made he was retested with the 'area test series' in the same manner as before.

The function trained was that of estimating areas from 10 to 100 sq. cm. with the aid of the correction of wrong tendencies supplied by ascertaining the real area after each judgment. We will call this 'function a.' A certain improvement was noted. What changes in the efficiency of closely allied functions are brought about by this improvement? Does the improvement in this function cause equal improvement (1) in the function of estimating areas of similar size but different shape without the correction factor? or (2) in the function of estimating identical areas without the correction factor? (3) In any case how much improvement was there? (4) Is there as much improvement in the function of estimating dissimilar shapes as similar? The last is the most important question.

We get the answer to 1 and part of 3 by comparing in various ways the average errors of the test areas of dissimilar shape in the before and after tests. These are given in Table I. The average errors for the last trial of the areas in the training series similar in size to the test series are given in the same table.

The function of estimating series 2 (same sizes, different shapes) failed

TABLE I.

Subject	Test series 2		Training series
	Av. error before training	Av. error after training	Av. error at end of training
T.	15.8	11.1	2.3
Be.	28.0	5.2	3.1
Br.	22.5	18.7	3.3
J. W.	12.7	21.0	1.5 approx.
W. (2)	17.0	20.0	4.0 approx.
E. B.	10.5	7.9	0.4

evidently to reach an efficiency equal to that of the function trained. Did it improve *proportionately* as much?

This is a hard question to answer exactly, since the efficiency of 'function *a*' increases with great rapidity during the first score or so of trials, so that the average error of even the first twenty estimates made is below that of the first ten, and that again is below that of the first five. Its efficiency at the start depends thus on what you take to be the start. The fact is that the first estimate of the training series is not an exercise of 'function *a*' at all and that the *correction* influence increases up to a certain point which we cannot exactly locate. The fairest method would seem to be to measure the improvement in 'function *a*' from this point and compare with that improvement the improvement in the other function or functions in question. This point is probably earlier in the series than would be supposed. If found, it would probably make the improvement in 'function *a*' greater than that given in our percentages.

The proportion of average error in the after test to that in the before test is greater in the case of the test series than in the case of the first and last estimations of the areas of the same size in the training series, save in the case of *Be.* The proportions are given in [Table II].

TABLE II.

Subject	Proportion of error after to error before training	
	Test series 2	Training series
T.	.70	.575
Be.	.19	.56
Br.	.83	.53
J. W.	1.75	.77 approx.
W. (2)	1.18	.83 approx.
E. B.	.75	.13

Question 2 is answered by a comparison of the average errors, before and after the training, of Series I. (identical areas) given without the correction factor. The efficiency reached in estimating without the correction factor (see column 2 of Table III.) is evidently below that reached

TABLE III.

Subject	Av. error before training of series 1	Av. error after training of series 1	Av. error after training of same sizes in training series	Av. error after training of series 2	Proportion of error after to error before training		
					Series 2	Series 1	Areas of training series identical with series 1
T.	9.0	6.0	2.1	11.1	.70	.67	.31
Be.	21.9	6.4	1.8	5.2	.19	.29	.45
Br.	24.2	14.7	3.7	18.7	.83	.61	.37
J. W.	7.7	8.6	1.5 app.	21.0	1.75	1.11	.77 app.
W. (2)	11.6	3.3 app.	4.0 app.	20.0	1.18	.28 app.	.83 app.
E. B.	9.8	4.1	0.4	7.9	.75	.42	.08

in 'function *a*.' The results there in the case of the same areas are given in column 3. The function of estimating an area while in the frame of mind due to being engaged in estimating a limited series of areas and seeing the extent of one's error each time, is evidently independent to a large extent of the function of judging them after the fashion of the tests.

If we ask whether the function of judging without correction improved proportionately as much as 'function *a*,' we have our previous difficulty about finding a starting point for *a*. Comparing as before the first 100 estimates with the last 100 we get the proportions in the case of the areas identical with those in the test. These are given in column 7. The proportions in the case of the test areas (series 1; same shape) are given in column 6. A comparison of columns 6 and 7 thus gives more or less of an answer to the question, and column 6 gives the answer to the further one: "How much improvement was there?"

We can answer question 4 definitely. Column 5 repeats the statement of the improvement in the case of the test areas of different shape, and by comparing column 6 with it we see that in every case save that of *Be*. there was more improvement when the areas were similar in shape to those of the training series. This was of course the most important fact to be gotten at.

To sum up the results of this experiment, it has been shown that the improvement in the estimation of rectangles of a certain shape is not equalled in the case of similar estimations of areas of different shapes. The function of estimating areas is really a function-group, varying according to the data (shape, size, etc.). It has also been shown that even after mental standards of certain limited areas have been acquired, the function of estimating with these standards constantly kept alive by noticing the real area after each judgment is a function largely independent of the function of estimating them with the standards fully acquired by one to two thousand trials, *but not constantly renewed by so noticing the real areas.* Just what happened in the training was the partial formation of a number of associations. These associations were between sense impressions of particular sorts in a particular environment coming to a person in a particular mental attitude or frame of mind, and a number of ideas or impulses.

What was there in this to influence other functions, other processes than these particular ones? There was first of all the acquisition of certain improvements in mental standards of areas. These are of some influence in judgments of different shapes. We think, "This triangle or circle or trapezoid is about as big as such and such a rectangle, and such a rectangle would be 49 sq. cm." The influence is here by means of an idea that may form an identical element in both functions. Again, we may form a particular habit of making a discount for a tendency to a constant error discovered in the training series. We may say, "I tend to judge with a minus error," and the habit of thinking of this may be beneficial in all cases. The habit of bearing this judgment in mind or of unconsciously making an addition to our first impulse is thus an identical element of both functions. This was the case with *Be.* That there was no influence due to a mysterious transfer of practice, to an unanalyzable property of mental functions, is evidenced by the total lack of improvement in the functions tested in the case of some individuals.

101   IVAN PETROVICH PAVLOV (1849–1936) ON CONDITIONED REFLEXES, 1904

I. P. Pavlov, *Dvadtzatiletni opit obektivnogo izucheniya vischei nervnoi deyatelnosti (povendeniya) zhivotnikh* (Moscow and Leningrad: State Publishing House, 1923; 3rd ed., 1925). Third edition translated by W. Horsley Gantt as *Lectures on Conditioned Reflexes* (New York: International Publishers, 1928), pp. 76–80.

Like Ebbinghaus and Thorndike, Pavlov created a distinctive experimental approach to the classical doctrine of association. It was his unique contribution to bring associationism and reflexology together for the first time by force of empirical evidence. Hartley, in the eighteenth century (see No. 70), had tried to accomplish this end by force of argument. Pavlov's contemporary and countryman V. M. Bekhterev proposed a similar blending of these two traditions, but Pavlov's thorough and extensive experimentation has won for him the approval of posterity. Pavlov and Bekhterev both worked within the framework laid down by their Russian predecessor, Sechenov (see No. 63).

While studying digestive reflexes in dogs, Pavlov noticed that alimentary secretion was often triggered by inappropriate stimuli, that is to say, by stimuli acting in other than their proper or normal manner, as when the sight and sound of an attendant approaching the dog sets off salivation. Other physiologists had been plagued by this phenomenon before,

but Pavlov saw it as a way to study higher nervous functions experimentally. He embarked upon a series of experiments that lasted for more than three decades, and that were concerned with the way in which other than adequate stimuli acquired the power to elicit reflex responses. He found that the sole requirement was that the inappropriate stimulus be paired, or associated, with the adequate stimulus. Nowadays, this form of learning is often called Pavlovian conditioning.

In 1904, Pavlov was awarded the Nobel prize for his physiological work on the digestive reflexes. The present selection—one of his earliest published accounts of conditioning—is from the address delivered on that occasion.

During the study of the gastric glands, I became more and more convinced, that the appetite acts not only as a general stimulus of the glands, but that it stimulates them in different degrees according to the object upon which it is directed. For the salivary glands the rule obtains that all the variations of their activity observed in physiological experiments are exactly duplicated in the experiments using a psychical stimulation, *i.e.,* in those experiments in which the stimulus is not brought into direct contact with the mucous membrane of the mouth, but attracts the attention of the animal from some distance. Here are examples of this. The sight of dry bread calls out a stronger salivary secretion than the sight of meat, although the meat, judging by the movement of the animals, excites a much livelier interest. On teasing the dog with meat or other foods, there flows from the submaxillary glands a concentrated saliva rich in mucus (lubricating saliva); on the contrary, the sight of a disagreeable substance produces from these same glands a secretion of very fluid saliva which contains almost no mucus (cleansing saliva). In brief, the experiments with psychical stimuli represent exact miniatures of the experiments with physiological stimulations by the same substances.

Thus, psychology, in relation to the work of the salivary glands, occupies a place close to that of physiology. And even more! On first view the psychological explanation of the activity of the salivary glands seems to be as incontrovertible as the physiological. When any object from a distance attracting the attention of the dog produces a flow of saliva, one has ground for assuming that this is a psychical and not a physiological phenomenon. When, however, after the dog has eaten something or has had something forced into his mouth, saliva flows, it is necessary to prove that in this phenomenon there is actually present a physiological cause, and not only a purely psychical one which, owing to the special conditions, is perhaps reinforced. From the following experiment this conception is seen to correspond in a remarkable way with reality. Most substances which during eating or forceful introduction into the mouth produce a flow of saliva, evoke a secretion after severance of all the sensory nerves of the tongue similar to that which they evoked before this operation. One must resort to more radical measures, such as poisoning of the animal or extirpation of the higher parts of the central nervous system, in order to convince oneself that between a substance stimulating the oral cavity and the salivary glands there exists not only a psychical but a purely physiological connection. Thus we have two series of apparently entirely different phenomena. How must the physiologist treat these psychical phenomena? It is impossible to neglect them, because they are closely bound up with purely physiological phenomena and determine the work of the whole organ. If the physiologist decides to study them, he must answer the question, How?

Following the examples of the study of the lowest representatives of the animal kingdom, and naturally not desiring to abandon physiology for psychology—especially after an entirely unsuccessful trial in this direction—we chose to maintain in our experiments with the so-called psychical phenomena a purely objective position. Above all, we endeavoured to discipline our thoughts and our speech about these phenomena, and not to concern ourselves with the imaginary mental state of the animal; and we limited our task to exact observation and description of the effect on the secretion of the salivary glands of the object acting from a distance. The results corresponded to our expectations—the relations we observed between the external phenomena and the variations in the work of the salivary glands appeared quite regularly, could be reproduced at will again and again as usual physiological phenomena, and were capable of being definitely systematised. To our great joy, we are convinced that we have started along the path which leads to a successful goal. I shall give some examples of the constant relations which have been established by the aid of this new method of research.

If the dog is repeatedly excited by the sight of substances calling forth a salivary secretion from a distance, the reaction of the salivary glands after each stimulation becomes weaker and weaker, and finally falls to zero. The shorter the intervals between separate stimulations, the quicker the reaction reaches zero, and vice versa. These rules are fully manifested only when the conditions of the experiment do not change. The identity of the conditions, however, need be only relative; it may be limited to those phenomena of the outer world with which had been associated the acts of eating or the forceful introduction of the corresponding substances into the animal's mouth; the variation of other conditions may remain without any effect. This relative identity can be easily attained by the experimenter, so that an experiment in which a stimulus is repeatedly applied from a distance gradually loses its effect, can be readily demonstrated in the lecture hall. If a substance, owing to its repeated employment as a distant stimulus, has become ineffective, the influence of other stimulating substances is not thereby annihilated: if milk from a distance ceases to stimulate the salivary glands, the distant action of bread remains clearly effective. After this has lost its influence by repetition, showing the dog acid will produce again a full effect on the salivary glands. These relations also explain the real meaning of the abovementioned identity of the experimental conditions; every detail of the surrounding objects appears as a new stimulus. When a certain stimulus has lost its efficacy due to repetition, then its action after a certain interval of minutes or of hours is restored without fail.

The effect when temporarily lost, can be restored at any given time, however, by special means. If bread repeatedly shown to the dog fails to stimulate the salivary glands, it is only necessary to give it to the dog to eat and thereupon the full effect of the bread at a distance is at once restored. The same result is obtained when the dog receives some other food. And even more. When some substance producing a salivary secretion, for example, acid, is forced into the dog's mouth, the original distant effect of the bread previously lost is again fully manifested. In general, everything that stimulates the salivary glands restores the lost reaction, and the more fully, the greater has been their activity.

Our reaction can be inhibited by certain influences with the same regularity; if, for example, some stimulus which evokes in the animal a definite motor reaction acts on the eye or ear of the dog.

For the sake of brevity, I shall limit myself to the above mentioned material, and now pass on to theoretical considerations of the experiments. Our given facts can readily be included in a framework of physiological description. The effects we produced on the salivary glands from a distance may properly be considered and termed reflexes. It is impossible not to see, by close attention, that the activity of the salivary glands, when present, is always excited by some external phenomenon; *i.e.,* in the same way as the usual physiological salivary reflex, it is always produced by an external stimulus. The difference consists chiefly in that the usual reflex is determined by the stimulation from the mouth cavity, whereas the new reflexes are evoked by stimulation of the eye, ear, etc. A further essential difference between the old and the new reflexes is that the former are constant and unconditional, while the latter are subject to fluctuation, and dependent upon many conditions. They, therefore, deserve the name of "conditioned."

Considering the phenomena more closely, I can not fail to see the following distinction between these two kinds of reflexes: in the *unconditioned* reflex, those properties of the substance to which the saliva is physiologically adapted act as the stimulus, for example, the hardness, the dryness, the definite chemical properties, etc.; in the *conditioned* reflex, on the other hand, those properties which bear no direct relation to the physiological rôle of the saliva act as stimuli, for example, colour, form, and the like. These last properties evidently receive their physiological importance as *signals* for the first ones, *i.e.,* for the essential properties. In their response one can not but notice a further and more delicate adaptation of the salivary glands to the external world. This is seen in the following case. We prepare to put acid into the dog's mouth, and the dog sees it. In the interest of the integrity of the buccal mucous membrane, it is highly desirable that before the acid comes into the mouth, there should be some saliva present; on the one hand, the saliva will hinder the direct contact of the acid with the mucous membrane, and, on the other hand, will serve to dilute the acid and thus weaken its injurious effect. But, of course, in reality the signals can have only a conditional significance, they are readily subject to change, as, for example, when the signalling objects do not come into contact with the mucous membrane. In this way the finer adaptation is based on the fact that the properties of the substances which serve as signals, now stimulate (*i.e.,* call out the reflex), now lose their exciting action. This is what occurs in reality. Any given phenomenon can be made a temporary signal of the object which stimulates the salivary glands, if the stimulation of the mucous membrane by the object has been once or several times associated simultaneously with the action of the stimulating phenomenon on another receptor surface of the body. We are now trying in our laboratory, with great success, to apply many such, and even highly paradoxical, combinations.

On the other hand, closely related and stable signals can be deprived of their stimulating action if they are often repeated without bringing the corresponding object into contact with the mucous membrane. If any food is shown to a dog for days or weeks without giving it to the animal it finally completely loses its distant stimulating effect on the salivary glands. The mechanism of the stimulation of the salivary glands through the signalising properties of objects, *i.e.,* the mechanism of the *"conditioned stimulation,"* may be easily conceived of from the physiological point of view as a function of the nervous system. As we have just said, at the basis of each conditioned reflex, *i.e.,* a stimulation through the signalising properties of an object, there lies an unconditioned reflex, *i.e.,* a stimulation through the essential attributes of the object. Then it must be assumed that the point of the central nervous system which during the unconditioned reflex becomes strongly stimulated, attracts to itself weaker impulses arriving simultaneously from the outer or internal worlds at other points of this system, *i.e.,* thanks to the unconditioned reflex, there is opened for all these stimulations a temporary path leading to the point of this reaction. The circumstances influencing the opening or closing of this path in the brain are the internal mechanism of the action or of the inaction of the signalising properties of the objects, and they represent the physiological basis of the finest reactivity of the living substance, the most delicate adaptation of the animal organism, to the outer world.

I desire to express my deep conviction that physiological research in the direction which I have briefly outlined, will be highly successful and will help us to make great advances.

Only one thing in life is of actual interest for us—our psychical experience. Its mechanism, however, has been, and remains, wrapped in deep mystery. All human resources—art, religion, literature, philosophy, historical science—all these unite to cast a beam of light into this mysterious darkness. Man has at his disposal one more powerful ally—biological science with its strictly objective methods. This study, as we all see and know, is making great advances every day. The facts and conceptions which I have given at the close of this lecture are typical of numerous trials to make use of systematic application of a purely naturalistic method of thinking in the study of the mechanism of the highest vital expression of the dog—this faithful and friendly representative of the animal world.

102   WOLFGANG KÖHLER (1887–) ON THE INSIGHT OF APES, 1917

Wolfgang Köhler, "Intelligenzprüfungen an Anthropoiden," *Abhandlungen der königlich preussischen Akademie der Wissenschaften,* 1917, no. 1, pp. 149–174. Translated for this book by Edwin G. Boring.

While the associationistic doctrine was finding experimental fulfillment in the research initiated by Ebbinghaus, Thorndike, and Pavlov, a countermovement was developing, which held that the mind is capable of more than associations of elements. As early as 1901, the British sociologist L. T. Hobhouse, in *Mind in Evolution,* objected to Thorndike's method of investigation and to his idea that learning was a mere "stamping in" of connections. The work by Köhler on anthropoid apes, done on the Spanish island of

Teneriffe off the coast of West Africa, gave this sort of dissenting opinion a mass of supporting evidence. On the basis of these experiments Köhler argued, in opposition to Thorndike, that an ape learns relations among stimuli and not only the connection between stimulus and response, that it can modify its behavior merely by perceiving a situation in a novel way and not only from the effect of its actions, and that learning sometimes proceeds by discontinuous improvements in performance and not only by the gradual building up of correct connections. This last process was called "insight" by Köhler.

In the case of the detour experiment, a practical success, one that consists of putting together by chance single and separate parts of the action, must be sharply distinguished from a genuine solution, which is usually characterized by an abrupt break in the smooth, continuous course of previous behavior. In the genuine solution the process corresponds as a whole to the structure of the situation and to the relation of its parts to one another. For example, the desired object may be visible on free ground behind an obstacle of some sort, and then suddenly there comes a smooth and unchecked movement along the course of action that leads to the solution. Thus we find ourselves obliged to believe that the course of this behavior is as a whole adequate from its start to the solution, which emerges from a complete survey of the entire situation. (Chimpanzees, whose behavior is ever so much more expressive than hens', show by their careful looking around that they actually begin with something very much like an inventory of the situation and that it is their survey that gives rise to the behavior required for the solution.)

In our own experiences we can readily distinguish between, on the one hand, the kind of behavior that from its very beginning arises out of our consideration of the structure of the situation and, on the other hand, the kind that does not. Only in the case of the former do we speak of insight, and the only behavior of animals that definitely appears to us as intelligence is one that takes account of the lay of the land from the very start and then deals with the situation along a single, continuous, and definite course. Here, then, is the criterion of insight: insight is shown by *the appearance of a solution complete with reference to the layout of the entire field*. The contrast between this theory and the one of parts put together by chance is absolute. If the natural parts of the action are not related to one another or to the structure of the situation, then the course of the solution must come into being through its dependence on itself and on the total situation as it is visually perceived.

. . . . .

One realizes that our theory requires that the field structure in its entirety and the relations of the members of the situation be bound together to produce a solution. We need here only to exclude the idea that the behavior of the animals can be explained by assuming that the solution is achieved without regard to the structure of the situation, but as a chance sequence of its parts, that is to say, without intelligence.

In the description of these experiments it should have been apparent that the chief essential is lacking for an explanation by chance actions, that is to say, the means by which the solution is composed out of chance parts are not apparent. Certainly it is not a characteristic of the chimpanzee, when he is brought into an experimental situation, that he should make chance movements out of which, among other things, a nongenuine solution could arise. Very seldom is a chimpanzee seen to attempt any action that would have to be considered accidental in relation to the situation—except, of course, when his interest is diverted from the desired object to other things. As long as his efforts are directed to the objective, all distinguishable stages of his behavior—as indeed with human beings in similar situations—tend to appear as complete attempts at solutions, of which none appears as the result of accidentally arranged parts. This is most certainly the pattern when the solution is finally successful. Indeed, success often follows upon a period of perplexity or quiet, a pause that is often also a period of survey. Never, in real and convincing cases, does the solution emerge from the disorder of blind impulses. The action is smooth and continuous and can be resolved into parts only *by the abstract thinking* of the observer. In *reality* the parts do *not* appear independently. Thus, without renouncing what is considered to be its chief merit, our theory cannot permit the supposition that, in the many "genuine" cases which have been described, the solutions that came as wholes could possibly have arisen from mere chance.

.　　.　　.　　.　　.

This sort of question can best be answered by observing the actual facts upon which this theory bases its claim to explain *all* of our experiments. Thus by his own observation one makes oneself more capable of judgment. Behavior suitable for such an examination occurred in building with boxes. Here in the solution in which, when taken *en bloc,* the rule of "higher box on top" was quite clear, the final result was achieved only by chance and after an almost entirely unintelligent muddling around. This procedure happened so often and so uniformly with all the animals used in the experiments that I can claim to know exactly that the process demanded by the theory of chance is general. *It must, therefore, be asserted even more strongly that a very striking difference exists between this kind of conduct, which is obviously ruled by chance, and the behavior described as "genuine" in achieving clear solutions.* The descriptions of these experiments have shown, moreover, how unwillingly a chimpanzee embarks on this procedure, for which the general outline comes to him as a genuine solution but the more detailed execution of which he must attempt by random trials, leaving the solution to chance. The animals would never have hit on making this kind of trial had an attempt *genuine* in outline not first placed them in a position where they encountered special conditions with which they were unable to deal. The fact that animals on such an occasion make these blind movements does not in any way contravene the assertion that, as a rule and under reasonable testing conditions, disordered impulses are not observed.

In these experiments the fact is always mentioned when chance may have brought about or favored a solution. In complicated experimental conditions . . . such cases become more frequent, and it must be said that even then the course of an experiment may not always agree with the theoretical interpretation. In the first place, it may happen that the animal will attempt a solution which, while it may not result in success, yet has in it some meaning with respect to the situation. Random trying then consists in attempts at solution in a *half-understood* situation, and the real solution may easily arise by some chance outcome of the trials; that is to say, it will not arise from chance impulses but rather from chance actions, which, because they are *au fond* sensible, are great aids to chance. Second, a lucky accident may occur in some action, one that has nothing to do with the objective. Here again there is no question of there being a meaningless impulse—as already noted, the chimpanzee gives way to these only when driven to them—but the activity is in a way intelligent even though it has nothing to do with the objective. This is what probably happened when Sultan discovered the way to put two sticks together. Only a philistine would call his playing with these sticks "meaningless impulses" merely because the play had no practical purpose. That an accident helped him is not the most important fact in either case. The important thing is how the experiment proceeds thereafter. We know from the case of man that even an accident may lead to *intelligent* further work (or intelligent repetition), especially in scientific discoveries . . . Thus Sultan's behavior, after he had once carried out with both the bamboo rods his usual play of "put stick in hole," was exactly the same as if he had discovered the new procedure in a genuine solution. After this there could be no doubt that he made intelligent use of the double-stick technique. The accident seems to have acted merely as an aid—a fairly strong one, to be sure—which led him at once to "insight."

If one does not watch attentively, the crude stupidities of these animals, already referred to on several occasions, might be taken as proofs that the chimpanzee does after all perform senseless actions, which could by a chance sequence result in an apparent solution.

.   .   .   .   .

For one who has watched the actual experiments, discussions like the foregoing have something comic about them. For instance, when one has seen for oneself how, in the first experiment in her life . . . it did not for hours dawn upon Tschego to push the obstructing box out of the way, how she merely stretched out her arm uselessly or else sat quietly until, fearing the loss of her food, she suddenly seized the obstacle and pushed it aside, thus solving the problem in a second—when one has watched all that, then it seems almost pedantic to speak of "securing these facts against misinterpretation." It is, however, so hard to reproduce the living impression that many a question can be raised about the words of a report that would not even occur to one after a little observation. Nevertheless it may be that, after this discussion, the description of another experiment, carried out as a model, will be particularly instructive. The experiment is characterized both by its simplicity and by the clarity of its relation to several different theories.

A heavy box stands upright some distance away on the other side of the bars that must so often be mentioned. One end of a stout cord is attached to the box, and the cord itself is laid down obliquely so that its free end lies between the vertical bars of the railing. Halfway between the box and the bars some fruit is tied to the cord (see Fig. 1), where it cannot be reached from the bars as it lies but can be if the cord is laid out straight . . . First of all, Chica pulls in the direction in which the cord lies, so hard that the board of the box breaks, the cord is freed,

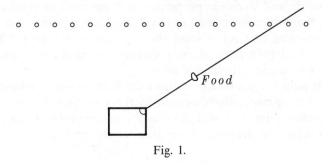

Fig. 1.

and the desired object can be pulled within reach. The box is next re-placed by a heavy stone and the cord is tied around it. Now that the simple solution of pulling has become impossible, Chica takes the cord in one hand, passes it around the bar to the other hand, which she has out through the next space, and carries on so, passing the cord along until it is at right angles to the bars, whereupon she can seize the ob-jective.

Grande at first seems not to see the cord, which is gray and lies on a gray ground. She drags stones about senselessly, an aftereffect of previous experiments . . . and then she tries to detach an iron rod from the wall, presumably wanting to use it as a stick. At last she sees the cord. There-after the experiment runs along as it did with Chica, a solution without any hesitiations.

Rana first pulls twice in the direction in which the cord lies. Then suddenly she changes the direction completely, trying to pull the cord to the spot just opposite the point at which it is tied (see Fig. 2). All the while she stands opposite the point of attachment herself and keeps

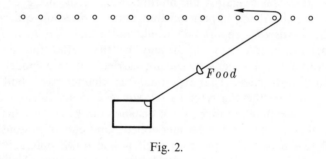

Fig. 2.

looking at the desired object and pulling the cord parallel to the bars. She makes this vain attempt twice in succession on separate occasions and then replaces it with the proper solution, like Grande's and Chica's. This experiment shows that the task consists of two parts. One, crude in its geometric and dynamic properties, is "turn cord at right angles to bars so that objective comes nearer." The other is the more refined special problem that arises from the structure of the bars. Chica and Grande solved both parts at once. Grande solved the first one quickly but the second one only after a delay.

Sultan pulls for a moment as Rana did (Fig. 2) but completely solves the problem immediately afterward in the way that the others did. It thus becomes quite clear that the crude dynamic problem can be solved with no regard to the special one (the second problem), which in this case seemed to be noticed only through nonsuccess. Similar effects had been encountered in building with boxes.

Tercera cannot be cajoled into taking part in the experiment. Tschego and Konsul show—a fact which must not be overlooked—that the solution is not obvious, for neither of them gets any further than pulling the cord in the direction in which it was lying.

The experiment is next repeated with Chica, but this time the cord lies on the floor *turned in the opposite direction.* The animal does not now pull at all in the direction of the cord, but starts in at once with the hand-over-hand process in the direction *opposite* from the previous experiment until the goal is reached. After this I did not think it necessary to make the same experiment with the other animals.

It need hardly be pointed out again, after the foregoing explanations, that experiments like the one just described give us *better* information about the chimpanzee than do the usual animal tests with such devices as complicated locked doors. It must also be realized that so simple and clear an experiment as this contains the whole problem to be considered.

If anyone is still of the opinion that such simple solutions are obvious and have nothing to do with intelligence, I can only invite him to show definitely and precisely how this successful procedure comes into being. No psychologist, I fear, will at the present time be able to come forward with such an explanation.

I have separated the two parts of this problem, which, as we have seen, are independent of each other. Now let me consider only the crude one and its solution. This solution is simply characterized in the sketch (Fig. 3), which does not show how the animal, at the first moment of the solution, actually performs the arrow movement in detail, whether it takes the bars into consideration or not.

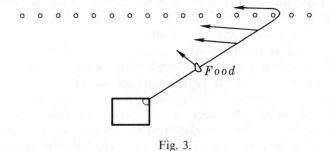

Fig. 3.

Do the animals arrive at the solution in accordance with the theory we have discussed? If so, we should expect in all cases to find the oc-

575

currence of a large number of impulses which might, in some of the chimpanzees at least, accidentally contain the right "fragments" in the right order. Actually Grande was the only animal that did anything senseless, and that was just a habit-stupidity that intervened before she had surveyed the possibilities of the problem. When she saw the cord, a new stage of behavior set in and immediately afterward she achieved a perfectly clear solution. Altogether only two movements really took place with regard to the objective . . . These two movements are:

1. *Pulling in the direction of the cord,* that is, a sensible proceeding, of which Chica once proved the practicability. No man, still less a chimpanzee, can find out otherwise whether the cord will really come loose from the box or the stone.

2. Pulling at the cord, or the continuous passing of it hand-over-hand —in both cases *in the right direction for a solution* (see the arrows in Fig. 3).

In not a single animal was there any pulling in a direction that came near being midway between the two directions observed. Much less was there a third, quite new direction of pulling. When the more primitive tendency (pulling in the direction of the cord) appeared first, the jump to the other direction was made quite abruptly.

Everyone must feel, it seems to me, that we have here a very clear, if peculiar, occurrence, one that has nothing at all to do with the postulates of our theory. Are we to squeeze the facts in order to force them to fit the theory, just in order to suit the principle of scientific economy, as it is called? In this case the observer is forced to conclude that attempts 1 and 2, appearing as wholes and yet each on its own, are a direct result of a visual survey of the situation. There is a certain scientific attitude— we might formulate it as a principle, the "principle of maximal scientific fertility"—that would lead us to begin the theoretical consideration with the character of the observations and not to eliminate it at any cost after the manner of the theory of chance.

There would be no need to discuss our theory further if the previous life of the chimpanzees we have tested were known to us in all details from birth to the moment of testing. Unfortunately it is not. Even if the possibility of the solutions' arising by chance were excluded in the *experiment,* still, within the scope of the theory, the possibility remains that they might have been developed by accident before the experiment, that they were then repeated and so improved that they now seem to appear as genuine solutions.

It is always difficult to refute arguments that lie outside the realm of possible proof. In the present case, however, we may overstep the bounds of experience without weakening the argument, for of course the chimpanzees we tested had spent several years as uncontrolled lively animals in the jungle of the West Coast, and while there they came into contact with objects similar to those employed in some of our experiments. Thus it becomes necessary to consider whether this circumstance influences the significance and factual value of our experiments.

Two points must be kept fixed in mind if the purpose of our discussion is not to be lost from sight:

1. The fact that the animals have had to deal with single objects or situations prior to our experiments need not have any direct connection with our problem. It is only when, exactly according to our theory, meaningless but successful chains of actions, externally like the behavior observed here, have been formed during this previous period—formed accidentally and then selected by success—that "previous experience" tells against the value of these experiments. I am far from asserting that the animals tested . . . have never had a stick or anything like one in their hands before the experiment. On the contrary, I take it for granted that every chimpanzee above a certain very low age has had some such experience: he will have seized a branch in play, scratched on the ground with it, and so on. Exactly the same thing is frequently observed in small children less than a year old; they too have had their "experience" with sticks before they use them as implements to pull toward them things that they could not reach otherwise. By no means, however, does this fact prove that children become accustomed to the use of implements in the mere chances of play and quite without insight, reproducing the action again without insight at two, four, or twenty years of age, nor should such a conclusion follow for the chimpanzee, whose test stick is not the first he ever held in his hand.

2. Nor am I by any means trying in this work to prove that the chimpanzee is a marvel of intelligence. On the contrary, the narrow limits of his powers, as compared with man's, have often been demonstrated. All that we have to decide is whether *any* of his actions ever have the characteristics of insight, and the answer to this question of principle is at present far more important than is an exact determination of his *degree* of intelligence. On the other hand, the theory of chance action, which is discussed here as a general principle of interpretation, is not concerned with the mere *diminution* of the *number* of intelligent acts in an experiment, for this theory, to be convincing, must explain consistently with itself *all* the tests without exception. It fails when, even though some observed results are explained by it, others are not. In the latter case, when the universal application of the theory collapses, there is less temptation to explain as products of accident those kinds of behavior which, by their nature, invite no such interpretation, even though it may be possible to force them under the theory of chance.

.   .   .   .   .

In discussing the chance theory I did not express my attitude toward the general theory of association. At the very beginning I pointed out that the question to be answered in this monograph might be affirmed or denied without thereby affirming anything about the relation of the experiments to the doctrine of association. For the time being we may assume that the two are independent. If we accept the principle of chance, we shall also have to agree that the animals have no insight whatsoever, and that presumption touches the very core of this investigation. Association theorists know and recognize what one calls insight in man, and they contend that they can explain this phenomenon just as well by their principles as they can the simplest association or reproduction by contiguity. The consequence for animal behavior is that, whenever it has an intelligent character, they treat it in the same way, but not at all as if the animal lacks what is usually called insight in man. Thus I can dispense with any more specific elaboration of this point. Let me merely observe here that the first and essential condition of a satisfactory associative explanation of intelligent behavior would be the following achievement by the theory of association, to wit: the principle of association must show in all strictness just what is meant by the grasp of a *material inner* relation of two things to each other (more universally, the grasp of the structure of a situation). Here the "relation" means an *interconnection based on the properties of these things themselves,* not a mere "frequent following of each other" or an "occurring together." This problem is the first that requires solution, because it is these "relations" that represent the most elementary participation in specifically intelligent behavior. There can be no doubt that these relations, along with other factors, regularly determine the chimpanzee's behavior. They are not facts like "sensations" and the like, or further associable elements; yet it can be very definitely proved . . . that by their functional properties they determine in very marked degree the chimpanzee's behavior, his inner processes. Either the association theory is capable of explaining clearly the relations "smaller than," "farther away than," "pointing straight toward," and the like, according to their true meanings as mere associations—and, if it can, then all is well—or else the theory may not be used for an explanation, as becomes the case if it cannot account for these factors that are so important for the chimpanzees (as they are for man). In this latter case the association principle can be allowed only as a *participation.* At the very least that other class of processes, those relations that are *not* external connections, must be recognized as an additional independent working principle.

XV   THE NATURE OF PSYCHOLOGY

For quite a long time now it has been conventional to treat the history of science in general and the history of psychology in particular as an institutionalized body of knowledge that grows continuously with the successive contributions made to it by Great Men. In such an account the contributions are sometimes factual, as when an investigator reports a discovery that has emerged from the use of the experimental method, and sometimes systematic, as when a theorist has clarified and successfully promoted the use of a new principle. The discovery that thought can proceed without imagery would be an example of the first; the establishment of the generalization that the facts of psychology can all be formulated in terms of the behavior of an organism would be an example of the second.

The modern view of the history of science, however, diminishes the importance of the Great Man theory and shows science for the most part as a continuous process of evolution with few sudden mutations. Almost every great discovery in science turns out to have been anticipated by others; it is assigned to the Great Man whose name it bears because his experiments or his arguments were more convincing that his predecessors'. New principles emerge in the same way and also in many instances because the times have become ready for an innovation in thinking that the generally accepted premises of an earlier period would have rejected, or perhaps actually did reject. These Great Men are eponyms, that is to say, they have their importance reinforced by the fact that their names have been given to periods or schools or principles or laws,

thus creating the delusion that important discovery or insight is sudden, whereas normally it has been quite gradual. In this way eponymy distorts history, although it has the use of simplifying thought for the historian.

A source book in the history of science is necessarily eponymous. It picks out representative samples from the writings of Great Men. In the present volume we have sought to diminish this inevitable distortion by presenting chronologically the evolution of each of fourteen basic trends in the history of psychology, choosing in general for reprinting the words of the Great Men, but not always the particular words that made them great. In this chapter we aim to give the reader a more conventional view of psychology's modern history by asking somewhat less than a score of Great Psychologists for their opinions as to the nature of psychology. When a brief opinion is available, we quote it. When the Great Psychologist is too prolix for brief quotation, we name him and say in a few words what place he occupied in the development under discussion. In this way we bring into mention by excerpt or abstract a number of very important psychologists who, by and large, contributed more to the philosophy of psychology than to factual discovery.

It is useful for the purposes of this chapter to think of the history of modern psychology as divided into three periods: the philosophical, from Descartes (1650) to Fechner (1860); the institutional or systematic, from Wundt (1874) to McDougall (1923); and the specific and factual, from Tolman (1932) on. In the philosophical period psychology was not a separate institutionalized discipline and the question whose answer now interests modern psychologists was: "What is the nature of mind?" Here, however, we venture on no such broad inquiry, but cite only the philosophers who led straight on to the period when psychology had become institutionalized and the new generation of experimental psychologists, having partly (as in Germany) or wholly (as in America) asserted freedom from the parent philosophy, had undertaken to defend the new discipline by defining the nature of psychology and supporting their definitions in their systematic treatises and texts. It can be argued that the last systematic texts to be written, texts that attempted to cover the whole range of psychology as it was historically laid out, were Titchener's, Watson's, and McDougall's. The third period also contains systematic texts, but they make no attempt to cover all the conventional topics of psychology. Tolman's was the first of these, and others have been Koffka's, Hull's, and Skinner's. Our volume does not, however, penetrate so far into the present.

The evolution of an institutionalized discipline of psychology was in part a change from one sort of question to another. The philosophers were, at first, asking, "What is mind?" Later, their intellectual descendants were asking, "How should mind be studied?" The first question is metaphysical—it asks about an ultimate reality; the second is epistemological—it asks about the sources and limits of a type of knowledge. Like most evolutionary processes, this one did not reveal itself suddenly or without ambiguity, for some of the early thinkers were interested in method, and some of the later ones in metaphysics. Nevertheless, the shift in emphasis is unmistakable when one examines two periods sufficiently separate, like Descartes's and Wundt's.

Although the answer to "How should mind be studied?" seems to depend upon the answer to "What is mind?" institutional psychology arose without the benefit of a metaphysical resolution. Fechner's psychophysical methods answered the epistemological question for the nineteenth century, but actually said nothing about the ultimate nature of mind—this in spite of the fact that Fechner was trying to answer the metaphysical question and even believed that his findings supported a panpsychism. His successors, however, were more than content to adopt his methods and neglect his metaphysics. It has often been true that sciences have developed with an unabashed disregard for the philosophical questions that got them under way.

The key problem in the metaphysical study of the mind was to distinguish it from the body and from matter in general. For hundreds of years Christian philosophy seemed to hold the clue to the distinction, which was to be found in the immortality of the soul. It was Descartes who first offered an alternative distinction and in so doing marked out the road leading to modern psychology.

The lifetime of René Descartes (1596–1650) spans an era remarkably rich in scientific discovery. In it William Gilbert (1544–1603) published his monumental study of magnetism, which founded this branch of modern physics and remains one of the earliest triumphs of the experimental method. Johannes Kepler (1571–1642) formulated his three great laws of planetary motion. Galileo Galilei (1564–1642) discovered in the science of mechanics the concepts which, with Kepler's laws, were to lead Newton to his greatest work, the *Principia* (1687). William Harvey (1578–1657) showed the heart to be a pump, and the blood vessels a hydraulic system. And, as the self-conscious spokesman for this new age, Francis Bacon (1561–

1626) expounded the philosophy of scientific induction and glorified the experimental method.

Yet if one man could stand out in an era as full as this, it was Descartes. For it was his philosophic contribution more than anyone else's that freed thought from the rigid dogmas, both religious and classical, that had bound it for centuries. Not since Aristotle had a philosopher constructed a new and influential system of thought that took into account the sum of knowledge, which in two thousand years had grown so significantly. Influence does not, however, mean general acceptance, for Descartes's era was a time of skepticism and questioning. Rather, it means that thoughtful men found in his works a fresh start, a new classification of knowledge, a new set of questions.

It may properly be said that modern psychology began in Descartes's conception of the duality of mind and matter. Although philosophers before him had also spoken of this duality, Descartes's version contributed something new. Mind, for him, was to be distinguished from everything else by its capacity to think, to be conscious of itself. That was no minor capacity, for it provided the central principle of the whole Cartesian system, the famous "I think, therefore I am." Unlike earlier dualistic philosophies, this one rested upon a purely psychological characteristic of the mind. To be sure, Descartes preserved some of the traditional distinctions between mind and matter, such as the indestructibility of the mind (soul), as contrasted with the impermanence of matter, but it is the psychological distinction that set this new dualism apart and made it especially important.

Man, then, has a mind and a body, and to study the two, said Descartes, requires different methods. The mind, manifested in thought, must be studied by introspection, while the body, manifested in action, is to be studied by the methods of natural science. Later on, psychologists, depending upon whether they were concerning themselves with thought or with action, were to find here the philosophical justification for both subjectivism and objectivism. The opposition between these two venerable traditions in psychology has been no less bitter for their having a common ancestor.

Descartes said that the mind is free because it can initiate action but that the body is governed by the operation of physical laws. But how is the freedom of the mind to be exercised if the body is bound by physical law? To solve the problem, Descartes suggested that the mind can interfere with the machinelike operation of the body when it wills to do so. By movements of the pineal body, the mind is able to affect the flow of animal spirits, which, in turn, governs the movements of the muscles. Descartes chose the pineal body as the point of interaction between mind and body because it was strategically located in the center of the head and also because he thought that only human beings possessed it. It seemed correct that animals, who lacked rational minds in the Cartesian sense, should also lack this key structure. Soon, however, physiologists identified the pineal body in lower animal forms, and this finding was not the only way in which Cartesian dualism was undermined.

As scientific physics developed, it soon became evident that Cartesian dualism, in which mind and body interacted, violated the principle of conservation of momentum since it implied that the mind, an incorporeal substance, could alter the direction of movement of the corporeal body. For some of Descartes's followers, this problem was solved by denying altogether the existence of a free, incorporeal soul. These materialists strove

581

to include the functions of the soul within the Cartesian view of the body as a machine. For others, the solution was to postulate a different sort of dualism, one in which the body and the mind exist as separate entities without ever interacting. Such psychophysical parallelism holds that mind and body are correlated with each other, but instead of interaction there is between them a pre-established harmony, as if there were two perfect clocks running side by side, a conception proposed by the Belgian philosopher Arnold Geulincx (1624–1669) and developed by the influential German philosopher Gottfried Wilhelm Leibnitz (1646–1716). If this view seemed implausible, it had at least the virtue that it was free of the logical and scientific difficulties of interactionism, a virtue important enough to establish it soon after Descartes as the more acceptable version of dualism.

The excerpt from Descartes contains his description of the nature of the interaction between mind and body (and is repeated here because indispensable in two places; see No. 44). He thought of the body as a machine operating in accordance with physical principles, and of the mind, or soul, as an unextended substance having its seat in the "little gland" in the brain. The body moves in response to the flow of animal spirits from the cavities in the brain, via the nerves, into the muscles. When the flow of the animal spirits is being guided by movements of the pineal body, the mind is influencing action. But the body also influences the mind, for the pineal gland is moved in a characteristic way by each pattern of activity in the animal spirits, and from these movements the mind perceives what is happening to the body and in the physical world beyond.

103   RENÉ DESCARTES (1596–1650)

René Descartes, *Les Passions de l'âme* (Amsterdam, 1650), pt. 1, art. 34. Translated by E. S. Haldane and G. R. T. Ross in *The Philosophical Works of Descartes* (Cambridge, Eng.: University Press, 1931), I, 347.

ARTICLE XXXIV

How the soul and the body act on one another

Let us then conceive here that the soul has its principal seat in the little gland which exists in the middle of the brain, from whence it radiates forth through all the remainder of the body by means of the animal spirits, nerves, and even the blood, which, participating in the impressions of the spirits, can carry them by the arteries into all the members. And, recollecting what has been said above about the machine of our body, i.e., that the little filaments of our nerves are so distributed in all its parts, that on the occasion of the diverse movements which are there excited by sensible objects, they open in diverse ways the pores of the brain, which causes the animal spirits contained in these cavities to enter in diverse ways into the muscles, by which means they can move the members in all the different ways in which they are capable of being moved; and also that all the other causes which are capable of moving the spirits in diverse ways suffice to conduct them into diverse muscles; let us here add that the small gland which is the main seat of the soul is so suspended between the cavities which contain the spirits that it can be moved by them in as many different ways as [it] may also be moved in diverse ways by the soul, whose nature is such that it receives in itself as many diverse impressions, that is to say, that it possesses as many diverse perceptions as there are diverse movements in this gland. Reciprocally, likewise, the machine of the body is so formed that from the simple fact that this gland is diversely moved by the soul, or by such other cause, whatever it is, it thrusts the spirits which surround it towards the pores of the brain, which conduct them by the nerves into the muscles, by which means it causes them to move the limbs.

The sturdy heritage of psychological speculation that operated on Descartes's thinking undoubtedly included the ancient Platonic doctrine of innate ideas, the view that the mind contains notions that are independent of experience. It is not surprising, then, that, in his attempt to classify the activities of the mind, Descartes spoke of innate ideas, which, he said, consisted of the notions of God, self, mathematics, geometry, space, and time. Nor would it be correct to place this bit of psychological speculation at the same high level as his dualistic thesis or his concept of the reflex (see No. 57), for it was neither so influential nor so original. What is most important about Descartes's doctrine of innate ideas is his equivocation in respect of it. In some of his writings, he states the doctrine firmly; in others, he plays it down.

Nevertheless, in those days it was radical to argue, as did Thomas Hobbes (1588–1679), that the mind contains only ideas drawn from experience and nothing innately. Hobbes's defense of this new view was, however, neither so exhaustive nor so influential (see No. 56) as the arguments put forth soon after him by John Locke (1632–1704). Locke was primarily a political philosopher, one who had the rare good fortune to see his beliefs mirrored in practical affairs, when the English monarchy was liberalized under William and Mary; but, in addition to a political theory, Locke advanced a theory of knowledge that established the important philosophic movement known as British empiricism. Like most philosophers of that era, Locke took as his starting point the Cartesian outlook. He differed from his great predecessor with regard to innate ideas, for he argued that experience is the sole source of mental content. Notwithstanding this difference, Locke was very much in the Cartesian tradition. Descartes had said that the essense of mind is thought; Locke, accepting this key principle, was defining thought by saying that its objects are ideas, which, in turn, are derived solely from observation either of the external world or of the mind itself in operation. These two sources of thought—sensation and reflection—provide the building blocks out of which, Locke said, all mental activity is composed.

583

104   JOHN LOCKE (1632–1704)

John Locke, *An Essay concerning Humane Understanding: In Four Books* (London, 1690), bk. II, chap. 1.

1. *Idea is the Object of Thinking.* Every man being conscious to himself, That he thinks, and that which his Mind is employ'd about whilst thinking, being the *Ideas,* that are there, 'tis past doubt, that Men have in their Minds several *Ideas,* such as are those expressed by the words, *Whiteness, Hardness, Sweetness, Thinking, Motion, Man, Elephant, Army, Drunkenness,* and others: It is in the first place then to be enquired, How he comes by them? I know it is a received Doctrine, That Men have native *Ideas,* and original Characters stamped upon their Minds, in their very first being. This Opinion I have at large examined already; and, I suppose, what I have said in the fore-going Book, will be much more easily admitted, when I have shewed, whence the Understanding may get all the *Ideas* it has, and by what ways and degrees they may come into the Mind; for which I shall appeal to every ones own Observation and Experience.

2. *All Ideas come from Sensation or Reflexion.* Let us then suppose the Mind to be, as we say, white Paper, void of all Characters, without any *Ideas;* How comes it to be furnished? Whence comes it by that vast store, which the busie and boundless Fancy of Man has painted on it, with an almost endless variety? Whence has it all the materials of Reason and Knowledge? To this I answer, in one word, From *Experience:* In that, all our Knowledge is founded; and from that it ultimately derives it self. Our observation employ'd either about *external, sensible Objects; or about the internal Operations of our Minds, perceived and reflected on by our selves, is that, which supplies our Understandings with all the materials of thinking.* These two are the Fountains of Knowledge, from whence all the *Ideas* we have, or can naturally have, do spring.

3. *The Objects of Sensation one Source of Ideas.* First, *Our Senses,* conversant about particular, sensible Objects, do *convey into the Mind,* several distinct *Perceptions* of things, according to those various ways, wherein those Objects do affect them: And thus we come by those *Ideas,* we have of *Yellow, White, Heat, Cold, Soft, Hard, Bitter, Sweet,* and all those which we call sensible qualities. This great Source, of most of the *Ideas* we have, depending wholly upon our Senses, and derived by them to our Understanding, I call, SENSATION.

4. *The Operations of our Minds about Sensible Ideas, the other Source of them.* Secondly, The other Fountain, from which Experience furnisheth the Understanding with *Ideas,* is the *Perception of the Operations of our own Minds* within us, as it is employ'd about the *Ideas* it has got; which

Operations, when the Soul comes to reflect on, and consider, do furnish the Understanding with another sett of *Ideas,* which could not be had from things without; and such are, *Perception, Thinking, Doubting, Believing, Reasoning, Knowing, Willing,* and all the different actings of our own Minds; which we being conscious of, and observing in our selves, do from these receive into our Understanding, as distinct *Ideas,* as we do from Bodies affecting our Senses. This Source of *Ideas,* every Man has wholly in himself: And though it be not Sense, as having nothing to do with external Objects; yet it is very like it, and might properly enough be call'd internal Sense. But as I call the other *Sensation,* so I call this REFLECTION, the *Ideas* it affords being such only, as the Mind gets by reflecting on its own Operations within it self. By REFLECTION then, in the following part of this Discourse, I would be understood to mean, that notice which the Mind takes of its own Operations, and the manner of them, by reason whereof, there come to be *Ideas* of these Operations in the Understanding. These two, I say, *viz.* External, Material things, as the Objects of SENSATION; and the Operations of our own Minds within, as the Objects of REFLECTION, are, to me, the only Originals, from whence all our *Ideas* take their beginnings.

The famous school of philosophy that Locke founded with his theory of knowledge was dominant in Great Britain and, to a lesser extent, in France until the middle of the nineteenth century. Eventually, it was to provide modern psychology with its major underlying assumptions and, perhaps, its prejudices, but this is not to say that the men who developed empiricism saw themselves as part of a succession of like-minded thinkers. On the contrary, Locke's successors more often regarded themselves as critics, not as followers, of their philosophical progenitor.

Soon after Locke's *Essay* appeared, a young Irish cleric, George Berkeley, published *A New Theory of Vision* (1709), a work in which he applied the empiristic philosophy to an ancient problem: how can man perceive a three-dimensional world when his eyes mediate only a two-dimensional picture of it? This is a hard problem

to the empiristic thinker, who is restrained from assuming that space is innately perceived. Berkeley's solution, based on Lockian principles, is a genetic account of space perception, in which the idea of space is compounded out of experience with visual and tactual sensations. Berkeley argued that it is the correlations between things seen and things touched that add the third dimension to perceived space (see No. 68).

Although in one sense Berkeley merely applied Lockian principles to a specific problem, in another he changed the quality of empiristic thought. Locke had pictured the mind as gaining knowledge only through experience but nevertheless gaining knowledge of the reality outside oneself. Berkeley, however, recognized the subjectivism tacit in empiricism: that one can never transcend the information that one's senses provide. Any reality outside oneself was, for Berkeley,

585

unknowable. He escaped solipsism by religious faith, for he believed that God arranges our sensations to be constant and dependable, as if the world outside were real and not a capricious illusion.

With David Hume (1711–1776), the next in this line of empiricists, the implicit solipsistic step was taken. For Hume, the existence of God cannot be logically demonstrated, and the skepticism that Berkeley just managed to avoid now became a characteristic of empiristic philosophy. It may be said that, as a philosophical system, empiricism found its logical completion in the writing of Hume (see No. 69), but as psychology it was just beginning to be exploited. Hume's contemporary David Hartley in 1749 wrote a compendium (see Nos. 59, 70) that was almost as psychological as philosophical in outlook. In respect of philosophy, there is little new in Hartley, certainly much less than in his three great predecessors. In this work, however, Hartley showed how empiristic philosophy can be wrought into a psychological system, one which could conceivably be coordinated with the physiology of the central nervous system. Hartley, trained as a physician, offered physiological hypotheses, none of which has survived except as history.

Empiricism is not the only important philosophical tradition to have shaped modern psychology. There is, in addition, the school of thought associated with the German philosopher Immanuel Kant (1724–1804), whose epistemological doctrines were formed as a refutation of Hume's empiricism. In one of his works, Kant said: "I confess frankly, it was the warning voice of David Hume that first, years ago, roused me from dogmatic slumbers." The warning voice had argued that causality as such can never be perceived, that it is but an illusion occasioned when events follow each other with regularity. There is, of course, no place for the idea of causality in empiricism: if all ideas arise in experience, then the only basis for the impression of causality is invariable sequentiality. Kant's refutation of Hume had to be, therefore, a refutation of the all-inclusiveness of the empiristic philosophy.

The refutation of Hume turned out to be the vastly influential *Critique of Pure Reason* (*Kritik der reinen Vernunft*), the work that initiated the school of German idealism and led ultimately, if tortuously, to much of Germany's nineteenth-century philosophy, including the Marxist system. Kant's answer to Hume was to include the concept of causality among the dozen innate capacities of the mind for perceiving relations among experiences. First, however, he had to justify his conviction in innate capacities, and this justification is the primary substance of his monumental work. In addition to the innate capacities, Kant spoke of the innate intuitions, which order experiences in respect of time and space. In the passage that follows, Kant outlined the argument in support of the innate, or a priori, nature of space and time.

105   IMMANUEL KANT (1724–1804)

Immanuel Kant, *Kritik der reinen Vernunft* (Riga, 1781), bk. I, pt. 1. Translated by Max Müller as *Critique of Pure Reason* (London, 1881).

Whatever the process and the means may be by which knowledge reaches its objects, there is one that reaches them directly, and forms the ultimate material of all thought, viz. intuition. This is possible only when the object is given, and the object can be given only (to human beings at least) through a certain affection of the mind.

This faculty (receptivity) of receiving representations, according to the manner in which we are affected by objects, is called sensibility.

Objects therefore are given to us through our sensibility. Sensibility alone supplies us with intuitions. These intuitions become thought through the understanding, and hence arise conceptions. All thought therefore must, directly or indirectly, go back to intuitions, i.e. to our sensibility, because in no other way can objects be given to us.

The effect produced by an object upon the faculty of representation, so far as we are affected by it, is called sensation. An intuition of an object, by means of sensation, is called empirical. The undefined object of such an empirical intuition is called phenomenon.

In a phenomenon I call that which corresponds to the sensation its *matter;* but that which causes the manifold matter of the phenomenon to be perceived as arranged in a certain order, I call its *form.*

Now it is clear that it cannot be sensation again through which sensations are arranged and placed in certain forms. The matter only of all phenomena is given us *a posteriori;* but their form must be ready for them in the mind *a priori,* and must therefore be capable of being considered as separate from all sensations.

I call all representations in which there is nothing that belongs to sensation, *pure* (in a transcendental sense). The pure form therefore of all sensuous intuitions, that form in which the manifold elements of the phenomena are seen in a certain order, must be found in the mind *a priori.* And this pure form of sensibility may be called the pure intuition.

Thus, if we deduct from the representation of a body what belongs to the thinking of the understanding, viz. substance, force, divisibility, etc. and likewise what belongs to sensation, viz. impermeability, hardness, colour, etc., there still remains something of that empirical intuition, viz. extension and form. These belong to pure intuition, which *a priori,* and even without a real object of the senses or of sensation, exist in the mind as a mere form of sensibility.

.   .   .   .   .

In Transcendental Aesthetic therefore we shall first isolate sensibility, by separating everything which the understanding adds by means of its concepts, so that nothing remains but empirical intuition.

Secondly, we shall separate from this all that belongs to sensation, so that nothing remains but pure intuition or the mere form of the phenomena, which is the only thing which sensibility *a priori* can supply. In the course of this investigation it will appear that there are, as prin-

587

ciples of *a priori* knowledge, two pure forms of sensuous intuition, namely, *Space* and *Time.*

Immanuel Kant's nativism was not merely a revival of the Cartesian notion of innate ideas, which pictured man as being born with substantive ideas. Rather, it pictured man as being born with principles of ordering such that, when sensations are elicited, they are automatically interwoven in specific ways. To psychology, Kant contributed the philosophical authority for nativistic theories of perception, which, like the modern Gestalt theory, hold that perception is, to some extent, based on man's tendency to order simple sensations in characteristic ways.

The two philosophical approaches to thought—the empiristic and the nativistic—epitomize many of the controversies in psychology. The opposition between Helmholtz and Hering, between Wundt and Stumpf, between behaviorists and Gestaltists, all echo the opposition between Hume and Kant, Locke and Descartes. Experience versus inheritance has not, however, been the only issue separating these disputants, for the two philosophical traditions gradually accumulated additional properties, which were to distinguish their adherents further and to intensify their disagreement. Those in the empiristic tradition were to become analytic, reductionistic, and, ultimately, positivistic in their views, while those in the nativistic tradition were to become phenomenological and rationalistic as opposed to experimental. These additional properties are not linked to the primary issue by logical necessity; they are linked more subtly and less surely, perhaps by virtue of the sort of mentality that is attracted to one theory of knowledge or the other.

In seeing the differences between these two schools of philosophy, one is likely to lose sight of the similarities. They both owe to Descartes the subjective outlook toward knowledge implicit in the *cogito,* and, with this subjective outlook, both schools gravitated toward psychological formulations. Both recognized the importance of sensory experience in governing thought, differing only in the degree to which they imputed additional powers to the mind. In view of all these similarities, a philosopher, as opposed to a psychologist, might well group together Descartes, Hume, and Kant as men who sought the key to knowledge by way of psychological speculation.

With the coming of the nineteenth century the physiology of the senses began to emerge and then to flourish. It must have affected James Mill, who was writing his treatise just after the establishment of the law of the spinal nerve roots by Charles Bell and François Magendie and just after Johannes Müller had undertaken his argument for the doctrine of the specific energies of nerves.

It was natural for adherents of empiristic philosophy to become especially interested in a psychology that emphasizes the study of the senses. James Mill's *Analysis of the Phenomena of the Human Mind* (1829) foreshadowed the metamorphosis of empiristic philosophy into academic psychology (see No. 72). He wrote it as a philosopher, but his text is really a psychology, for it discusses each sense, as well as association, language, memory, imagination, will, and feeling. To be sure, it is not a scientific psychology, since it relies on casual instead of systematic observation, but it shows, nevertheless, how empiristic philosophy could so readily give way to the systematic, scientific psychology soon to be founded in Germany by Wundt.

Johannes Müller, sometimes called the father of experimental physiology, wrote between 1833 and 1840 his definitive seven-book compendium of what was then the new physiology. In dealing with the brain, it was proper for him to say something about the mind for which the brain functions.

He was in these matters a good Cartesian dualist: the mind, he noted, is affected by the action of the brain and yet, being utterly different in nature, is largely independent of it. Actually he was more than a dualist, for he was also a vitalist, believing that life could not be reduced to physicochemical processes—a triplist, as it were. His dualism was interactionistic. Psychophysical parallelism had not quite yet received its powerful new support from the theory of conservation of energy.

106 JOHANNES MÜLLER (1801–1858)

Johannes Müller, *Handbuch der Physiologie des Menschen* (Coblenz, 1840), bk. VI, sect. 1, chap. 2, "Wirkung des Gehirns beim Seelenleben," pp. 516–517. Translated by William Baly as *Elements of Physiology*, vol. II (London, 1842).

Action of the brain in the production of mental phenomena.—The peculiar quality of the mind is "consciousness," a something of which no further definition can be given, and which admits of description as little as the states or properties called sound, blue, red, and bitter. Just as it is the property of a specific nerve connected with the sensorium to have sensibility, so it is the property of the brain to have consciousness. The conception of ideas, thought, and emotion, or the affections, are modes of consciousness. There is no sufficient reason for admitting the existence of special organs or regions set apart in the brain for the different acts of the mind, or for regarding these as distinct powers or functions. They are, in fact, as we shall presently show, merely different modes of action of the same power. Moreover, although the clearness of the ideas conceived, and of the thoughts, and the intensity of the emotions are affected by structural changes of the brain; and although the integrity of that organ is quite essential to consciousness, yet the action of the mind cannot be explained as the result of material processes, but must rather be regarded as in its essence independent of all material relations, though influenced in its clearness and intensity of the condition of the organ which is its seat. It is true that the mind is rendered conscious of external impressions only through the medium of the nerves of sense, and their action on the brain: but the retention and reproduction of the mental images of external objects of sense, exclude altogether the notion of particular orders of ideas being fixed in particular parts of the brain; for example, in the ganglionic corpuscules of the grey substance. For the thoughts accumulated in the mind become associated in the most various manners, in a chronological succession, according to the relation of simultaneous occurrence, or according to their similarity or contrariety; and these relations of the ideas or thoughts to each other change every moment. It is certainly correct that structural changes of the brain are sometimes followed by partial loss of memory; facts relating to certain periods of time, or certain kinds of names, substantives, or adjectives being forgotten. But the occurrence of partial loss of memory of the first kind could only be used as an argument for impressions being successively fixed in the order of time in stratified parts of the brain,—a supposition which cannot be for a moment entertained. And if the perception of mental impressions and thought were ascribed in a general way to the ganglionic corpuscules; if the direction of the mind from individualities to general ideas, or from general ideas to individualities, were supposed to depend on the relative exaltation of action of the peripheral part of a ganglionic corpuscule above its nucleus, or of the nucleus above the peripheral part; if the combination of ideas in an act of thought, or a proposition, for which the conception of object, predicata and copula, at the same time is necessary, were explained by a reciprocal reaction of ganglionic corpuscules through the medium of intermediate processes acting as copulæ; or, lastly, if the association of ideas successively in the order of their first conception, or simultaneously according to their previous existence were admitted to be attended with a successive action of certain united corpuscules, or a simultaneous action of several corpuscules,—if all these suppositions were adopted, we should be merely indulging in vague and groundless hypotheses.

It is best, therefore, to limit ourselves to the supposition that the clearness and distinctness of our ideas depend on the intensity of the organic actions of the grey globules or nucleated corpuscules of the brain.

Just after Johannes Müller had finished with his *Handbook* came the decade when the principle of conservation of energy was established, an outstanding event in the history of physics, contributed to by J. R. Mayer (1842), J. P. Joule (1840–1845), the great Helmholtz (1847), and Lord Kelvin (1851–1854). It was at this time that Johannes Müller's brilliant pupils Ernst Brücke (Freud's teacher), Emil du Bois-Reymond, Carl Ludwig, and Helmholtz joined together in declaring that they would work to show that all life comes under physicochemical law or other laws "of equal dignity," a faith that their vitalist master could not accept. Presently the new doctrine of energy was to force psychologist-dualists away from the principle of psychophysical interactionism to parallelism.

In 1852 Hermann Lotze's *Medicinische Psychologie, oder Physiologie der Seele* appeared. Lotze was never a physiologist, always a philosopher, but as a philosopher he was intensely interested in the new movement toward an experimental physiological psychology—in much the same way as the philosophically oriented William James was thirty years later. Whether Lotze was influenced by the theory of conservation of energy is not clear. He does not allude to it when he discusses the mind-body problem. He was a dualistic parallelist, holding it obvious that body and mind are constituted of entirely different kinds of stuff. He was an early anticipator of psychophysiological isomorphism in that he discussed how the order of change in a perception would imply a corresponding order of change in the underlying brain process (see No. 31). Unfortunately he is too prolix as he dwells upon these matters for us to quote him here.

Fechner may be said to have been an epistemological dualist and a metaphysical monist, which is a way of saying that he believed in the double-aspect theory of mind and brain. He was fighting nineteenth-century materialism and believed that, by estab-

lishing the law of the interrelation between sensation and brain process, he could demonstrate the mutuality of the two and confirm, as he hoped, not the dependence of mind upon brain, but a panpsychism, the fact that physical events are indissolubly identified with mental. The law of this relation was what he called Weber's law (actually his own Fechner's law), and his effect upon contemporary thinking was to establish the belief in the dependence of mind upon brain, not the other way around, as he wanted it. As it was, the formula for the relation between mind and body furnished just as good support for psychophysical parallelism as for the double-aspect view. An interesting thing about the following excerpt is that it shows Fechner supporting his metaphysical belief by saying that the difference between mind and matter, the alleged dualism, is only a difference in the point of view from which the psychophysical entity is observed. Presently we shall find Mach, Avenarius, Wundt, and Titchener all arguing that the difference between psychology and physics is a difference given by point of view.

107   GUSTAV THEODOR FECHNER (1801–1887)

G. T. Fechner, *Elemente der Psychophysik* (Leipzig, 1860), vol. I, chap. 1. Translated by Helmut E. Adler as *Elements of Psychophysics,* ed. Edwin G. Boring and Davis Howes (New York: Holt, Rinehart and Winston, in preparation).

When standing inside a circle, its convex side is hidden, covered by the concave side; conversely, when outside, the concave side is covered by the convex. Both sides belong together as indivisibly as do the mental and material sides of man and can be looked upon as analogous to his inner and outer sides. It is just as impossible, standing in the plane of the circle, to see both sides of the circle simultaneously as it is to see both sides of man from the plane of human existence. Only when we change our standpoint, is the side of the circle which we view changed, so that we now see the hidden side behind the one we had seen before. The circle is, however, only a metaphor and what counts is a question of fact . . .

Let me add a second illustrative example to the first. The solar system offers quite a different aspect as seen from the sun than as observed from the earth. One is the Copernican, the other the Ptolemaic world. It will be impossible for all time for the same observer to perceive both world systems simultaneously, in spite of the fact that both belong quite indivisibly together and, just like the concave and convex sides of the circle, are basically only two different modes of appearance of the same matter from different standpoints. Here again one needs but to change his standpoint in order to make evident the one world instead of the other . . .

What will appear to you as your mind from the internal standpoint, where you yourself are this mind, will on the other hand appear from the outside point of view as the material basis of this mind. There is a difference whether one thinks with the brain or examines the brain of a thinking person . . . These activities appear to be quite different, but the standpoint is quite different too, for here one is an inner, the other an outer point of view. They are even more immensely different than in the previous examples, and for that reason the differences between the modes of their appearance are immensely greater. For the twofold mode of appearance of the circle or the planetary system was after all basically gained by taking two different external standpoints; whether within the circle or on the sun, the observer remained outside the sweep of the circles, outside the planets. The appearance of the mind to itself, on the other hand, is gained from a truly internal standpoint of the underlying being with respect to itself, at a point of coincidence with itself, whereas the appearance of the material state belonging to it derives from a standpoint which is truly external, a state of non-coincidence.

Now it becomes obvious why no one could ever observe mind and body simultaneously even though they are so inextricably united. It is just impossible for anyone to be inside and outside of the same thing at one time.

Alexander Bain's (1818–1903) work was done in Great Britain long before the universities had established their chairs of psychology. Thus, he occupied the chair of logic at the University of Aberdeen, where he made his important contributions in psychology. His two books, *The Senses and the Intellect* (1855) and *The Emotions and the Will* (1859), were definitive texts in Great Britain for nearly a half century. They show the influence of his close friend John Stuart Mill, who was twelve years his senior, and whose outlook toward psychological topics he absorbed. With Bain, it may safely be said, British empiricism and associationism had become more a school of psychology than of philosophy.

The following excerpt comes from *Mind and Body* (1873), which was Bain's attempt to settle the problem of dualism. His solution was not unlike Fechner's, for he too argued that mind and body are metaphysically inseparable. He has been called a psychophysical parallelist, but it would be incorrect to equate his views with those of either Leibnitz or Hartley, the parallelists of the preceding century. Bain did not espouse the implausible view that mind and body are in concert only because of the arbitrary whim of a preordained harmony. He believed, rather, that the parallel courses of mind and body are really but one stream, a stream that can be observed objectively as matter or subjectively as mind. Like Fechner, then, he was a metaphysical monist and an epistemological dualist.

108   ALEXANDER BAIN (1818–1903)

Alexander Bain, *Mind and Body* (London, 1873), chap. 6.

The doctrine of two substances—a material united with an immaterial in a certain vaguely defined relationship—which has prevailed from the time of Thomas Aquinas to the present day, is now in course of being modified, at the instance of modern physiology. The dependence of purely intellectual operations, as memory, upon the material processes, has been reluctantly admitted by the partisans of an immaterial principle; an admission incompatible with the isolation of the intellect in Aristotle and in Aquinas. This more thorough-going connexion of the mental and the physical has led to a new form of expressing the relationship, which is nearer the truth, without being, in my judgment, quite accurate. It is now often said that *the mind and the body act upon each other;* that neither is allowed, so to speak, to pursue its course alone; there is a constant interference, a mutual influence between the two. This view is liable to the following objections:—

In the first place, it assumes that we are entitled to speak of mind apart from body, and to affirm its powers and properties in that separate capacity. But of mind apart from body we have no direct experience, and absolutely no knowledge. The wind may act upon the sea, and the waves may react upon the wind; yet the agents are known in separation, they are seen to exist apart before the shock of collision; but we are not allowed to perceive a mind acting apart from its material companion.

In the second place, we have every reason for believing that there is, in company with all our mental processes, *an unbroken material succession.* From the ingress of a sensation, to the outgoing responses in action, the mental succession is not for an instant dissevered from a physical succession. A new prospect bursts upon the view; there is a mental result of sensation, emotion, thought—terminating in outward displays of speech or gesture. Parallel to this mental series is the physical series of facts, the successive agitation of the physical organs, called the eye, the retina, the optic nerve, optic centres, cerebral hemispheres, outgoing nerves, muscles, &c. While we go the round of the mental circle of sensation, emotion, and thought, there is an unbroken physical circle of effects. It would be incompatible with everything we know of the cerebral action, to suppose that the physical chain ends abruptly in a physical void, occupied by an immaterial substance; which immaterial substance, after working alone, imparts its results to the other edge of the physical break, and determines the active response—two shores of the material with an intervening ocean of the immaterial. There is, in fact, no rupture of nervous continuity. The only tenable supposition is, that mental and physical proceed together, as undivided twins. When, therefore, we speak of a mental cause, a mental agency, we have always a *two-sided cause;* the effect produced is not the effect of mind alone, but of mind in company with body. That mind should have operated on the body, is as much as to say, that a two-sided phenomenon, one side being bodily, can influence the body; it is, after all, body acting upon body. When a shock of fear paralyses digestion, it is not the emotion of fear, in the abstract, or as a pure mental existence, that does the harm; it is the emotion in company with a peculiarly excited condition of the brain and nervous system; and it is this condition of the brain that deranges the stomach. When physical nourishment, or a physical stimulant, acting through the blood, quiets the mental irritation, and restores a cheerful tone, it is not a bodily fact causing a mental fact by a direct line of causation: the nourishment and the stimulus determine the circulation of blood to the brain, give a new direction to the nerve currents; and the mental condition corresponding to this particular mode of cerebral action henceforth manifests itself. The line of mental sequence is thus, not mind causing body, and body causing mind, but mind-body giving birth to mind-body; a much more intelligible position. For this double, or conjoint causation, we can produce evidence; for the single-handed causation we have no evidence.

It was after Fechner that psychophysical parallelism became the standard faith of the nineteenth-century experimental psychologists. When pressed they turned out mostly to be metaphysical monists, for they were all empiricists who must admit that physics and psychology as sciences are both derived from experience. As a rule, however, they eschewed metaphysics and thought in terms of observed fact. A fact is a function of method, and the sciences seem to be distinguished by the objects with which they deal. Paradoxical as it may seem, objects turn out to be subjective, for their character depends on the method of their observation. Epistemology is at bottom the theory of objects, as the German equivalent (*Gegenstandstheorie*) shows. The objects of introspection (sensations, images, thoughts) seemed so different from the objects of physics (weight, heat, light, electricity) that epistemological dualism emerged as almost the final dichotomy, even when experience was accepted as the ultimate stuff upon which all knowledge is based. Physics and psychology deal with the same stuff, though their objects are so very different. Parallelism applies to the objects, setting up an epistemological dualism.

Thus Fechner supported parallelism in spite of his panpsychism. Helmholtz went along with the others, as might be expected of one of the entrepreneurs of the conservation of energy. He was, in fact, more concerned about the defense of Lotze's and Wundt's empiricism against Hering's and Stumpf's nativism. Helmholtz's epistemological dualism is clear in his discussion of the resonance theory of hearing (see No. 12) and in his account of the nature of perception and the operation of unconscious inference in perception (see No. 40).

Wundt was a clear parallelist. He held that psychology by the method of introspection takes experience immediately as it is given to the observing person, whereas physics deals mediately with experience, forming it into stable objects by inferential procedures. This view appears explicitly in the fourth edition of Wundt's classical handbook, *Grundzüge der physiologischen Psychologie* (1893), but we have chosen the simpler exposition from Wundt's *Grundriss* of 1896. Although Wundt did not agree with the complicated Avenarius, whom we shall mention presently, and polemized at length against him, this excerpt shows Wundt finding that the difference between psychology and physics, between mind and matter, between immediate and mediate experience, lies in the point of view from which experience is observed.

109 WILHELM WUNDT (1832–1920)

Wilhelm Wundt, *Grundriss der Psychologie* (Leipzig, 1896), introduction to sect. 1, "Aufgabe der Psychologie," para. 2. Translated for this book by Edwin G. Boring.

The expressions "outer experience" and "inner experience" do not indicate different objects, but *different points of view* from which we take under consideration the scientific treatment of a unitary experience. We come naturally to these points of view, because every concrete experience immediately divides into *two factors*—a *content* that is presented to us, and our *apprehension* of this content. The first of these factors we designate as *objects of experience*, whereas the second is the *experiencing subject*. Such a division indicates two modes of treatment of the experience. One is the mode of the *natural sciences*, which concern themselves with the *objects* of experience, which in turn are regarded as independent of the subject. The other is the mode of *psychology*, which investigates the whole content of experience in its relations to the subject and also in respect of the attributes that this content derives directly from the subject. The point of view of natural science may, accordingly, be designated as yielding *mediate experience,* since it becomes possible only after abstracting from the subjective factor that is present in all actual experience. The point of view of psychology, on the other hand, may be designated as that of *immediate experience,* since it intentionally does away with this abstraction and all its consequences.

In actuality Ernst Mach comes ahead of Wundt, but we put him here because he and Avenarius, and not Wundt, fixed the ground upon which Titchener argued for the difference by point of view between dependent and independent experience, between psychology and physics, and Titchener is the master expositor in the matter of the differentiation of the sciences by point of view. If Mach is not clear, let the reader look ahead to Titchener's explanation of the matter. Mach was primarily concerned with showing that all sciences, since their data lie only in experience, have the same subject matter, and that physics, like psychology, begins with sensory observation. He emphasized his point by calling experience sensations, confusing as that usage is to the un-prepared reader. Although Mach's chief concern in this famous work is to convince his readers that psychology and physics are both derived from sensations, it is important to note that he does indeed mention the difference between the two fields as depending on point of view or the "direction of investigation." In psychology one regards experience as dependent on the body that mediates it (for color, on the properties of the retina). That was in 1886, two years before Avenarius had formulated his schema of dependent and independent vital events, which led to Titchener's basic assumptions. In subsequent editions of his book Mach proclaimed his agreement with Avenarius and welcomed Avenarius' support of this view.

110   ERNST MACH (1838–1916)

E. Mach, *Beiträge zur Analyse der Empfindungen* (Jena, 1886), chap. 1, "Antimetaphysische Vorbemerkungen," sects. 2, 3, 8, 12. Translated by C. M. Williams as *Contributions to the Analysis of the Sensations* (Chicago, 1897).

2

Colors, sounds, temperatures, pressures, spaces, times, and so forth, are connected with one another in manifold ways; and with them are associated moods of mind, feelings, and volitions. Out of this fabric, that which is relatively more fixed and permanent stands prominently forth, engraves itself in the memory, and expresses itself in language. Relatively greater permanency is exhibited, first, certain *complexes* of colors, sounds, pressures, and so forth, which are connected in time and space, receive special names, and are designated *bodies*. Absolutely permanent such complexes are not.

My table is now brightly, now dimly lighted. Its temperature varies. It may receive an ink stain. One of its legs may be broken. It may be repaired, polished, and replaced part for part. But for me, amid all its changes, it remains the table at which I daily write . . .

Our greater intimacy with this sum-total of permanency, and its preponderance as contrasted with the changeable, impel us to the partly instinctive, partly voluntary and conscious economy of mental representation and designation, as expressed in ordinary thought and speech. That which is perceptually represented in a single image receives *a single* designation, *a single* name.

As relatively permanent, there are exhibited, further, those complexes of memories, moods, and feelings, joined to a particular body (the human body), which are denominated the "I" or "Ego." I may be engaged upon this or that subject, I may be quiet or animated, excited or ill-humored. Yet, pathological cases apart, enough durable features remain to identify the ego. Of course, the ego is only of relative permanence.

·   ·   ·   ·   ·

3

The useful habit of designating such relatively permanent compounds by *single* names, and of apprehending them by *single* thoughts, without going to the trouble each time of an analysis of their component parts, is apt to come into strange conflict with the tendency to isolate the component parts. The vague image which we have of a given permanent complex's being an image which does not perceptibly change when one or another of the component parts is taken away, gradually establishes itself as something which exists *by itself*. Inasmuch as it is possible to take away *singly* every constituent part without destroying the capacity of the image to *stand for* the totality and of being recognised again, it is imagined that it is possible to subtract *all* the parts and to have something still remaining. Thus arises the monstrous notion of a *thing in itself*, unknowable and different from its "phenomenal" existence.

·   ·   ·   ·   ·

8

The traditional gulf between physical and psychological research, accordingly, exists only for the habitual stereotyped method of observation. A color is a physical object so long as we consider its dependence upon its luminous source, upon other colors, upon heat, upon space, and so forth. Regarding, however, its dependence upon the retina . . . it becomes a psychological object, a sensation. Not the subject, but the direction of our investigation, is different in the two domains.

. . . . .

12

Bodies do not produce sensations, but complexes of sensations (complexes of elements) make up bodies. If, to the physicist, bodies appear the real, abiding existences, whilst sensations are regarded merely as their evanescent, transitory show, the physicist forgets, in the assumption of such a view, that all bodies are but thought-symbols for complexes of sensations (complexes of elements). Here, too, the *elements* form the real, immediate, and ultimate foundation, which it is the task of physiological research to investigate. By the recognition of this fact, many points of psychology and physics assume more distinct and economical forms, and many spurious problems are disposed of.

It was in 1888 and 1890 that Richard Avenarius, a philosopher at Zürich, wrote two ponderous volumes of about 170,000 words, which he titled *Kritik der reinen Erfahrung.* He died six years later, exhausted, it is said, by his labor. Avenarius was essentially a psychophysical dualist: he believed that the events that make up experience depend for their occurrence on the functioning of a proper portion of the nervous system, and, since he was not sure just which nervous processes condition consciousness, he avoided the issue in a circular manner by calling the part of the central nervous system that underlies experience "System C." The organism, he held, lives a perpetual life of readjustment, being thrown out of equilibrium by stimulation and regaining equilibrium by its own natural processes. These changes go on in System C, and Avenarius called the disequilibria "vital differences" and the changes in them "vital series." Since the vital series occur in System C, they necessarily

appear as experience that is inseparable from events in System C. System C is, nevertheless, something more permanent and stable than evanescent experience. How can this seeming contradiction be resolved?

Avenarius argued that the difference is only apparent and arises because there are two points of view that one can take toward the events in System C, which are also items of experience. If one regards experience as dependent upon the changing events in System C, then one comes away with a description of mental events, but, if one takes the view that these events exist in their own right independently of System C, then one comes away with the description of the more stable events and objects with which physics deals. Put thus, this view comes close to Wundt's that psychology deals with immediate and physics with mediate experience, but Wundt disagreed with Avenarius on many other matters, and polemized against him at length. It is not fair to Avenarius to attempt so simple an epitome of his views, but the purpose of this paragraph is simply to show where Titchener got the terms *dependent* and *independent,* which he applied to experience in order to distinguish psychology from physics.

Titchener's exposition of experience as the common subject matter of psychology and physics, and of the difference of the two as dependent and independent experience, is so simple and clear as not to need introductory comment. What Titchener says constitutes the culmination of this view of the nature of psychology, for after him the importance of introspection and consciousness in psy-chology waned. Introspection was attacked in America by the behaviorists and in Germany by the Gestalt psychologists. Titchener took his stand with epistemological psychophysical parallelism. In this respect he was like most of the experimental psychologists, content with epistemological dualism. To inquire about ultimate identification of the mental process with its underlying neural substrate, to choose a double-aspect theory or an identity theory of mind and body, was to press the problem toward its metaphysical ultimate, and most psychologists then and now have eschewed metaphysics.

111   EDWARD BRADFORD TITCHENER (1867–1927)

E. B. Titchener, *A Text-Book of Psychology* (New York: Macmillan, 1910), pp. 2–8, 13, 16, 19–27.

First of all, then, it is plain that all the sciences have the same sort of subject-matter; they all deal with some phase or aspect of the world of human experience. If we take a mere fragment of this world,—say, our own experience during a single day,—we find it a rather hopeless mixture. Our lawn-sprinkler obeys the third law of motion, while our pleasure in possessing it is a fact for psychology; the preparation of our food is an applied chemistry, its adulteration depends upon economic conditions, and its effect upon health is a matter of physiology; our manner of speech is governed by phonetic laws, while the things we say reflect the moral standards of the time: in a word, one science seems to run into another science as chance may decide, without order or distinction. If, however, we look over the world as a whole, or examine historically any long period of human existence, the survey is less bewildering. The world of nature breaks up at once, as we inspect it, into living objects, the objects that change by growth, and non-living objects, the objects that change only by decay. And living objects divide, again, into objects that grow in one place, the plants, and objects that move about as they grow, the animals. Here, almost at the first glance, we have distinguished the raw materials of three different sciences: geology, botany, zoology. Now let us turn to some stage of human evolution: we may choose the social life of mankind before the dawn of civilisation.

Primitive man was required, by the necessities of his case, to make himself weapons; to hunt animals for food; to protect himself by clothing and shelter, and to avoid eating or drinking from poisonous or tainted sources. If he ventured upon the water, he must steer his course by the stars; if he banded with his fellows, he must hold to the code of honour of the tribe. He dreamed, and told his dreams; when he was glad, or angry, or afraid, he showed his feeling in gesture or by the expression of his face. Doubtless, his daily experience, if he ever thought about it, seemed to him as chaotic as our own has just appeared to us. But we, who have a larger vision of that experience, can see that it contained the natural germs of many sciences: mechanics, zoology and physiology,— astronomy, ethics and psychology.

.   .   .   .   .

Experience, we have seen, presents itself under different aspects. The differences are roughly outlined, but are definite enough to serve as a starting-point. These different aspects engage the attention of different men. Division of labour is necessary, if the whole of experience is to be brought within the sphere of science; and men's interests are so various that every aspect of experience is sure, in the long run, to find a student. As scientific investigation proceeds, and as the number of scientific men increases, more and more aspects of experience are revealed, and the sciences multiply. They do not exist independently, side by side, as accounts of separate portions of the world or of separate regions of experience; they overlap and coincide, describing one and the same world of experience as it appears from their special standpoints. They are not like blocks of knowledge, which when cut to the proper size and properly fitted together will give us a map of the universe; they are rather like the successive chapters of a book which discusses a large topic from every possible point of view. Some chapters are long, and some are short; some are general; and some are special: this depends upon the sort of attitude which a given science takes towards experience. But all the chapters, or sciences, deal with the same world under its various aspects.

.   .   .   .   .

. . . If it is true that all the sciences have the same sort of subject-matter, there can be no essential difference between the raw materials of physics and the raw materials of psychology. Matter and mind, as we call them, must be fundamentally the same thing. Let us find out, now, whether this statement is really as paradoxical as at first thought it appears.

600

All human knowledge is derived from human experience; there is no other source of knowledge. But human experience, as we have seen, may be considered from different points of view. Suppose that we take two points of view, as far as possible apart, and discover for ourselves what experience looks like in the two cases. First, we will regard experience as altogether independent of any particular person; we will assume that it goes on whether or not anyone is there to have it. Secondly, we will regard experience as altogether dependent upon the particular person; we will assume that it goes on only when someone is there to have it. We shall hardly find standpoints more diverse. What are the differences in experience, as viewed from them?

Take, to begin with, the three things that you first learn about in physics: space, time and mass. Physical space, which is the space of geometry and astronomy and geology, is constant, always and everywhere the same. Its unit is 1 cm., and the cm. has precisely the same value wherever and whenever it is applied. Physical time is similarly constant; and its constant unit is the 1 sec. Physical mass is constant; its unit, the 1 gr., is always and everywhere the same. Here we have experience of space, time and mass considered as independent of the person who experiences them. Change, then, to the point of view which brings the experiencing person into account. The two [horizontal] lines in Fig. 1 are physically equal; they measure alike in units of 1 cm. To

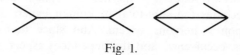

Fig. 1.

you, who see them, they are not equal. The hour that you spend in the waiting-room of a village station and the hour that you spend in watching an amusing play are physically equal; they measure alike in units of 1 sec. To you, the one hour goes slowly, the other quickly; they are not equal. Take two circular cardboard boxes of different diameter (say, 2 cm. and 8 cm.), and pour sand into them until they both weigh, say, 50 gr. The two masses are physically equal; placed on the pans of a balance, they will hold the beam level. To you, as you lift them in your two hands, or raise them in turn by the same hand, the box of smaller diameter is considerably the heavier. Here we have experience of space, time and mass considered as dependent upon the experiencing person. It is the same experience that we were discussing just now. But our first point of view gives us facts and laws of physics; our second gives us facts and laws of psychology.

·   ·   ·   ·   ·

Common sense says that we cry because we are sorry, laugh because we are amused, run because we are frightened; that we feel gloomy and morose because we do not digest our food, go insane from softening of the brain, lose consciousness because we have inhaled ether. Mind influences body, and body influences mind. Our own position has been that mind and body, the subject-matter of psychology and the subject-matter of physiology, are simply two aspects of the same world of experience. They cannot influence each other, because they are not separate and independent things. For the same reason, however, wherever the two aspects appear, any change that occurs in the one will be accompanied by a corresponding change in the other. Your view of a town from the east cannot influence your view of the same town from the west; but as your view from the east differs in sunlight and moonlight, so correspondingly will your view from the west differ. This doctrine of the relation of mind to body is known as the doctrine of psychophysical parallelism: the common-sense doctrine is that of interaction.

. . . . .

We have defined mind as the sum-total of human experience considered as dependent upon the experiencing person. We have said, further, that the phrase 'experiencing person' means the living body, the organised individual; and we have hinted that, for psychological purposes, the living body may be reduced to the nervous system and its attachments. Mind thus becomes the sum-total of human experience considered as dependent upon a nervous system. And since human experience is always process, occurrence, and the dependent aspect of human experience is its mental aspect, we may say, more shortly, that mind is the sum-total of mental processes. All these words are significant. 'Sum-total' implies that we are concerned with the whole world of experience, not with a limited portion of it; 'mental' implies that we are concerned with experience under its dependent aspect, as conditioned by a nervous system; and 'processes' implies that our subject-matter is a stream, a perpetual flux, and not a collection of unchanging objects.

. . . . .

. . . Scientific method may be summed up in the single word 'observation'; the only way to work in science is to observe those phenomena which form the subject-matter of science. And observation implies two things: attention to the phenomena, and record of the phenomena; that is, clear and vivid experience, and an account of the experience in words or formulas.

. . . . .

The method of psychology, then, is observation. To distinguish it from the observation of physical science, which is inspection, a looking-at, psychological observation has been termed introspection, a looking-within. But this difference of name must not blind us to the essential likeness of the methods.

.   .   .   .   .

In principle, then, introspection is very like inspection. The objects of observation are different; they are objects of dependent, not of in-dependent experience; they are likely to be transient, elusive, slippery. Sometimes they refuse to be observed while they are in passage; they must be preserved in memory, as a delicate tissue is preserved in harden-ing fluid, before they can be examined. And the standpoint of the ob-server is different; it is the standpoint of human life and of human interest, not of detachment and aloofness. But, in general, the method of psychology is much the same as the method of physics.

.   .   .   .   .

. . . If mind is the sum-total of human experience considered as de-pendent upon the experiencing person, it follows that each one of us can have direct acquaintance only with a single mind, namely, with his own. We are concerned in psychology with the whole world of human experience; but we are concerned with it solely under its dependent aspect, as conditioned by a nervous system; and a nervous system is a particular thing, possessed by a particular individual. In strictness, there-fore, it is only his own mind, the experience dependent upon his own nervous system, that each of us knows at first-hand; it is only to this limited and individual subject-matter that the method of experimental introspection can be directly applied. How, then, is a scientific psychology possible? How can psychology be anything more than a body of personal beliefs and individual opinions?

The difficulty is more apparent than real. We have every reason to believe, not only in general that our neighbours have minds like our own, that is, are able like ourselves to view experience in its dependent aspect, but also in detail that human minds resemble one another pre-cisely as human bodies do. Within a given race there is much apparent diversity of outward form: differences in height and figure, in colour of hair and eyes, in shape of nose and mouth. We notice these differences, because we are obliged in everyday life to distinguish the persons with whom we come in contact. But the resemblances are more fundamental than the differences. If we have recourse to exact measurements, we find that there is in every case a certain standard or type to which the

603

individual more or less closely conforms and about which all the individuals are more or less closely grouped. And even without measurement we have evidence to the same effect: strangers see family likenesses which the members of the family cannot themselves detect, and the units in a crowd of aliens, Chinese or Negroes, look bewilderingly alike.

.　　.　　.　　.　　.

If, however, we attribute minds to other human beings, we have no right to deny them to the higher animals. These animals are provided with a nervous system of the same pattern as ours, and their conduct or behaviour, under circumstances that would arouse certain feelings in us, often seems to express, quite definitely, similar feelings in them. Surely we must grant that the highest vertebrates, mammals and birds, have minds. But the lower vertebrates, fishes and reptiles and amphibia, possess a nervous system of the same order, although of simpler construction. And many of the invertebrates, insects and spiders and crustaceans, show a fairly high degree of nervous development. Indeed, it is difficult to limit mind to the animals that possess even a rudimentary nervous system; for the creatures that rank still lower in the scale of life manage to do, without a nervous system, practically everything that their superiors do by its assistance. The range of mind thus appears to be as wide as the range of animal life.

The development of thought from Helmholtz and Wundt to Titchener exhibits the standard belief of the orthodox experimental psychologists for the half century between 1860 and 1910. These psychologists were dualists who wished to set psychology off from physics, to distinguish clearly between mind and matter while maintaining the ultimate identity of the two as founded in experience. They believed in introspection as the observational technique for finding out about mind. Introspection, however, was developing as an analytical method that reduces consciousness to elements—sensations, images, and feelings.

The opposition that grew up to this line of description also believed in a difference between mind and matter, or at least between mind and not-mind, and believed that mind is necessarily observed as immediate experience, but in general this opposition did not use the term *introspection* because it was making the point that the stream of consciousness cannot be analyzed into those elements that conventional introspection requires. Consciousness, these men thought, consists of conscious functions or acts or phenomena, not of sensations, images, and feelings. In general, these psychologists were more philosophically oriented than were the tough-minded experimentalists of the Wundtian school, and this tradition of "act psychology" was philosophically very respectable because it had its roots deep in scholastic philosophy. The first representative whom we cite here is Brentano, whose enormously influential book was published in 1874, the same year as the first edition of Wundt's *Grundzüge der physiologischen Psychologie,* to which it became a counterpoise.

Brentano held that mind consists of acts that are phenomena characterized by their immanent objectivity, that is to say, these acts have what he called *intentional inexistence,* an object that inexists intentionally within every act. "I see a red." Seeing is the act. Red is the object inexisting within the act. The red is the physical object, within which nothing ever inexists intentionally. Here is what Brentano says about the psychic, the physical, and the distinction between them in terms of intentional inexistence.

112   FRANZ BRENTANO (1838–1917)

Franz Brentano, *Psychologie vom empirischen Standpunkte* (Leipzig, 1874), bk. II, chap. 1, sect. 9, pp. 126–127. Translated for this book by Edwin G. Boring.

In conclusion let us bring together the results of our discussion of the difference between the psychic and the physical. First we demonstrate the characteristics of both of these classes by examples. Then we designate as psychic phenomena ideas and such phenomena as are based upon ideas; everything else belongs to the physical. In this connection we speak of the attribute of extension, which psychologists insist upon as a peculiarity of all physical phenomena; everything psychic lacks it. This assertion has not, however, gone unopposed, and later investigation has not yet come to a decision about it. All that has been established is that psychic phenomena really appear altogether without extension. Next we find that the discriminating peculiarity of all psychic phenomena is their *intentional inexistence,* their relation to something as an object. Nothing in physical events shows any similarity to this. Further, we assert that psychic phenomena are the exclusive objects of inner perception; they alone are perceived (*wahrgenommen*) by immediate data; indeed they alone are accepted as true (*wahrgenommen*) in the full sense of that word. To them, moreover, the further specification is added that they alone are the phenomena from which intentional as well as actual existence is externally derived. In addition, we have shown, as a final difference, that the psychic phenomena which any one perceives always appear to him, in spite of all their diversification, as unitary, whereas the physical phenomena, which perhaps he perceives at the same moment, do not all present themselves in the same manner but occur as parts of a single phenomenon.

Shortly after Franz Brentano had advanced the cause of phenomenology in Germany, James Ward, an English philosopher, did so in England.

Ward's ideas first appeared in the scholarly ninth edition (1886) of the *Encyclopædia Britannica,* for which he wrote the article on psychology. For the eleventh edition (1911), he wrote a revision that has become a classic of systematic thought in psychology.

Brentano and Ward shared an antipathy toward Wundt's conception of the mind as passive, describable in terms of its sensory and affective contents. Both of them gave the mind an active role in perceiving, judging, acting, and so on. For neither of them was the doctrine of association the all-inclusive, fundamental principle that it was for Wundt, since they endowed the mind with other powers as well. The similarity between Brentano and Ward is, however, limited. The background for Brentano's thinking was the scholasticism of Catholic philosophy, whereas for Ward's it included Darwin's theory of evolution. Ward was countering the psychological atomism of Wundt and his followers by insisting that mental processes have evolved from an undifferentiated to a differentiated state. While Wundt was saying that sensory and affective elements are the building blocks of the mind, Ward was saying that these elements emerge slowly out of an undifferentiated mentality by a process of evolution.

113   JAMES WARD (1843-1925)

James Ward, "Psychology," *Encyclopædia Britannica,* 9th ed. (New York, 1886), XX, 45-46.

What is implied in this process of differentiation or mental growth and what is it that grows or becomes differentiated?—these are the questions to which we must now attend. Psychologists have usually represented mental advance as consisting fundamentally in the combination and recombination of various elementary units, the so-called sensations and primitive movements, or, in other words, in a species of "mental chemistry." If we are to resort to physical analogies at all—a matter of very doubtful propriety—we shall find in the growth of a seed or an embryo far better illustrations of the unfolding of the contents of consciousness than in the building up of molecules: the process seems much more a segmentation of what is originally continuous than an aggregation of elements at first independent and distinct. Comparing higher minds or stages of mental development with lower—by what means such comparison is possible we need not now consider—we find in the higher conspicuous differences between presentations which in the lower are indistinguishable or absent altogether. The worm is aware only of the difference between light and dark. The steel-worker sees half a dozen tints where others see only a uniform glow. To the child, it is said, all faces are alike; and throughout life we are apt to note the general, the points of resemblance, before the special, the points of difference. But, even when most definite, what we call a presentation is still part of a larger whole. It is not separated from other presentations, whether simultaneous or successive, by something which is not of the nature of presentation, as one island is separated from another by the intervening sea, or one note in a melody from the next by an interval of silence. In our search for a theory of presentations, then, it is from this "unity of consciousness" that we must take our start. Working backwards from this as we find it now, we are led alike by particular facts and general considerations to the conception of a *totum objectivum* or objective continuum which is gradually differentiated, thereby becoming what we call distinct presentations, just as with mental growth some particular presentation, clear as a whole, as Leibnitz would say, becomes a complex of distinguishable parts. Of the very beginning of this continuum we can say nothing: absolute beginnings are beyond the pale of science. Actual presentation consists in this continuum being differentiated; and every differentiation constitutes a new presentation. Hence the commonplace of psychologists:—We are only conscious as we are conscious of change.

But "change of consciousness" is too loose an expression to take the place of the unwieldy phrase differentiation of a presentation-continuum, to which we have been driven. For not only does the term "consciousness" confuse what exactness requires us to keep distinct, an activity and its object, but also the term "change" fails to express the characteristics which distinguish presentations from other changes. Differentiation implies that the simple becomes complex or the complex more complex; it implies also that this increased complexity is due to the persistence of former changes; we may even say such persistence is essential to the very idea of development or growth. In trying, then, to conceive our psychological individual in the earliest stages of development we must not picture it as experiencing a succession of absolutely new sensations, which, coming out of nothingness, admit of being strung upon the "thread of consciousness" like beads picked up at random, or cemented into a mass like the bits of stick and sand with which the young caddis covers its nakedness. The notion, which Kant has done much to encourage, that psychical life begins with a confused manifold of sensations not only without logical but without psychological unity is one that becomes more inconceivable the more closely we consider it. An absolutely new presentation, having no sort of connexion with former presentations till the subject has synthesized it with them, is a conception for which it would be hard to find a warrant either by direct observation, by inference from biology, or in considerations of an *a priori* kind. At any given moment we have a certain whole of presentations, a "field of consciousness" psychologically one and continuous; at the next we have not an entirely new field but a partial change within this field. Many who would allow this in the case of representations, *i.e.,* where idea succeeds idea by the workings of association, would demur to it in the case of primary presentations or sensations. "For," they would say, "may not silence be broken by a clap of thunder, and have not the blind been made to see?" To urge such objections is to miss the drift of our discussion, and to answer them may serve to make it clearer. Where silence can be broken there are representations of preceding sounds and in all probability even subjective presentations of sound as well; silence as experienced by one who has heard is very different from the silence of Condillac's statue before it had ever heard. The question is rather whether such a conception as that of Condillac's is possible; supposing a sound to be, qualitatively, entirely distinct from a smell, could a field of consciousness consisting of smells be followed at once by one in which sounds had part? And, as regards the blind coming to see, we must remember not only that the blind have eyes but that they are descended from ancestors who could see. What nascent presentations of sight are thus involved it would be hard to say; and the problem of heredity is one that we have for the present left aside.

The view here taken is (1) that at its first appearance in psychical life a new sensation or so-called elementary presentation is really a partial modification of some pre-existing presentation, which thereby becomes as a whole more complex than it was before; and (2) that this complexity and differentiation of parts never become a plurality of discontinuous presentations, having a distinctness and individuality such as the atoms or elementary particles of the physical world are supposed to have. Beginners in psychology, and some who are not beginners, are apt to be led astray by expositions which begin with the sensations of the special senses, as if these furnished us with the type of an elementary presentation. The fact is we never experience a mere sensation of colour, sound, touch, and the like; and what the young student mistakes for such is really a perception, a sensory presentation combined with various sensory and motor presentations and with representations—and having thus a definiteness and completeness only possible to complex presentations. Moreover, if we could attend to a pure sensation of sound or colour by itself, there is much to justify the suspicion that even this is complex and not simple, and owes to such complexity its clearly marked specific quality. In certain of our vaguest and most diffused organic sensations, in which we can distinguish little besides variations in intensity and massiveness, there is probably a much nearer approach to the character of the really primitive presentations.

Brentano's intentional psychology has often been characterized by the German word *Funktion,* the English translation of which is "function" as a process or action. Thus the word is appropriate, as it opposes the more static approach of the Wundtians. But the correspondence between this psychology of *Funktion* and the American functional psychology that came into prominence around the turn of the century is not so direct and simple as the use of cognate words suggests. American functionalism was not only process-oriented and action-oriented, it was also purposive. It included that connotation of the word "function" evident in phrases like "a function is served." For the Americans, it was function in the Darwinian sense, which would have the mind a useful, not merely an active, characteristic of the organism.

The father, if not the leader, of American functionalism was William James. In the passage that follows, James equated psychology with purposivism—the mark of mentality is direction toward a goal. The debt to evolutionary theory is obvious and explicit in all of James's psychological writing. Yet the connection between his functionalism and Brentano's is also evident. For these two, as for Ward, and in contrast with Wundt and his followers, the study of the mind was necessarily concerned with acts and dynamisms.

114 WILLIAM JAMES (1842–1910)

William James, *The Principles of Psychology* (New York, 1890), I, 6–8.

If some iron filings be sprinkled on a table and a magnet brought near them, they will fly through the air for a certain distance and stick to its surface. A savage seeing the phenomenon explains it as the result of an attraction or love between the magnet and the filings. But let a card cover the poles of the magnet, and the filings will press forever against its surface without its ever occurring to them to pass around its sides and thus come into more direct contact with the object of their love. Blow bubbles through a tube into the bottom of a pail of water, they will rise to the surface and mingle with the air. Their action may again be poetically interpreted as due to a longing to recombine with the mother-atmosphere above the surface. But if you invert a jar full of water over the pail, they will rise and remain lodged beneath its bottom, shut in from the outer air, although a slight deflection from their course at the outset, or a re-descent towards the rim of the jar when they found their upward course impeded, would easily have set them free.

If now we pass from such actions as these to those of living things, we notice a striking difference. Romeo wants Juliet as the filings want the magnet; and if no obstacles intervene he moves towards her by as straight a line as they. But Romeo and Juliet, if a wall be built between them, do not remain idiotically pressing their faces against its opposite sides like the magnet and the filings with the card. Romeo soon finds a circuitous way, by scaling the wall or otherwise, of touching Juliet's lips directly. With the filings the path is fixed; whether it reaches the end depends on accidents. With the lover it is the end which is fixed, the path may be modified indefinitely.

Suppose a living frog in the position in which we place our bubbles of air, namely, at the bottom of a jar of water. The want of breath will soon make him also long to rejoin the mother-atmosphere, and he will take the shortest path to his end by swimming straight upwards. But if a jar full of water be inverted over him, he will not, like the bubbles, perpetually press his nose against its unyielding roof, but will restlessly explore the neighborhood until by re-descending again he has discovered a path round its brim to the goal of his desires. Again the fixed end, the varying means!

Such contrasts between living and inanimate performances end by leading men to deny that in the physical world final purposes exist at all. Loves and desires are to-day no longer imputed to particles of iron or of air. No one supposes now that the end of any activity which they may display is an ideal purpose presiding over the activity from its outset and soliciting or drawing it into being by a sort of *vis a fronte.* The end, on the contrary, is deemed a mere passive result, pushed into being *a tergo,* having had, so to speak, no voice in its own production. Alter the pre-existing conditions, and with inorganic materials you bring forth each time a different apparent end. But with intelligent agents, altering the conditions changes the activity displayed, but not the end reached; for here the idea of the yet unrealized end co-operates with the conditions to determine what the activities shall be.

The pursuance of future ends and the choice of means for their attainment are thus the mark and criterion of the presence of mentality in a phenomenon. We all use this test to discriminate between an intelligent and a mechanical performance. We impute no mentality to sticks and stones, because they never seem to move for *the sake of* anything, but always when pushed, and then indifferently and with no sign of choice. So we unhesitatingly call them senseless.

Just so we form our decision upon the deepest of all philosophic problems: Is the Kosmos an expression of intelligence rational in its inward nature, or a brute external fact pure and simple? If we find ourselves, in contemplating it, unable to banish the impression that it is a realm of final purposes, that it exists for the sake of something, we place intelligence at the heart of it and have a religion. If, on the contrary, in surveying its irremediable flux, we can think of the present only as so much mere mechanical sprouting from the past, occurring with no reference to the future, we are atheists and materialists.

During the first ten years of the new century Brentano's influence was to be seen in the growing importance of phenomenology among German psychologists of distinction. The term "phenomenology" itself was old but had been given new significance by Edmund G. Husserl in his *Logische Untersuchungen,* of which the first volume appeared in 1900. Husserl had been a student first of Brentano's and then of Stumpf's. His phenomenology was a description "of pure consciousness by immanent inspection." Even for Husserl it was not psychology. Husserl's psychology of act resembled Brentano's very closely. Phenomenology was the most primitive kind of observation of experience that it is possible for man to achieve, but no psychologist who took over the term ever intended anything quite so inchoate.

Carl Stumpf, who had published the two volumes of his classical *Tonpsychologie* in 1883 and 1890 before he came to the chair at Berlin, a treatment of psychology along the lines orthodox for the experimental psychologists of that period, was taking a fresh look at psychology in this first decade of the new century and deciding that Brentano and Husserl were closer to the proper nature of psychology than Helmholtz and Wundt. In 1907 he published two of his addresses before the Berlin Akademie der Wissenschaften: *Zur Einteilung der Wissenschaften* and *Erscheinungen und psychische Funktionen.* In the first he distinguished between phenomenology and psychology, deciding that his *Tonpsychologie* had really

been phenomenology and that psychology properly deals with psychical functions, as Brentano had said. Stumpf characterized phenomenology as a propaedeutic science (*Vorwissenschaft*), the discipline that considers bare experience out of which all sciences are derived.

Meanwhile, at Würzburg a comparable change of heart was going on with Oswald Külpe, Wundt's one-time assistant at Leipzig, who in 1893 wrote a *Grundriss der Psychologie* much along Wundtian lines, although he did not accept Wundt's dictum that thought cannot be examined by the experimental method. The experimental attack on the nature of thought at Würzburg in 1901–1908 under Külpe's direction brought out the fact that Wundtian introspection is not adequate to a description of thinking. Thought is something more than the associative course of sensations and images. Külpe believed that he had discovered a new kind of conscious element, not images but an impalpable awareness, and eventually he found himself moving over to Brentano's side of the great dichotomy and deciding that these new elements that emerge in systematic introspective description are conscious functions. He argued not for two disciplines, phenomenology and psychology, but for a bipartite psychology, which included in its materials, on the one side, contents (sensations, images, feelings) and, on the other, psychical functions (acts). Külpe, who died in 1915 without having published his new views, left the manuscripts of lectures which, although incomplete, became available in 1920 as his *Vorlesungen über Psychologie.*

Since it is not possible to find brief excerpts of these psychologies of function, either in Stumpf's papers of 1907 or in Külpe's posthumously printed lectures, we must content ourselves with these brief notes, enough to establish the continuity of the development away from a psychology that consists entirely of content (phenomena) to one that includes psychical functions (acts). Both Stumpf and Külpe undertook to show how content and function are differentiated by their independent variability, how, in the seeing of red, the function can change from noticing to judging while the content remains constant, or the content can change from red to green with the seeing constant.

That Brentano should have won out over Wundt in the thinking of such important psychologists as Stumpf and Külpe, at the same time and without obvious collaboration between them, is an example of how the spirit of the times can affect thought and opinion, especially with younger men, for Wundt was too old and too committed to change. Phenomenology was in the air, and it had at the same time another effect in another quarter.

While Wundtian analytical introspection had worked well for the psychophysical problems of sensation in establishing the relations of sensory quality and intensity to the physical dimensions of the stimulus, to many this analysis seemed artificial when applied to the description of perception. Perception had characteristics that could not be adequately reduced to sensations and their four attributes of quality, intensity, extension, and duration. Phenomenology, on the other hand, placed no such restrictions on the description of consciousness. One could use whatever terminology

one needed to tell about what one saw or heard or felt. Thus there grew up what has sometimes been called an experimental phenomenology, first at Göttingen in G. E. Müller's famous laboratory and then elsewhere as Gestalt psychology emerged under the stimulus of Wertheimer and his associates Köhler and Koffka.

The important Göttingen monograph, important in part because it preceded Wertheimer's paper that initiated Gestalt psychology, appeared in 1911 and was by David Katz: *Die Erscheinungsweisen der Farben und ihre Beeinflussung durch die individuelle Erfahrung.* This work is an elaborate study of the phenomena of color. It did not call itself phenomenology, but it made frequent reference to Hering, who described color in similar ways and mentioned an indebtedness to Husserl. Katz began with a description of the primary modes of appearance of colors: film colors (as in the spectrum), surface colors, transparent, reflected, and volumic colors, luster, and luminous and glowing colors. These modes of appearance are phenomenologically primary. They are not reducible to sensations and their conventional attributes, and thus they illustrate the freedom of description that phenomenology affords. Since Katz did not discuss these general principles, but merely illustrated them in his extensive descriptions, it is impracticable to excerpt him here.

There was one other volume in the phenomenology of visual perception from G. E. Müller's laboratory at Göttingen, Edgar Rubin's *Synsoplevede Figurer* of 1915. In it the description of figure and ground in visual perception is demonstrated and discussed, but this research came after Wertheimer's on seen movement and thus does not antedate Gestalt psychology. In fact, the Gestalt psychologists hailed it as a contribution to their movement and were pleased to see it translated from the Danish into German in 1921.

It was Max Wertheimer's study of seen movement that started Gestalt psychology off in 1912. One can see movement of an object from one spot to another even when the stimulus does not move but consists in the discrete displacement of the object pattern, as in motion pictures. Movement is a clear visual phenomenon, and Wertheimer appropriately named it Phi—the Phi phenomenon. Phi was plainly a piece of experience that would not reduce to Wundtian visual attributes. It had, for instance, no color. Having already cited Wertheimer twice, we need not quote him again. See his discussion of the difference between the phenomenal world and the Wundtians' mosaic of elements (No. 43) and his parallelistic account of ismorphism (No. 55).

Certainly Wertheimer founded Gestalt psychology in 1912, yet it is interesting to note how many influences were at work to prepare the way. Wertheimer admitted owing a great deal to the nativist Hering, who counts as a phenomenologist even though he made his contributions so early. Wertheimer worked at Berlin with Schumann and Stumpf, and Schumann, the last of the school of form quality (*Gestaltqualität*), was at Frankfurt when Wertheimer made his discovery of the Phi phenomenon there. Although the Gestalt psychologists had all worked at Berlin, they did not acknowledge Stumpf's influence on their movement. Nevertheless, at the very least it can be said that Berlin saw Stumpf espouse one kind of phenomenology, whereas the Gestalt psychologists chose another kind, one not so different, only a little later. Nor may one forget that Wertheimer took his doctorate *summa cum*

laude with Külpe at Würzburg in 1904 at the time that Külpe was moving over from the Wundtian position to that of Brentano.

Robert Sessions Woodworth's presidential address to the American Psychological Association in 1914 was entitled "A Revision of Imageless Thought" (*Psychological Review 22,* 1915), and was a more radical departure from the Wundtian position than Külpe or his students had been able to conceive. Woodworth argued that what was called imageless thought is really a type of reaction that is neither sensory nor motor, but purely mental. By this argument, Woodworth put himself in the German tradition of *Funktion,* but he was also very much in the American tradition of functional psychology, being one of its major spokesmen for almost sixty years.

Unlike the functionalists who preceded him, James and Dewey, for example, he was reacting against more than Wundtian structuralism. By 1918, when he published the book from which the following excerpt comes, America had already heard, and reacted to, Watson's call for behaviorism (see No. 94), the definition of psychology as the study of behavior rather than the study of the mind. Woodworth, then, was searching for a synthesis that would take account of behaviorism as well as of the older psychologies. The synthesis has been called eclectic. Woodworth argued for a psychology that would take its data from any source that offered itself. Introspection and rigorous observation, human sensory processes and animal behavior, were all to be included if they could shed light on the causal mechanisms of psychology; and a psychology that deals with the causes of thoughts and actions can well be named *dynamic,* so Woodworth thought.

115 ROBERT SESSIONS WOODWORTH (1869–1962)

Robert Sessions Woodworth, *Dynamic Psychology* (New York: Columbia University Press, 1918), pp. 34–36.

A beginner in psychology, approaching the subject from the side of common interests and unworried as yet by controversies within the ranks of psychologists, would be inclined to suppose that the aim of the science was fairly clear, and to express it as an attempt to understand the 'workings of the mind.' He wishes to be informed how we learn and think, and what leads people to feel and act as they do. He is interested, namely, in cause and effect, or what may be called dynamics.

. . . . .

What is meant by a study of cause and effect—since we no longer hope to discover ultimate causes—is an attempt to gain a clear view of the action or process in the system studied, both in its minute elements and in its broad tendencies, noting whatever uniformities occur, and what laws enable us to conceive the whole process in an orderly fashion. Now neither consciousness nor behavior provides a coherent system of processes for causal treatment. Consciousness is not a coherent system, because much of the process that is partly revealed in consciousness goes on below the threshold of consciousness; and behavior, considered as a series of motor reactions to external stimuli, is incoherent because it leaves out of account the process intervening between the stimulus and the reaction. Nor do consciousness and behavior taken together provide a coherent system, since much of the internal process intervening between stimulus and reaction is unconscious. We shall undoubtedly have to look to brain physiology for a minute analysis of the process; but until brain physiology is able to give us such an analysis, and probably even after it has done so, we shall derive some satisfaction from the coarser analysis which we can derive from the introspective and behavioristic methods of psychology. But the essential thing is to keep the dynamic point of view, and to be working always toward a clearer view of the mental side of vital activity, refusing to be contented with the fragmentary views offered us by the exclusive students of either consciousness or behavior, but endeavoring to utilize the results of both these parties, and the results of brain physiology as well, for an understanding of the complete processes of mental activity and development.

William McDougall, an important Englishman who wrote his systematic text at Harvard, was promoting a purposive psychology, one that was in some ways derived from Ward and thus remotely from Brentano. He had said earlier that psychology is the science of behavior, but now refused to accept the title of behaviorist because he found himself in polemical opposition to John B. Watson, the founder of behaviorism in America. McDougall believed that purposiveness can be seen in animal behavior, and he listed the evidence for it in his seven marks of purposive behavior which the following excerpt gives. In a sense he provides, therefore, an intermediate position between a conscious functional psychology, opposed to Wundtian orthodoxy, and a nonconscious behaviorism, also opposed to Wundtian orthodoxy—two opponents of the Wundtian tradition that were themselves opposed.

116   WILLIAM McDOUGALL (1871–1938)

William McDougall, *Outline of Psychology,* copyright 1923 Charles Scribner's Sons; renewal copyright 1951 Anne A. McDougall. Pages 3–6, 43–46, 56–57. Reprinted with the permission of Charles Scribner's Sons, New York, and Methuen & Co., Ltd., London.

Each of us has no direct or immediate acquaintance with minds other than his own. Each one of us experiences pain and pleasure and various emotions, thinks and strives, remembers and expects and resolves. And it is generally agreed that all such experiences are manifestations of his mind or mental capacities. By reflection upon such experiences a man may form some notion of what his mind does and can do. And, by comparing notes with other men, he learns that they have similar experiences upon similar occasions, and infers that they have minds not unlike his own. Such observation of the varieties of one's own experience is called *introspection* . . . [which] has for a long time been a well-recognized method; it has in fact often been declared to be the sole practical method of psychological study, the only legitimate and effective method of obtaining knowledge of the mind.

· · · · ·

The introspective method has . . . peculiar difficulties and limitations; yet, in spite of these, it is possible for it to achieve a generalized description of types of experience. It could and did achieve in this way a certain stage of psychological science, namely, the descriptive classificatory stage, which is but the first stage of the development of a science. But even this could be achieved only by taking note of the conditions under which we enjoy the experiences that we more or less successfully describe in words . . . Some of these conditions are facts of the outer world, some are facts of experience; and by noting systematically such occasions or conditions of various types of experience, it is possible to establish a certain number of empirical rules which raise to the explanatory stage the purely descriptive psychology attainable by introspection alone . . .

A third great type of observation enables us to carry yet further our understanding of our experience, and at the same time raises another group of problems. This is the observation of conduct or behavior, both our own and that of other persons.

· · · · ·

By "behavior" we commonly mean the action or actions of some living thing, [for] behavior . . . is peculiar to living things. When an animal is dead, its corpse does not "behave"; it has become inert, the sport of forces that play upon it from without. This indicates one of the marks of behavior, namely, *a certain spontaneity of movement.* In behaving, an animal is not simply pushed or pulled by forces external to itself; but, if it actively resists the push or pull, it is behaving. It is true that the behavior of an animal often appears to be a response or reaction to some sense-impression, a sound, or a touch, or a ray of light. And some of the mechanists dogmatically lay down the law that every movement is a response to some such impression . . . Whether this assumption is well founded we cannot at present say. But, even if it be true that every instance of behavior is initiated by a "stimulus," it is evident that the movement or train of behavior, once initiated, often continues independently of the initiating stimulus. A momentary noise, such as the snapping of a twig, may send a rabbit scurrying to his burrow, put to flight a flock of birds, and throw the timid deer into the attitude and motions of alert watchfulness . . . This is the second mark of behavior; namely the *persistence of activity independently of the continuance of the impression which may have initiated it.*

An inanimate object, when set in motion, continues to move in the same direction, if not acted upon by any forces which deflect or arrest it . . . Its movements and changes are in principle strictly predictable according to physical laws . . . But, when an animal persists in the movements initiated by a sense-impression, its movements are not predictable in detail . . . When, for example, [an animal], like a rabbit, belongs to some timid species which normally shelters itself in holes in the earth, we may predict that, if it is set running by a sudden noise, it will continue to run until it finds such shelter, and that, if the course, it first takes, leads to no such shelter, it will dodge hither and thither until such shelter is found. Such *variation of direction of persistent movements* is a third mark of behavior.

The movements of an animal are commonly continued, with more or less variation of direction, until they bring about that kind of change in its situation which, as we have noted, is predictable in general terms from a knowledge of the species; and when that new situation is achieved, the train of activity commonly ceases, perhaps giving place to some activity of an altogether different kind . . . This *coming to an end of the animal's movements as soon as they have brought about a particular kind of change in its situation* is a fourth mark of behavior.

Again, we may often observe that, while the animal's movements are maintained, they seem to show in some degree preparation for, or anticipation of, the new situation which will bring them to an end or will give rise to a new and very different train of movements . . . The cat, aroused by the squeak of the mouse behind the wainscot, stealthily approaches the hole and there lies in wait in the attitude of preparation for the spring upon the prey. Such *preparation for the new situation toward the production of which the action contributes* is a fifth mark of behavior.

.   .   .   .   .

A sixth mark of behavior, which is less easy to observe, has been very commonly accepted as the most trustworthy indication of mental life; namely, *some degree of improvement in the effectiveness of behavior, when it is repeated by the animal under similar circumstances* . . . No doubt, when such improvement may be observed, it provides the surest criterion; but, without this sixth mark, we may infer mental activity from the other five. And it is to be noted that this sixth mark implies the others; if the train of movements did not present those characters, we should not be able to infer Mind from the sixth alone.

.   .   .   .   .

In contrasting reflex action with purposive action or behavior, we must take notice of yet another distinction of great importance, which perhaps deserves to rank as a seventh objective mark of behavior; namely, a reflex action is always a partial reaction, but a *purposive action is a total reaction of the organism* . . .

In purposive action . . . the whole organism is commonly involved; the processes of all its parts are subordinated and adjusted in such a way as to promote the better pursuit of the natural goal of the action. If, while you amuse yourself by repeatedly exciting the scratch-reflex in your dog, some sound excites him to behavior, then, even though the behavior consists in nothing more than assuming an alert attitude with eyes and ears directed toward the disturbing object, your stimulation of his flank becomes ineffective . . . If the sound is followed by the appearance of a stranger (dog or man) your dog springs to his feet with every muscle and organ at work in preparation for attack . . . That is the type of the total reaction.

Nowadays it can be seen that the end of the period of systematic handbooks is well represented by the texts of Titchener, McDougall, and Watson, each so different in purpose from the other two, yet all alike in making the attempt to define psychology and to cover the entire range of what was at the time recognized as psychology. Thus the revolt of John B. Watson against introspectionism (Wundt, Titchener) and against mentalism (Brentano, McDougall) is his founding of behaviorism. It belongs here. The revolt was announced by Watson in 1913 (see No. 94) just before the First World War, and his systematic text came in 1919, *Psychology from the Standpoint of a Behaviorist,* just after the war. Watson deplored the unreliability and inadequacy of the introspective method and advocated limiting psychology to the observation of the actual objective behavior of men and animals. He did not deny the existence of consciousness, and is said by a colleague to have had excellent visual imagery himself, but he undertook to suggest how certain topics in psychology that had not yet yielded to objective observation could properly be handled by behaviorism. Thinking, he proposed, might be observed in faint laryngeal or other vocimotor movements; feeling might be found in glandular response; and association was already being taken care of by Pavlov's conditioned response.

It was this Watsonian revolt that initiated the positivistic phase in American psychology, fifty years of it at the least. Watson himself was not temperamentally suited to provide a sound philosophical foundation for behaviorism, but E. B. Holt shortly remedied this deficiency in 1915 (see No. 42), and there followed the enthusiastic formulation of many different kinds of behaviorism, the general American acceptance of logical positivism and operationism, and the theoretical contributions of leaders like E. C. Tolman, C. L. Hull, and B. F. Skinner, all of which lies beyond the proper limit of the present volume.

Shall we take a moment to review what we have been saying? Descartes began it all with a thoroughgoing metaphysical dualism. Matter is extended substance, mind is unextended. Some ideas are innate. Locke accepted dualism, but argued for the origin of all ideas in experience, the beginning, except for Hobbes, of psychological empiricism. Berkeley held that experience is primary and that the concept of matter is derived from experience. He avoided solipsism by an appeal to divine interposition, but Hume, rejecting this view, left the business unresolved in his skepticism. Hartley gave us a straightforward psychophysical parallelism freed of metaphysical implication.

Kant, rectifying Hume, validated mind with his intuitions, which are reminiscent of Descartes's innate ideas. He was primarily responsible for the nativism of the vitalist Johannes Müller. From that ancestry sprang the nativistic theories of the perception of space propounded by Hering and Stumpf. On the other side, empiricism found support in Lotze's genetic theory of space perception, a view which was elaborated by Helmholtz and Wundt, both of them in the empiristic tradition. Lotze was a parallelist in his discussion of the relation of mind to body.

There follow Fechner, Bain, Wundt, Mach, Avenarius, and Titchener, all of whom as empiricists believed that experience is the foundation of all science and that psychology and physics are to be distinguished by the point of view from which experience is regarded. Fechner was arguing for a metaphysical monism and demonstrating the usefulness of an epistemological dualism. Bain, having regard to the need for the brain's physical events' constituting a closed system, was for an epistemological psychophysical parallelism. Wundt held that psychical processes are immediately given in experience, whereas physical ones are mediate or derived. Mach stressed the identity of the basic subject matters of psychology and physics, since both are forms of experience, whereas Avenarius drew the distinction by point of view, holding that mental events are seen when experience is regarded as dependent upon its being experienced, while physical events appear when experience is taken as occurring in-dependently. Titchener followed both Mach and Avenarius, who agreed with each other.

Outside of this trend is the functional movement that held that mind is active, as did Brentano, and that it is to be also understood in respect of its purpose or use, as did William James. Ward's views were similar to James's. Stumpf and then Külpe followed the tradition of Brentano that immediate observation reveals mind as active. McDougall, writing in the tradition of Ward, argued that the criterion of mind, be it in immediate or mediate events, is purposive.

Watson's role is outside of both these traditions. He revolted against the use of immediate experience and, founding behaviorism, initiated the American positivistic trend in psychology, although he was by no means the very first to attempt to make psychology objective. Descartes in his views about animal action, the French materialists, and certain of the nineteenth-century students of animal behavior were all ancestors of Watsonian behaviorism.

LIST OF EXCERPTED WORKS

The translators of foreign works are cited unless the translation was made especially for this book. The parenthetical numbers following citations refer to the numbers of the excerpts in the text.

Angell, J. R., "The province of functional psychology," *Psychological Review 14* (1907). Published by the American Psychological Association. (No. 93)

Aristotle, *De anima*, ca. 350 B.C. Translated by J. A. Smith in W. D. Ross, ed., *The Works of Aristotle*, vol. III, Oxford: Clarendon Press, 1931. By permission. (No. 1)

——— *De memoria et reminiscentia*, ca. 350 B.C. Translated by J. I. Beare in W. D. Ross, ed., *The Works of Aristotle*, vol. III, Oxford: Clarendon Press, 1931. By permission. (No. 65)

Bain, Alexander, *Mind and Body*, London, 1873. (No. 108)

Baldwin, J. M., *Mental Development in the Child and the Race*, New York, 1895. (No. 92)

Bell, Charles, *Idea of a New Anatomy of the Brain: Submitted for the Observation of His Friends*, London, 1811. (Nos. 6, 8)

Berkeley, George, *An Essay towards a New Theory of Vision*, Dublin, 1709. (Nos. 28, 36)

——— *The Theory of Vision, or Visual Language, Shewing the Immediate Presence and Providence of a Deity, Vindicated and Explained*, London, 1733. (No. 68)

Binet, Alfred, and Victor Henri, "La Psychologie individuelle," *L'Année psychologique 2*, 1895. By permission. (No. 81)

Bouguer, Pierre, *Traité d'optique sur la gradation de la lumière*, Paris, 1760. (No. 15)

Brentano, Franz, *Psychologie vom empirischen Standpunkte*, Leipzig, 1874. By permission of Dunker & Humblot Verlagsbuchhandlung, Berlin. (No. 112)

Broca, Paul, "Remarques sur le siège de la faculté du langage articulé, suivies d'une observation d'aphémie," *Bulletin de la Société Anatomique de Paris* [2] 6 (1861). (No. 47)

Brown, Thomas, *Lectures on the Philosophy of the Human Mind*, vols. I and II, Edinburgh, 1820. (Nos. 38, 71)

Calkins, M. W., "Association: an essay analytic and experimental," *The Psychological Review Monograph Supplements 1*, no. 2 (1896). Published by the American Psychological Association. (No. 96)

Cattell, J. McK., "Mental tests and measurements," *Mind 15* (1890). By permission. (No. 80)

Darwin, Charles, *On the Origin of Species by Means of Natural Selection, or the Preservation of the Favoured Races in the Struggle for Life*, London, 1859. (No. 77)

Delboeuf, J. R. L., *Examen critique de la loi psychophysique: sa base et sa signification*, Paris, 1883. (No. 20)

Delezenne, C. E. J., "Sur les valeurs numériques des notes de la gamme," *Recueil des travaux de la Société des Sciences, de l'Agriculture et des Arts de Lille*, 1827. (No. 16)

Descartes, René, *La Dioptrique*, Leiden, 1638. From Victor Cousin, ed., *Oeuvres de Descartes*, vol. V, Paris, 1824. (No. 27)

―――― *L'Homme*, 1664. From Victor Cousin, ed., *Oeuvres de Descartes,* vol. IV, Paris, 1824. (No. 57)

―――― *Les Passions de l'âme,* Amsterdam, 1650. Translated by E. S. Haldane and G. R. T. Ross, *The Philosophical Works of Descartes* (Cambridge, Eng.: University Press, 1931). By permission. (Nos. 44, 103)

Dewey, John, "The reflex arc concept in psychology," *Psychological Review 3* (1896). Published by the American Psychological Association. (No. 64)

Ebbinghaus, Hermann, *Ueber das Gedächtnis,* Leipzig, 1885. Translated by H. A. Ruger and Clara Bussenius as *Memory,* Educational Reprint No. 3, New York: Teachers College, Columbia University, 1913. By permission. (No. 95)

―――― "Ueber eine neue Methode zur Prüfung geistiger Fähigkeiten und ihre Anwendung bei Schulkindern," *Zeitschrift für Psychologie und Physiologie der Sinnesorgane 13* (1897). Translated by William H. Everhardy at the National Institutes of Health, Bethesda, Maryland, as "A new method for testing mental capacities and its use with school children." By permission. (No. 82)

Epicurus, Letter to Herodotus, from the original text of Diogenes Laertius, ca. 300 B.C. Translated by Cyril Bailey in W. J. Oates, ed., *The Stoic and Epicurean Philosophers: Complete Extant Writings of Epicurus, Epictetus, Lucretius, and Marcus Aurelius,* New York: Random House, 1940. By permission of the Clarendon Press, Oxford. (No. 22)

Fechner, G. T., *Elemente der Psychophysik,* 2 vols., Leipzig, 1860. Translated by Herbert S. Langfeld in Benjamin Rand, ed., *Classical Psychologists,* Boston: Houghton Mifflin, 1912; by permission (No. 18). Translated by Helmut E. Adler as *Elements of Psychophysics,* ed. Edwin G. Boring and Davis Howes, New York: Holt, Rinehart and Winston (A Henry Holt Edition), in preparation; by permission. (No. 107)

Flourens, P. J. M., *Recherches expérimentales sur les propriétés et les fonctions du système nerveux dans les animaux vertébrés,* Paris, 1824. (No. 46)

Franz, S. I., "Variations in the distribution of the motor centers," *Psychological Monographs 19,* no. 81 (1915). Published by the American Psychological Association. (No. 50)

Frey, Max von, *Vorlesungen über Physiologie,* Berlin: Springer, 1904. By permission. (No. 13)

Fritsch, Gustav, and Eduard Hitzig, "Ueber die elektrische Erregbarkeit des Grosshirns," *Archiv für Anatomie, Physiologie, und wissenschaftliche Medicin,* 1870. (No. 48)

Gall, F. J., *Sur les fonctions du cerveau et sur celles de chacune de ses parties . . .,* vols, IV, VI, Paris, 1825. Translated by Winslow Lewis, *Gall's Works,* vols. IV, VI, Boston, 1835. (No. 45)

Galton, Francis, *Hereditary Genius: An Inquiry into Its Laws and Consequences,* London, 1869. (No. 78)

―――― *Inquiries into Human Faculty and Its Development,* London, 1883. (No. 79)

Hall, Marshall, *New Memoir on the Nervous System,* London, 1843; *Synopsis of the Diastaltic Nervous System,* London, 1850. (No. 62)

Hartley, David, *Observations on Man, His Frame, His Duty, and His Expectations,* London and Bath, 1749. (Nos. 59, 70)

Head, Henry, *Aphasia and Kindred Disorders of Speech,* vol. I, New York: Macmillan, 1926. By permission of Cambridge University Press. (No. 52)

Helmholtz, H. L. F. von, *Handbuch der physiologischen Optik,* vols. II and III,

Leipzig, 1860 and 1866. Translated by J. P. C. Southall as *Helmholtz' Treatise on Physiological Optics,* vols. II and III, [Rochester, N. Y.]: Optical Society of America, 1924 and 1925. By permission. (Nos. 11, 34, 40)

—— *Die Lehre von den Tonempfindungen,* Brunswick, 1863; 3rd ed., 1870. Third edition translated by A. J. Ellis as *Sensations of Tone,* London, 1875. (No. 12)

Henri, Victor, *see* Binet

Hering, Ewald, *Beiträge zur Physiologie: Zur Lehre vom Ortsinn der Netzhaut,* Leipzig, 1861–1864, pt. 5. (No. 33)

—— *Zur Lehre Vom Lichtsinne,* Vienna, 1878. By permission of C. Gerold's Sohn Verlagsbuchhandlung, Vienna. (No. 53)

Hitzig, Eduard, *see* Fritsch

Hobbes, Thomas, *Leviathan, or the Matter, Forme and Power of a Commonwealth Ecclesiasticall and Civill,* London, 1651. (No. 66)

Holt, E. B., *The Freudian Wish and Its Place in Ethics,* New York: Henry Holt, 1915, which reprints "Supplement: Response and cognition," *Journal of Philosophy, Psychology and Scientific Methods 12* (1915). By permission. (No. 42)

Hume, David, *A Treatise of Human Nature; being An Attempt to introduce the experimental Method of Reasoning into Moral Subjects,* vol. I, London, 1739; *Philosophical Essays concerning Human Understanding,* 2nd ed., London, 1751. (No. 69)

Jackson, J. Hughlings, *The Croonian Lectures on the Evolution and Dissolution of the Nervous System,* London, 1884. (No. 49)

James, William, *The Principles of Psychology,* vols. I and II, New York, 1890. (Nos. 75, 91, 114)

Jennings, H. S., *Behavior of the Lower Organisms,* New York: Columbia University Press, 1906. (No. 90)

Kant, Immanuel, *Kritik der reinen Vernunft,* Riga, 1781. Translated by John Watson in *The Philosophy of Kant as Contained in Extracts of His Own Writings,* New York, 1888 (No. 30). Translated by F. M. Müller as *Critique of Pure Reason,* London, 1881. (No. 105)

Kepler, Johannes, *Ad Vitellionem paralipomena, quibus astronomiae pars optica traditur,* Frankfurt, 1604. Translated by Alistair C. Crombie in I. B. Cohen and René Taton, eds., *Mélanges Alexandre Koyré: L'Aventure de la science,* Paris: Hermann, 1964. By permission. (No. 23)

Köhler, Wolfgang, "Intelligenzprüfungen an Anthropoiden," *Abhandlungen der königlich preussischen Akademie der Wissenschaften,* 1917, no. 1. By permission of Springer-Verlag, Berlin-Göttingen-Heidelberg. (No. 102)

—— *Die physische Gestalten in Ruhe und im stationären Zustand,* Brunswick: Vieweg, 1920. By permission. (No. 56)

La Mettrie, J. O. de, *L'Homme machine,* Leiden, 1748. Translated by G. C. Bussey and M. W. Calkins as *Man a Machine,* Chicago and London: Open Court Publishing House, 1927. By permission. (No. 58)

Lashley, K. S., *Brain Mechanisms and Intelligence: A Quantitative Study of Injuries to the Brain,* Chicago: University of Chicago Press, 1929. Copyright 1929 by The University of Chicago. By permission. (No. 51)

Locke, John, *An Essay concerning Humane Understanding: In Four Books,* London, 1690 (Nos. 5, 104); 4th ed., London, 1700. (No. 67)

Loeb, Jacques, *Einleitung in die vergleichende Gehirnphysiologie und vergleichende Psychologie mit besonderer Berücksichtigung der wirbellosen Thiere,* Leipzig,

1899. Translated by A. L. Loeb, with changes by the author, as *Comparative Physiology of the Brain and Comparative Psychology,* New York: Putnam, 1900. (No. 89)

Lotze, R. H., *Medicinische Psychologie, oder Physiologie der Seele,* Leipzig, 1852. (No. 31)

McDougall, William, *Outline of Psychology,* New York: Scribners; London: Methuen, 1923. By permission. (No. 116)

Mach, Ernst, *Beiträge zur Analyse der Empfindungen,* Jena, 1886. Translated by C. M. Williams as *Contributions to the Analysis of the Sensations,* Chicago, 1897. By permission of Gustav Fischer Verlag, Jena. (No. 110)

Magendie, François, "Expériences sur les fonctions des racines des nerfs rachidiens," and "Expériences sur les fonctions des racines des nerfs qui naissent de la moëlle épinière," *Journal de physiologie expérimentale et pathologique 2* (1822). (No. 7)

Mill, James, *Analysis of the Phenomena of the Human Mind,* London, 1829. (No. 72)

Mill, J. S., *An Examination of Sir William Hamilton's Philosophy,* London, 1865. (No. 39)

—— *A System of Logic, Ratiocinative and Inductive, being a connected View of the Principles of Evidence, and the Methods of Scientific Investigation,* London, 1843. (No. 73)

Molyneux, William, *Dioptrica Nova: A Treatise of Dioptrics,* London, 1692. (No. 24)

Morgan, C. Lloyd, *An Introduction to Comparative Psychology,* London, 1894. By permission. (No. 88)

Müller, G. E., "Zur Psychophysik der Gesichtsempfindungen." *Zeitschrift für Psychologie 10* (1896). By permission of J. A. Barth, Leipzig. (No. 54)

Müller, Johannes, *Handbuch der Physiologie des Menschen,* bks. V and VI, Coblenz, 1838 and 1840. Translated by William Baly as *Elements of Physiology,* vol. II, London, 1842. (Nos. 9, 106)

—— *Zur vergleichenden Physiologie des Gesichtssinnes,* Leipzig, 1826. (No. 25)

Newton, Isaac, "An hypothesis explaining the properties of light . . .," 1675, in Thomas Birch, *History of the Royal Society of London,* vol. III, London, 1757. (No. 2)

—— *Opticks,* London, 1704. (No. 3)

—— *See also* Young

Pavlov, I. P., *Dvadtzatiletni opit obektivnogo izucheniya vischei nervnoi deyatelnosti (povendeniya) zhivotnikh,* Moscow and Leningrad: State Publishing House, 1923; 3rd ed., 1925. Third edition translated by W. G. Gantt as *Lectures on Conditioned Reflexes,* New York: International Publishers, 1928. By permission. (No. 101)

Plateau, J. A. F., "Sur la mesure des sensations physiques, et sur la loi qui lie l'intensité de ces sensations à l'intensité de la cause excitante," *Bulletins de l'Académie Royale des Sciences, des Lettres et des Beaux-Arts de Belgique* [2] *33* (1872). (No. 19)

Prochaska, George, *De functionibus systematis nervosi,* Prague, 1784. Translated by Thomas Laycock as *A Dissertation on the Functions of the Nervous System,* London, 1851. (No. 61)

Reid, Thomas, *Essays on the Intellectual Powers of Man,* Edinburgh, 1785. (No. 37)

Romanes, G. J., *Animal Intelligence,* London, 1882. (No. 87)

Sechenov, I. M., *Refleksy golovnogo mozga*, St. Petersburg, 1863. Translated by A. A. Subkov as "Reflexes of the Brain" in I. M. Sechenov, *Selected Works*, Moscow and Leningrad: State Publishing House for Biological and Medical Literature, 1935. By permission. (No. 63)

Sharp, S. E., "Individual psychology: a study in psychological method," *American Journal of Psychology 10* (1899). (No. 83)

Small, W. S., "Experimental study of the mental processes of the rat, II," *American Journal of Psychology 12* (1901). (No. 99)

Spearman, C. E., "'General intelligence,' objectively determined and measured," *American Journal of Psychology 15* (1904). (No. 85)

Spencer, Herbert, *The Principles of Psychology*, London, 1855. (No. 74)

Stern, William, *Die psychologische Methoden der Intelligenzprüfung*, Leipzig: Barth, 1912. Translated by G. M. Whipple as *The Psychological Methods of Testing Intelligence*, Baltimore: Warwick and York, 1914. (No. 86)

Stratton, G. M., "Vision without inversion of the retinal image," *Psychological Review 4* (1897). Published by the American Psychological Association. (No. 26)

Thorndike, E. L., "Animal intelligence: an experimental study of the associative processes in animals," *The Psychological Review Monograph Supplements 2*, no. 4; whole no. 8 (1898). Published by the American Psychological Association. (No. 97)

—— and R. S. Woodworth, "The influence of improvement in one mental function upon the efficiency of other functions," *Psychological Review 8* (1901). Published by the American Psychological Association. (No. 100)

Titchener, E. B., *Experimental Psychology*, vol. II, New York: Macmillan, 1905. (No. 21)

—— *An Outline of Psychology*, New York: Macmillan, 1896. (No. 14)

—— *A Text-Book of Psychology*, New York: Macmillan, 1910. By permission of the Estate of E. B. Titchener. (Nos. 41, 111)

Ward, James, "Psychology," *Encyclopædia Britannica*, 9th ed., New York, 1886. By permission. (No. 113)

Watson, J. B., "Psychology as the behaviorist views it," *Psychological Review 20* (1913). Published by the American Psychological Association. (No. 94)

Weber, E. H., *De pulsu, resorptione, auditu et tactu: annotationes anatomicae et physiologicae*, Leipzig, 1834. (No. 17)

—— "Der Tastsinn und das Gemeingefühl," in Rudolph Wagner, ed., *Handwörterbuch der Physiologie*, vol. III, Brunswick, 1846. (No. 10)

—— "Ueber den Raumsinn und die Empfindungskreise in der Haut und im Auge," *Berichte der königlich-sächsischen Gesellschaft der Wissenschaften zu Leipzig, mathematisch-physische Classe 4* (1852). (No. 32)

Wertheimer, Max, "Experimentelle Studien über das Sehen von Bewegung," *Zeitschrift für Psychologie 61* (1912). By permission of J. A. Barth, Leipzig. (Nos. 35, 55)

—— "Untersuchungen zur Lehre von der Gestalt," *Psychologische Forschung 4* (1923). By permission of Springer-Verlag, Berlin-Göttingen-Heidelberg. (No. 43)

Wheatstone, Charles, "Contributions to the physiology of vision: on some remarkable and hitherto unobserved phenomena of binocular vision," *Philosophical Transactions of the Royal Society of London*, 1838. (No. 29)

Whytt, Robert, *An Essay on the Vital and Other Involuntary Motions of Animals,* Edinburgh, 1751. (No. 60)

Wissler, Clark, "The correlation of mental and physical tests," *The Psychological Review Monograph Supplements 3,* no. 6; whole no. 16 (1901). Published by the American Psychological Association. (No. 84)

Woodworth, R. S., *Dynamic Psychology,* New York: Columbia University Press, 1918. By permission. (No. 115)

———— *See also* Thorndike

Wundt, Wilhelm, *Grundriss der Psychologie,* Leipzig, 1896 (No. 109). Translated by C. H. Judd as *Outlines of Psychology,* London and New York, 1897. (No. 76)

Yerkes, R. M., "The formation of habits in the turtle," *The Popular Science Monthly 58* (1901). (No. 98)

Young, Thomas, "On the theory of light and colours," *Philosophical Transactions of the Royal Society of London 92* (1802); also in George Peacock, ed., *Miscellaneous Works of the Late Thomas Young,* vol. I, London, 1855. Young quotes a letter of Isaac Newton of 18 November 1672, which was published in *Philosophical Transactions 7* (1672). (No. 4)

INDEX OF NAMES

The important items are keyed by subject and given first. Less important items and incidental mentions of a name are labeled "inc." and given next. Citations of the literature are labeled "ref." and given last. No "inc." item is included when there is an important subject item or a "ref." item for the same page; nor are the items in the "List of Excerpted Works" (pp. 620–625) included in this index.

INDEX OF SUBJECTS